VIBRATIONAL SPECTRA AND STRUCTURE OF SILICATES
KOLEBATEL'NYE SPEKTRY I STROENIE SILIKATOV
КОЛЕБАТЕЛЬНЫЕ СПЕКТРЫ И СТРОЕНИЕ СИЛИКАТОВ

VIBRATIONAL SPECTRA AND STRUCTURE OF SILICATES

A. N. Lazarev
I. V. Grebenshchikov Institute of Silicate Chemistry
Leningrad, USSR

Translated from Russian by
G. D. Archard

Translations editor
Victor C. Farmer
Department of Spectrochemistry
Macaulay Institute for Soil Research
Aberdeen, Scotland

c/b CONSULTANTS BUREAU • NEW YORK-LONDON • 1972

ADRIAN NIKOLAEVICH LAZAREV was born in Leningrad in 1928. He was graduated from Leningrad State University in 1951 and from that time has worked in the I. V. Grebenshchikov Institute of Silicate Chemistry of the Academy of Sciences of the USSR. Lazarev defended his dissertation for the degree of Candidate of Physicomathematical Sciences in 1963 and was awarded the degree of Doctor of Chemical Sciences in 1969. His major scientific interest lies in the field of molecular spectroscopy.

The original Russian text, published by Nauka Press in Leningrad in 1968, has been corrected and expanded by the author for the present edition. The English translation is published under an agreement with Mezhdunarodnaya Kniga, the Soviet book export agency.

А. Н. Лазарев

КОЛЕБАТЕЛЬНЫЕ СПЕКТРЫ И СТРОЕНИЕ СИЛИКАТОВ

Library of Congress Catalog Card Number 70-136984

ISBN 0-306-10856-9

A Division of Plenum Publishing Corporation

227 West 17th Street, New York, N. Y. 10011

United Kingdom edition published by Consultants Bureau, London

A Division of Plenum Publishing Company, Ltd.

Davis House (4th Floor), 8 Scrubs Lane, Harlesden, London, NW10 6SE, England

Published in the United States of America

FOREWORD TO THE ENGLISH EDITION

I am happy to have the opportunity of contributing this Foreword to the English edition in the hope that this book will make its own contribution to the cooperation of research workers and in the conviction that knowledge will ultimately always serve Mankind.

The Soviet version was mainly written in 1965-1966 and the new edition accordingly contains a number of changes. Errors have been eliminated and references to recent important investigations have been added. Of these we may specially mention a section concerning the use of solid-solution spectra in increasing the reliability of frequency calculations for complex anions and a completely rewritten chapter on the spectra of rare-earth silicates. These two features constitute shortened versions of the author's contributions to Russian-language collections just published; they contain the results of investigations carried out in cooperation with T. F. Tenisheva, I. S. Ignat'ev, and A. P. Mirgorodskii.

Such books "age" quickly; the author is in no way distressed by this, as it indicates steady progress in the corresponding field of knowledge.

It is to be hoped that the book will play its part in drawing attention to the prospects of using spectroscopy in the chemistry of inorganic oxides, although there is little doubt that the mutual benefit derived from the narrowing of the gap between inorganic chemistry and solid-state physics will soon lead to the necessity (and opportunity) of a deeper analysis of the vibrational spectra and dynamics of even such complex systems as silicates and their analogs.

FOREWORD

Silicates make up the greater part of the mass of the earth's core, and throughout the course of history they have constituted the greater part of industrial production. In recent years silicates and other oxide compounds have found new spheres of application and in a number of cases have taken the place of metals. Oxygen compounds of silicon, constituting one of the most important subjects of investigation, have secured widespread applications even in fields of technology traditionally associated with organic compounds: it is enough to mention organosilicon rubbers, oils, resins, electrical insulating materials, and so forth. Materials obtained by the grafting of organic radicals to a silicate base have also been created; these constitute an intermediate link between silicates and organosilicon oxygen-containing compounds.

For a long time the development of the crystal chemistry of silicates was chiefly associated with advances in x-ray structural analysis. Despite the considerable cognitive possibilities of this method, its uses were limited by practical difficulties, and conclusions as to the nature of chemical bonds based solely on data relating to the spatial arrangement of the atoms in the lattice were not always unambiguous. In the last 15 to 20 years considerable interest has been aroused in the vibrational spectra of silicates. This has been due to the development of numerous commercial spectrometers available to every laboratory and also to advances in the theory of the vibrational spectra of crystals containing complex ions. The accumulation of substantial experimental material now enables us to use spectroscopic data both for the identification of natural and artificial silicate minerals and also for the solution of many problems associated with their structure and chemical properties.

The aim of this book is to attempt a systematized exposition of methods of interpreting the vibrational spectra of silicates and similar compounds, and to present illustrations of certain applications of these methods to specific problems associated with studying the configuration of the complex anion, the mutual influence of the cation and anion (affecting particularly the individual bonds of the latter), the coordination of the atoms in the lattice, the mechanisms of phase transformations, and the relation between these and the dynamics of the lattice. In many cases it has seemed appropriate to consider the spectra of organosilicon compounds containing Si–O bonds in addition to the spectra of silicates. Special attention has been given to the information attainable by using the symmetry properties of the vibrations, since quantitative methods of calculating the frequencies in the vibrational spectra of silicates are still far from perfect. The use of this information, together with the results of other methods of investigation, is of practical interest even at the present time.

Further progress in the use of spectroscopic methods for studying the structure of silicates must clearly provide for a transition from qualitative methods, which are restricted to the comparatively high frequencies of the internal vibrations of the anions, to a complete analysis of all the normal lattice vibrations. The necessary prerequisites for solving this problem are being created by the rapid development of computing methods and high-speed electronic computers. Automation of the process of setting up and solving secular equations of high orders enables us to dispense with simplifying assumptions which ignore the interaction between the internal vibrations of the anion and the lattice vibrations. We can therefore obtain a fuller and more reliable set of coefficients characterizing the force field of the crystal. In the solution of this problem, a considerable part will have to be played by investigations into and calculations of the vibrational spectra of organosilicon compounds containing Si–O bonds; this is essential in order to determine the set of force constants describing the internal force field of the silicate anion.

The material presented in this book is largely based on the results of work carried out in the I. V. Grebenshchikov Institute of the Chemistry of Silicates of the Academy of Sciences of the USSR. The fullness of the exposition of individual problems has been chiefly dictated by the sphere of interests of the author; the bibliography is therefore not very extensive. These failings are partly compensated by the inclusion of sections of a compilative nature: Chapter I and a few paragraphs of Chapter IV.

The majority of the results here presented were secured in collaboration with T. F. Tenisheva. The author is greatly indebted to V. A. Kolesova and Ya. I. Ryskin, who took part in the discussion of a number of problems and made valuable comments. The author is also extremely grateful to V. V. Orlov for preparing a large number of the illustrations and schematic material.

CONTENTS

CHAPTER I

INTRODUCTION TO THE THEORY OF THE VIBRATIONAL SPECTRA OF MOLECULES AND COMPLEX IONS IN CRYSTALS

The study of the vibrational spectra of molecules in the gas and liquid phases has for a long time been one of the fundamental methods of structural chemistry. The application of vibrational spectroscopy to the study of crystals, particularly inorganic compounds, has only received wide attention in the last 15 years. This has been largely due to intensive development of the theory of the spectra of crystals, particularly those containing molecules and complex ions. On the other hand, progress has been made in experimental methods of studying the spectra of single crystals, and simple techniques that are applicable to a wide range of materials have been devised for obtaining the spectra of polycrystalline samples.

Theories and methods of molecular spectroscopy have been set out in a large number of textbooks and are well known to all chemists concerned with structural investigations. Problems relating to the spectroscopy of crystals have so far been mainly confined to specialized literature.† We shall therefore now give a brief review of methods of qualitatively interpreting the spectra of molecular crystals and crystals with complex ions, and also some reference material required when using these methods.

§1. Vibrational Spectra and the Symmetry of Crystals

Born's Theory. The basic principles of modern ideas relating to the dynamics of crystals were set out over 50 years ago by Born and Karman [17]. Born's theory considers a crystal as a mechanical system of mN atoms where m is the number of atoms in the unit cell and N is the number of unit cells in the crystal. In order to obtain the spectrum of normal vibrations of the crystal the classical equations of motion are introduced and their solution is considered in the form of plane waves periodic in time and space. The vibrational frequencies constitute 3m roots of the secular equation of the system, depending on the value of the wave

† General questions as to the dynamics of crystals are set out in a classical monograph [1] and also in [2]. The application of group theory to the spectra of crystals is considered in [3-6]. These problems, together with the mechanism underlying the interaction between the vibrations of the molecules in a crystal are treated in [7, 8]. A review of the spectra of crystals was published in [9]. The spectra of high polymer crystals are considered by Zbinden [9a]. The Raman spectra of light in crystals are discussed by Sushchinskii [9b]. Various "non-spectroscopic" methods of studying the vibrations of crystal lattices were considered in [10]. Reviews on the spectra of inorganic molecules and complexes were published in [11, 12]. Reference data regarding the infrared spectra of minerals are contained in the atlas [13] and bibliography [14]. The spectra of silicates were considered in [15, 16]. Some of the material in this chapter is also covered by Mitra [48], but with different examples and different emphasis.

vector $\vec{k}$, which is capable of assuming N discrete values. Of the 3m branches of normal vibrations, three acoustic branches give a zero frequency ν as $\vec{k} \to 0$ and this case corresponds to translation of the whole crystal. The remaining 3m − 3 optical branches have finite values of ν as $\vec{k} \to 0$, and the frequencies of these limiting vibrations may be observed in the infrared spectrum and in the Raman spectrum.

The main content of the theory may be illustrated by the analysis of a simple model, a linear chain of regularly alternating atoms of two kinds. The chain is characterized by the masses of the atoms, m_1 and m_2, the "lattice" constant $2a$ (twice the distance between neighboring atoms), and the elastic interaction constant between the atoms f. Let the atoms with masses m_1 and m_2 have even and odd numbers, respectively, in the sequential enumeration of the atoms in the chain. If we denote the displacement of the n-th atom along the axis of the chain by u_n, then the equation of motion (in the harmonic approximation) for atoms 2n and 2n+1 takes the form

$$\begin{aligned} m_1\ddot{u}_{2n} &= f\,(u_{2n+1} + u_{2n-1} - 2u_{2n}) \\ m_2\ddot{u}_{2n+1} &= f\,(u_{2n+2} + u_{2n} - 2u_{2n+1}) \end{aligned} \tag{1}$$

A system of coupled equations of this kind corresponds to the motions of the whole single-dimensional lattice. We shall seek the solutions in the form of the functions

$$\begin{aligned} u_{2n} &= u' e^{i(2\pi\nu t + 2nka)} \\ u_{2n+1} &= u'' e^{i[2\pi\nu t + (2n+1)ka]} \end{aligned} \tag{2}$$

where the parameters $2\pi\nu$ and ka reflect the time and space periodicity of these solutions.†

After substituting these solutions in the equations of motion we obtain a system of equations for the amplitudes u′ and u″; the condition for the compatibility of these equations is that the secular determinant should vanish:

$$\begin{vmatrix} 2f - 4\pi^2\nu^2 m_1 & -2f\cos ka \\ -2f\cos ka & 2f - 4\pi^2\nu^2 m_2 \end{vmatrix} = 0 \tag{3}$$

From this we may obtain a dispersion equation characterizing the frequency as a function of wave number

$$\nu^2 = \frac{1}{4\pi^2}\left[\frac{f}{\mu} \pm \left(\frac{f^2}{\mu^2} - \frac{4f^2\sin^2 ka}{m_1 m_2}\right)^{1/2}\right] \tag{4}$$

Here μ is the reduced mass of a single cell: $1/\mu = 1/m_1 + 1/m_2$. Depending on the sign in front of the root (+ or −), the dispersion formula describes two sets of possible values of $\nu(k)$, corresponding to the two branches of vibrations. If $k \to 0$ $(\lambda \to \infty)$, then one branch gives $u'/u'' = -m_2/m_1$ and the other $u'/u'' = 1$, i.e., in the first case the displacements of the atoms with masses m_1 and m_2 (both the neighboring and distant ones, in view of the spatial periodicity of the solutions) are opposite in direction, while the frequency of the vibrations for small k equals $\frac{1}{2\pi}\sqrt{f/\mu}$. This branch has become known as the optical branch. The other (acoustic) branch corresponds to displacements of the atoms in the same direction. For small k its frequency is

$$\nu = \frac{1}{2\pi}\left(\frac{2f}{m_1 + m_2}\right)^{1/2} ka \tag{5}$$

† $k = 2\pi/\lambda$ is the wave vector defining the length and direction of propagation of the elastic wave in the crystal.

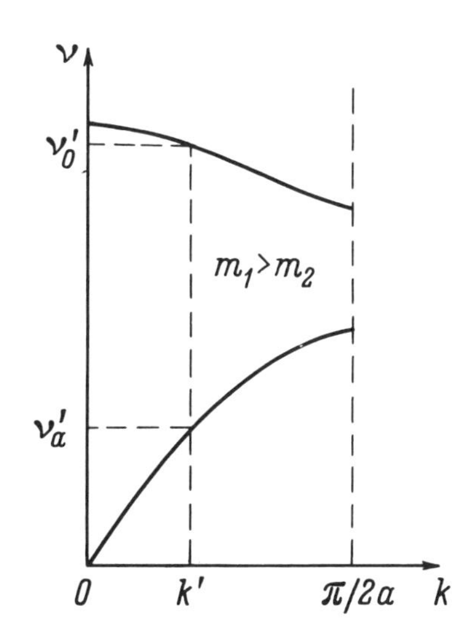

Fig. 1. Optical and acoustic branches of the vibrations of a diatomic linear chain.

Thus the frequency tends to zero as $k \to 0$, and for $k = 0$ this branch describes the translation of the whole one-dimensional crystal. For $k = \pi/2a$ the frequency of the acoustic branch is $\frac{1}{2\pi}\sqrt{2f/m_1}$ and that of the optical branch $\frac{1}{2\pi}\sqrt{2f/m_2}$ (Fig. 1). We note that since ν is a periodic function of ka with period π it is sufficient to consider $\nu(k)$ in the range between 0 and $\pm\pi/2a$.

The dimensions of a real crystal are always limited. In order to avoid a situation in which the characteristic vibrations of a large crystal depend on its surface area† Born introduced the following boundary conditions. Let the infinite lattice consist of an infinite number of successive sections of mN each (in our present model 2N atoms in each), the motions of all these parts being the same. In other words, the motions of the lattice are regarded as spatially periodic, with a period equal to the section $2Na$ or fractions of this containing a whole number of atoms: $\lambda = 2Na/n$ (n = whole number). Then the quantity $ka = \pi n/N$ cannot have any arbitrary values but takes a number of discrete values in the range $-\pi/2$ to $+\pi/2$, and no new types of motion correspond to values of $n > N$.‡ In all, a one-dimensional crystal of N cells with m atoms in the cell has $m-1$ optical and one acoustic branch corresponding to the mN normal vibrations of the system. It is clear from the foregoing that the physical sense of the cyclic boundary conditions in the Born theory reduces to the fact that the number of possible vibrations of the atoms in a finite crystal is limited to those which are repeated again and again in different parts of the crystal. The motions of a parallelepiped of N cells are the same in a crystal of arbitrary shape and finite dimensions as in an infinite crystal.

In order to study many physical properties of crystals, for example, in order to calculate the specific heat, it is important to know the way in which the normal vibrations are distributed over various values of the wave vector or frequency. The function $g(\nu)$, the density of the vibrational states, characterizing the number of states $g(\nu)\,d\nu$ in the range of ν values between ν

---

†In a crystal of limited dimensions, in addition to the volume (three-dimensional) waves described by the Born theory, there are also surface vibrations of the lattice, and under certain conditions these lead to extra optical effects. The introduction of cyclic boundary conditions eliminates the surface waves, for which the boundary conditions would have to have a different form.

‡The boundary conditions may conveniently be written in terms of the vector $\vec{b}$ of the reciprocal lattice in the form $\vec{k} = n/N\vec{b}$. The reciprocal lattice is constructed in such a way that $\vec{a}_i\vec{b}_j = \delta_{ij}$, where δ_{ij} is the Kronecker delta, i.e., $\delta = 1$ for $i = j$ and $\delta = 0$ for $i \neq j$, while $\vec{a}_i$ are the unit vectors of the direct lattice.

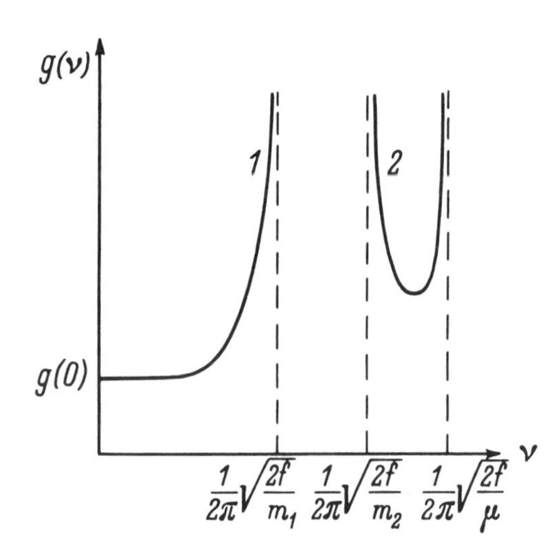

Fig. 2. Distribution function of the frequencies of a linear diatomic chain. 1) Acoustic branch; 2) optical branch.

and $\nu + d\nu$, may be obtained by means of the boundary conditions and dispersion relation. For the one-dimensional model under consideration the function $g(\nu)$ is shown in Fig. 2.

It follows from Figs. 1 and 2 that for frequencies in the range between $\frac{1}{2\pi}\sqrt{2f/m_1}$ and $\frac{1}{2\pi}\sqrt{2f/m_2}$ no solutions of the equations can be obtained for our one-dimensional lattice, i.e., near $k = \pi/2a$, this range corresponds to the forbidden band. We see from this that the range of k from 0 to $\pm\pi/2a$ determines the first Brillouin zone of the one-dimensional crystal.

Infrared radiation may excite optical vibrations for which the directions of the displacements of different types of atoms are in opposition. The wave vector of a light wave, having a wavelength much greater than the lattice constant, may be considered as very small compared with the maximum value of k for the vibrations of the lattice ($\pi/2a \approx 10^8$ cm^{-1}). Hence the limiting vibration with frequency $\frac{1}{2\pi}\sqrt{2f/\mu}$, corresponding to $k \approx 0$ will be excited optically. The vanishing of the wave vector in fact means that the vibrations of equivalent atoms are in phase in all the unit cells. Then the total electrical moment of a region of crystal containing a fairly large number of unit cells (but small in comparison with the light wavelength) will change during the vibrations, causing the absorption of light. For vibrations taking place with a phase shift between equivalent atoms of the lattice, the total change in the electrical moment of the whole region of the crystal will be zero, and no light absorption will occur. Analogous considerations are also applicable to the Raman scattering of light in the crystal; this may occur for nonzero changes in the polarizability of a fairly large part of the crystal.

Thus for the fundamental frequencies the optical spectrum of a crystal has a discrete character. As in the case of the spectra of molecules, anharmonicity of the vibrations makes it possible for overtones and combination frequencies to appear in the spectrum. There is no need for the phase of each of the combining vibrations to be the same in all the cells. It is sufficient that $\theta_{ij}(\nu_1) + \theta_{ij}(\nu_2) = 0$, where θ_{ij} are the phase differences of vibrations in the i-th and j-th cells, i.e., that the phase differences of the combining vibrations should be mutually compensated. Thus, for our linear model this means that the combinations $\nu_0' \pm \nu_a'$ or $2\nu_a'$, $2\nu_0'$ are optically active if both frequencies correspond to the same absolute value of the wave vector k′ (see Fig. 1). It follows from the foregoing argument that spectra of the second and higher orders may have a quasicontinuous character, while their specific form will depend on the frequency distribution function $g(\nu)$.

In the case of a three-dimensional crystal, the Born theory leads to solutions of the equations of motion in the form of superposed plane waves of the $\exp i(\omega t - \vec{k}\vec{r}_j)$ type. The energies of the 3mN normal vibrations of a crystal comprising N cells are quantized, and the term phonon is usually employed to denote the quantum of lattice vibrations,† in the same way as photon for the quantum of electromagnetic field. For the majority of crystals quantitative calculations present considerable difficulties and they have accordingly only been carried out for a relatively small number of the simplest cases.

The interpretation of the vibrational spectrum of a crystal usually bears a semiempirical character, at any rate for lattice vibrations, if a crystal containing molecules or complex ions is under consideration. Here the use of group-theory methods enables us to determine the number of different vibrations and establish the selection rules and polarization of the vibrations by considering the symmetry properties. After this, if necessary, some model of the force field may be chosen in order to make a quantitative estimate of the frequencies and study the force constants as well as the form of the vibrations and the intensity of the bands in the spectra.

Symmetry of Normal Vibrations. Selection Rules. The main characteristic of the crystalline state which affects crystal dynamics is the existence of translational symmetry, a spatial periodicity in the arrangement of the atoms, ions, or molecules. In addition to this, the arrangement of the atoms in a crystal also possesses symmetry properties not exclusive to infinite objects. Such symmetry operations as inversion, rotation (simple or with reflection), and mirror reflection are possible for both finite and infinite sets of material points (molecules and crystals, respectively). In the latter case the combination of translation with these operations may give new forms of symmetry operation only associated with crystals: the screw operation (rotation combined with translation along a screw axis), and the glide operation (reflection combined with translation parallel to a glide plane).

The set of all symmetry operations satisfied by the arrangement of the atoms in the crystal forms one of 230 space groups, since the number of possible ways of combining symmetry operations is finite. The order (number of symmetry elements) of a space group describing the symmetry properties of an infinitely extended crystal is infinite.

A real crystal is spatially limited. In addition to this, the application of group-theory methods to a study of the vibrations of crystals is simplified if the well-developed theory of the representations of finite groups may be employed. It is therefore convenient to introduce finite space groups, defined as follows. Let $\vec{a}_1$, $\vec{a}_2$, $\vec{a}_3$ be the principal vectors of the lattice defining the primitive cell of the crystal, N_1, N_2, and N_3 be fixed positive whole numbers, and n_1, n_2, and n_3 be arbitrary whole numbers. All the symmetry elements of an infinite space group, the result of applying which differs only in respect of a translation through $n_1N_1\vec{a}_1 + n_2N_2\vec{a}_2 + n_3N_3\vec{a}_3$ are considered as one and the same symmetry element of the finite space group. In other words, this group is the group of symmetry operations of lattices with constants $N_1\vec{a}_1$, $N_2\vec{a}_2$, $N_3\vec{a}_3$, and translations through these vectors are equivalent to identity. The application of such finite space groups to the study of crystal vibrations is equivalent to imposing the Born–Karman cyclic boundary conditions, providing for the vibrations of blocks of $N_1N_2N_3$ unit cells to have an in-phase character.

Any space group may be represented in the form of the product of an invariant subgroup of pure translations (finite or infinite) and a factor group. In other words, the subgroup of translations is formed by all the lattice vectors (or those of a block of $N_1N_2N_3$ unit cells) while the factor group is formed by all the symmetry elements transforming any vector of the translations group into any other vector of the same group. The factor group is isomorphic with one of the 32 crystallographic point groups, but in contrast to these may in-

† In molecular crystals phonon is usually taken to mean the quantum of the external vibration of the molecule in the lattice.involving a translational or restricted rotational motion of the molecule.

clude screw operations and glide operations in addition to purely point symmetry operations. The characters of the irreducible representations of the factor groups coincide with the well-known characters of the corresponding point groups. We also note that if the order of the factor group is H then, since the order of the finite group of translations equals $N_1N_2N_3$, the order of the finite space group is $N_1N_2N_3H$.

A particular normal vibration of the system may be observed in the infrared spectrum or the Raman spectrum if it is accompanied by a change in dipole or polarizability, respectively. In terms of group theory this means that a transition from state i to state j is allowed in the infrared spectrum if the representation of the product of the wave functions of the initial and final states $\psi_i^*\psi_j$ contains the representation of a vector, and allowed in the Raman spectrum if the representation of $\psi_i^*\psi_j$ contains the representation of a symmetrical tensor. Hence the problem of determining the selection rules for optical transitions in a crystal, as in the case of molecular spectra, reduces to determining the number of components of a symmetrical tensor (Raman scattering) or vector (infrared absorption) transforming by a specified irreducible representation.

For this purpose we may use the ordinary expressions for the characters [3, 18] of a symmetrical tensor

$$\chi(R) - 2\cos\varphi(R) | \pm 1 + 2\cos\varphi(R) | \tag{6}$$

and of a vector

$$\chi(R) = \pm 1 + 2\cos\varphi(R) \tag{7}$$

where the signs (+ or −) correspond to simple rotation and rotation with reflection.

The vectors and symmetrical tensors are invariant with respect to a translation. Hence the only active representation of the group of translations is its totally symmetric representation $\Gamma(0)$. In a finite space group only those functions which belong to representations of the factor group correspond to the totally symmetric subgroup of translations. Hence the $3mN_1N_2N_3$ motions of the block under consideration in the crystal reduce to the 3m motions of the smallest translationally nonequivalent cell (m is the number of atoms in the cell). In other words, if we exclude translations of the whole crystal, not more than $3m - 3$ fundamental vibrations may be observed in the crystal's optical spectrum. Thus, for the fundamental vibrations, analysis of the selection rules on the basis of a finite space group is equivalent to finding the active representations of the factor group. Hence it is sufficient to consider one representative each of the H sets of representations of the finite block of the crystal.

We note that the condition of total symmetry toward translations is entirely equivalent to the imposition of cyclic boundary conditions in Born's dynamics and leads to the same result: the limitation to in-phase vibrations of all the unit cells gives not more than $3m - 3$ optically active fundamental vibrations. The conditions of cyclic properties, requiring that

$$\vec{k} = \frac{n_1}{N_1}\vec{b}_1 + \frac{n_2}{N_2}\vec{b}_2 + \frac{n_3}{N_3}\vec{b}_3 \, (n_i = 0, 1, 2, \ldots, N_i - 1) \tag{8}$$

for the ground tone correspond to $N_i = 1$ ($\vec{k} = 0$).

In practice, an analysis of the selection rules for the fundamental vibrations may be carried out by the method of Bagavantam and Venkatarayudu [3], which is entirely analogous to the methods of obtaining the selection rules for an isolated molecule. For a crystal, the part of the latter is played by a set of atoms in the primitive cell, and instead of the point symmetry group the factor group of the space group appears. The slightly different way of analyzing molecular crystals or crystals with complex ions proposed by Halford and Winston [4, 19] enables one to consider first the change in the symmetry of the molecule in the internal field of the crystal and then to consider the influence of resonance interaction between several molecules in the unit cell, the Davydov effect [7, 8]. A more detailed exposition of these methods will be set out in the next section.

In determining the selection rules for the overtones and combination vibrations in the crystal we must discover whether the representation $\psi^*_{\text{init}} \cdot \psi_{\text{fin}}$ contains an active representation of the group in question. As in the case of the fundamental vibrations, a necessary condition for optical activity of a combination corresponding to the wave vectors $\vec{k}_i$ and $\vec{k}_j$ is

$$\Gamma^* (\vec{k}_i) \Gamma (\vec{k}_j) = \Gamma (0) \tag{9}$$

i.e., the combined representation must be a totally symmetric representation of the group of translations. Using the representations of the group of translations derived by Seitz [20] it may be shown that condition (9) is satisfied if

$$\vec{k}_i = \vec{k}_j \tag{10}$$

i.e., both vibrations should correspond to the same wave vector. Each of the representations $\Gamma(\vec{k}_i)$ and $\Gamma(\vec{k}_j)$ individually may not be totally symmetric with respect to the translation. This is again equivalent to the conclusion arising from Born's dynamics as to the optical activity of any combinations of different vibrational branches, and not only the limiting frequencies, if the phase differences of the corresponding vibrations between any elementary cells mutually compensate. As already mentioned, this leads to the possibility of spectra of higher order having a continuum character. More detailed study of the problem carried out by Winston and Halford [4], using finite space groups, showed that condition (10) was in essence the only strict selection rule for the combination frequencies and overtones in the spectrum of a crystal. This is associated with the fact that the overwhelming majority of wave vectors in the lattice are only invariant with respect to the identity operation and contain no other symmetry elements.

Thus a determination of the selection rules for combinations and overtones on the basis of the factor group of the space group of the crystal cannot give a full set of potentially active frequencies, and it is therefore desirable to consider a block comprising a fair number of elementary or unit cells. In a number of cases the observed overtones and combinations correspond only to transitions the symmetry of which corresponds to active representations of the factor group. Hence the selection rules based on the factor group may probably be used in order to gain an approximate initial idea of the strongest summation and difference bands.

§2. Method of Interpreting the Vibrational Spectra of Molecular Crystals and Crystals with Complex Ions†

The generally complex problem of interpreting the vibrational spectrum of a crystal may be simplified if the crystal contains a group of atoms (molecules or complex ions) and the forces of interaction between the atoms within such a group are much greater than the forces acting between the groups. Of the 3m optical and acoustic branches corresponding to the vibrations of m translationally nonequivalent atoms, two groups may be singled out: external and internal vibrations. The external vibrations correspond to rotational and translational motions of molecules and complex ions in the crystal lattice; their frequencies are low and depend considerably on the wave vector $\vec{k}$. The frequencies of the internal vibrations, in which there is

†Some of the procedures described in this and the following sections are simplified by the use of tables prepared by Adams and Newton [49].

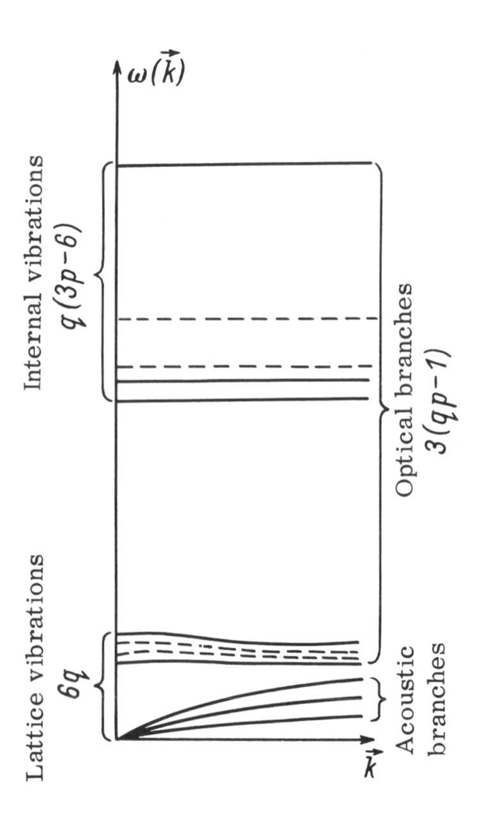

Fig. 3. Internal and external vibrations in a molecular crystal.

no motion of the center of gravity or rotation of the molecule as a whole, are usually much higher and as a result of the weak coupling between the vibrations of neighboring molecules depend little on the wave vector. These vibrations differ comparatively little from the vibrations of an isolated molecule in the spectrum of a gas and change little with temperature, external fields, or changes in the state of aggregation.

If the m atoms in the unit cell are distributed over q molecules, containing p atoms each, then 6q branches, including three acoustic branches, correspond to the lattice vibrations. Of the $6q-3$ optical lattice vibrations $3q-3$ correspond to translational and 3q to rotational vibrations of the molecules (Fig. 3).

It follows from the foregoing that $(3p-6)q$ vibrations of the crystal correspond to the $3p-6$ vibrations of the isolated molecule. The relative stability of these vibrations in the spectra of the molecules in various states of aggregation, and the fact that the number of vibrations observed in the spectrum of the crystal is in the majority of cases much smaller than expected, indicate the desirability of dividing these into groups, each of which corresponds to an internal vibration of a single molecule. The influence of the mutual interaction between the molecules in the crystal reduces to a certain splitting and displacement of the frequencies, and to a change in the selection and polarization rules. The comparatively small extent of these effects has made it possible to develop a theory of the spectra of molecular crystals in which the interaction between the molecules is treated by the perturbation method [7].

We shall now give an elementary exposition of methods of interpreting the spectra of molecular crystals and crystals containing complex ions by considering the symmetry properties of these systems. The sequence of the exposition corresponds to the recommended sequence of analyzing experimental results. First the usually more striking changes in the spectrum of a complex ion caused by the internal field of the crystal are described. Then, if the number of observed frequencies exceeds that expected, the possibility of resonance interactions taking place between the vibrations of the molecules is considered. Then attention is paid to overtones and combinations, including the interaction between internal and external vibrations.

The symmetry of the position of the molecule or ion in the lattice is considered by expanding the irreducible representations of the point symmetry group of the molecule or complex ion with respect to the irreducible representations of the so-called site symmetry group corresponding to the crystal site on which the center of gravity of the molecule is situated [19]. Each point of the crystal remains invariant or coincides with itself after the operation of some of the operations of the symmetry space group of the crystal. (If the point lies on none of the symmetry elements of the crystal, i.e., if it occurs in a "general position," to use crystallographic terminology, it remains invariant with respect to the identity operation and has a trivial C_1 symmetry.) The set of all such symmetry operations, i.e., the symmetry elements on the intersection of which the point under consideration lies, forms a certain group which we shall in future call the site symmetry group. This group naturally only contains operations of point symmetry (since no point of the crystal remains invariant to glide or screw operations), and is a subgroup of the symmetry group of the molecule and simultaneously a subgroup of the factor group of the space group of the crystal. Several points with the same site symmetry may form a set of points any of which may be transformed into all the remaining points of the set by using the symmetry operations of the factor group not forming a part of this site group. Despite the fact that the site groups of equivalent points of the set are isomorphic, this does not mean that they are necessarily formed by the same elements of the factor group. It is not difficult to show that, if the order of a site group is h and the multiplicity factor of the set equals n, then the order of the factor group is

$$H = nh \tag{11}$$

The essence of the approximation described reduces to the assumption that the effective symmetry of the molecule in the crystal is described by the site symmetry group. Physically this is equivalent to considering the vibrational states of the molecule in a certain force field, the symmetry of which corresponds to the site symmetry group. In other words, the use of the site symmetry for analyzing the vibrational spectrum of the molecule is analogous to the method of analyzing the electron terms of the atoms in a crystal in accordance with the symmetry of their positions in the lattice [21].

The transition from the symmetry group of the molecule to the site group enables us to predict whether there is any possibility of removing or reducing the degree of degeneracy (splitting bands of degenerate vibrations), and changing the selection rules and polarization of the bands. Since the site group is a subgroup of the molecular group, the degree of degeneracy cannot be increased or an active vibration become inactive in this way. Of course, we still cannot draw any conclusion as to the extent of the splitting of degenerate vibrations or the intensity of transitions becoming active, without making further assumptions as to the strength and anisotropy of the internal field of the crystal.

Practically, the problem of passing from the symmetry group of the molecule to the site group reduces to establishing relationships between the irreducible representations of these groups. Since the site group is a subgroup of the molecular group, any irreducible representation of the latter is a representation of the site group, in general irreducible. Any representation Γ of a finite group G may be expressed in the form

$$\Gamma = \sum n_i \Gamma_i \quad (i = 1,\ 2,\ 3 \ldots p) \tag{12}$$

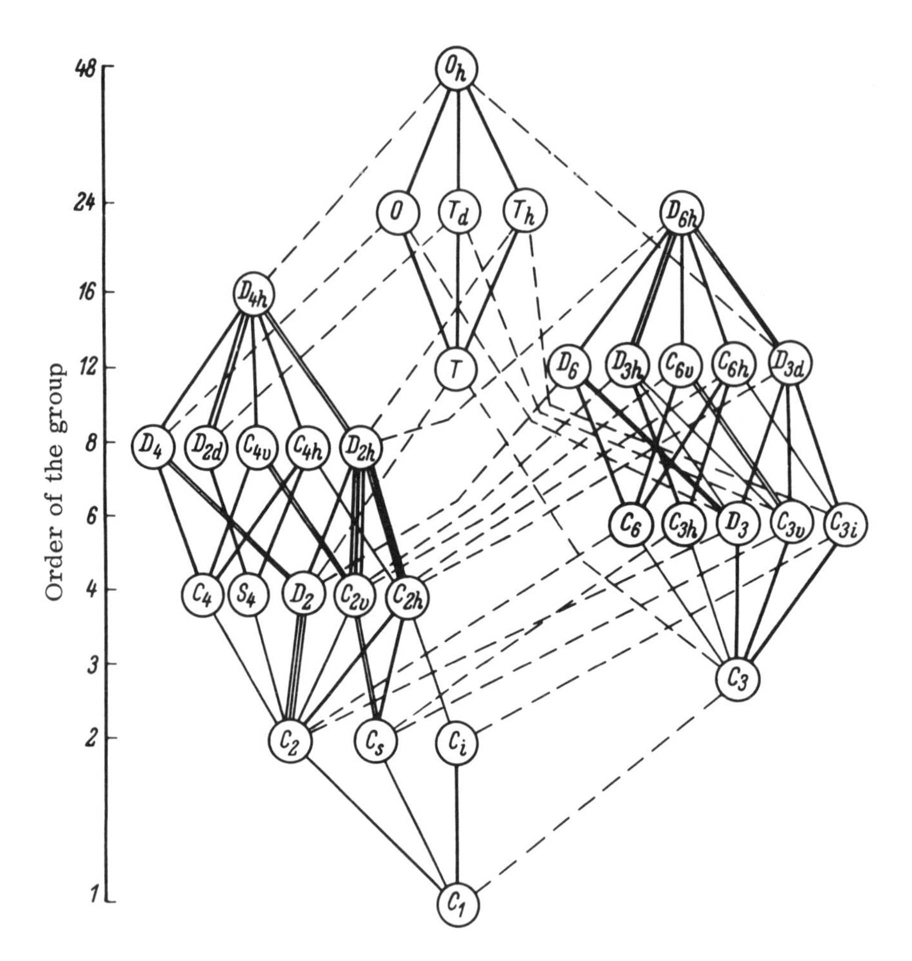

Fig. 4. Relations between the groups and subgroups of the 32 crystallographic point groups.

We note that p, i.e., the number of irreducible representations of the group, is equal to the number of conjugate classes† of the elements of the group.

If R is some transformation of the group G and $\chi'(R)$ and $\chi_i(R)$ are, respectively, the characters of the transformation R in the representation Γ and the irreducible representation Γ_i, relation (12) may be rewritten in the form of a relation between the characters of the representations:

$$\chi'(R) = \sum_{i=1}^{p} n_i \chi_i(R) \tag{13}$$

† A conjugate class of a group is the name given to a set of elements of the group which are conjugate relative to one another, i.e., are formed from a single element s by using the operations $t^{-1}st$, where t is an element of the same group. The number of conjugate elements is called the order of the conjugate class and is usually indicated in tables of group characters together with an element of the particular conjugate class. A single element of a group (identity operation) always forms a conjugate class of order 1. An example of conjugate classes is provided by the two C_3 axes, the three C_2 axes, or the three σ_v planes of the point group D_{3d}.

The coefficients n_i may be determined from the formula

$$n_i = \frac{1}{N} \sum h_\rho \chi'_\rho (R) \chi_i (R) \tag{14}$$

where N is the order of the group G, h_ρ is the order of the ρ-th conjugate class. The summation is carried out over all p conjugate classes of the group. The index ρ in $\chi'_\rho(R)$ indicates that the operation R belongs to the ρ-th conjugate class.

The relations between the groups and subgroups (and their orders) for the 32 crystallographic points groups are shown in the illustration (Fig. 4).

The transformation from the irreducible representations of the molecular group to the irreducible representations of the site group may also be effected without any calculations based on formulas (12) to (14) if we refer to the so-called correlation tables of the irreducible representations of points groups and their subgroups, which are given in many reference books on molecular spectroscopy (see, for example, [18]). It should be noted that in many cases the establishment of correlations between the irreducible representations of the molecular and site groups may be carried out in several different ways and will only lead to unambiguous results for a specific mutual orientation of the symmetry elements of the two groups. These additional conditions are also indicated in correlation tables. In practice this means that, in addition to the symmetry of the molecules, we must know not only the site symmetry group of the molecule in the crystal but also the orientation of the molecule relative to the symmetry elements of the site group.

Where the mutual orientation of the symmetry elements of the molecular and site groups is unimportant or, conversely, where the position and orientation of the molecules are unambiguously established, complete x-ray structural data are not essential in order to carry out an analysis of the symmetry properties of the vibrations in accordance with the site symmetry of the molecule. It is sufficient to know the space group of the crystal, the symmetry of the molecule, and the number of molecules in a unit cell. Then by using Table 1 we may establish the site symmetry of the molecule. In so doing, however, we must remember the possibility of there being complicating circumstances: identical molecules may lie on several sets of equivalent points with different site symmetries; again, since more than one molecule can lie on a given plane (or axis) of symmetry, the number of molecules per unit cell with this site symmetry may be two or more times the number of planes (or axes) of symmetry in a set. In these cases more detailed information is required as to the structure of the crystal, just as when there are several sets of sites with symmetry appropriate to the molecule and with multiplicities equal to the number of molecules in the cell.

In using Table 1 we must remember that the multiplicities of sets of points with the same site symmetry are given for crystallographic unit cells. Such cells correspond to the primitive (i.e., the smallest translationally nonidentical) cell of the crystal only for simple Bravais lattices. In body- and A, B, or C face-centered lattices, the volume of the crystallographic unit cell is twice that of the primitive cell and in face-centered lattices four times this. In Table 1 these are easily distinguished if we note that, for the primitive cell, the multiplicity factor of the general positions C_1 equals the order of the factor group.

We also note that, according to the definition of a site symmetry group, its order cannot be higher than the order of the point group of the isolated molecule or ion. Comparatively rare exceptions to this rule are of particular interest, since they give a direct indication of the free rotation of the molecule in the crystal relative to one or several axes, which also leads to an increase in its "effective" symmetry. Examples of this kind are crystals of ammonium chloride, where the NH_4^+ ions occupy positions with site symmetry O_h, higher than the T_d symmetry of an isolated ion.

Table 1. Number, Symmetry, and Multiplicity Factor of Sets of Particular Positions in the 230 Space Groups

Space group	Number and multiplicity factor of the sets of positions
$P1 - C_1^1$	C_1
$P\bar{1} - C_i^1$	$C_1\,(2),\ 8\mathbf{C}_i$
$P2 - C_2^1$	$C_1\,(2),\ 4C_2$
$P2_1 - C_2^2$	$C_1\,(2)$
$B2 - C_2^3$	$C_1\,(4),\ 2C_2\,(2)$
$Pm - C_s^1$	$C_1\,(2),\ 2C_s$
$Pb - C_s^2$	$C_1\,(2)$
$Bm - C_s^3$	$C_1\,(4),\ C_s\,(2)$
$Bb - C_s^4$	$C_1\,(4)$
$P2/m - C_{2h}^1$	$C_1\,(4),\ 8\mathbf{C}_{2h},\ 4C_2\,(2),\ 2C_s\,(2)$
$P2_1/m - C_{2h}^2$	$C_1\,(4),\ 4\mathbf{C}_i\,(2),\ C_s\,(2)$
$B2/m - C_{2h}^3$	$C_1\,(8),\ 4\mathbf{C}_{2h}\,(2),\ 2\mathbf{C}_i\,(4),\ 2C_2\,(4),\ C_s\,(4)$
$P2/b - C_{2h}^4$	$C_1\,(4),\ 4\mathbf{C}_i\,(2),\ 2C_2\,(2)$
$P2_1/b - C_{2h}^5$	$C_1\,(4),\ 4\mathbf{C}_i\,(2)$
$B2/b - C_{2h}^6$	$C_1\,(8),\ 4\mathbf{C}_i\,(4),\ C_2\,(4)$
$P222 - D_2^1$	$C_1\,(4),\ 8\mathbf{D}_2,\ 12C_2\,(2)$
$P222_1 - D_2^2$	$C_1\,(4),\ 4C_2\,(2)$
$P2_12_12 - D_2^3$	$C_1\,(4),\ 2C_2\,(2)$
$P2_12_12_1 - D_2^4$	$C_1\,(4)$
$C222_1 - D_2^5$	$C_1\,(8),\ 2C_2\,(4)$
$C222 - D_2^6$	$C_1\,(8),\ 4\mathbf{D}_2\,(2),\ 7C_2\,(4)$
$F222 - D_2^7$	$C_1\,(16),\ 4\mathbf{D}_2\,(4),\ 6C_2\,(8)$
$I222 - D_2^8$	$C_1\,(8),\ 4\mathbf{D}_2\,(2),\ 6C_2\,(4)$
$I2_12_12_1 - D_2^9$	$C_1\,(8),\ 3C_2\,(4)$
$Pmm2 - C_{2v}^1$	$C_1\,(4),\ 4\mathbf{C}_{2v},\ 4C_s\,(2)$
$Pmc2_1 - C_{2v}^2$	$C_1\,(4),\ 2C_s\,(2)$
$Pcc2 - C_{2v}^3$	$C_1\,(4),\ 4C_2\,(2)$
$Pma2 - C_{2v}^4$	$C_1\,(4),\ 2C_2\,(2),\ C_s\,(2)$
$Pca2_1 - C_{2v}^5$	$C_1\,(4)$
$Pnc2 - C_{2v}^6$	$C_1\,(4),\ 2C_2\,(2)$
$Pmn2_1 - C_{2v}^7$	$C_1\,(4),\ C_s\,(2)$
$Pba2 - C_{2v}^8$	$C_1\,(4),\ 2C_2\,(2)$
$Pna2_1 - C_{2v}^9$	$C_1\,(4)$
$Pnn2 - C_{2v}^{10}$	$C_1\,(4),\ 2C_2\,(2)$
$Cmm2 - C_{2v}^{11}$	$C_1\,(8),\ 2C_{2v}\,(2),\ C_2\,(4),\ 2C_s\,(4)$
$Cmc2_1 - C_{2v}^{12}$	$C_1\,(8),\ C_s\,(4)$
$Ccc2 - C_{2v}^{13}$	$C_1\,(8),\ 3C_2\,(4)$
$Amm2 - C_{2v}^{14}$	$C_1\,(8),\ 2C_{2v}\,(2),\ 3C_s\,(4)$
$Abm2 - C_{2v}^{15}$	$C_1\,(8),\ 2C_2\,(4),\ C_s\,(4)$
$Ama2 - C_{2v}^{16}$	$C_1\,(8),\ C_2\,(4),\ C_s\,(4)$
$Aba2 - C_{2v}^{17}$	$C_1\,(8),\ C_2\,(4)$
$Fmm2 - C_{2v}^{18}$	$C_1\,(16),\ C_{2v}\,(4),\ C_2\,(8),\ 2C_s\,(8)$
$Fdd2 - C_{2v}^{19}$	$C_1\,(16),\ C_2\,(8)$
$Imm2 - C_{2v}^{20}$	$C_1\,(8),\ 2C_{2v}\,(2),\ 2C_s\,(4)$
$Iba2 - C_{2v}^{21}$	$C_1\,(8),\ 2C_2\,(4)$
$Ima2 - C_{2v}^{22}$	$C_1\,(8),\ C_2\,(4),\ C_s\,(4)$
$Pmmm - D_{2h}^1$	$C_1\,(8),\ 8\mathbf{D}_{2h},\ 12C_{2v}\,(2),\ 6C_s\,(4)$
$Pnnn - D_{2h}^2$	$C_1\,(8),\ 4\mathbf{D}_2\,(2),\ 2\mathbf{C}_i\,(4),\ 6C_2\,(4)$
$Pccm - D_{2h}^3$	$C_1\,(8),\ 4\mathbf{C}_{2h}\,(2),\ 4\mathbf{D}_2\,(2),\ 8C_2\,(4),\ C_s\,(4)$
$Pban - D_{2h}^4$	$C_1\,(8),\ 4\mathbf{D}_2\,(2),\ 2\mathbf{C}_i\,(4),\ 6C_2\,(4)$
$Pmma - D_{2h}^5$	$C_1\,(8),\ 4\mathbf{C}_{2h}\,(2),\ 2C_{2v}\,(2),\ 2C_2\,(4),\ 3C_s\,(4)$
$Pnna - D_{2h}^6$	$C_1\,(8),\ 2\mathbf{C}_i\,(4),\ 2C_2\,(4)$
$Pmna - D_{2h}^7$	$C_1\,(8),\ 4\mathbf{C}_{2h}\,(2),\ 3C_2\,(4),\ C_s\,(4)$
$Pcca - D_{2h}^8$	$C_1\,(8),\ 2\mathbf{C}_i\,(4),\ 3C_2\,(4)$
$Pbam - D_{2h}^9$	$C_1\,(8),\ 4\mathbf{C}_{2h}\,(2), 2C_2\,(4),\ 2C_s\,(4)$
$Pccn - D_{2h}^{10}$	$C_1\,(8),\ 2\mathbf{C}_i\,(4),\ 2C_2\,(4)$
$Pbcm - D_{2h}^{11}$	$C_1\,(8),\ 2\mathbf{C}_i\,(4),\ C_2\,(4),\ C_s\,(4)$
$Pnnm - D_{2h}^{12}$	$C_1\,(8),\ 4\mathbf{C}_{2h}\,(2),\ 2C_2\,(4),\ C_s\,(4)$
$Pmmn - D_{2h}^{13}$	$C_1\,(8),\ 2C_{2v}\,(2),\ 2\mathbf{C}_i\,(4),\ 2C_s\,(4)$
$Pbcn - D_{2h}^{14}$	$C_1\,(8),\ 2\mathbf{C}_i\,(4),\ C_2\,(4)$
$Pbca - D_{2h}^{15}$	$C_1\,(8),\ 2\mathbf{C}_i\,(4)$
$Pnma - D_{2h}^{16}$	$C_1\,(8),\ 2\mathbf{C}_i\,(4),\ C_s\,(4)$
$Cmcm - D_{2h}^{17}$	$C_1\,(16),\ 2\mathbf{C}_{2h}\,(4),\ C_{2v}\,(4),\ \mathbf{C}_i\,(8),\ C_2\,(8),\ 2C_s\,(8)$
$Cmca - D_{2h}^{18}$	$C_1\,(16),\ 2\mathbf{C}_{2h}\,(4),\ \mathbf{C}_i\,(8),\ 2C_2\,(8),\ C_s\,(8)$
$Cmmm - D_{2h}^{19}$	$C_1\,(16),\ 4\mathbf{D}_{2h}\,(2),\ 2\mathbf{C}_{2h}\,(4),\ 6C_{2v}\,(4),\ C_2\,(8),\ 4C_s\,(8)$
$Cccm - D_{2h}^{20}$	$C_1\,(16),\ 2\mathbf{D}_2\,(4),\ 4\mathbf{C}_{2h}\,(4),\ 5C_2\,(8),\ C_s\,(8)$

Note. The figure to the left of the sight group symbol indicates the number of sets of this symmetry in the space group; the multiplicity factor of the set is shown in brackets. Sets containing points for which all the coordinates are fixed and hence capable of being only singly-occupied are denoted by heavy print.

Table 1 (cont.)

Space group	Number and multiplicity factor of the sets of positions
$Cmma - D_{2h}^{21}$	$C_1\,(16),\ 2\boldsymbol{D}_2\,(4),\ 2\boldsymbol{C}_{2h}\,(4),\ C_{2v}\,(4),\ 5C_2\,(8),\ 2C_s\,(8)$
$Ccca - D_{2h}^{22}$	$C_1\,(16),\ 2\boldsymbol{D}_2\,(4),\ 2\boldsymbol{C}_i\,(8),\ 4C_2\,(8)$
$Fmmm - D_{2h}^{23}$	$C_1\,(32),\ 2\boldsymbol{D}_{2h}\,(4),\ 3\boldsymbol{C}_{2h}\,(8),\ \boldsymbol{D}_2\,(8),\ 3C_{2v}\,(8),\ 3C_2\,(16),\ 3C_s\,(16)$
$Fddd - D_{2h}^{24}$	$C_1\,(32),\ 2\boldsymbol{D}_2\,(8),\ 2\boldsymbol{C}_i\,(16),\ 3C_2\,(16)$
$Immm - D_{2h}^{25}$	$C_1\,(16),\ 4\boldsymbol{D}_{2h}\,(2),\ 6C_{2v}\,(4),\ \boldsymbol{C}_i\,(8),\ 3C_s\,(8)$
$Ibam - D_{2h}^{26}$	$C_1\,(16),\ 2\boldsymbol{D}_2\,(4),\ 2C_{2h}\,(4),\ \boldsymbol{C}_i\,(8),\ 4C_2\,(8),\ C_s\,(8)$
$Ibca - D_{2h}^{27}$	$C_1\,(16),\ 2C_i\,(8),\ 3C_2\,(8)$
$Imma - D_{2h}^{28}$	$C_1\,(16),\ 4\boldsymbol{C}_{2h}\,(4),\ C_{2v}\,(4),\ 2C_2\,(8),\ 2C_s\,(8)$
$P4 - C_4^1$	$C_1\,(4),\ 2C_4,\ C_2\,(2)$
$P4_1 - C_4^2$	$C_1\,(4)$
$P4_2 - C_4^3$	$C_1\,(4),\ 3C_2\,(2)$
$P4_3 - C_4^4$	$C_1\,(4)$
$I4 - C_4^5$	$C_1\,(8),\ C_4\,(2),\ C_2\,(4)$
$I4_1 - C_4^6$	$C_1\,(8),\ C_2\,(4)$
$P\bar{4} - S_4^1$	$C_1\,(4),\ 4\boldsymbol{S}_4,\ 3C_2\,(2)$
$I\bar{4} - S_4^2$	$C_1\,(8),\ 4\boldsymbol{S}_4\,(2),\ 2C_2\,(4)$
$P4/m - C_{4h}^1$	$C_1\,(8),\ 4\boldsymbol{C}_{4h},\ 2\boldsymbol{C}_{2h}\,(2),\ 2C_4\,(2),\ C_2\,(4),\ 2C_s\,(4)$
$P4_2/m - C_{4h}^2$	$C_1\,(8),\ 4\boldsymbol{C}_{2h}\,(2),\ 2\boldsymbol{S}_4\,(2),\ 3C_2\,(4),\ C_s\,(4)$
$P4/n - C_{4h}^3$	$C_1\,(8),\ 2\boldsymbol{S}_4\,(2),\ C_4\,(2),\ 2\boldsymbol{C}_i\,(4),\ C_2\,(4)$
$P4_2/n - C_{4h}^4$	$C_1\,(8),\ 2\boldsymbol{S}_4\,(2),\ 2\boldsymbol{C}_i\,(4),\ 2C_2\,(4)$
$I4/m - C_{4h}^5$	$C_1\,(16),\ 2\boldsymbol{C}_{4h}\,(2),\ \boldsymbol{C}_{2h}\,(4),\ \boldsymbol{S}_4\,(4),\ C_4\,(4),\ \boldsymbol{C}_i\,(8),\ C_2\,(8),\ C_s\,(8)$
$I4_1/m - C_{4h}^6$	$C_1\,(16),\ 2\boldsymbol{S}_4\,(4),\ 2\boldsymbol{C}_i\,(8),\ C_2\,(8)$
$P422 - D_4^1$	$C_1\,(8),\ 4\boldsymbol{D}_4,\ 2\boldsymbol{D}_2\,(2),\ 2C_4\,(2),\ 7C_2\,(4)$
$P42_12 - D_4^2$	$C_1\,(8),\ 2\boldsymbol{D}_2\,(2),\ C_4\,(2),\ 3C_2\,(4)$
$P4_122 - D_4^3$	$C_1\,(8),\ 3C_2\,(4)$
$P4_12_12 - D_4^4$	$C_1\,(8),\ C_2\,(4)$
$P4_222 - D_4^5$	$C_1\,(8),\ 6\boldsymbol{D}_2\,(2),\ 9C_2\,(4)$
$P4_22_12 - D_4^6$	$C_1\,(8),\ 2\boldsymbol{D}_2\,(2),\ 4C_2\,(4)$
$P4_322 - D_4^7$	$C_1\,(8),\ 3C_2\,(4)$
$P4_32_12 - D_4^8$	$C_1\,(8),\ C_2\,(4)$
$I422 - D_4^9$	$C_1\,(16),\ 2\boldsymbol{D}_4\,(2),\ 2\boldsymbol{D}_2\,(4),\ C_4\,(4),\ 5C_2\,(8)$
$I4_122 - D_4^{10}$	$C_1\,(16),\ 2\boldsymbol{D}_2\,(4),\ 4C_2\,(8)$
$P4mm - C_{4v}^1$	$C_1\,(8),\ 2C_{4v},\ C_{2v}\,(2),\ 3C_s\,(4)$
$P4bm - C_{4v}^2$	$C_1\,(8),\ C_4\,(2),\ C_{2v}\,(2),\ C_s\,(4)$
$P4_2cm - C_{4v}^3$	$C_1\,(8),\ 2C_{2v}\,(2),\ C_2\,(4),\ C_s\,(4)$
$P4_2nm - C_{4v}^4$	$C_1\,(8),\ C_{2v}\,(2),\ C_2\,(4),\ C_s\,(4)$
$P4cc - C_{4v}^5$	$C_1\,(8),\ 2C_4\,(2),\ C_2\,(4)$
$P4nc - C_{4v}^6$	$C_1\,(8),\ C_4\,(2),\ C_2\,(4)$
$P4_2mc - C_{4v}^7$	$C_1\,(8),\ 3C_{2v}\,(2),\ 2C_s\,(4)$
$P4_2bc - C_{4v}^8$	$C_1\,(8),\ 2C_2\,(4)$
$I4mm - C_{4v}^9$	$C_1\,(16),\ C_{4v}\,(2),\ C_{2v}\,(4),\ 2C_s\,(8)$
$I4cm - C_{4v}^{10}$	$C_1\,(16),\ C_4\,(4),\ C_{2v}\,(4),\ C_s\,(8)$
$I4_1md - C_{4v}^{11}$	$C_1\,(16),\ C_{2v}\,(4),\ C_s\,(8)$
$I4_1cd - C_{4v}^{12}$	$C_1\,(16),\ C_2\,(8)$
$P\bar{4}2m - D_{2d}^1$	$C_1\,(8),\ 4\boldsymbol{D}_{2d}\ 2\boldsymbol{D}_2\,(2),\ 2C_{2v}\,(2),\ 5C_2\,(4),\ C_s\,(4)$
$P\bar{4}2c - D_{2d}^2$	$C_1\,(8),\ 4\boldsymbol{D}_2\,(2),\ 2\boldsymbol{S}_4\,(2),\ 7C_2\,(4)$
$P\bar{4}2_1m - D_{2d}^3$	$C_1\,(8),\ 2\boldsymbol{S}_4\,(2),\ C_{2v}\,(2),\ C_2\,(4),\ C_s\,(4)$
$P\bar{4}2_1c - D_{2d}^4$	$C_1\,(8),\ 2\boldsymbol{S}_4\,(2),\ 2C_2\,(4)$
$P\bar{4}m2 - D_{2d}^5$	$C_1\,(8),\ 4\boldsymbol{D}_{2d},\ 3C_{2v}\,(2),\ 2C_2\,(4),\ 2C_s\,(4)$
$P\bar{4}c2 - D_{2d}^6$	$C_1\,(8),\ 2\boldsymbol{D}_2\,(2),\ 2\boldsymbol{S}_4\,(2),\ 5C_2\,(4)$
$P\bar{4}b2 - D_{2d}^7$	$C_1\,(8),\ 2\boldsymbol{S}_4\,(2),\ 2\boldsymbol{D}_2\,(2),\ 4C_2\,(4)$
$P\bar{4}n2 - D_{2d}^8$	$C_1\,(8),\ 2\boldsymbol{S}_4\,(2),\ 2\boldsymbol{D}_2\,(2),\ 4C_2\,(4)$
$I\bar{4}m2 - D_{2d}^9$	$C_1\,(16),\ 4\boldsymbol{D}_{2d}\,(2),\ 2C_{2v}\,(4),\ 2C_2\,(8),\ C_s\,(8)$
$I\bar{4}c2 - D_{2d}^{10}$	$C_1\,(16),\ 2\boldsymbol{D}_2\,(4),\ 2\boldsymbol{S}_4\,(4),\ 4C_2\,(8)$
$I\bar{4}2m - D_{2d}^{11}$	$C_1\,(16),\ 2\boldsymbol{D}_{2d}\,(2),\ \boldsymbol{D}_2\,(4),\ \boldsymbol{S}_4\,(4),\ C_{2v}\,(4),\ 3C_2\,(8),\ C_s\,(8)$
$I\bar{4}2d - D_{2d}^{12}$	$C_1\,(16),\ 2\boldsymbol{S}_4\,(4),\ 2C_2\,(8)$
$P4/mmm - D_{4h}^1$	$C_1\,(16),\ 4\boldsymbol{D}_{4h},\ 2\boldsymbol{D}_{2h}\,(2),\ 2C_{4v}\,(2),\ 7C_{2v}\,(4),\ 5C_s\,(8)$
$P4/mcc - D_{4h}^2$	$C_1\,(16),\ 2\boldsymbol{D}_4\,(2),\ 2\boldsymbol{C}_{4h}\,(2),\ \boldsymbol{C}_{2h}\,(4),\ \boldsymbol{D}_2\,(4),\ 2C_4\,(4),\ 4C_2\,(8),\ C_s\,(8)$
$P4/nbm - D_{4h}^3$	$C_1\,(16),\ 2\boldsymbol{D}_4\,(2),\ 2\boldsymbol{D}_{2d}\,(2),\ 2\boldsymbol{C}_{2h}\,(4),\ C_4\,(4),\ C_{2v}\,(4),\ 4C_2\,(8),\ C_s\,(8)$
$P4/nnc - D_{4h}^4$	$C_1\,(16),\ 2\boldsymbol{D}_4\,(2),\ \boldsymbol{D}_2\,(4),\ \boldsymbol{S}_4\,(4),\ C_4\,(4),\ \boldsymbol{C}_i\,(8),\ 4C_2\,(8)$
$P4/mbm - D_{4h}^5$	$C_1\,(16),\ 2\boldsymbol{C}_{4h}\,(2),\ 2\boldsymbol{D}_{2h}\,(2),\ C_4\,(4),\ 3C_{2v}\,(4),\ C_s\,(8)$
$P4/mnc - D_{4h}^6$	$C_1\,(16),\ 2\boldsymbol{C}_{4h}\,(2),\ \boldsymbol{C}_{2h}\,(4),\ \boldsymbol{D}_2\,(4),\ C_4\,(4),\ 2C_2\,(8),\ C_s\,(8)$
$P4/nmm - D_{4h}^7$	$C_1\,(16),\ 2\boldsymbol{D}_{2d}\,(2),\ C_{4v}\,(2),\ 2\boldsymbol{C}_{2h}\,(4),\ C_{2v}\,(4),\ 2C_2\,(8),\ 2C_s\,(8)$
$P4/ncc - D_{4h}^8$	$C_1\,(16),\ \boldsymbol{D}_2\,(4),\ \boldsymbol{S}_4\,(4),\ C_4\,(4),\ \boldsymbol{C}_i\,(8),\ 2C_2\,(8)$
$P4_2/mmc - D_{4h}^9$	$C_1\,(16),\ 4\boldsymbol{D}_{2h}\,(2),\ 2\boldsymbol{D}_{2d}\,(2),\ 7C_{2v}\,(4),\ C_2\,(8),\ 3C_s\,(8)$
$P4_2/mcm - D_{4h}^{10}$	$C_1\,(16),\ 2\boldsymbol{D}_{2h}\,(2),\ 2\boldsymbol{D}_{2d}\,(2),\ \boldsymbol{D}_2\,(4),\ \boldsymbol{C}_{2h}\,(4),\ 4C_{2v}\,(4),\ 3C_2\,(8),\ 2C_s\,(8)$
$P4_2/nbc - D_{4h}^{11}$	$C_1\,(16),\ 3\boldsymbol{D}_2\,(4),\ \boldsymbol{S}_4\,(4),\ \boldsymbol{C}_i\,(8),\ 5C_2\,(8)$
$P4_2/nmm - D_{4h}^{12}$	$C_1\,(16),\ 2\boldsymbol{D}_{2d}\,(2),\ 2\boldsymbol{D}_2\,(4),\ 2\boldsymbol{C}_{2h}\,(4),\ C_{2v}\,(4),\ 5C_2\,(8),\ C_s\,(8)$
$P4_2/mbc - D_{4h}^{13}$	$C_1\,(16),\ 2\boldsymbol{C}_{2h}\,(4),\ \boldsymbol{S}_4\,(4),\ \boldsymbol{D}_2\,(4),\ 3C_2\,(8),\ C_s\,(8)$
$P4_2/mnm - D_{4h}^{14}$	$C_1\,(16),\ 2\boldsymbol{D}_{2h}\,(2),\ \boldsymbol{C}_{2h}\,(4),\ \boldsymbol{S}_4\,(4),\ 3C_{2v}\,(4),\ C_2\,(8),\ 2C_s\,(8)$
$P4_2/nmc - D_{4h}^{15}$	$C_1\,(16),\ 2\boldsymbol{D}_{2d}\,(2),\ 2C_{2v}\,(4),\ \boldsymbol{C}_i\,(8),\ C_2\,(8),\ C_s\,(8)$
$P4_2/ncm - D_{4h}^{16}$	$C_1\,(16),\ \boldsymbol{D}_2\,(4),\ \boldsymbol{S}_4\,(4),\ 2\boldsymbol{C}_{2h}\,(4),\ C_{2v}\,(4),\ 3C_2\,(8),\ C_s\,(8)$
$I4/mmm - D_{4h}^{17}$	$C_1\,(32),\ 2\boldsymbol{D}_{4h}\,(2),\ \boldsymbol{D}_{2h}\,(4),\ \boldsymbol{D}_{2d}\,(4),\ C_{4v}\,(4),\ \boldsymbol{C}_{2h}\,(8),\ 4C_{2v}\,(8),\ C_2\,(16),\ 3C_s\,(16)$

Table 1 (cont.)

Space group	Number and multiplicity factor of the sets of positions
$I4/mcm - D_{4h}^{18}$	$C_1(32)$, $\mathbf{D}_4(4)$, $\mathbf{D}_{2d}(4)$, $\mathbf{C}_{4h}(4)$, $\mathbf{D}_{2h}(4)$, $\mathbf{C}_{2h}(8)$, $C_4(8)$, $2C_{2v}(8)$, $2C_2(16)$, $2C_s(16)$
$I4_1/amd - D_{4h}^{19}$	$C_1(32)$, $2\mathbf{D}_{2d}(4)$, $2\mathbf{C}_{2h}(8)$, $2C_2(16)$, $C_s(16)$
$I4_1/acd - D_{4h}^{20}$	$C_1(32)$, $\mathbf{S}_4(8)$, $\mathbf{D}_2(8)$, $\mathbf{C}_i(16)$, $3C_2(16)$
$P3 - C_3^1$	$C_1(3)$, $3C_3$
$P3_1 - C_3^2$	$C_1(3)$
$P3_2 - C_3^3$	$C_1(3)$
$R3 - C_3^4$	$C_1(3)$, C_3
$P\bar{3} - C_{3i}^1$	$C_1(6)$, $2\mathbf{C}_{3i}$, $2C_3(2)$, $2\mathbf{C}_i(3)$
$R\bar{3} - C_{3i}^2$	$C_1(6)$, $2\mathbf{C}_{3i}$, $C_3(2)$, $2\mathbf{C}_i(3)$
$P312 - D_3^1$	$C_1(6)$, $6\mathbf{D}_3$, $3C_3(2)$, $2C_2(3)$
$P321 - D_3^2$	$C_1(6)$, $2\mathbf{D}_3$, $2C_3(2)$, $2C_2(3)$
$P3_112 - D_3^3$	$C_1(6)$, $2C_2(3)$
$P3_121 - D_3^4$	$C_1(6)$, $2C_2(3)$
$P3_212 - D_3^5$	$C_1(6)$, $2C_2(3)$
$P3_221 - D_3^6$	$C_1(6)$, $2C_2(3)$
$R32 - D_3^7$	$C_1(6)$, $2\mathbf{D}_3$, $C_3(2)$, $2C_2(3)$
$P3m1 - C_{3v}^1$	$C_1(6)$, $3C_{3v}$, $C_s(3)$
$P31m - C_{3v}^2$	$C_1(6)$, C_{3v}, $C_3(2)$, $C_s(3)$
$P3c1 - C_{3v}^3$	$C_1(6)$, $3C_3(2)$
$P31c - C_{3v}^4$	$C_1(6)$, $2C_3(2)$
$R3m - C_{3v}^5$	$C_1(6)$, C_{3v}, $C_s(3)$
$R3c - C_{3v}^6$	$C_1(6)$, $C_3(2)$
$P\bar{3}1m - D_{3d}^1$	$C_1(12)$, $2\mathbf{D}_{3d}$, $2\mathbf{D}_3(2)$, $C_{3v}(2)$, $2\mathbf{C}_{2h}(3)$, $C_3(4)$, $2C_2(6)$, $C_s(6)$
$P\bar{3}1c - D_{3d}^2$	$C_1(12)$, $3\mathbf{D}_3(2)$, $\mathbf{C}_{3i}(2)$, $2C_3(4)$, $\mathbf{C}_i(6)$, $C_2(6)$
$P\bar{3}m1 - D_{3d}^3$	$C_1(12)$, $2\mathbf{D}_{3d}$, $2C_{3v}(2)$, $2\mathbf{C}_{2h}(3)$, $2C_2(6)$, $C_s(6)$
$P\bar{3}c1 - D_{3d}^4$	$C_1(12)$, $\mathbf{D}_3(2)$, $\mathbf{C}_{3i}(2)$, $2C_3(4)$, $\mathbf{C}_i(6)$, $C_2(6)$
$R\bar{3}m - D_{3d}^5$	$C_1(12)$, $2\mathbf{D}_{3d}$, $C_{3v}(2)$, $2\mathbf{C}_{2h}(3)$, $2C_2(6)$, $C_s(6)$
$R\bar{3}c - D_{3d}^6$	$C_1(12)$, $\mathbf{D}_3(2)$, $\mathbf{C}_{3i}(2)$, $C_3(4)$, $\mathbf{C}_i(6)$, $C_2(6)$
$P6 - C_6^1$	$C_1(6)$, C_6, $C_3(2)$, $C_2(3)$
$P6_1 - C_6^2$	$C_1(6)$
$P6_5 - C_6^3$	$C_1(6)$
$P6_2 - C_6^4$	$C_1(6)$, $2C_2(3)$
$P6_4 - C_6^5$	$C_1(6)$, $2C_2(3)$
$P6_3 - C_6^6$	$C_1(6)$, $2C_3(2)$
$P\bar{6} - C_{3h}^1$	$C_1(6)$, $6\mathbf{C}_{3h}$, $3C_3(2)$, $2C_s(3)$
$P6/m - C_{6h}^1$	$C_1(12)$, $2\mathbf{C}_{6h}$, $2\mathbf{C}_{3h}(2)$, $C_6(2)$, $2\mathbf{C}_{2h}(3)$, $C_3(4)$, $C_2(6)$, $2C_s(6)$
$P6_3/m - C_{6h}^2$	$C_1(12)$, $3\mathbf{C}_{3h}(2)$, $\mathbf{C}_{3i}(2)$, $2C_3(4)$, $\mathbf{C}_i(6)$, $C_s(6)$
$P622 - D_6^1$	$C_1(12)$, $2\mathbf{D}_6$, $2\mathbf{D}_3(2)$, $C_6(2)$, $2\mathbf{D}_2(3)$, $C_3(4)$, $5C_2(6)$
$P6_122 - D_6^2$	$C_1(12)$, $2C_2(6)$
$P6_522 - D_6^3$	$C_1(12)$, $2C_2(6)$
$P6_222 - D_6^4$	$C_1(12)$, $4\mathbf{D}_2(3)$, $6C_2(6)$
$P6_422 - D_6^5$	$C_1(12)$, $4\mathbf{D}_2(3)$, $6C_2(6)$
$P6_322 - D_6^6$	$C_1(12)$, $4\mathbf{D}_3(2)$, $2C_3(4)$, $2C_2(6)$
$P6mm - C_{6v}^1$	$C_1(12)$, C_{6v}, $C_{3v}(2)$, $C_{2v}(3)$, $2C_s(6)$
$P6cc - C_{6v}^2$	$C_1(12)$, $C_6(2)$, $C_3(4)$, $C_2(6)$
$P6_3cm - C_{6v}^3$	$C_1(12)$, $C_{3v}(2)$, $C_3(4)$, $C_s(6)$
$P6_3mc - C_{6v}^4$	$C_1(12)$, $2C_{3v}(2)$, $C_s(6)$
$P\bar{6}m2 - D_{3h}^1$	$C_1(12)$, $6\mathbf{D}_{3h}$, $3C_{3v}(2)$, $2C_{2v}(3)$, $3C_s(6)$
$P\bar{6}c2 - D_{3h}^2$	$C_1(12)$, $3\mathbf{D}_3(2)$, $3\mathbf{C}_{3h}(2)$, $3C_3(4)$, $C_2(6)$, $C_s(6)$
$P\bar{6}2m - D_{3h}^3$	$C_1(12)$, $2\mathbf{D}_{3h}$, $2\mathbf{C}_{3h}(2)$, $C_{3v}(2)$, $2C_{2v}(3)$, $C_3(4)$, $3C_s(6)$
$P\bar{6}2c - D_{3h}^4$	$C_1(12)$, $\mathbf{D}_3(2)$, $3\mathbf{C}_{3h}(2)$, $2C_3(4)$, $C_2(6)$, $C_s(6)$
$P6/mmm - D_{6h}^1$	$C_1(24)$, $2\mathbf{D}_{6h}$, $2\mathbf{D}_{3h}(2)$, $C_{6v}(2)$, $2\mathbf{D}_{2h}(3)$, $C_{3v}(4)$, $5C_{2v}(6)$, $4C_s(12)$
$P6/mcc - D_{6h}^2$	$C_1(24)$, $\mathbf{D}_6(2)$, $\mathbf{C}_{6h}(2)$, $\mathbf{D}_3(4)$, $\mathbf{C}_{3h}(4)$, $C_6(4)$, $\mathbf{D}_2(6)$, $\mathbf{C}_{2h}(6)$, $C_3(8)$, $3C_2(12)$, $C_s(12)$
$P6_3/mcm - D_{6h}^3$	$C_1(24)$, $\mathbf{D}_{3h}(2)$, $\mathbf{D}_{3d}(2)$, $\mathbf{C}_{3h}(4)$, $\mathbf{D}_3(4)$, $C_6(4)$, $\mathbf{C}_{2h}(6)$, $C_{2v}(6)$, $C_3(8)$, $C_2(12)$, $2C_s(12)$
$P6_3/mmc - D_{6h}^4$	$C_1(24)$, $\mathbf{D}_{3d}(2)$, $3\mathbf{D}_{3h}(2)$, $2C_{3v}(4)$, $\mathbf{C}_{2h}(6)$, $C_{2v}(6)$, $C_2(12)$, $2C_s(12)$
$P23 - T^1$	$C_1(12)$, $2\mathbf{T}$, $2\mathbf{D}_2(3)$, $C_3(4)$, $4C_2(6)$
$F23 - T^2$	$C_1(48)$, $4\mathbf{T}(4)$, $C_3(16)$, $2C_2(24)$
$I23 - T^3$	$C_1(24)$, $\mathbf{T}(2)$, $\mathbf{D}_2(6)$, $C_3(8)$, $2C_2(12)$
$P2_13 - T^4$	$C_1(12)$, $C_3(4)$
$I2_13 - T^5$	$C_1(24)$, $C_3(8)$, $C_2(12)$
$Pm3 - T_h^1$	$C_1(24)$, $2\mathbf{T}_h$, $2\mathbf{D}_{2h}(3)$, $4C_{2v}(6)$, $C_3(8)$, $2C_s(12)$
$Pn3 - T_h^2$	$C_1(24)$, $\mathbf{T}(2)$, $2\mathbf{C}_{3i}(4)$, $\mathbf{D}_2(6)$, $C_3(8)$, $2C_2(12)$
$Fm3 - T_h^3$	$C_1(96)$, $2\mathbf{T}_h(4)$, $\mathbf{T}(8)$, $\mathbf{C}_{2h}(24)$, $C_{2v}(24)$, $C_3(32)$, $C_2(48)$, $C_s(48)$
$Fd3 - T_h^4$	$C_1(96)$, $2\mathbf{T}(8)$, $2\mathbf{C}_{3i}(16)$, $C_3(32)$, $C_2(48)$
$Im3 - T_h^5$	$C_1(48)$, $\mathbf{T}_h(2)$, $\mathbf{D}_{2h}(6)$, $\mathbf{C}_{3i}(8)$, $2C_{2v}(12)$, $C_3(16)$, $C_s(24)$
$Pa3 - T_h^6$	$C_1(24)$, $2\mathbf{C}_{3i}(4)$, $C_3(8)$
$Ia3 - T_h^7$	$C_1(48)$, $2\mathbf{C}_{3i}(8)$, $C_3(16)$, $C_2(24)$
$P432 - O^1$	$C_1(24)$, $2\mathbf{O}$, $2\mathbf{D}_4(3)$, $2C_4(6)$, $C_3(8)$, $3C_2(12)$
$P4_232 - O^2$	$C_1(24)$, $\mathbf{T}(2)$, $2\mathbf{D}_3(4)$, $3\mathbf{D}_2(6)$, $C_3(8)$, $5C_2(12)$
$F432 - O^3$	$C_1(96)$, $2\mathbf{O}(4)$, $\mathbf{T}(8)$, $\mathbf{D}_2(24)$, $C_4(24)$, $C_3(32)$, $3C_2(48)$
$F4_132 - O^4$	$C_1(96)$, $2\mathbf{T}(8)$, $2\mathbf{D}_3((16)$, $C_3(32)$, $2C_2(48)$
$I432 - O^5$	$C_1(48)$, $\mathbf{O}(2)$, $\mathbf{D}_4(6)$, $\mathbf{D}_3(8)$, $\mathbf{D}_2(12)$, $C_4(12)$, $C_3(16)$, $3C_2(24)$

Table 1 (cont.)

Space group	Number and multiplicity factor of the sets of positions
$P4_332 - O^6$	$C_1\,(24),\ 2\mathbf{D}_3\,(4),\ C_3\,(8),\ C_2\,(12)$
$P4_132 - O^7$	$C_1\,(24),\ 2\mathbf{D}_3\,(4),\ C_3\,(8),\ C_2\,(12)$
$I4_132 - O^8$	$C_1\,(48),\ 2\mathbf{D}_3\,(8),\ 2\mathbf{D}_2\,(12),\ C_3\,(16),\ 3C_2\,(24)$
$P\bar{4}3m - T_d^1$	$C_1\,(24),\ 2\mathbf{T}_d,\ 2\mathbf{D}_{2d}\,(3),\ C_{3v}\,(4),\ 2C_{2v}\,(6),\ C_2\,(12),\ C_s\,(12)$
$F\bar{4}3m - T_d^2$	$C_1\,(96),\ 4\mathbf{T}_d\,(4),\ C_{3v}\,(16),\ 2C_{2v}\,(24),\ C_s\,(48)$
$I\bar{4}3m - T_d^3$	$C_1\,(48),\ \mathbf{T}_d\,(2),\ \mathbf{D}_{2d}\,(6),\ C_{3v}\,(8),\ \mathbf{S}_4\,(12),\ C_{2v}\,(12),\ C_2\,(24),\ C_s\,(24)$
$P\bar{4}3n - T_d^4$	$C_1\,(24),\ \mathbf{T}\,(2),\ \mathbf{D}_2\,(6),\ 2\mathbf{S}_4\,(6),\ C_3\,(8),\ 3C_2\,(12)$
$F\bar{4}3c - T_d^5$	$C_1\,(96),\ 2\mathbf{T}\,(8),\ 2\mathbf{S}_4\,(24),\ C_3\,(32),\ 2C_2\,(48)$
$I\bar{4}3d - T_d^6$	$C_1\,(48),\ 2\mathbf{S}_4\,(12),\ C_3\,(16),\ C_2\,(24)$
$Pm3m - O_h^1$	$C_1\,(48),\ 2\mathbf{O}_h,\ 2\mathbf{D}_{4h}\,(3),\ 2C_{4v}\,(6),\ C_{3v}\,(8),\ 3C_{2v}\,(12),\ 3C_s\,(24)$
$Pn3n - O_h^2$	$C_1\,(48),\ \mathbf{O}\,(2),\ \mathbf{D}_4\,(6),\ \mathbf{C}_{3i}\,(8),\ \mathbf{S}_4\,(12),\ C_4\,(12),\ C_3\,(16),\ 2C_2\,(24)$
$Pm3n - O_h^3$	$C_1\,(48),\ \mathbf{T}_h\,(2),\ \mathbf{D}_{2h}\,(6),\ 2\mathbf{D}_{2d}\,(6),\ \mathbf{D}_3\,(8),\ 3C_{2v}\,(12),\ C_3\,(16),\ C_2\,(24),\ C_s\,(24)$
$Pn3m - O_h^4$	$C_1\,(48),\ \mathbf{T}_d\,(2),\ 2\mathbf{D}_{3d}\,(4),\ \mathbf{D}_{2d}\,(6),\ C_{3v}\,(8),\ \mathbf{D}_2\,(12),\ C_{2v}\,(12),\ 3C_2\,(24),\ C_s\,(24)$
$Fm3m - O_h^5$	$C_1\,(192),\ 2\mathbf{O}_h\,(4),\ \mathbf{T}_d\,(8),\ \mathbf{D}_{2h}\,(24),\ C_{4v}\,(24),\ C_{3v}\,(32),\ 3C_{2v}\,(48),\ 2C_s\,(96)$
$Fm3c - O_h^6$	$C_1\,(192),\ \mathbf{O}\,(8),\ \mathbf{T}_h\,(8),\ \mathbf{D}_{2d}\,(24),\ \mathbf{C}_{4h}\,(24),\ C_{2v}\,(48),\ C_4\,(48),\ C_3\,(64),\ C_2\,(96),\ C_s\,(96)$
$Fd3m - O_h^7$	$C_1\,(192),\ 2\mathbf{T}_d\,(8),\ 2\mathbf{D}_{3d}\,(16),\ C_{3v}\,(32),\ C_{2v}\,(48),\ C_s\,(96),\ C_2\,(96)$
$Fd3c - O_h^8$	$C_1\,(192),\ \mathbf{T}\,(16),\ \mathbf{D}_3\,(32),\ \mathbf{C}_{3i}\,(32),\ \mathbf{S}_4\,(48),\ C_3\,(64),\ 2C_2\,(96)$
$Im3m - O_h^9$	$C_1\,(96),\ \mathbf{O}_h\,(2),\ \mathbf{D}_{4h}\,(6),\ \mathbf{D}_{3d}\,(8),\ \mathbf{D}_{2d}\,(12),\ C_{4v}\,(12),\ C_{3v}\,(16),\ 2C_{2v}\,(24),\ C_2\,(48),\ 2C_s\,(48)$
$Ia3d - O_h^{10}$	$C_1\,(96),\ \mathbf{C}_{3i}\,(16),\ \mathbf{D}_3\,(16),\ \mathbf{D}_2\,(24),\ \mathbf{S}_4\,(24),\ C_3\,(32),\ 2C_2\,(48)$

Interaction of the Vibrations of Several Molecules or Ions in the Unit Cell. Allowance for the site symmetry of the molecules or complex ions in the crystal enables us to explain the splitting of degenerate vibrations and certain changes in the selection rules. However, this method provides no explanation for the splitting of nondegenerate internal vibrations of molecules, not infrequently observed in the spectra of crystals. Equally, conclusions as to the polarization of the bands and also the selection rules, particularly for overtones and combination frequencies, drawn on the basis of this method, are often at variance with experimental facts.

A. S. Davydov [7, 8] developed a theory of the vibrational spectra of molecular crystals enabling him to explain the splitting of nondegenerate vibrations of the molecules and predict the polarization of the components formed by the splitting. The theory is based on the concept of the collective nature of the excitation of a crystal by light and the resonance character of the interaction of the molecular vibrations in the crystal. As a result of the forces of interaction between the molecules in the crystal, the excitation of one molecule is not localized in the latter but is transferred to other molecules; this may be described as the propagation of waves of excitation or quasiparticles (excitons) in the crystal. The effective mass of an exciton may be expressed in terms of the matrix element of the transfer of excitation between neighboring molecules M in the direction of the lattice vector a along which the exciton is propagating, $m^* = h^2/2Ma^2$.

If there are q translationally nonequivalent positions for molecules in the crystal (q molecules in the unit cell), q bands of excited states of the crystal correspond to one nondegenerate excited state of an isolated molecule; the moments of the transitions into these states are different in magnitude and orientation. In an ideal crystal, when the molecules maintain fixed positions in the lattice, transitions under the influence of light satisfy the law of conservation of momentum in the form $\vec{k} = \vec{Q}$, where $\vec{k}$ and $\vec{Q}$ are the wave vectors of the exciton and absorbed light. Since $Q = 2\pi/\lambda$ (λ is the wavelength of the light) and k is of the order of $1/a$ (a is the basic vector of the lattice), the condition $\vec{k} = \vec{Q}$ means that $\vec{k} \approx 0$. In other words, only limiting vibrations are excited by light, and the exciton mechanism of the excitation of a molecular crystal leads to the splittings of the bands of intramolecular transitions into q discrete bands.

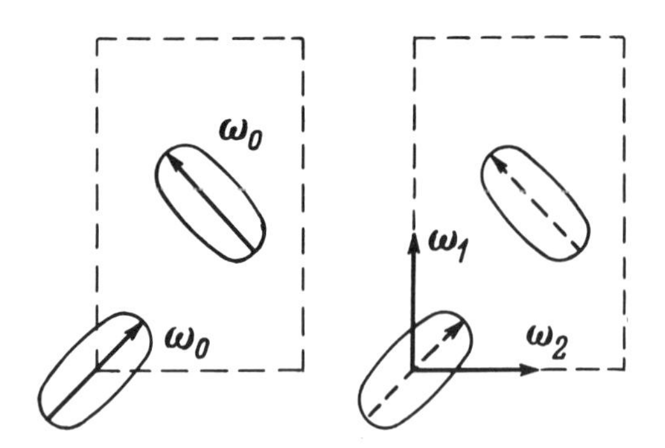

Fig. 5. Schematic representation of the transition from the vibrations of noninteracting molecules ω_0 to coupled molecular vibrations in the crystal, ω_1, ω_2.

The scheme indicated in Fig. 5 illustrates the essence of the transition from the model of oriented noninteracting molecules in the crystal to the model used in the Davydov theory. For two molecules in the unit cell, in the first case the directions of the dipole moments of the transition are independent and correspond to the same frequency ω_0; in the second case there should be two different frequencies ω_1 and ω_2 corresponding to the collective excitations of the molecule, with dipole (transition) moments corresponding to the sum and difference of the moments of individual molecular transitions. Thus two absorption bands with mutually perpendicular polarization should be observed.

The extent of the splitting of the intramolecular vibrations is determined by the matrix elements of the transfer of excitation between two molecules in the theory. For the calculation of these the Coulomb interaction of the electrons and the nuclei of the two molecules is expanded in series in reciprocal powers of the intermolecular distances. The terms of this series characterize the interaction of multipole fields of various orders. The greatest splitting should be shown by transitions active in the infrared spectrum associated with a change in dipole moment, while a considerably smaller splitting is expected for inactive transitions. To a first approximation the extent of the splitting is proportional to the square of the derivative of the dipole moment with respect to the normal coordinate $\nu\,\Delta\nu \sim (\partial\mu/\partial Q)^2$. Without further considering the methods of theoretically estimating the extent of the resonance splitting, we simply mention Davydov's review of these questions [8] and also references [22, 25].

The resonance character of the splitting of the frequencies may in some cases be particularly convincingly proved by comparing the spectra of a purely molecular crystal and a dilute solid solution of the same molecules in a matrix crystal with isotopically substituted molecules, if the two crystals are isomorphous. In the spectrum of the solid solution, resonance splitting of the internal vibrations of the molecules of low concentration should vanish. At the same time the splitting of the degenerate vibrations of the molecule due to its site symmetry in the crystal should be preserved. The structure of the band of the degenerate bending vibration of $CH_3(CD_3)$ in CH_3I and CD_3I crystals and their solid solutions was studied in some detail from these points of view in [22]. It was later shown [23] that a correct interpretation of the structure of this band also requires allowance for intramolecular resonance (Fermi), the Davydov splitting of the band being more substantial than the splitting due to site symmetry in the crystal. We note that from general considerations the influence of site symmetry on the spectrum should be more considerable for crystals with complex ions, while the Davydov effect, for which the "equilibrium" value of μ is less important than $\partial\mu/\partial Q$, may play a considerable part even in crystals with nonpolar molecules.

For the majority of real systems the theoretical calculation of the energy and wave functions of the exciton states of a crystal presents considerable difficulties. However, for the purpose of interpreting experimental data it is often sufficient to be able to determine the selection rules (the number of active components of any particular intramolecular vibration in

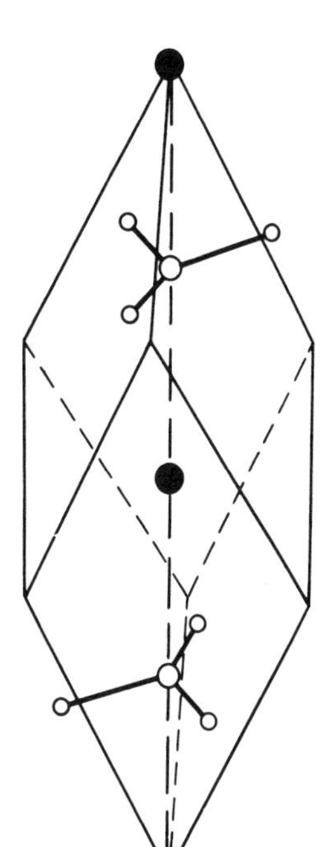

Fig. 6. Structure of the unit cell of $NaNO_3$.

the Raman or infrared absorption spectrum) and the expected polarization of the bands. This may be effected by group-theory methods, by comparing the symmetry properties of the molecule and the crystal. If the influence of the site symmetry on the vibrations of the molecule in the crystal is already taken into account, then allowing for the Davydov effect reduces to passing from the representations of the site group to those of the factor group of the space group of the crystal. This problem, like the transition from the molecular symmetry group to the site group, may be solved by means of formulas (12) to (14) or correlation tables of the representations of groups and their subgroups. In many cases, as already mentioned, it is sufficient to know the symmetry of the molecules, their number in the unit cell, and the space group of the crystal.

We may note one useful special result: it is not difficult to show that, in a centrosymmetric crystal, in which centrosymmetric molecules occupy positions on centers of symmetry of the crystal, the internal vibrations of the molecule cannot be both infrared and Raman active (the mutual exclusion rule).

Let us illustrate the idea of passing from the molecular to the site and then to the factor group as set out in this section by a specific example. Calcite and sodium nitrate crystals, $CaCO_3$ and $NaNO_3$, belong to the D_{3d}^6 space group and contain two formula units in the unit cell. This cell is the smallest translationally nonidentical cell, since (Table 1) the multiplicity factor of the general position equals 12, which corresponds to the order of the factor group of a crystal isomorphic to D_{3d}^6. This space group, like many other groups of the trigonal system, may be described not only by the rhombohedral cell shown in Fig. 6 but also by a hexagonal cell of three times the volume. Since the intrinsic symmetry of an isolated complex anion XO_3 is D_{3h}, it is easy to convince oneself with the aid of the scheme of Fig. 4 and the list of particular positions in Table 1 that the only possible site symmetry group of these ions is D_3. Another set of particular positions with multiplicity factor 2 in fact corresponds to the site group $S_6 = C_{3i}$, not constituting a subgroup of the molecular group. Moreover, the positions of the two monatomic cations are also established unambiguously as S_6, since the other set with multiplicity factor 2 is already "occupied." This set has fixed values of all three coordinates, as we may easily convince ourselves with the help of the coordinates of particular positions in [24], and cannot be occupied twice.

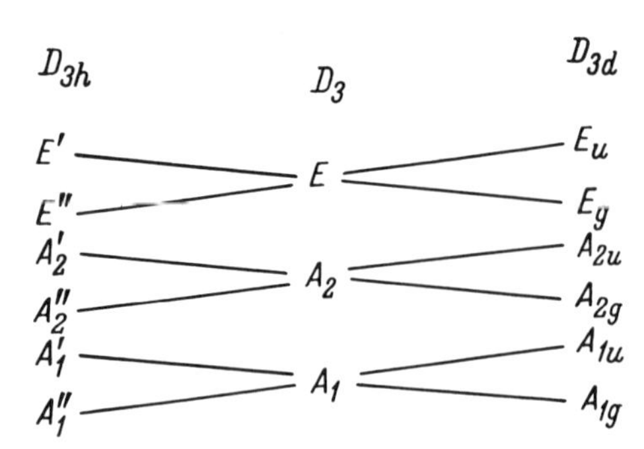

Fig. 7. Scheme of correlations between the irreducible representations of the intrinsic symmetry group of the NO_3 ion, its site group in the $NaNO_3$ crystal, and the factor group of the crystal with symmetry D_{3d}^6.

Let us consider the changes in the internal vibrations of the anions allowing for their site symmetry. A direct comparison of the tables of characters of the groups D_{3h} and D_3 enables us, if we wish, to use formulas (12) and (14) and thus express the irreducible representations of the first group in terms of the irreducible representations of the second

$$E' = E,\ E'' = E,\ A_2'' = A_2,\ A_2' = A_2,\ A_1'' = A_1,\ A_1' = A_1.$$

The six normal vibrations of the four-atom anion XO_3 are distributed in the following way with respect to the irreducible representations of the D_{3h} group: three stretching vibrations relating to types A_1' and E', of which the totally symmetric one is only active in the Raman spectrum, while the two degenerate ones are active in the infrared spectrum also; two deformation vibrations belong to the degenerate representation E' and one to A_2'', only active in the infrared spectrum. The selection rules, the orientations of the dipole moments, and the polarization of the Raman lines suffer no changes on passing to the site group in the present case.

On passing to the factor group of the space group, we may analogously express the representations of the site group D_3 in terms of the irreducible representations of the D_{3d} group

$$E = E_u + E_g,\ A_1 = A_{1g} + A_{1u},\ A_2 = A_{2g} + A_{2u}.$$

This result may also be obtained without using formulas (12) to (14) by using the correlation tables in [18].

Thus the interaction of the vibrations of two translationally nonequivalent XO_3 groups should lead to the splitting of the internal vibrations into two components with different selection rules. The whole sequence of operations here described may conveniently be written in the form of a scheme of correlations between the molecular, site, and factor group of the space group (Fig. 7). It follows immediately from this scheme that the degenerate vibrations of the anion retain degeneracy (as a result of the trigonal site symmetry), but each couples with the corresponding vibration of the other anion in the set to give two degenerate components, one of which is active in the infrared spectrum and the other in the Raman spectrum. Each of the two nondegenerate vibrations also couples to give two components: only one component of the nondegenerate deformation vibration is active in the infrared spectrum, and both are inactive in the Raman spectrum. The totally symmetric vibration remains inactive in absorption.

External Vibrations of Molecules in a Crystal. Classification of All Fundamental Vibrations of a Crystal.† Up to the present we have only considered the internal vibrations of molecules and complex ions in a crystal. Let us now turn to their external degrees of freedom, i.e., to "lattice vibrations," so as to be able to classify all $3m-3$ optical branches in the spectrum of the crystal. By analogy with the approximate divi-

† The tables of Adams and Newton [49] are particularly useful in this context.

sion into internal and external vibrations it is in practice useful to divide the latter into translational and rotation vibrations.

The main sources of information regarding the external vibrations of molecules in crystals, the frequencies of which are very low (this of course does not always apply to crystals containing complex ions), were until recently Raman spectra. It is therefore useful to mention the semiempirical rule [3] according to which the lines of rotational vibrations are usually stronger in the Raman spectrum than lines of translational vibrations.

The application of group-theory methods to the analysis of the motions of the 3m translationally nonequivalent atoms of a crystal, using the factor group of the space group, enables us to obtain the distribution of these normal vibrations with respect to its irreducible representations quite quickly [3]. For this purpose we must use formula (14), which enables us to determine how many times the irreducible representation Γ_i is contained in the representation Γ of a finite group if the characters of these representations are known. In the case under consideration the characters $\chi_i(R)$ of the irreducible representations Γ_i of the factor group coincide with the characters of the representations of the point group isomorphic with it, while the characters $\chi'_\rho(R)$ of the representation Γ are determined by the rules about to be given.

If Γ is a representation determined by all 3m Cartesian coordinates of m nonequivalent atoms, then its character

$$\chi'_\rho(R) = U_R(\pm 1 + 2\cos\varphi_R) \tag{15}$$

where U_R is the number of atoms invariant to the transformation R, while the signs (+ and −) as earlier depend on whether R is a simple or mirror rotation through an angle φ. Putting (15) into (14) gives n_i, a number which in the present case corresponds to the number of all the normal vibrations, internal and external, including acoustic vibrations, transforming by the irreducible representation Γ_i of the factor group. In order to use (14) for obtaining T, the number of translations, i.e., of acoustic branches in the representation Γ_i, we must clearly take $U_R = 1$ for any R in (15).

If the m atoms may be divided into q groups, molecules or ions (single- or polyatomic), in order to calculate T′, the number of optical external vibrations of the translational type, we must take

$$\chi'_\rho(R) = [U_R(q) - 1](\pm 1 + 2\cos\varphi_R) \tag{16}$$

where $U_R(q)$ corresponds to the number of groups (out of the total number q) invariant relative to R.

In order to obtain R′, the number of external rotational vibrations, the character $\chi'_\rho(R)$ is taken in the form

$$\chi'_\rho(R) = U_R(q - v)(1 \pm 2\cos\varphi_R) \tag{17}$$

Here v is the number of monatomic groups not possessing rotational degrees of freedom, and hence $U_R(q-v)$ is the number of polyatomic groups (of the total number q − v in the cell) invariant to the operation R.

The number n'_i, i.e., the number of internal vibrations of the q − v molecules, may be obtained by subtracting from n_i the number of acoustic and external optical (translational and rotational) vibrations, or directly by substituting

$$\chi'_\rho(R) = [U_R - U_R(q)](\pm 1 + 2\cos\varphi_R) - U_R(q - v)(1 \pm 2\cos\varphi_R) \tag{18}$$

into (14).

Table 2. Normal Vibrations of a Calcite Crystal

Space group	Characters of factor group						Number of normal vibrations					Selection rules
D_{3d}^6	E	$2S_6$	$2C_3$	i	$3\sigma_d$	$3C_2$	n_i	T	T'	R'	n_i'	
A_{1g}	1	1	1	1	1	1	1	0	0	0	1	$\alpha_{xx}+\alpha_{yy},\ \alpha_{zz}$
A_{2g}	1	1	1	1	−1	−1	3	0	1	1	1	—
A_{1u}	1	−1	1	−1	−1	1	2	0	1	0	1	—
A_{2u}	1	−1	1	−1	1	−1	4	1	1	1	1	M_z
E_g	2	−1	−1	2	0	0	4	0	1	1	2	$(\alpha_{xx}-\alpha_{yy},\ \alpha_{xy}),\ (\alpha_{yz},\ \alpha_{zx})$
E_u	2	−1	−1	−2	0	0	6	1	2	1	2	$(M_x,\ M_y)$
$U_R(n_i)$	10	2	4	2	0	4						
$U_R(q)$	4	2	4	2	0	2						
$U_R(q-v)$	2	0	2	0	0	2						
$h_\rho\chi_\rho'(n_i)$	30	0	0	−6	0	−12						
$h_\rho\chi_\rho'(T)$	3	0	0	−3	3	−3						
$h_\rho\chi_\rho'(T')$	9	0	0	−3	−3	−3						
$h_\rho\chi_\rho'(R')$	6	0	0	0	0	6						

Let us use these methods to analyze the vibrations of the calcite cell, the internal vibrations of the anions of which were considered earlier. For two formula-units in the cell m = 10, q = 4, and v = 2. Since the analysis is carried out with the aid of the factor group of the crystal, an atom must be regarded as invariant toward a symmetry operation as a result of which it remains in its former place or is transferred to a translationally equivalent position. An atom is noninvariant toward an operation transferring it to another point of the set. Referring to Fig. 6, we may establish (as also follows from the site symmetry S_6) the invariance of the Ca^{2+} ions to the operations E, S_6, C_3, i.

The C atoms (and the CO_3^{2-} ions as whole groups) in positions D_3 are invariant to E, C_3, C_2. The six O atoms lying in points with site symmetry C_2, are invariant toward E; however, to each of the three C_2 axes only two of them are invariant.† From the foregoing considerations the values of $U_R(n_i)$, $U_R(q)$ and $U_R(q-v)$ are immediately calculated for all R of the factor group. This enables us to calculate n_i, T, T′, R′, n_i' by means of (14) to (18). The results are given in Table 2 together with the selection rules (the components of the dipole moment and polarization tensor transforming by the irreducible representation in question are indicated). It follows in particular from Table 2 that, in the calcite crystal, two simple and two doubly degenerate rotational vibrations of the CO_3 ions and three simple and three doubly degenerate translational vibrations correspond to the lattice vibrations.

It should be noted that the division of the external vibrations of the molecules in the crystal into translational and rotational in general bears an approximate character. More detailed theoretical investigations of the external vibrations of molecules in a crystal carried out by Ansel'm and Porfir'eva [26] showed that the separation of translational and orientational vibrations only occurred for a specific symmetry of the crystal and only for the frequencies of the

† We should direct attention to the fact that the ion CO_3 itself is invariant as a whole to all three C_2 axes, while for each of the operations of class C_2 only one of its oxygen atoms transforms into itself.

limiting vibrations. Of course, the classification of crystal vibrations into external and internal degrees of freedom bears no less an arbitrary character. In crystals with molecules linked to each other by relatively weak Van der Waals forces, the greater part of the internal vibrations of the molecules actually possess considerably higher frequencies than the frequencies of the lattice vibrations. As regards crystals with complex ions, this is only valid, strictly speaking, in those cases in which the frequencies of the vibrations of the cation and anion sublattices are determined by purely electrostatic interactions and are fairly low compared with the frequencies of the vibrations of the essentially covalent bonds in the complex ion. In other cases such a classification can only be applied with great caution, and its validity demands special justification for every new group of subjects.

Overtones and Combination Frequencies. Interaction of Internal and External Vibrations. The preceding subsections have, generally speaking, exhausted the uses of the symmetry properties of molecular crystals in interpreting the spectrum in the frequency range of the fundamental vibrations. Here we shall briefly consider overtones and combination frequencies, in particular, combinations of internal and external vibrations, in relation to the possibility that such interactions may exert an influence on the interpretation of the results obtained when studying the fundamental vibrations. The determination of the selection rules for overtones or combinations in the case of the limiting frequencies causes no difficulty and is effected by expanding the direct product of the corresponding representations of the factor group of the crystal with respect to its irreducible representations, exactly as in the case of the spectra of isolated molecules [18]. The characters χ_{ij} are determined in this case by the ordinary rule

$$\chi_{ij}(R)=\chi_i(R)\chi_j(R) \tag{19}$$

It has already been mentioned however, that, in principle, any combinations satisfying the requirement $\vec{k}_i = \vec{k}_j$ may be active in the spectrum of the crystal. Hence the use of the selection rules obtained by means of the factor group of the crystal may only give rough information suitable for the interpretation of the strongest overtones and combinations. When studying the infrared spectra of polycrystalline samples in the form of thin layers of crystals dispersed in a suitable medium, the observed frequencies may usually be reduced to transitions corresponding to the active representations of the factor group of the crystal [29].

The greatest practical interest among the overtones and combinations appearing in the spectra of molecular crystals lies in combinations of branches corresponding to the internal vibrations of the molecules with branches of lattice vibrations. In view of the low frequencies of the latter such combination frequencies may be observed close to the frequency of the corresponding intramolecular vibration, resulting in the development of a complex structure of the band and sometimes an apparent breach of the selection rules if the actual intramolecular vibration is forbidden. Physically, transitions of this type correspond to the simultaneous excitation of a molecule by light and the displacement of its equilibrium position in the lattice, i.e., the simultaneous development of an exciton and one or several phonons. Then, in contrast to the ideal case of molecules fixed in the lattice considered earlier, the conditions for the activity of combination transitions in the spectrum may be written both as a law of conservation of energy in the form

$$h\nu=\sum_i(\pm h\nu_i) \tag{20}$$

where† ν is the frequency of the light wave, ν_i are the frequencies of the combining branches,

† The signs (+ and −) correspond to the emission and absorption of light. In view of the low energy of the lattice levels, even at ordinary temperatures the population of the excited states is quite large, so that there is point in considering both possibilities.

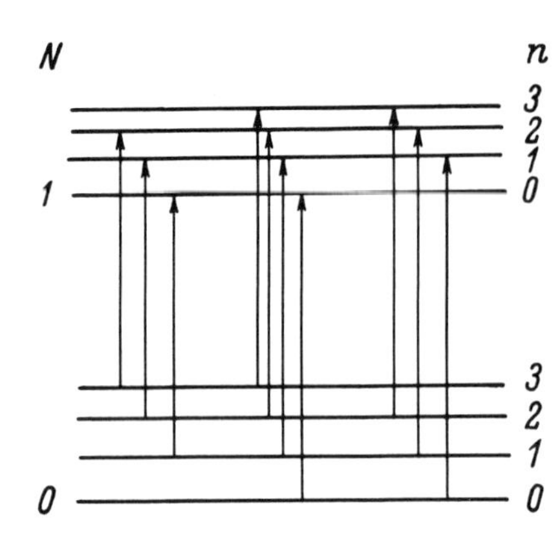

Fig. 8. Interaction of internal and external vibrations of the molecules in a crystal. N and n are quantum numbers; the transitions are $\Delta n = 0, \pm 1$, $\Delta N = 1$.

and also as a law of conservation of momentum in the form

$$\vec{Q} = \sum_i \vec{k}_i + n\vec{b}. \tag{21}$$

where $\vec{Q}$ is the wave vector of the light wave, $\vec{k}_i$ are the wave vectors of the combining vibrations (internal and external), $\vec{b}$ is the reciprocal lattice vector, and n is any whole number. Thus, on satisfying certain conditions of polarization and (20) or (21), all 3mN normal vibrations of a crystal of N cells may take part in optically active combinations. Moreover, in addition to the optical branches, acoustic branches (inactive in the form of fundamental vibrations) may participate in the combinations. This does not mean that combinations with acoustic vibrations, or quite generally with any branches of the lattice vibrations, will only lead to continuous absorption, since the existence of singular points in the frequency distribution function $g(\nu)$ may cause the development of absorption maxima. Hence we must remember that analysis of the experimentally observed structure of the bands formed by transitions $\nu_{int} \pm \nu_{ext}$ may lead to values of ν_{ext} (for optical vibrations) differing from the limiting frequencies obtained from first-order spectra. In solving the phonon structure of the bands of internal transitions, valuable information regarding the phonon spectrum may be given by not only Raman and long-wave infrared spectra but also inelastic neutron scattering, the elastic constants of the crystals, and the specific heat, particularly if we are concerned with optically inactive acoustic phonons.

Relatively few investigators have attempted to allow for the contributions of all the vibrational branches (not only the limiting vibrations) to the observed intensity distribution in spectra of higher orders. The first efforts in this direction related to the Raman spectrum of rock salt, which contains no active fundamental vibrations. The second-order spectrum, calculated by means of the frequency-distribution function, agreed satisfactorily with the very weak and diffuse Raman scattering pattern from this crystal. The diamond spectrum was considered in an analogous manner. Recently the phonon structure of the infrared spectrum of semiconductors has been frequently discussed.

The majority of investigations into combinations involving internal and external vibrations of molecules and complex ions in a crystal are chiefly limited to considering the interaction of the limiting vibrations. Figure 8 shows an idealized scheme for the intramolecular transition N $0 \rightarrow 1$, accompanied by a simultaneous change in the energy state of the lattice described by the quantum number n ($\Delta n = \pm 1$); these can also originate from an excited lattice level, i.e., for $n \neq 0$. The use of this kind of scheme for interpreting experimental results in addition to the foregoing provisions usually involves additional complications on account of, firstly, the existence of several external vibrations (with quantum numbers n_1, n_2, n_3, ...) interacting with the internal vibration, and, secondly, the possibility that transitions with $\Delta N = 1$ and $|\Delta n| > 1$ may appear in the spectrum. In relation to the latter possibility it is useful to consider a

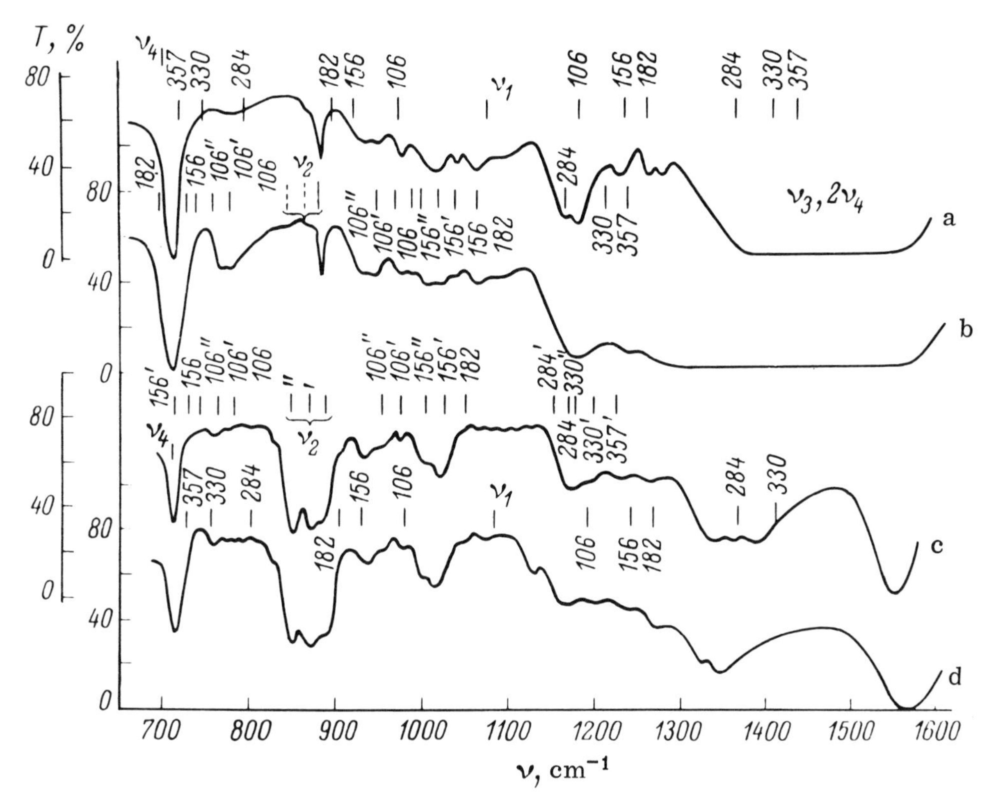

Fig. 9. Infrared spectrum of crystals of calcite $CaCO_3$ [31]. a) Sample with plane perpendicular to the optic axis (thickness 0.12 mm) at -160°; b) the same at 20°; c) sample with plane parallel to the optic axis (thickness 0.10 mm) at -160°; d) the same at 20°. Primes (') and ('') indicate frequencies reckoned from the corresponding components of vibration ν_2.

simple model of a diatomic vibrator executing librational vibrations of small amplitude around the equilibrium position [30] under the influence of harmonic forces. Let the vibrational states of the vibrator be described by quantum numbers N, while quantum numbers n = 0, 1, 2, 3 ... and |m| = n, (n−2), (n−4) ... correspond to the librations. The selection rules for transitions in this kind of scheme have the following form

$$\text{for} \quad \Delta m = 0 \quad \text{(parallel bands)}$$
$$\Delta n = 0, \ \pm 2, \ \pm 4, \ \pm 6 \ldots$$
$$\text{for} \quad \Delta m = \pm 1 \quad \text{(perpendicular bands)}$$
$$\Delta n = \pm 1, \ \pm 3, \ \pm 5 \ldots$$

Thus in addition to the strongest band of the purely vibrational transition Δn = 0 there may be a complex system of transitions reminiscent of the rotational structure of the bands in the spectra of gases. A change in the moment of inertia of the molecule in the excited (N ≠ 0) state and an increase in the anharmonicity of the lattice vibrations in the presence of intramolecular excitation, as noted by Hexter and Dows, lead to a displacement of the transitions N 0 → 1, Δn = 0 (n ≠ 0) to frequencies lower than the "origin" of the band, that is, than the frequency of the transition N 0 → 1, Δn = 0 (n = 0). The effective value of the lattice force constant may either rise or fall on internal excitation of the molecule and hence cause a displacement of the transitions to high or low frequencies compared with the zero band.

Figure 9 shows the infrared spectrum of a calcite single crystal illustrating the complex structure of the bands of combinations involving lattice vibrations. A study of such transitions requires fairly thick layers in the form of oriented sections of single crystals (in view of the obvious dichroism of the bands). The use of low temperatures enables us to obtain a sharper pattern and also to distinguish sum and difference frequencies by the different temperature dependences of the band intensities. Figure 9 also shows lines separated from the centers of the bands of fundamental vibrations of the anion by distances corresponding to the lattice-vibration frequencies, identified from the Raman spectrum and by the method of residual rays [32]. The arrangement of these lines satisfactorily reflects the intensity distribution of the combination bands; however, as we should expect, there is not perfect agreement.

In accordance with the results of the foregoing analysis of the selection rules based on the crystal factor group, the totally symmetric vibration ν_1 of the CO_3^{2-} ion is absent from the absorption spectrum.† Unexpected is the complex structure of the band of the nondegenerate deformation vibration ν_2, observed in the form of a triplet ν_2, ν_2', ν_2'', although of the two components of ν_2 only one is active, according to the selection rules, in the infrared spectrum.‡ It is possible that the special features in the structure of this band may be explained by resonance interactions of the vibrations in the excited lattice [31], the structure and symmetry of which may differ from the ideal symmetry (approached only at absolute zero) used to obtain the selection rules. This is the more probable in view of the fact that, among the normal vibrations of the CO_3^{2-} ion the ν_2 vibration has the greatest dipole moment and is therefore the most sensitive to interactions.

The possibility that excitation of lattice vibrations could modify the effective symmetry and also the internal vibrations of molecules in a crystal was studied theoretically by Hornig [5]. This problem was considered with due allowance for the site symmetry of the molecule only, and hence only concerned the possibility of the splitting of the degenerate vibrations of the molecules by the lattice vibrations. This presentation of the problem permits use of the analogy with the John–Teller dynamic effect [34], the splitting of degenerate electron states of the molecule by the nontotally symmetric vibrations of its nuclei. It is clear that the excitation of some of the external vibrations of the molecule in the crystal may change its site symmetry. This of course refers to optical vibrations, or acoustic vibrations of short wavelength, since for the particular model acoustic vibrations of long wavelengths are equivalent to a simple translation of the molecule and the whole of its immediate surroundings. In a considerable proportion of cases the frequencies of the lattice vibrations are far lower than those of the internal vibrations of the molecules, and the excited states are substantially populated even at room temperatures. Hence a high-frequency internal vibration, a large number of cycles of which are executed for practically unvarying external coordinates, will take place in a field of low effective symmetry; this may lead to splitting or broadening of the bands of the degenerate internal vibrations.

Table 3. Splitting of the Degenerate Internal Vibrations of the Molecules by the Crystal Lattice Vibrations [5]

Molecule site-symmetry	Internal vibration symmetry-species	Lattice vibrations causing splitting
C_3	E	$T(E)$; $R(E)$
C_{3v}	E	$T(E)$; $R(E)$
C_{3h}	E', E''	$T(E')$
D_3	E	$T(E)$; $R(E)$
D_{3d}	E_g, E_u	$R(E_g)$
S_4	E	$T(B)$
D_{2d}	E	$T(B_2)$
T	E	Not split
	F	$T(F)$; $R(F)$
T_d	E	Not split
	F_1, F_2	$T(F_2)$
T_h	E_g, E_u	Not split
	F_g, F_u	$R(F_g)$

Note. T = translational; R = rotational.

Thus if a tetrahedral group (for example, the NH_4^+ ion) is situated in the center of a cubic cell and the direction of the bonds coincides with the three-fold axes, then

† It is interesting to note that the reduction of the site symmetry of the CO_3^{2-} ion in calcite crystals under a pressure of over 10,000 atm leads to the appearance of the vibration ν_1 in the infrared spectrum [33].

‡ The component ν_2'' (850 cm^{-1}) has been identified as due to $^{13}CO_3^{2-}$ in natural abundance [50].

the site symmetry of the group is T_d. An external (translational) triply degenerate vibration corresponds to displacements of the center of gravity of the ion in directions normal to the cube faces. For each of these vibrations the site symmetry falls to C_{2v}, which leads to the removal of degeneracy for internal vibrations of type F_2. A more detailed analysis, entirely analogous to that carried out by Jahn and Teller for the electron states of molecules, enables us to indicate which of the degenerate internal vibrations of the molecule in the crystal may be split and which external vibrations are responsible for this splitting (Table 3) for any site symmetry.

It follows from a consideration of this table that, for example, the triply degenerate vibrations of the NH_4^+ ion in the NH_4Cl crystal may be split while the doubly degenerate may not. For a site symmetry corresponding to the groups C_4, C_{4v}, C_{4h}, D_4, D_{4h}, C_6, C_{6v}, C_{6h}, D_6, D_{6h}, O, and O_h there is no splitting of the degenerate internal vibrations by the vibrations of the lattice.

We note that splitting arising in this manner, in contrast to the removal of degeneracy as a result of a *static* reduction in the symmetry of the molecule in the crystal, should possess a distinct temperature dependence and be "frozen" at fairly low temperatures. Up to the present time, no examples of cases in which this mechanism of removing degeneracy has been proved experimentally have appeared in the literature.

A considerable proportion of investigations into the band structure of the internal vibrations of molecules and complex ions in crystals and their interactions with lattice vibrations have been associated with studying the character of the rotational vibrations of molecules in crystals. In this respect the example most studied (even spectroscopically) has been the rotation of the H_2 molecule in solid hydrogen. Many molecular crystals and crystals with complex ions (NH_4^+, NO_3^-) exhibit specific-heat anomalies at certain points of the temperature scale, the so-called λ points, the existence of which is associated with the character of the rotational motions of the molecules or ions. According to the view set forward by Pauling and then developed by Fowler, solid-phase transitions of this kind are due to a transition from the librations of molecules around an equilibrium position to free or almost free rotation. From Frenkel's point of view developed by Landau in the theory of phase transformations of the second kind, the motion of molecules has the character of restricted rotational oscillations both above and below the transformation temperature, and the transformation itself is associated with "orientational melting," i.e., with a loss of long-range order in the orientations of the molecules relative to one or several axes.

In spectroscopical investigations of ammonium halides no gas-type rotational structure of the bands of internal vibrations of the NH_4^+ ions were observed; moreover analysis of the combination transitions established the existence of an external torsional vibration (libration) above the transformation temperature [35]. A libration of the NO_3^- ion was identified in the spectrum of $NaNO_3$ crystals by Hexter [36], who denied the possibility of a free rotation above the λ point, although some Japanese authors [37], studying the contour of the $\nu_1 + \nu_4$ band of the NO_3^- ion in these crystals, concluded that there was a transition to a free rotation around the three-fold axis. Here we may also mention another paper [38] in which an attempt was made to interpret some of the combination bands in the infrared spectra of carbonates and nitrates as combinations of the internal vibrations of the anions with series of librational transitions. Another group of samples which have been thoroughly studied in respect of the combinations of internal vibrations of the anions with lattice vibrations in their spectra are the hydroxides of the alkali and alkaline-earth metals with simple diatomic $[OH]^-$ anions. In the spectrum of brucite $Mg(OH)_2$ a very complex band structure was observed near the frequency of the OH stretching vibration [39]. The interpretation of this structure by way of the foregoing simple scheme of libration-vibrational transitions of a two-atom vibrator considered by Hexter and Dows [30] gave only a partial explanation of the observed picture; however, in certain respects the results disagreed considerably with experimental data. For this reason attempts were made in a number of papers [40, 41] to explain this structure by considering the possibility of combinations taking place between νOH and various lattice vibrations, including those of the acoustic kind. Mitra [9] gave a fairly detailed review of investigations into the structure of the spectra of $Mg(OH)_2$ crystals and other hydroxides of metals belonging to the first and second groups of the Periodic System.

§3. Interpretation of the Vibrational Spectra of Polymeric Chain or Layer Molecules and Complex Ions

In the crystals of many organic and inorganic compounds the molecules (complex ions) have a considerable extension in one of two directions, many times exceeding the lattice constant. In an ideal, defect-free crystal their dimensions correspond to the macroscopic dimensions of the crystal, while the composition and structure remain strictly periodic. If the forces of interaction between macromolecules in a crystal are considerably weaker than the interaction between the atoms, then, as in the case of small molecules, it is useful as a first approximation to consider the spectrum of this kind of isolated system and then to allow for possible complications of the spectrum arising from intermolecular interactions. The existence of translational symmetry in one or two directions enables us to apply the Born theory in order to determine the optical vibrations of such systems, i.e., to consider them as infinitely extended one- or two-dimensional crystals. In the following paragraphs we shall consider methods of allowing for the symmetry properties of chain and layer structures in order to determine the number of normal vibrations and the selection rules in vibrational spectra.

One-Dimensional Crystals. By this term we mean any structures characterized by the existence of translational symmetry in one direction only, this being called the chain axis. The chain itself may contain arbitrarily complex side radicals or consist of a finite number of parallel chains joined together, such as for example, the ribbon anions well known in the crystal chemistry of silicates. The problem of interpreting the spectra of such structures originally arose in connection with investigations into crystals of organic high polymers; it has been discussed in detail in a number of papers [42-44].

An infinite one-dimensional crystal, containing (according to chemical composition) p atoms in the unit link and q such links in a one-dimensional cell, i.e., in the identity period of the chain, possesses 3pq−4 branches of internal vibrations (apart from the three translations we must also exclude rotation around the chain axis). The Born boundary conditions for such a system have the form

$$\vec{k} = \frac{n}{N}\vec{b}\ (n = 0,\ 1,\ 2,\ 3, \ldots, N-1). \tag{22}$$

As in the three-dimensional case, N = 1 corresponds to consideration of the fundamental vibrations. Then the boundary conditions may also be written in the form of the following relations for the phase shift θ of the optically active vibrations between neighboring chemical unit links of the chain

$$\theta = (2\pi/q)\, r\ \ (r = 0,\ 1,\ 2,\ \ldots,\ q-1). \tag{23}$$

Thus, for example, in the plane chain of a paraffin hydrocarbon $[CH_2CH_2]_\infty$ with q = 2 θ may be equal to 0 and π. The vibrations of the hydrocarbon skeleton or framework of such a chain, including four zero vibrations, is shown schematically in Fig. 10. The first and third from the top clearly correspond to the nonzero, i.e., optical vibrations of the chain.

In order to study the symmetry properties of the normal vibrations of such systems it is convenient to use the representations of one-dimensional space groups. These

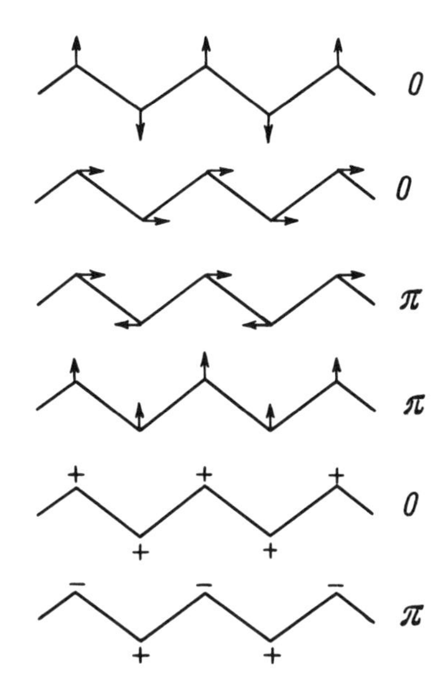

Fig. 10. Normal vibrations of a plane hydrocarbon chain [43].

groups may be obtained by considering three-dimensional groupings of different forms arranged with periodical repetitions along a specific direction. Of course the groupings themselves need not be on the chain axis. The symmetry of such structures ("rods") has been repeatedly described in the literature. Possible symmetry operations, apart from those usual for point groups, include translation along the chain axis, gliding reflection in a plane passing through the chain axis, and screw displacement along the chain axis with rotation through $2\pi/n$ in each move. In the latter case, in contrast to the screw axes of three-dimensional space groups, n may be equal not only to 2, 3, 4, and 6, but to any other number (not only whole numbers). Simple rotation axes C_n in a direction coinciding with the axis of the translations may also correspond to an arbitrarily large n. In a direction perpendicular to the chain axis, only two-fold symmetry axes can occur.

Thus one-dimensional space groups may be constructed on the basis of the points groups C_n, D_n, C_{ni}, C_{nh}, C_{nv}, D_{nd} and D_{nh} by combining the symmetry elements of these point groups with translations. If we confine ourselves to values of n equal to 1, 2, 3, 4, and 6, we may derive 75 one-dimensional space groups [45].

Since vectors and symmetrical tensors are totally symmetric with respect to translations, the optically-active representations of a one-dimensional space group are the representations of its factor group. The latter is determined in the same way as for three-dimensional space groups, and moreover is isomorphic with respect to one of the point groups. Subsequent establishment of the number of normal vibrations and selection rules is carried out by the normal methods. The total number of normal vibrations in the i-th irreducible representation of the factor group (excluding translations of the unit cell as a whole) is determined in analogy with formula (14) by the expression

$$n_i = \frac{1}{N} \sum h_\rho (U_R - 1)(\pm 1 + 2\cos\varphi_R)\chi_i(R) \tag{24}$$

The types of symmetry associated with the rotations, including the rotation around the chain axis, which has to be excluded from the list of optical vibrations, have to be found separately for every specific factor group. We note, incidentally, that rotations of a unit link around the two other axes involve stretching and bending of bonds in the chain, i.e., they constitute internal degrees of freedom of the chain.

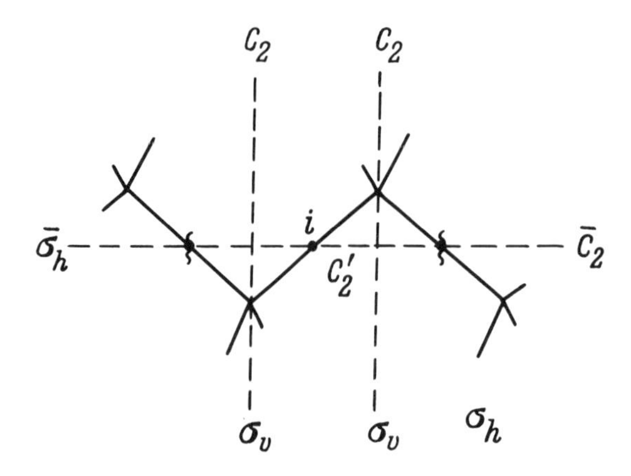

Fig. 11. Symmetry elements of the polyethylene chain.

Where the observed spectrum of the crystal cannot be completely explained by considering the vibrations of an isolated chain, it is appropriate to consider the site symmetry of the molecule in the crystal and the possibility of resonance interactions occurring between the vibrations of several molecules in the three-dimensional unit cell. This may be carried out by the methods described in the previous section. However, instead of the point group of site symmetry we must consider the one-dimensional site group, since an infinitely extended chain molecule in the crystal may be invariant not only toward certain point operations but also toward operations including translations. The set of these operations determines the one-dimensional site group, the factor group of which is used for finding the selection rules for the fundamental vibrations. The transition to the factor group of the three-dimensional space group of the crystal is completely analogous to the transition described earlier for crystals with molecules of finite dimensions, since the identity period of the chain coincides with the dimensions of the unit cell of the crystal along one of its axes. The only difference lies in that some of the lattice vibrations of a crystal with small molecules become internal vibrations of the chain in a polymer crystal.

The method of analysis described has been successfully used for many polymeric organic compounds in the crystalline state. By way of example, let us consider the case of polyethylene [42, 46]. The infinite plane zigzag chain $CH_2{-}CH_2{-}CH_2{-}CH_2$ shown in Fig. 11 possesses the following symmetry elements: identity E, symmetry plane σ_h containing the carbon atoms, two-fold axes C_2 along the bisectrices of the HCH angles, symmetry plane σ_v coinciding with the HCH planes, two-fold axes C_2' normal to σ_h and passing through the middles of the C–C bonds, centers of inversion i on the intersection of C_2' with σ_h, two-fold screw axis $\bar{C}_2$ along the chain axis, glide reflection plane $\bar{\sigma}_h$ (containing the chain axis) perpendicular to σ_h and σ_v, and simple translation along the chain axis. The unit cell of this chain contains two CH_2 links.

On considering the factor group of the one-dimensional space group, i.e., taking a translation along the chain axis as an identity transformation, and the $\bar{C}_2$ axis and $\bar{\sigma}_h$ plane as a simple two-fold axis and mirror plane, we obtain the following multiplication table for the elements of this factor group (Table 4). This table is obtained by the successive application of all possible pairs of operations of the one-dimensional space group to the atoms of the unit cell, any transformation transferring all the atoms to the neighboring cell being taken as the identity operation E. It follows di-

Table 4. Multiplication Table for the Elements of the Factor Group of the One-Dimensional Space Group of the Polyethylene Chain

	E	σ_v	σ_h	$\bar{\sigma}_h$	i	$\bar{C}_2$	C_2'	C_2
E	E	σ_v	σ_h	$\bar{\sigma}_h$	i	$\bar{C}_2$	C_2'	C_2
σ_v	σ_v	E	C_2	C_2'	$\bar{C}_2$	i	$\bar{\sigma}_h$	σ_h
σ_h	σ_h	C_2	E	$\bar{C}_2$	C_2'	$\bar{\sigma}_h$	i	σ_v
$\bar{\sigma}_h$	$\bar{\sigma}_h$	C_2'	$\bar{C}_2$	E	C_2	σ_h	σ_v	i
i	i	$\bar{C}_2$	C_2'	C_2	E	σ_v	σ_h	$\bar{\sigma}_h$
$\bar{C}_2$	$\bar{C}_2$	i	$\bar{\sigma}_h$	σ_h	σ_v	E	C_2	C_2'
C_2'	C_2'	$\bar{\sigma}_h$	i	σ_v	σ_h	C_2	E	$\bar{C}_2$
C_2	C_2	σ_h	σ_v	i	$\bar{\sigma}_h$	C_2'	$\bar{C}_2$	E

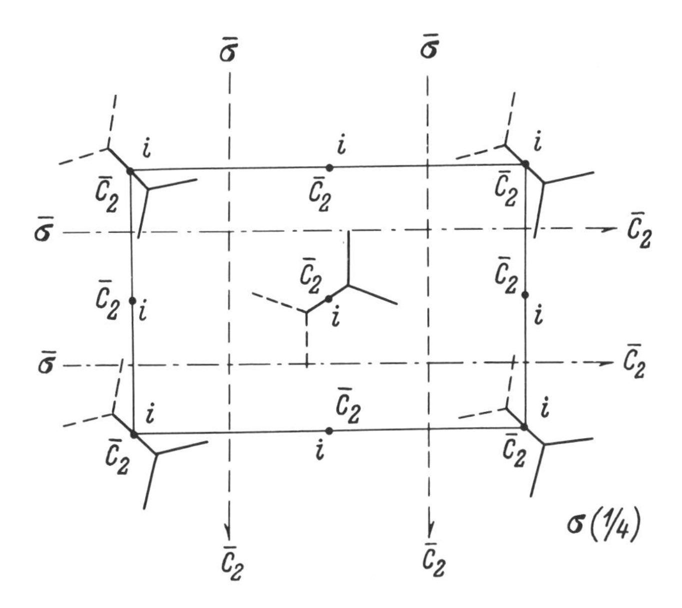

Fig. 12. Schematic representation of the structure of the unit cell and symmetry elements of a polyethylene crystal.

rectly from the table that the factor group of the one-dimensional space group considered is isomorphic to the point group $D_{2h} \equiv V_h$. We note that, in contrast to the corresponding point group, not all the axes and symmetry planes of the one-dimensional space group intersect at one point, the inversion center. This is due to the presence of combined symmetry elements, a screw axis and a glide plane.

We may now use ordinary methods (i.e., the characters of the point group D_{2h}) to determine by what irreducible representations the components of the dipole moment and polarizability tensor (and also translations and rotations of the molecule) transform. The number of normal vibrations of each type of symmetry is established in the same way as for molecules of finite size, nevertheless remembering that rotations of a unit link around axes normal to the chain axis constitute internal vibrations of the macromolecule (Table 5).

We may then proceed from the consideration of an isolated chain to its vibrations in the crystal with space group D_{2h}^{16}, containing unit sections of two chains in a three-dimensional cell. It follows from Fig. 12 that, among the symmetry operations of the factor group of the space group of the crystal, each chain is invariant toward operations of mirror reflection, screw displacement, and inversion. It is by these symmetry elements that the site group (the factor group of which is isomorphic with C_{2h}) is defined. We may now construct a scheme of correlations between the irreducible representations of the factor groups of the one-dimensional site and three-dimensional space groups, as shown in Fig. 13. The distribution of the 36 branches of the vibrational spectrum of the polyethylene crystal (three acoustic, three optical translational and two rotational lattice vibrations, and 28 internal vibrations of the chains) with respect to the irreducible representations, which may easily be obtained by the methods described previously, is given in Table 5, together with a classification of the vibrations of a single chain relative to its intrinsic and site symmetry.

We see from Table 5 and the correlation scheme (Fig. 13) that the number of internal vibrations of the chain is doubled in the spectrum of the crystal. According to the selection rules, both Davydov components of the vibrations with a moment perpendicular to the chain axis are active in the infrared spectrum and possess different polarization. For vibrations with a moment along the axis of the chain only one of the components is active.

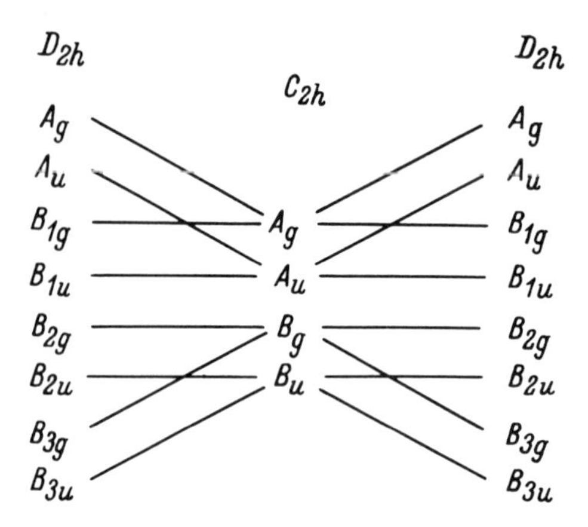

Fig. 13. Scheme of correlations of the irreducible representations of the factor groups of the one-dimensional space, site, and three-dimensional symmetry groups of crystalline polyethylene.

The results of the analysis presented here enable us, for example, to explain the splitting of the rocking mode (ρ) of the CH_2 groups in the infrared spectra of paraffin hydrocarbons with long chains on passing from the liquid to the crystal. In accordance with the splitting scheme

$$\underset{\text{chain}}{B_{2u}} \rightarrow \underset{\text{crystal}}{B_{2u} + B_{3u}}$$

the components of the bands of ρCH_2 at 725 to 731 cm^{-1} are sharply polarized along the *a* and b axes of the crystal (Fig. 14). The resonance character of the splitting of this band in the spectrum of the crystal is confirmed by comparison with the spectrum of the dilute solid solution in the corresponding deuterated hydrocarbon (Fig. 15). As we should expect, with decreasing concentration the doublet rapidly vanishes, although its contraction ought to have been more abrupt if the assumption as to the exclusive role of interaction between neighboring molecules had been correct.

Table 5. Characters and Selection Rules for the D_{2h} Group and Classification of the Vibrations of the Polyethylene Chain by Means of the Factor Group of the One-Dimensional Space Group, the Site Group, and the Factor Group of the Crystal

Chain D_{2h}	Characters E	σ_v	σ_h	$\bar{\sigma}_h$	i	$\bar{C}_2$	C_2'	C_2	Selection rules	Rotations and translations	Internal vibrations of the $[Ch_2CH_2]_\infty$ chain	Symmetry types: site group[1] C_{2h}	factor group[2] D_{2h}^{16}
A_g	+1	+1	+1	+1	+1	+1	+1	+1	$\alpha_{\bar{C}_2\bar{C}_2}$, $\alpha_{C_2'C_2'}$, $\alpha_{C_2C_2}$	—	ν_{CH_2}, δ_{CH_2}, ν_{CC}	$3A_g$	$3(A_g, B_{1g})$
A_u	+1	−1	−1	−1	−1	+1	+1	+1	—	—	τ_{CH_2}	A_u	A_u, B_{1u}
B_{1g}	+1	+1	−1	−1	+1	+1	−1	−1	$\alpha_{C_2'C_2}$	$R_{\bar{C}_2}$	ν_{CH_2}, ρ_{CH_2}	$2A_g$	$2(A_g, B_{1g})$
B_{1u}	+1	−1	+1	+1	−1	+1	−1	−1	$M_{\bar{C}_2}$	$T_{\bar{C}_2}$	γ_{CH_2}	A_u	A_u, B_{1u}
B_{2g}	+1	−1	+1	−1	+1	−1	+1	−1	$\alpha_{\bar{C}_2C_2}$	$(R_{C_2'})$	γ_{CH_2}, ν_{CC}	$2B_g$	$2(B_{2g}, B_{3g})$
B_{2u}	+1	+1	−1	+1	−1	−1	+1	−1	$M_{C_2'}$	$T_{C_2'}$	ν_{CH_2}, ρ_{CH_2}	$2B_u$	$2(B_{2u}, B_{3u})$
B_{3g}	+1	−1	−1	+1	+1	−1	−1	+1	$\alpha_{\bar{C}_2C_2'}$	(R_{C_2})	τ_{CH_2}	B_g	B_{2g}, B_{3g}
B_{3u}	+1	+1	+1	−1	−1	−1	−1	+1	M_{C_2}	T_{C_2}	ν_{CH_2}, δ_{CH_2}	$2B_u$	$2(B_{2u}, B_{3u})$

Note. 1. Site group formed by symmetry elements E, $\bar{C}_2$, i, σ_v. 2. Acoustic lattice vibrations of types B_{1u}, B_{2u}, B_{3u}; optical, translational A_u, B_{2u}, B_{3u}, rotational A_{1g}, B_{1g}.

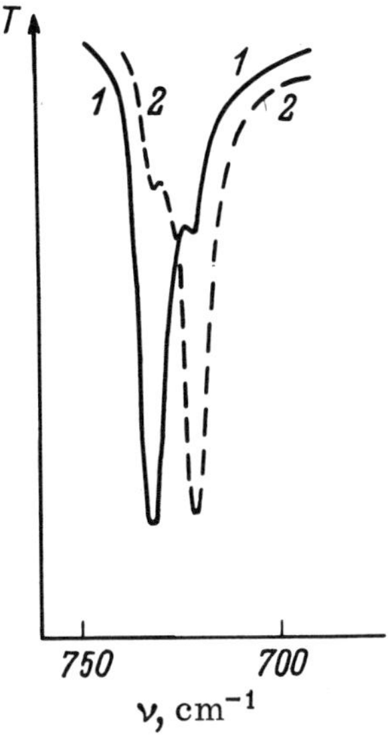

Fig. 14. Infrared spectrum of crystalline (monoclinic) n-$C_{36}H_{74}$, the ρCH_2 band in polarized light [46]. 1) Electric vector along the a axis; 2) electric vector along the b axis.

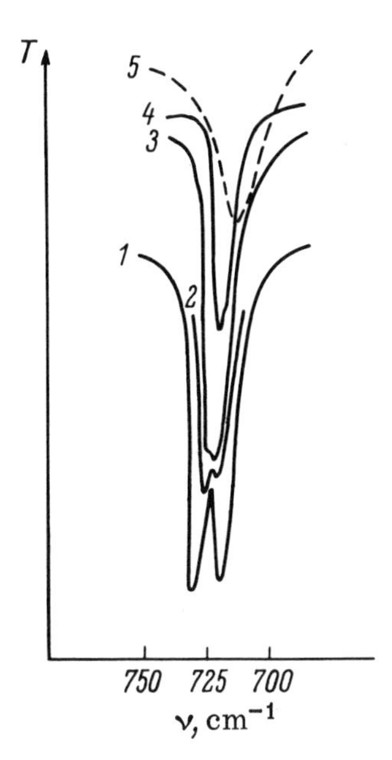

Fig. 15. Infrared spectra of solid solutions [46]. Ratio of the components $C_{64}H_{130}/C_{100}D_{202}$: 1) 1:0; 2) 1:1; 3) 1:3; 4) 1:20; 5) polyethylene melt.

Overtones and Combinations. Disordered State of the Single-Dimensional Crystal. It was noted earlier that the overwhelming majority of wave vectors in the crystal are only invariant with respect to identity operations. Hence the only strict limitation imposed on the possibility that an overtone or combination may appear in the spectrum lies in the condition that both components should correspond to the same wave vector. This condition, also valid for a one-dimensional crystal, may be supplemented in the present case by some additional limitations; this is associated with coincidence between the directions of the direct and reciprocal lattice. In chains of fairly high symmetry, the wave vectors are often invariant toward a whole series of symmetry operations. Thus in the example considered here of a polyethylene chain all the wave vectors† are invariant toward the operations E, σ_h, $\bar{\sigma}_h$ and $\bar{C}_2$. The set of these operations forms a group isomorphic with respect to C_{2v}. Hence for determining the symmetry properties of the combinations and overtones we must expand the direct product of the corresponding representations of the group C_{2v} with respect to the irreducible representations of the factor group D_{2h}.

Using the scheme of correlations between D_{2h} and C_{2v}

D_{2h}	A_g	B_{1u}	A_u	B_{1g}	B_{2g}	B_{3u}	B_{2u}	B_{3g}
C_{2v}	A_1		A_2		B_1		B_2	

we may immediately establish, for example, that the torsional vibrations of the CH_2 groups of symmetry type A_u may give combinations (active in the infrared spectrum and the Raman spectrum) with the branch of translations along C_2' (B_{2u})

$$A_u \to A_2, \ B_{2u} \to B_2; \ A_2 \times B_2 = B_1, \ B_1 \to B_{2g}, \ B_{3u}$$

† The wave vector with its end lying at the boundary of the first Brillouin zone is invariant toward all operations of the factor group of the chain [42].

In an analogous manner, on passing to the one-dimensional site symmetry group in a crystal the factor group of which is isomorphic with respect to C_{2h}, we may readily convince ourselves that any wave vector is invariant toward operations of the subgroup isomorphic with respect to C_2.

Let us now consider possible changes in the vibrational spectrum when the periodicity of the structure of the one-dimensional crystal is disrupted. In practice, the most important deviations from periodicity encountered in real crystals containing chain molecules or ions reduce to truncations of the chains and relative disordering of the chains in the crystal.

If the length of a chain is fairly small and constitutes only a few identity periods of an infinite chain, the strict selection rules may be obtained by considering the point symmetry group of the molecule. For a fairly extensive length of the chain the following approximate approach to the question is possible [44]. Let us consider any normal vibration of the one-dimensional unit cell of the chain. If all such cells vibrate in phase, the vibration is potentially active, i.e., it is observed in the spectrum if the factor group of the corresponding infinite chain belongs to an active irreducible representation. The intensity I is proportional to $(nA)^2$, where n is the number of unit cells in the finite chain and A is the maximum amplitude of the variation in the dipole moment (polarizability) of one cell for the vibration in question. If for the same vibration of the cell the following cells do not vibrate in phase and all possible phase relations between the cells are considered, a "branch" is formed. The vibrations forming this branch may be active, but with a considerably lower intensity, if a particular phase condition is satisfied, i.e., if an odd number of half waves is superimposed on the length of the finite chain. The intensity I′ of the vibrations satisfying this condition is approximately proportional to $\{nA/[2(2q+1)]\}^2$, where $q = 0, 1, 2 \ldots$. Correspondingly $I'/I = 1/[2(2q+1)]^2$, in other words, in addition to the limiting vibration the representation of which is a representation of the factor group of the infinite chain, a series of bands of decreasing intensity (or a broadening of the bands) may be observed in the spectrum of the finite chain. In addition to this, the spectrum of the finite chain may exhibit bands of vibrations which are inactive according to the selection rules for an infinite chain.‡

The analysis of the changes in the spectrum in the case of disordering in the relative disposition and shape of the chains involves great difficulties. It has been established experimentally that, on passing from the spectrum of a crystalline polymer to that of the amorphous polymer, certain bands vanish and certain others lose their multiplet structure. At the same time some bands are only observed in the spectrum of the amorphous polymer. We may suppose that the bands (or multiplets) only observed in the spectrum of the crystal are associated with resonance interactions of the vibrations of the chains. The bands exclusively belonging to the spectrum of the amorphous phase of the polymer may be due either to the disruption of the selection rules acting in the crystal or to the existence of different conformations of the chain.

Vibrations of Layer Structures. In many crystalline silicates and compounds of similar structure, complex anions possessing translational symmetry in two dimensions may be distinguished in the form of layers. In order to analyze the spectra of such structures, by way of a first approximation it is desirable to use a model of the two-dimensional crystal analogous to the one-dimensional model described in the foregoing for chain-type molecules and ions.

Let us start with a description of the symmetry properties of layer structures, for which it is useful to consider two-dimensional space groups, reflecting the symmetry of "patterns" of three-dimensional groupings repeating themselves periodically in two directions.

‡Questions as to the dynamics and method of calculating the normal vibrations of chains of finite length are considered in [47].

Symmetry of this kind has been described in the literature [3]. Apart from point-group symmetry operations, possible operations include: translations along the two principal directions of the two-dimensional lattice, a screw translation along each of these directions, and a gliding reflection with displacement in the plane of the translational symmetry axes. The direction of the displacement in the last operation need not coincide with any of the axes; the screw axes cannot have an order of greater than two. Simple rotation axes lying in the plane of the axes of translational symmetry can also only be of the second order, while the C_n axes perpendicular to this plane are limited, as in the three-dimensional case, to n = 1, 2, 3, 4, 6. Earlier [3] 80 two-dimensional space groups were derived and expressions were given for the transformations of the coordinates corresponding to the generating elements of the four crystallographic systems possible for these lattices. It should be noted that two-dimensional space groups of the hexagonal system do not contain any screw axes or glide planes.

Analysis of the symmetry properties of the normal vibrations and the determination of the selection rules for the spectra of such structures are completely analogous to those already described for three- and one-dimensional systems. If we introduce the Born boundary conditions and consider only the in-phase vibrations of translationally equivalent atoms in order to obtain the selection rules for the fundamental frequencies, then the representations of such vibrations are representations of the factor group of the two-dimensional space group. The number of internal vibrations of a layer with m atoms in the unit cell is $3m-3$, since only the three translations have to be excluded. All forms of rotations of the two-dimensional unit cell correspond to changes in the interatomic distances and angles within the limits of the actual layer. In other words, in a three-dimensional crystal formed from such layers their external vibrations can only have a translational character.

We also note that, as in the three-dimensional case, analysis of the spectrum on the basis of the factor group of the layer requires choice of the smallest (noncentered) unit cell. Further consideration of the spectrum may be based on the usual principles as follows. First we separate out the group of operations of the three-dimensional crystal with respect to which the layer is invariant. The factor group of this group is an analog of the site group. Then, after this, we may pass to the factor group of the three-dimensional space group in order to elucidate the possibility of resonance splittings of the internal vibrations of the layer taking place.

References

1. M. Born and K. Huang, Dynamic Theory of Crystal Lattices, University Press, Oxford (1954).
2. A. A. Maradudin, E. W. Montroll, and G. H. Weiss, Theory of Lattice Dynamics in the Harmonic Approximation, Academic Press, New York (1963).
3. S. Bagavantam and T. Venkatarayudu, Group Theory and Its Application to Physical Problems, Academic Press, London (1969).
4. H. Winston and R. S. Halford, J. Chem. Phys., 17:607 (1949).
5. D. F. Hornig, J. Chem. Phys., 16:1063 (1948).
6. J.-P. Mathieu, Spectres, Vibration, et Symétrie des Molecules et des Cristaux, Hermann et Cie., Paris (1945).
7. A. S. Davydov, Theory of the Absorption of Light in Molecular Crystals, Izd. AN Ukrainian SSR, Kiev (1951).
8. A. S. Davydov, Usp. Fiz. Nauk, 82:393 (1964).
9. S. S. Mitra, Vibration Spectra of Solids. Solid State Physics, Vol. 13, Academic Press, New York-London (1962), pp. 2-53.

9a. R. Zbinden, Infrared Spectroscopy of High Polymers, Academic Press, New York-London (1964).

9b. M. M. Sushchinskii, Raman Spectra of Molecules and Crystals, Nauka, Moscow (1969).
10. W. Cochran, Lattice Vibrations. Reports on Progress in Physics, Vol. 26 (1963), p. 1.
11. K. Nakamoto, Infrared Spectra of Inorganic and Coordination Compounds, John Wiley, New York (1963); 2nd edn (1970).
11a. H. Siebert, Anwendungen der Schwingungsspektroskopie in der Anorganischen Chemie, Springer Verlag, Berlin-Heidelberg-New York (1966).
12. K. E. Lawson, Infrared Absorption Spectra of Inorganic Substances, Reinhold, New York (1961).
13. H. Moenke, Mineralspektren I und II, Acad. Verl., Berlin (1962 and 1966).
14. R. J. P. Lyon, Minerals in the Infrared, Stanford Research Inst., Menlo Park, California (1962).
15. H. Moenke, Spektralanalyse von Mineralien und Gesteinen. Eine Anleitung zur Emissions und Absorptionsspektroskopie, Akad. Verlag., Leipzig (1962).
16. P. Tarte, Étude Expérimentale et Interprétation du Spectre Infrarouge des Silicates et des Germanates. Application à des Problèmes Structuraux Relatifs à l'État Solide, Acad. Royale de Belgique, Mémoires, Vol. 35 (4a, 4b) Brussels (1965).
17. M. Born and T. von Karman, Z. Phys., 13:297 (1912).
18. E. B. Wilson, J. C. Decius, and P. C. Cross, Molecular Vibrations: The Theory of Infrared and Raman Spectra, McGraw-Hill, New York (1955).
19. R. S. Halford, J. Chem. Phys., 14:8 (1946).
20. F. Seitz, Ann. Math., 37:17 (1936).
21. H. Bethe, Ann. Phys., 3:133 (1929), see also M. A. El'yashevich, Spectra of the Rare Earths, Gostekhteorizdat, Moscow (1953).
22. R. M. Hexter, J. Chem. Phys., 33:1833 (1960).
23. R. Kopelman, J. Chem. Physics, 44:3547 (1966).
24. International Tables for X-Ray Crystallography, Vol. 1. Symmetry Groups (N. F. M. Henry and K. Lonsdale, eds., Kynoch, Birmingham (1952).
25. W. Vedder and D. F. Hornig, "Infrared Spectra of Crystals," in Advances in Spectroscopy, Vol. II (H. W. Thompson, ed.), Intersci. Publ., New York, London (1961), pp. 189-262.
26. A. I. Ansel'm and N. N. Porfir'eva, Zh. Éksp. Teor. Fiz., 19:438 (1949).
27. N. N. Porfir'eva, Zh. Éksp. Teor. Fiz., 19:692 (1949); 20:97 (1950).
28. S. Asano and Y. Tomishima, J. Phys. Soc. Japan, 12:346, 890, 900 (1957).
29. C. J. H. Schutte, Z. Phys. Chem., 39:241 (1963).
30. R. M. Hexter and D. A. Dows, J. Chem. Phys., 25:504 (1956).
31. A. A. Shultin, Dokl. Akad. Nauk SSSR, 125:767 (1959).
32. A. K. Ramdas, Proc. Indian Acad. Sci., 37A:441 (1953).
33. C. E. Weir, E. R. Lippincott, A. Van Valkenberg, and E. N. Bunting, J. Res. Nat. Bur. Standards, 63A:55 (1959).
34. H. A. Jahn and E. Teller, Proc. Roy. Soc., A161:220 (1937).
35. E. L. Wagner and D. F. Hornig, J. Chem. Phys., 18:296, 305 (1950).
36. R. M. Hexter, Spectrochimica Acta, 10:291 (1958).
37. Y. Sato, K. Gesi, and Y. Takagi, J. Phys. Soc. Japan, 19:449 (1964).
38. R. A. Schroeder, C. E. Weir, and E. R. Lippincott, J. Res. Nat. Bur. Standards, 66A:407 (1962).
39. R. T. Mara and G. B. B. M. Sutherland, J. Opt. Soc. America, 43:1100(1953); 46:462 (1956).
40. P. A. Buchanan and H. H. Caspers, J. Chem. Phys., 36:2665 (1962); 38:1025 (1963).
41. S. S. Mitra, J. Chem. Phys., 39:3031 (1963).
42. M. C. Tobin, J. Chem. Phys., 23:891 (1955).
43. C. J. Liang, S. Krimm, and G. B. B. M. Sutherland, J. Chem. Phys., 25:543 (1956).
44. C. J. Liang, J. Molecular Spectr., 1:61 (1957).

45. C. Hermann, Z. Kristallogr., 69:250 (1929).
46. S. Krimm, C. J. Liang, and G. B. B. M. Sutherland, J. Chem. Phys., 25:549 (1956).
47. H. Matsuda, K. Okada, T. Takase, and T. Yamamoto, J. Chem. Phys., 41:1527 (1964).
48. S. S. Mitra, in: Optical Properties of Solids (S. Nudelman and S. S. Mitra, eds.), Plenum Press, New York (1969), pp. 333-451.
49. D. M. Adams and D. C. Newton, J. Chem. Soc. A (London), 2822 (1970); Tables for Factor Group and Point Group Analysis, Beckman-RIIC Ltd, 4 Bedford Park, Croydon, CR9 3LG, England (1970).
50. W. Sterzel and E. Chorinsky, Spectrochim. Acta, 24A:353 (1968).

CHAPTER II

VIBRATIONAL SPECTRA AND STRUCTURE OF COMPLEX ANIONS IN SILICATES. "ISLAND" STRUCTURES

Silicate crystals are characterized by the existence of a large number of complex anions distinguished both by the degree of condensation of the silicon–oxygen tetrahedra (from an "isolated" SiO_4 group to highly condensed three-dimensional frameworks) and also by the configuration, i.e., the relative dispositions of the tetrahedra in an anion of a particular composition. Judgment as to the structure of a complex anion is usually based on the results of a complete x-ray diffraction solution of the crystal structure or else on indirect considerations, which are not always reliable. It is therefore interesting to study the correlation between the vibrational spectrum and the structure of a complex silicate anion in order to elucidate the possibility of identifying different types of anions by spectroscopic methods. The use of the methods set out in the preceding chapter also justifies us in expecting to be able to obtain information regarding the dispositions of these anions in the crystal lattice. Here we shall consider the spectra of silicates containing anions incorporating a finite number of silicon–oxygen tetrahedra, the spectra of "model" organosilicon compounds, and general questions as to the interpretation and calculation of silicate spectra.

§1. Spectroscopic Evidence for Special Features in the Electron Structure of Silicon–Oxygen Bonds

Before proceeding to discuss the interpretation of the spectra of silicates, it is perhaps appropriate to consider some of the expected differences in the force constants of Si–O bonds in relation to the nature of the substituents adjacent to the silicon and oxygen. The main sources of information of this kind are x-ray and electron-diffraction data regarding interatomic distances and valence angles and the results of spectroscopic investigations into organosilicon molecules containing Si–O bonds, which will be briefly summarized in subsequent Sections.†

Orders of the Si–O Bonds, Interatomic Distances, and Force Constants. Formally, the electron configuration of a neutral silicon atom ($1s^2$, $2s^2$, $2p^6$, $3s^2$, $3p^2$) leads to the development of sp^3 hybridization when forming bonds, giving tetrahedral angles between the bonds, as in the case of the carbon atom. However, many characteristics of the chemical bonds formed by silicon cannot be explained simply by differences in the electronegativity of Si and C, and this has led to the view that the vacant 3d orbitals of silicon may be used in order to create donor–acceptor $p\pi$–$d\pi$ bonds with atoms possessing unshared pairs of elec-

† Information regarding the nature of these bonds based on thermodynamic data, physicochemical constants, and the mechanisms of reactions may be found in a number of monographs on the chemistry of organosilicon compounds [1-4].

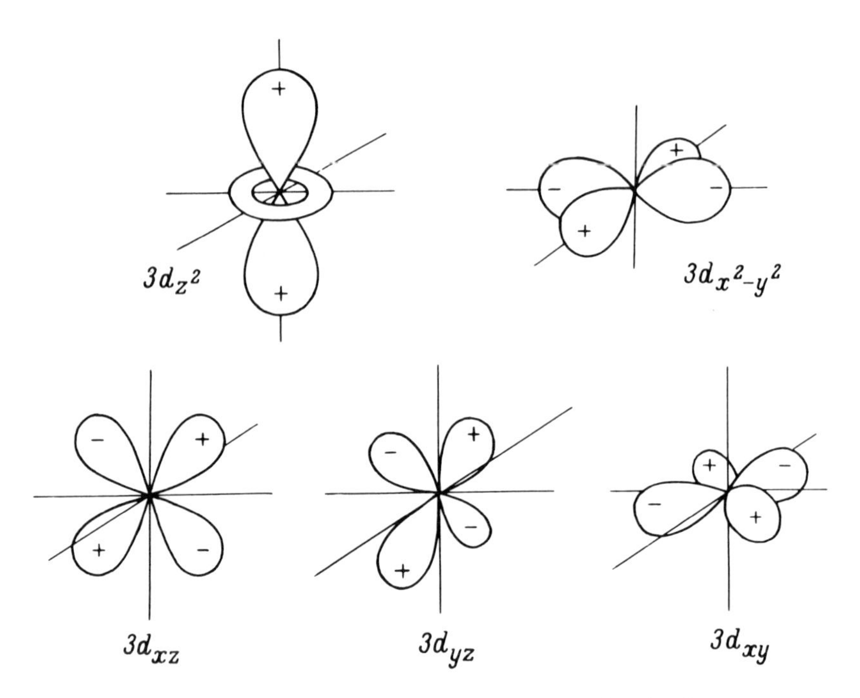

Fig. 16. Configuration of the 3d orbitals of silicon.

trons. Thus the interatomic Si–O distances in the majority of cases are no greater than 1.65 Å, which is smaller than the sum of the covalent radii of Si and O (1.83 Å) and the expected Si–O bond length (1.76 Å) allowing for corrections associated with the difference in electronegativity [5]. The possibility of using the 3d orbitals of silicon and other elements of the third period (P, S, Cl) in the formation of bonds involving occupation of these hybridized orbitals by unshared pairs of p electrons of neighboring atoms has been considered theoretically in a whole series of papers [6-9]. The five linearly independent d orbitals schematically depicted in Fig. 16 differ substantially from the p orbitals in symmetry properties. In contrast to the axially symmetric p orbitals, regions with alternating signs are divided, in the d orbitals, by two nodal planes. It follows from geometrical considerations that the possibility of strong overlapping taking place depends substantially on the spatial disposition of the d orbitals relative to the σ bonds: for effective overlapping between the d orbitals of one partner and the p orbitals of the other, their nodal plane must coincide and the σ bond must coincide with the plane in which the vectors representing the maximum densities of the d orbital lie. From this point of view it is improbable that there will be a simultaneous formation of four strong tetrahedrally oriented $d\pi-p\pi$ bonds. It has also been established that appreciable overlap integrals can only be obtained in the presence of a formal positive charge on the atom possessing the d orbitals, and so we may expect a strengthening of the $d\pi-p\pi$ bonds on increasing the electronegativity of neighboring atoms.

Detailed analysis of the interatomic distances in the isoelectronic ions XO_4^{n-} (X = Si, P, S, Cl) and more complex ions formed on the condensation of the tetrahedra was first carried out by Cruickshank [9], starting from an assumption as to the formation of two strongly bonding molecular orbitals obtained from the $3d_{z^2}$ and $3d_{x^2-y^2}$ orbitals of X and combinations of the $1/2\ (p_1+p_2+p_3+p_4)$ and $1/2\ (p_1'+p_2'+p_3'+p_4')$ type of the 2p orbitals of oxygen. It was concluded that there was an approximately linear relationship between the interatomic distances P–O, S–O, and the orders of the π bonds calculated by the valence-bond method; for example, for the XO_4^{n-} ion the order of the X–O bond is $1/4+1/4=1/2$. The comparatively small range of Si–O distances obtained in the as-yet small number of x-ray diffraction analyses of silicates carried out to a reasonable accuracy is ascribed to the partial compensation of the differences in the orders of the π bonds by electrostatic effects. Electron-diffraction determinations of the bond lengths in organo-silicon molecules embrace only a few substances [10-13], and the difference in the Si–O distances is hardly greater than experimental error.

Despite the indefinite nature of the absolute magnitudes associated with the choice of a means for describing the force field, the force constants are a more sensitive characteristic of changes in the orders of the bonds than the interatomic distances. Thus on passing from a simple C–C bond to a triple C≡C bond the interatomic distance changes by 22%, while the change in the force constant is 180%. Hence for an error of 0.02 to 0.03 Å in determining the interatomic distances, i.e., 1.5 to 2% and an error of ± 5% in calculating the force constants, the latter gives at least a two-fold advantage in sensitivity. The order of the bond may be found spectroscopically both by establishing correlations between the force constants and interatomic distances, on the one hand, and the interatomic distances and the orders on the other, and also by establishing a direct correlation between the force constant and the order of the bond. The first of these versions involves the use of the Badger [14] formula or other similar empirical formulas, and thus makes the accuracy of determining the order of the bond dependent on the accuracy of determining the interatomic distances. The second approach was employed by Siebert, who proposed an empirical formula not containing any dependence of the force constant on the interatomic distance. In a somewhat altered form [16] (compared with the original), this formula determines the force constant of a single X–Y bond in relation to the atomic numbers z, the principal quantum numbers u of the X and Y atoms, and the number of unshared pairs of electrons n in the two atoms

$$K'_{X-Y} = 7.20 \frac{z_X z_Y}{u_X^3 u_Y^3} \frac{6}{n+6} 10^5 \text{ dyn/cm}$$

In deriving the formula it was assumed that, in the molecules of the methyl derivatives $X(CH_3)_n$, the order of the C–X bonds is unity.

The order of the bond N is determined by means of the relations

$$N = \frac{K^N_{X-Y}}{K'_{X-Y}} \text{ for } 0 < N < 1.5 \text{ and}$$

$$N = 0.69 \frac{K^N_{X-Y}}{K'_{X-Y}} + 0.37 \text{ for } N > 1.5$$

On comparing the orders of the bonds formed between oxygen and elements of the third period in ions of the XO_4 type (calculated by Siebert [17]) we find a certain proportion of double-bondedness in the SiO_4 ions, in accordance with the X–O bond lengths.

	SiO_4	PO_4	SO_4
K^N_{X-O}	4.44	5.65	6.52 × 10^5 dyn/cm
K'_{X-O}	3.73	4.00	4.27 × 10^5 dyn/cm
N_{X-O} (by Siebert)	1.19	1.41	1.53
r'_{X-O} (calculated by the Stevenson–Schomaker method)	1.76	1.71	1.69 Å
r_{X-O} (experimental value [9])	1.63	1.54	1.49 Å
Shortening ($r'_{X-O} - r_{X-O}$)	0.13	0.17	0.20 Å

In subsequent discussions of spectroscopic data we shall use the Siebert approach, despite the slight uncertainty in the concept of bond order employed in this method.

Recently correlations between the force constants, the interatomic distances, and the orders of the bonds formed by silicon, phosphorus, sulfur, and chlorine with oxygen have been studied by a number of authors [9, 18-22]. The existence of considerable discrepancies in the proposed relationships is attributable both to the different methods of determining the order of the bond and also to the insufficient reliability of the force constants, particularly for the oxygen compounds of silicon.

Calculations of the frequencies of organosilicon molecules with Si–O bonds have usually been carried out allowing for only some of the vibrational coordinates, and only a small number of parameters has been used to describe the force field. The accuracy of the resultant force constants is insufficient for making any judgement as to the relatively small changes in bond order. Thus the force constant of the Si–O bond in the hexamethyldisiloxane molecule obtained by means of the triatomic model equals $4.65 \cdot 10^5$ dyn/cm [23], while on allowing for a large number of vibrational degrees of freedom the value $4.288 \cdot 10^5$ dyn/cm is obtained [24]. The differences in the methods of describing the force field exacerbate the difficulties of comparing the force constants of the Si–O bonds in molecules of differing structure. In what follows we shall briefly consider conclusions of a qualitative character, such as may be drawn by comparing the spectra of molecules with Si–O bonds, and also results obtained by the author together with Kr. Poiker and T. F. Tenisheva in connection with a complete calculation of the frequencies of some of the most important representatives of this class of compounds.†

Changes in the Nature of the Si–O Bonds in Relation to the Substituent Next to the Silicon and Oxygen Atoms. These changes may be explained as being due principally to changes in the electronegativities of the R and R′ atoms (groups) in the $R_3Si–OR'$ system, which influence the efficiency of the $p\pi–d\pi$ interactions between O and Si [25, 25a-25e]. It is also important to allow for the possibility of analogous interactions occurring in the Si–R bonds if the R atom possesses unshared pairs of electrons; in such cases the resultant "competition" may clearly prevent a rise in the order of the Si–O bond.

Let us consider the force constants of the Si–O bonds in $R_3Si–OR'$ molecules, the spectra of which are calculated with due allowance for all the vibrations of the system:

R	R'	K_{SiO}	K^*_{SiO}
Cl	CH_3	$9.00 \cdot 10^6$ cm^{-2}	$8.23 \cdot 10^6$ cm^{-2}
CH_3	CH_3, CD_3	6.75	6.17
CH_3	H, D	7.20	6.92
CH_3	Na	8.75	8.66

† The use of data relating to the spectra of alkyl halide silanes as calculated by Kovalev [31-33], and also data relating to the spectra of the deutero-analogs of these compounds, enabled the constants of the quadratic potential function of general form to be determined with an adequate objectivity. It should be noted that, in using this function containing cross terms $K_{ij}\Sigma Q_i Q_j$, the diagonal force constants K_{ii} do not, strictly speaking, constitute an individual characteristic of the elasticity of the particular bond in the molecule. Hence, in future, together with the coefficients K_{ii}, we also give the so-called "independent" force constants K^*_{ii}, which constitute a more objective measure of the elasticity of the bond being distorted. The method of determining K^*_{ii} from known K_{ii} and K_{ij} is described in [34]. As in other sections of this book, the force constants are here given using 1/16 of the mass of an oxygen atom as mass unit and 1 Å as length unit, in contrast to the scale frequently used in the spectroscopy of organic compounds, in which the unit of length is the length of the C–H bond and the unit of mass the "spectroscopic" mass of a hydrogen atom.

For calculating the spectra of $(CH_3)_3SiOCH_3$ and $(CH_3)_3SiOCD_3$ the original experimental data are used.

The smallest value of K^*_{SiO} is obtained for R = R′ = CH_3. Arbitrarily taking this value as the force constant of the Si–O single bond, we may use the Siebert method‡ to estimate the change in the order of the bond for various R and R′. The resultant values of bond orders may be a little too low, since even in $(CH_3)_3SiOCH_3$ the order of the Si–O bond may be slightly greater than unity.

The increase in the electronegativity of R as a result of the transition $(CH_3)_3SiOCH_3 \rightarrow Cl_3SiOCH_3$ leads to a 33% rise in K^*_{SiO}. Experimentally this appears as a rise in the frequency νSiO from 720 to 810 cm^{-1} [26-30]. The change in K^*_{SiO} may be partly attributed to a change in the polarity of the σ bond of Si–O. However, an analogous structural transition involving a bond between silicon and an atom not having unshared pairs has less effect on the value of the force constant. According to the results of Kovalev [31-33] K_{SiC} increases by only 15% from $(CH_3)_3Si-CH_3$ to Cl_3Si-CH_3, while transformation to the "independent" force constant K^*_{SiC} reduces this difference to 9%. This suggests that the strengthening of the $(p \rightarrow d)\,\pi$ bond of Si–O on replacing R = CH_3 by R = Cl raises K^*_{SiO} by no less than 20 to 25%. In the $Si(OCH_3)_4$ molecule the force constant of the Si–O bond is close to that for trimethylmethoxysilane (see Chap. 2, §3), which is probably associated with the fact that the influence of the increasing positive charge of Si (on replacing CH_3 by OCH_3) is compensated by the "competition" between several Si–O bonds having a common Si atom. It also clearly follows from this that the role of the $d\pi-p\pi$ interactions in the formation of the Si–Cl bonds of the Cl_3SiOCH_3 molecule is insignificant. The sequence of changes in the frequencies of the Si–O vibrations in the $Cl_{4-n}Si(OCH_3)_n$ series agrees with these considerations. The frequencies and the symmetry species (allowing for the symmetry of the $Cl_{4-n}SiO_n$ group of atoms only) are taken from [28]. The frequencies of molecules with n = 2, 3 have not yet been calculated.

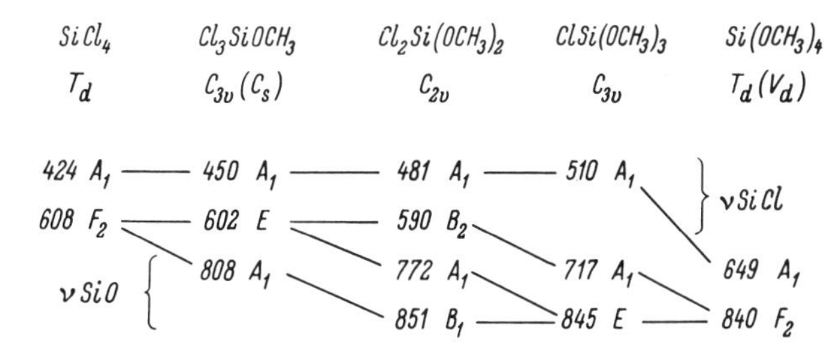

The force constant of the Si–O bond is greatest for n = 1 and, judging from the frequencies νSiO it changes little in the transition $Cl_3SiOCH_3 \rightarrow Cl_2Si(OCH_3)_2$. However, it clearly falls on passing to molecules with three and four methoxy groups, which enables us to associate this reduction not only with the smaller electronegativity of the OCH_3 groups as compared with the Cl atoms but also with competition between the Si–O bonds, i.e., the necessity of "sharing" the two 3d orbits of Si between three and four such bonds.

The same dependence of K^*_{SiO} on the electronegativity of R is found also for R′ = SiR_3, i.e., in $R_3SiOSiR_3$ molecules; the order of the bonds, according to Siebert, equals 1.13, 1.33, and 1.49 for R = CH_3, H, and Cl, respectively (see p. 39). The transition $R_3Si-OCH_3 \rightarrow R_3Si-O-SiR_3$ moreover, leads to the same rise in K^*_{SiO} (~13%) both for R = CH_3 and for R = Cl. A slightly different sequence in the changes of K^*_{SiO} occurs in the series of $(CH_3)_3Si-OR'$ compounds on

‡ Siebert's earlier-proposed force constant for the Si–O single bond differs little from the value here used and equals $6.35 \cdot 10^6$ cm^{-2}, while passing to the "independent" force constant of this bond determined from the spectrum of the SiO_4 ion (allowing for the Siebert bond order in this ion) gives $6.24 \cdot 10^6$ cm^{-2}.

replacing R′. We may expect a rise in K^*_{SiO} as the electropositivity of R′ increases; hence the rise in K^*_{SiO} in the ratio 1.00 : 1.12 : 1.40 for the series R′ = $CH_3(CD_3)$ → H(D) → Na indicates that the electropositivity of the H connected to the electronegative O atom is greater than that of the CH_3 group (in all the R_3Si-OR' compounds considered the signs of the polarity of the Si–R and O–R′ bonds are opposite). This result also agrees closely with the results of Cruickshank and Robinson [22], who came to the conclusion that the order of the O–H bonds in the POH and SOH systems was 0.65 to 0.70 of the order of the O–R bonds in analogous systems.

Apart from the frequency and force constant of the Si–O(H) bond in the isolated molecule, the changes in these characteristics on association of the molecules by means of hydrogen bonds are also very interesting. The νOH frequency of isolated silanols has been repeatedly found to be higher than that of carbinols or primary alcohols [35-41] and this has been considered as due to the electronegativity of silicon being less than that of carbon; in agreement with this, the dissociation constant of silicic acid is less than that of carbonic acid. On the other hand, silanols often behave as if more acidic than carbinols, and the reduction in the νOH frequency on formation of hydrogen bonds is about 400 cm^{-1} for the former as against 250 to 300 cm^{-1} for the latter. The stronger hydrogen bonds of the silanols are probably due to the possibility of the electron charge released in the partial ionization of the hydrogen finding an energetically favorable location in the π bond between the O and Si. The corresponding strengthening of the Si–O(H) bond is reflected in the 50-to-60-cm^{-1} higher νSi–O(H) frequency of the $(C_2H_5)_3SiOH$ molecule on forming hydrogen bonds with strong proton acceptors [42-45]† while the νC–OH frequency of methyl alcohol is practically unaltered on passing from the isolated to the associated molecule. We note that a comparison of NMR data [46] for trimethylsilanol and trimethylcarbinol led to the conclusion that there was less screening of the proton in the silanol (contrary to expectations based on the relation between the electronegativities of the Si and C), which also supported the idea of attraction of some of the electronic charge of the H into the O–Si bond.

An increase in the electronegativity of R in the R_3SiOH molecules, as in the alkoxyl silanes, evidently promotes a strengthening of the Si–O bond; the hydrogen bonds strengthen considerably in the transition $(CH_3)_3SiOH$ → Cl_3SiOH, judging from the νOH frequencies.

The earlier-mentioned increase in K^*_{SiO} in the transition $(CH_3)_3SiOH$ → $(CH_3)_3SiONa$ is reflected in the appearance of strong bands near 950 cm^{-1} in the spectra of silanolates studied by Tatlock and Rochow [37] and interpreted by Ryskin. The order of the Si–O bond in $(CH_3)_3SiONa$, higher than that obtained by Siebert for the $[SiO_4]^{4-}$ ion, may be associated both with the presence of only *one* Si–O bond in the silanolate molecule and also with the fact that Siebert used averaged data for the orthosilicates, including those with less basic cations than alkali metals.

Order of the Si–O–Si Bonds and Valence Angles of the Oxygen Atoms. These quantities are strictly related, as indicated in Fig. 17 [48]. If the bonds of the O atom in the Si–O–Si bridge are simple p bonds, then for the formation of the (p→d) π bonds with Si only one of the p orbitals of O can be used. The increase in the SiOSi angle corresponding to a transition to sp^3, sp^2, and sp (for ∠SiOSi = 180°) configurations of the O atom leads in the limiting case to the appearance of two equivalent p orbitals occupied by unshared pairs, the orientation of which is favorable for interaction with the d orbitals of Si. It follows

† For $(CH_3)_3SiOH$ the picture is somewhat more complicated: as Kriegsmann showed in [47], in addition to the ν and δOH frequencies, two other frequencies, at about 780 and 900 cm^{-1}, are associated with the vibrations of the SiOH group. It follows from our calculation of the spectrum that this is due to the strong interaction between the νSiO vibration and a rocking vibration of CH_3; hence the contribution of Si–O bond stretching is about equal for both vibrations.

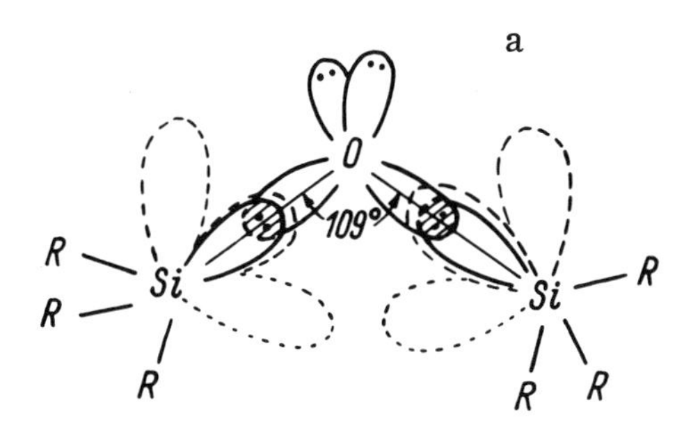

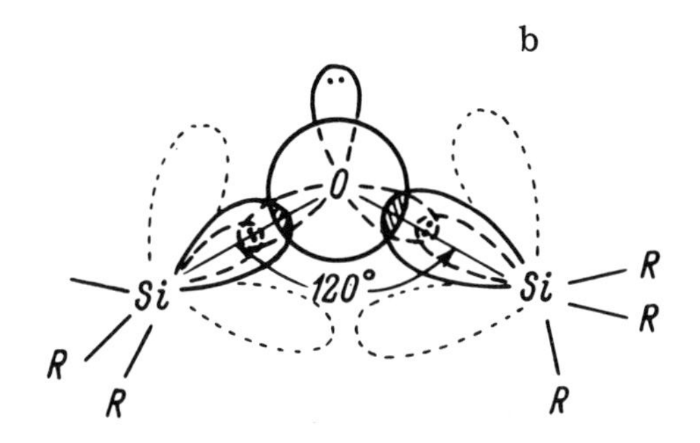

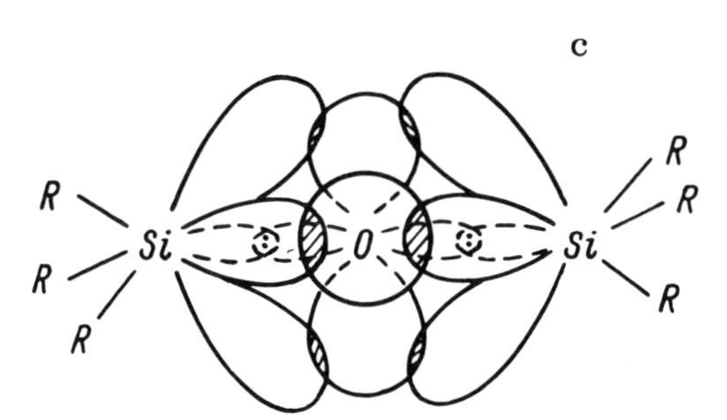

Fig. 17. Scheme of the electron structure of the Si–O–Si bridge for various configurations of the O atom. a) sp^3; b) sp^2; c) sp.

from these considerations, in particular, that on increasing ∠SiOSi (strengthening the $p\pi-d\pi$ interaction) the proton acceptor capacity of oxygen should be weakened. Actually the chemical shift of the magnetic resonance of the proton in solutions of chloroform in ethers, alkoxysilanes, and siloxanes changes successively from $\delta = -0.740$ for ethers to $\delta < -0.057$ for siloxanes, in the opposite sense to that expected on the basis of the electronegativity of the atoms connected to the O: i.e., Si–O–Si, Si–O–C, C–O–C [49]. The relatively high value of δ for hexamethylcyclotrisiloxane (-0.3) may also be associated with the smaller value of ∠SiOSi in the plane trisiloxane ring: for ∠SiOSi ≃ 120 to 130° and a nearly sp^2 configuration of the O atom, the existence of an unshared pair in an extended orbital makes it possible for a $\begin{matrix}Si\\Si\end{matrix}\!\!>\!O\ldots H$ bond to be formed. A study of the displacement of the νOH bands in the infrared spectra of proton donors in similar systems [50] led to analogous results.

Another important conclusion to be drawn from the scheme of Fig. 17 is that the angle SiOSi is highly sensitive to the electronegativity of R in $R_3SiOSiR_3$ molecules. As the positive charge on the Si increases, we may expect an increase in the SiOSi angle, not only as a result of electrostatic interaction, but also on account of the strengthening of the $(p \rightarrow d)\,\pi$ bonds (coinciding with a rise in the Si–O force constant). Let us compare some spectroscopic (and electron-diffraction) data† regarding the dynamics and structure of the $R_3SiOSiR_3$ molecules.

		$(CH_3)_3SiOSi(CH_3)_3$ [$(CD_3)_3SiOSi(CD_3)_3$]	$H_3SiOSiH_3$ [$D_3SiOSiD_3$]	$Cl_3SiOSiCl_3$
ν_{as} SiOSi (infrared spectrum, vapor)		$1073\ cm^{-1}$ [$1065\ cm^{-1}$]	$1107\ cm^{-1}$ [$1094\ cm^{-1}$]	$1131\ cm^{-1}$
ν_s SiOSi (Raman spectrum, liquid)†		$520\ cm^{-1}$ [$467\ cm^{-1}$]	$606\ cm^{-1}$ [$555\ cm^{-1}$]	$718\ cm^{-1}$ ‡
in $10^6\ cm^{-2}$	K_{SiO}	7.55	8.75	9.60
	K^*_{SiO}	6.96	8.17	9.18
	$H_{SiO,\ SiO}$	$\leqslant 0$	$\simeq 0.5$	$\simeq 1.0$
	K_{SiOSi}	0.5 — 1.0	0.2	0.0 — 0.2
$\angle$ SiOSi by a spectroscopic estimate		135 — 140°	150° [55]	160 — 170°
Electron-diffraction data	$\angle$ SiOSi	130±10° [10]	144.1±0.8° [12]	175±5° [52]
	r_{SiO} (Å)	1.63 [10]	1.634±0.002 [22]	1.64 [10]

The values of K^*_{SiO} and $\angle$ SiOSi given here agree with the foregoing considerations, in particular confirming the rise in the order of the Si–O (Si) bond on increasing the angle SiOSi, while the values of r_{SiO}, which have been determined fairly accurately only for $H_3SiOSiH_3$, show no strengthening in the series R = CH_3, H, Cl. Despite the much lower accuracy in estimating the interaction constant of the Si–O bonds with a common O atom ($H_{SiO,\ SiO}$), this undoubtedly increases with rising K^*_{SiO}. Since the coefficient $H_{SiO,\ SiO}$ characterizes (for $H_{SiO,\ SiO} > 0$) the strengthening of one bond when the other is being weakened, its intensification may be regarded as a measure of the delocalization of the π electrons in the Si–O–Si bridge as the electron structure of this group gradually approaches that of the cumulative system in which both unshared pairs of electrons of the oxygen are attracted into each of its (p → d) π bonds with the two Si atoms. We should also notice the fact that the rigidity of the Si–O–Si bond angle characterized by the bending force constant K_{SiOSi} (the accuracy in estimating this is also not very high) falls with increasing $\angle$ SiOSi and K^*_{SiO}. The dependence of this flexibility on the electron structure of the birdge and its effect on the properties of compounds containing such bridges is considered in Chapter VI, Section 3.

The frequencies of the two stretching vibrations of the bridge (ν_{as} SiOSi and ν_s SiOSi), given here increase with K^*_{SiO}, independently of any change in the masses of R and the shape of the molecules. (For constant K^*_{SiO}, an increase in the masses of R reduces both frequencies, and chiefly ν_s SiOSi, while a rise in $\angle$ SiOSi reduces ν_s SiOSi but raises ν_{as} SiOSi). This enables us (with some caution) to use the frequencies of the corresponding vibrations in order to judge changes in the bond strengths of Si–O–Si (and similar) bridges in molecules for which the spectra have not been calculated. A sympathetic rise in the ν_{as} SiOSi and ν_s SiOSi frequencies, unambiguously indicating the strengthening of the Si–O bonds, was observed by Kriegsmann [56] in the spectra of $X(CH_3)_2SiOSi(CH_3)_2X$ compounds for X = CH_3, H, C_6H_5, Cl; an analogous result was obtained for the Si–N bonds in $X(CH_3)_2SiNHSi(CH_3)_2X$ molecules [57]. Existing data regarding the structure of the $H_3SiSSiH_3$ [58] and also the results of spectroscopic investigations of $H_3SiSSiH_3$ [59], $(CH_3)_3SiSSi(CH_3)_3$ [23, 62], $H_3SiCH_2SiH_3$ [60, 61], $(CH_3)_3SiCH_2Si(CH_3)_3$ [23], $H_3GeOGeH_3$, and $H_3GeSGeH_3$ [63] give no indication of a rise in the orders of the Si–S, Si–C, Ge–O, and Ge–S bonds in these compounds. It should also be noted that the very fact of the much narrower range

† For hexamethyl- and hexachlorodisiloxane we use the original data for the vibrational spectra. The frequencies in the spectra of $H_3SiOSiH_3$ and $D_3SiOSiD_3$ are taken from [51].

‡ The intense polarization of the 353 cm^{-1} line earlier [53, 54] ascribed to the ν_s SiOSi vibration is due to Fermi resonance between the ν_s $SiCl_3$ vibration and the overtone of the deformation vibration of the $SiCl_3$ group, or is associated with the contamination of $Cl_3SiOSiCl_3$ by a trace of $Cl_3SiSiCl_3$, as indicated in [54a].

of COC or SiCSi angles encountered in different compounds serves as an indication of the absence of any tendency towards a straightened configuration similar to that occurring in the Si–O–Si bridges. Thus the transition from the spectra of $(CH_3)_3SiCH_2Si(CH_3)_3$ to $H_3SiCH_2SiH_3$ and $Cl_3SiCH_2SiCl_3$ does not lead to any substantial change in the frequencies of the Si–C–Si vibrations and gives no indications of any appreciable changes in the SiCSi angle [53].

For the structures of silicates, in which in the majority of cases the same Si atoms form bonds of both the Si–O(Si) and $Si-O^-$ $(\ldots M^+)$ types, the question as to the possible mutual influence of the two types of bond is particularly important. In this respect it is interesting to compare the spectra of the $XO(CH_3)_2SiOSi(CH_3)_2OX$ compounds on passing from X = H to X = Li, Na, K [64]. On replacing the hydrogen atom by potassium, the infrared spectrum shows a rise in the frequencies of the stretching vibrations of the Si–O(H, K) bonds, although not so large as for $(CH_3)_3SiOH \rightarrow (CH_3)_3SiOK$, and at the same time a fall in the frequencies of the ν_s SiOSi and ν_{as} SiOSi vibrations. It follows from this that, on increasing the force constant (increasing the order) of the "end" Si–O bond, as a result of a rise in the negative charge on the O atom there is a fall in the order of the Si–O–Si bond not attributable to any other causes. The fact that the fall in the ν_{as} SiOSi frequency is much greater than in ν_s SiOSi suggests that the SiOSi valence angle also diminishes with the fall in the order of the Si–O–Si bond. The opposing effect of the two factors, dynamic and kinematic, leads to a comparatively small change in the frequency of the symmetrical Si–O–Si vibration. Figure 18 gives a schematic picture of the Si–O vibration frequencies in the spectra of $HO(CH_3)_2SiOSi(CH_3)_2OH$ and $KO(CH_3)_2SiOSi(CH_3)_2OK$, and also the results of a calculation of the frequencies of the stretching vibrations of the simplified five-atom model O\Si/O\Si/O, illustrating the foregoing considerations: thus the changes taking place in frequency on replacing hydrogen by potassium may be satisfactorily reproduced if we consider that the force constant of the Si–O(H, K) bond rises by 7% and the force constant of the Si–O(Si) bond falls by about the same amount, with a simultaneous reduction of 8 to 10° in the $\angle$SiOSi.

§ 2. Calculations of the Vibrational Frequencies of Complex Anions in Silicates

The most important problem in the spectroscopic study of silicates is that of establishing correlations between the vibrational spectrum and the structure of the complex anion, since

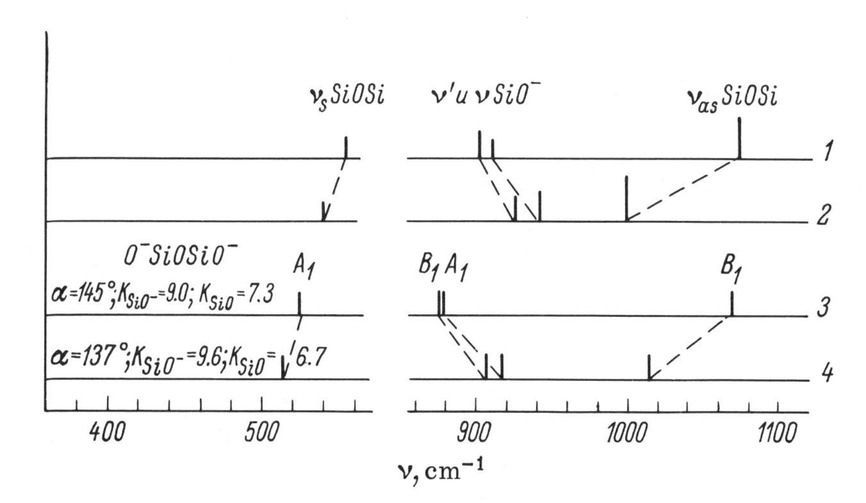

Fig. 18. Frequencies of the stretching vibrations of the Si–O bond in $XO(CH_3)_2SiOSi(CH_3)_2OX$ molecules. 1) X = H; 2) X = K (experimental data); 3) and 4) calculated data for a plane five-atom model of OSiOSiO (α – $\angle$SiOSi, OSiO angle tetrahedral).

the majority of crystalline silicates contain complex anions formed by the condensation of several (or an infinite number of) silicon–oxygen tetrahedra. At first, attempts were made to reduce the spectrum of the complex anion roughly to the vibrations of an isolated tetrahedron [65-73]; these not only were devoid of physical meaning, since they ignored both the dynamic and kinematic interaction of the vibrations of the coupled tetrahedra, but also frustrated the very presentation of the problem of establishing the configuration of the complex anion by spectroscopic methods. The first investigations in the field of the theoretical analysis and calculation of some of the vibrational frequencies of α quartz were carried out by Saksena [74-79]; Barriol [80] considered the vibrations of the more symmetrical lattice of β quartz; finally, Matossi [81] calculated the vibrations of Si_2O_7 groups, Si_3O_9 rings, chains of the $(Si_2O_6)_\infty$ type, and quartz.

The development of efficient methods of calculating the vibrational spectra of molecules [82-84] again eased the problem of using these fairly simple and universal procedures for calculating the vibration frequencies of silicate anions. The more systematic investigations in this direction were carried out by Stepanov and Prima, using the method of internal coordinates (changes in the bond lengths and valence angles) in order to establish the secular equations. In a number of papers [85-87] such secular equations were obtained for the vibrations of several typical structures of complex anions possessing translational symmetry in one, two, or three directions. In this way consideration was given to infinite chains of the pyroxene type, a layer similar to that found in structures of the talc and mica type, and the three-dimensional space lattices of β cristobalite and β quartz. In a later investigation Prima obtained and analyzed the secular equations of vibrations of plane silicate rings with D_{nh} symmetry [88]. In papers by Saksena [89, 90] the secular equations for a number of structures of silicate anions were compared, using Cartesian coordinates for the displacements of the atoms. However, a comparison of the calculated frequencies of the normal vibrations of the complex anions with the experimental spectra of the silicates by no means always led to a correct interpretation of the latter. The main cause of this disagreement was not errors in considering the symmetry properties of the systems under examination nor the rather idealized structure assumed in deriving the secular equations. Of far more significance were difficulties of a quite general nature arising from attempts at calculating the vibrational spectra of any inorganic crystals with complex anions. It is therefore desirable to present a brief discussion of the main factors limiting the applicability of existing computing methods to the vibrational spectra of silicates, and also certain features of the actual technique of setting up the secular equations.

Silicate Anions of the "Island" Type. In calculating the frequencies of internal vibrations these may be considered as isolated groups of atoms, so that the problem of setting up the secular equation for them differs in no way from the analogous problem for a molecule. If necessary, a possible reduction in the site symmetry of the anion in the lattice may be allowed for when passing from internal to symmetry coordinates. However, remembering the approximate nature of the calculation, in the majority of cases it is more convenient to confine oneself to calculating the frequencies of the more symmetrical model, and when comparing the calculated and experimental frequencies to allow for the possibility of a change in the selection rules and a reduction in the degree of degeneracy (or the removal of degeneracy). Resonance interactions between the vibrations of translationally nonequivalent complex anions in the lattice can of course not be allowed for when calculating the frequencies of one anion.

In solving the mechanical problem associated with calculating the vibrations of a complex anion (as in the case of molecules), the number of observed frequencies of the fundamental vibrations is much smaller than the number of potential-energy constants, which include both valence force constants and interaction force constants. In calculating the vibrational spectra of molecules of organic compounds this difficulty may be avoided either by using experimental data relating to the spectra of isotopically substituted molecules or else by assuming the constancy of a whole series of force constants in different molecules. The first method is in general only of interest for systems containing hydrogen atoms, since for other atoms the mass differences of different isotopes are too small. The second method also has a limited applicability in relation to silicate anions; this follows from the preceding discussion of the wide range of possible values of the force constants of organosilicon molecules containing Si–O bonds. This method cannot be effectively used until fairly reliable calculations of the spectra of many compounds of this class have been carried out.

The most serious difficulties arising in any attempt at calculating the vibrational frequencies of complex anions in silicates are associated with the failure to allow for the interaction of the internal vibrations of the anions with the vibrations of the cation-oxygen bonds, i.e., the lattice vibrations. Even in crystals only containing cations of the alkali metals, only the stretching vibrations of the anion can be considered (to a fairly close approximation) as not interacting with the lattice vibrations. Among the bending (deformation) vibrations of a complex anion, many are characterized by frequencies of the order of a few hundreds or even tens of reciprocal centimeters, and these frequencies cannot be satisfactorily calculated by means of the isolated-anion model. For this reason in some papers [86] only the stretching frequencies of the silicate anions have been calculated initially, allowing only for the stretching force constants, and then, in order to secure agreement with experimental data, introducing one or two more force constants characterizing valence angle deformations. Here, of course, the constants introduced really had a different physical meaning, since they included not only the interaction between stretching and bending vibrations but also the interaction with the vibrations of the lattice. The majority of experimental data regarding the vibrational spectra of silicates relate to crystals in which the part of cations is played by elements of the second, third, and subsequent groups of the periodic system. The interaction of these cations with the oxygen atoms of the complex anions has a partially covalent (directional) character and influences different internal vibrations of the anion in a very different manner. For the same reason there is hardly any sense in trying to improve the agreement between the calculated and experimental frequencies by considering the vibrations of the complex anion in some effective external field. It should also be noted that, even for purely electrostatic interactions between the cation and anion, their influence on the internal vibrations of the anion depends on the form of these vibrations. Hence, even in these cases, the displacement of different internal frequencies on passing from the isolated anion to the crystal differs considerably. This question is discussed in more detail in [91].

The interaction between the vibrations of similar but translationally nonequivalent groups of atoms in the lattice is considered as a small perturbation in Davydov's theory [92], which is reasonably valid with respect to molecules, but in structures with complex ions is only justified for crystals with predominantly electrostatic interactions between the cations and anions. It will subsequently be shown by a number of examples that the splitting of the nondegenerate internal vibrations of silicate anions in crystals increases considerably on increasing the proportion of covalence in the bonds between the ions. In the limiting case, the model of the isolated anion is unsuitable even for a qualitative interpretation of the stretching vibrations of the anion. For such cases one must consider all the normal vibrations of the crystal lattice, and there is no longer any point in dividing them into internal and external. The limits of applicability of such a classification also depend on the masses of the cations and on the frequencies of the internal vibrations of the anions. In addition to this, the coordination number of the cation is important (this partly characterizes the proportion of covalence in the cation/anion interaction). Table 6 approximately classifies various cations in relation to the suitability of the model of the isolated silicate anion for interpreting the spectra in the frequency range of its stretching vibrations. This classification naturally changes somewhat on passing, for example, to germanate anions, with lower characteristic frequencies, or, on the other hand, to phosphates or borates.

Table 6. Classification of the Cations in Silicates with Respect to Their Influence on the Vibrational Spectrum of a Complex Anion

Weak cations	Medium-strong cations	Strong cations
Na^+, K^+, Rb^+, Cs^+, Li^+, Ag^+, Cu^+, Ba^{2+}, Cu^{2+}, Sr^{2+}, Ca^{2+}, Ln^{2+}, Fe^{2+}, Mn^{2+}	Co^{2+}, Ni^{2+}, Cd^{2+}, Fe^{3+}, $[^{4,6}]Mg^{2+}$, $[^{4,6}]Zn^{2+}$, Sc^{3+}, Y^{3+}, Ln^{3+}, Bi^{3+}, Zr^{4+}, Pb^{2+}, $[^{6}]Ti^{4+}$, $[^{6}]Al^{3+}$	All remaining cations: Be^{2+}, B^{3+}, $[^{4}]Al^{3+}$ etc.

Note. Weak cations are those whose presence in the structure does not prevent the identification of the internal vibrations of the silicate anion, with frequencies above ~400 cm^{-1}; only in the presence of cations Na^+ to Cs^+ can the bending vibrations of the anion be calculated. The presence of medium-strong cations as a rule admits the possibility of identifying the frequencies of stretching vibrations of the complex anion; the calculation of the stretching vibrations of the anion involves considerable errors, while the bending vibrations of the complex anion mix strongly with the lattice vibrations. On introducing strong cations into the structure, the division of frequencies into internal and external loses all meaning.

Calculation of the Frequencies Representing the Normal Vibrations of Systems Possessing Translational Symmetry. The calculation of such frequencies in systems with translational symmetry in one, two, or three directions (chains, layers, or three-dimensional lattices of silicon–oxygen tetrahedra) differs somewhat from the calculation of the frequencies associated with molecules of finite size. Theoretical investigations into this question have been carried out by a number of authors [93-102]. In the majority of cases the aim has been to calculate and analyze the spectra of molecules in the crystals of high-molecular organic compounds. Greatest attention has therefore been paid to structures of the chain type, the most widespread form of which is the helix. Since many other forms of chains may be considered as special cases of helices, we shall discuss the question of calculating the frequencies of the normal vibrations in accordance with the exposition of Higgs [101].

An infinite helix formed of consecutively connected, identical groups of atoms is characterized by two parameters,† l and ψ. Each group of atoms transforms into its neighbor by the operation $H(l, \psi)$, consisting of a translation l along the chain axis with a rotation through an angle ψ around it. The set of all the operations H^n (n is any whole number) transforming any of the groups of atoms into another forms a finite group $\mathbf{H}(l, \psi)$ isomorphic with respect to the cyclic group C_∞. All its irreducible representations are correspondingly unidimensional and may be denoted by $\Gamma(\theta)$, where θ is a parameter taking any values in the range $-\pi < \theta \leq \pi$. The corresponding characters are $\chi(\theta, H^n) = e^{in\theta}$. Each normal vibration of the chain belongs to one of the representations $\Gamma(\theta)$. This means that, if in a particular normal vibration some group of atoms vibrates in a particular fashion with amplitude A, then another such group of atoms lying at a distance of n such groups from the first vibrates in an analogous manner with an amplitude $Ae^{-in\theta}$. Each of the vibration frequencies of a single group ν thus engenders a branch $\nu(\theta)$, while θ characterizes the phase difference. For each frequency $\nu(\theta)$ there is a complex conjugate $\nu(-\theta)$ equal to it, since the frequencies are real. Hence the frequencies

† We note that no restrictions are placed on the value of ψ: π/ψ is any number, including irrational quantities. The latter may be interesting for some organic polymers.

belonging to the representations $\Gamma(\theta)$ and $\Gamma(-\theta)$ are degenerate in pairs (except for $\nu(\theta)$ and $\nu(\pi)$), while the complex normal vibrations may be transformed into pairs of real vibrations with amplitudes A cos (nθ) and A sin (nθ).

In setting up the secular equation, as usual we use internal vibrational coordinates, denoting these by q_{nk}, where the index n corresponds to the numeration of the "cells," groups of atoms, while k corresponds to the numeration of the coordinates of a particular cell. Here n is any whole number while k = 1, 2, ..., N, where N is the number of coordinates of the cell. The coordinates are chosen in the same way for any cell; hence any of the coordinates q_{nk} transforms into another corresponding to a different value of n by operations of the group H with a fixed k. The frequencies of the normal vibrations $\nu(\theta)$ may be calculated from the secular equation

$$|\boldsymbol{T}^{-1}\boldsymbol{U} - \lambda\boldsymbol{E}| = 0 \tag{1}$$

Here $\lambda = 4\pi^2\nu^2$, and E is the unitary matrix.

The potential energy U and kinetic energy T of the vibrations are associated with the matrices **U** and $\mathbf{T}^{-1}$ by the relations $U = \frac{1}{2}\boldsymbol{q}^*\boldsymbol{U}\boldsymbol{q}$ and $T = \frac{1}{2}\dot{\boldsymbol{q}}^*\frac{1}{\boldsymbol{T}^{-1}}\dot{\boldsymbol{q}}$. Since U and T are invariant with respect to operations of the group H, the matrix elements $U_{nk,\,ml}$ and $T^{-1}_{nk,\,ml}$ only depend on the difference in the values of n and m. If n − m = s, then we may write $U_{nk,\,ml} = {}^{(s)}U_{kl}$ and $T^{-1}_{nk,\,ml} = {}^{(s)}T^{-1}_{kl}$. The quantities ${}^{(s)}U_{kl}$ and ${}^{(s)}T^{-1}_{kl}$ respectively characterize the dynamic and kinematic interaction of the k-th coordinate of a particular cell with the l-th coordinate of the cell displaced by s units relative to the former.

Partial diagonalization of the matrices **U** and $\mathbf{T}^{-1}$ is effected by transforming to the symmetry coordinates

$$\left.\begin{aligned} S_k(\theta) &= (2\pi)^{-1/2}\sum_{n=-\infty}^{+\infty} q_{nk}e^{in\theta} \\ q_{nk} &= (2\pi)^{-1/2}\int_{-\pi}^{+\pi} S_k(\theta)e^{-in\theta}d\theta \end{aligned}\right\} \tag{2}$$

As a result of the transformation the following matrix elements not containing the index s may be obtained

$$\left.\begin{aligned} U_{kl}(\theta) &= \sum_{s=-\infty}^{+\infty} {}^{(s)}U_{kl}e^{-is\theta} \\ T^{-1}_{kl}(\theta) &= \sum_{s=-\infty}^{+\infty} {}^{(s)}T^{-1}_{kl}e^{-is\theta} \end{aligned}\right\} \tag{3}$$

and this enables us to pass from the secular equation (1) of infinite degree to an infinite set of equations of degree N

$$|\boldsymbol{T}^{-1}(\theta)\,\boldsymbol{U}(\theta) - \lambda(\theta)\,\boldsymbol{E}| = 0 \tag{4}$$

where $\lambda(\theta) = 4\pi^2\nu^2(\theta)$ and $-\pi < \theta \leqslant \pi$.

A study of the selection rules shows that in the Raman spectrum [101] only vibrations with $\theta = 0$ and $\theta = 2\psi$ are active, while in the infrared spectrum only vibrations with $\theta = 0$ and $\theta = \psi$ are allowed. Hence, in order to obtain the frequencies of optically active fundamental vibrations of the helix, it is sufficient, out of the whole set of equations (4), to solve three, with $\theta = 0,\ \psi,\ 2\psi$.

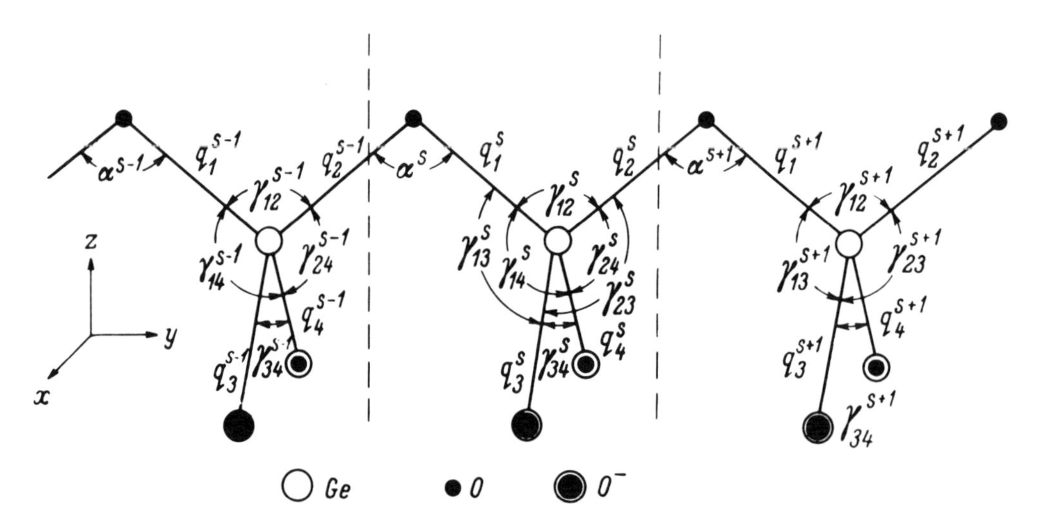

Fig. 19. Configuration of the $(GeO_3)_\infty$ chain in copper metagermanate. Notation and numeration of the internal vibrational coordinates, s being the number of the unit cell of the chain.

The meaning of the transformations described lies in the use of the translational symmetry of the system. The problem of setting up a secular equation for specific chains of comparatively uncomplicated structure presents no special difficulties, all the more so in view of the fact that, in order to analyze the frequencies of the fundamental vibrations in the infrared and Raman spectra it is only necessary to study vibrations with an infinite wavelength, i.e., in-phase vibrations of all the translationally identical parts of the chain.

In order to illustrate the foregoing it is useful to analyze an example relating to structures of present interest, confining attention to a consideration of the limiting vibrations from the very beginning. The simplest model we may use is an infinite chain of germanium–oxygen tetrahedra with one tetrahedron in the identity period ($\psi = 2\pi$), such as occur in copper metagermanate crystals (Fig. 19). The symmetry elements of the chain are symmetry planes σ_{xz} and σ_{yz} and a two-fold axis in the direction of the z axis; together with translations, these form a one-dimensional space group, the factor group of which is isomorphic with respect to the point group C_{2v}. The complete set of internal coordinates of the s-th cell of the chain includes the following (the intrinsic symmetry of the coordinates is shown in brackets): changes in the bond lengths Ge—O(Ge) — q_1^s, q_2^s (C_s), changes in the bond lengths Ge—O⁻ — q_3^s, q_4^s (C_s), changes in the angles O⁻GeO — γ_{14}^s, γ_{13}^s, γ_{24}^s, γ_{23}^s (C_1), the angle OGeO — γ_{12}^s (C_{2v}), the angle O⁻GeO⁻ — γ_{34}^s (C_{2v}), and the angle GeOGe — α^s (C_{2v}).

For an infinite chain we may set up an infinite kinematic-interaction matrix, calculating the elements of this by the usual method. Table 7 shows part of this matrix, including all the nonzero interactions of the coordinates of the cell s with the coordinates of the same cell and neighboring cells. Since, in the case of the vibrations with infinite wavelength of present interest, the motions of translationally equivalent atoms of all cells are in phase, for such vibrations (totally symmetric with respect to translations) the infinite matrix may be transformed into a finite number of similar and finite matrices by summing the columns (or lines) corresponding to equivalent coordinates of all the cells in the chain. Allowing for the symmetry of the factor group (by introducing coordinates of point symmetry, as in the case of a finite system) is the main problem in the subsequent transformation of the resultant eleventh-order matrix. Then the matrix may be reduced to quasi-diagonal form, the orders of the submatrices for symmetry types A_1, B_1, B_2, and A_2 being respectively equal to six, two, two, and one.

Table 7. Matrix of Kinematic Interaction for an Infinite $(GeO_3)_\infty$ Chain

	...	q_1^{s-1}	q_2^{s-1}	q_3^{s-1}, q_4^{s-1}	γ_{12}^{s-1}	γ_{13}^{s-1}, γ_{14}^{s-1}	γ_{23}^{s-1}	γ_{24}^{s-1}	γ_{34}^{s-1}	α^{s-1}	q_1^s	q_2^s	q_3^s	q_4^s	γ_{12}^s	γ_{13}^s	γ_{14}^s	γ_{23}^s	γ_{24}^s	γ_{34}^s	α^s	α^{s+1}	q_1^{s+1}	γ_{12}^{s+1}	γ_{13}^{s+1}	γ_{14}^{s+1}	...
...	...	...	...	...	...	...	...	...	...	...	...	...	...	...	...	...	...	...	...	...	...	...	...	...	...	...	...
q_1^s	...		$-v$		$-2f$		f	f			a	$-b$	$-b$	$-b$	$-2c$	$-2d$	$-2d$	e	e	$2d$	$-2f$	$-2c$					...
q_2^s	...										$-b$	a	$-b$	$-b$	$-2c$	e	e	$-2d$	$-2d$	$2d$	$-2c$	$-2f$	$-v$	$-2f$	f	f	...
q_3^s	...										$-b$	$-b$	a	$-b$	$2c$	$-2c$	e	$-2c$	e	$-2d$	c	c					...
q_4^s	...										$-b$	$-b$	$-b$	a	$2c$	e	$-2c$	e	$-2c$	$-2d$	c	c					...
γ_{12}^s	...		$-2f$		$2w$		$-w$	$-w$			$-2c$	$-2c$	$2c$	$2c$	g	h	h	h	h	$-2i$	k	k	$-2f$	$2w$	$-w$	$-w$	...
γ_{13}^s	...		f		—		$5w$	$-4w$			$-2d$	e	$-2c$	e	h	l	m	q	p	n	o	r					...
γ_{14}^s	...		f		$-w$		$-4w$	$5w$			$-2d$	e	e	$-2c$	h	m	l	p	q	n	o	r					...
γ_{23}^s	...										e	$-2d$	$-2c$	e	h	q	p	l	m	n	r	o	f	$-w$	$5w$	$-4w$	...
γ_{24}^s	...										e	$-2d$	e	$-2c$	h	p	q	m	l	n	r	o		$-w$	$-4w$	$5w$	...
γ_{34}^s	...										$2d$	$2d$	$-2d$	$-2d$	$-2i$	n	n	n	n	t	$-i$	$-i$					...
α^s	...	$-2c$	$-2f$	c	k	r	o	o	$-i$	y	$-2f$	$-2c$	c	c	k	o	o	r	r	$-i$	u	y					...
...	...	...	...	...	...	...	...	...	...	...	...	...	...	...	...	...	...	...	...	...	...	...	...	...	...	...	...

Note: $a = \varepsilon_1 + \varepsilon_2$; $b = \frac{1}{3}\varepsilon_1$; $c = \frac{\sqrt{2}}{3}\varepsilon_1\sigma_1$; $d = \frac{\sqrt{2}}{3}\varepsilon_1\sigma_2$; $e = \frac{\sqrt{2}}{3}\varepsilon_1(\sigma_1 + \sigma_2)$; $f = \frac{\sqrt{2}}{3}\varepsilon_2\sigma_1$; $g = \frac{8}{3}\varepsilon_1\sigma_1^2 + 2\varepsilon_2\sigma_1^2$; $h = -\frac{2}{3}\varepsilon_1\sigma_1(\sigma_1 - \sigma_2) - \frac{1}{2}\varepsilon_2\sigma_1^2$; $i = \frac{4}{3}\varepsilon_1\sigma_1\sigma_2$; $k = \frac{4}{3}(\varepsilon_1 + \varepsilon_2)\sigma_1^2$; $l = \varepsilon_1\left(\sigma_1^2 + \sigma_2^2 + \frac{2}{3}\sigma_1\sigma_2\right) + \varepsilon_2(\sigma_1^2 + \sigma_2^2)$; $m = \varepsilon_1\left(-\frac{1}{2}\sigma_1^2 - \frac{1}{3}\sigma_1\sigma_2 + \frac{5}{6}\sigma_2^2\right) - \frac{1}{2}\varepsilon_2\sigma_1^2$; $n = \frac{2}{3}\varepsilon_1\sigma_2(\sigma_1 - \sigma_2) - \frac{1}{2}\varepsilon_2\sigma_2^2$; $o = -\frac{1}{2}\varepsilon_1\sigma_1^2 - \frac{1}{6}\varepsilon_1\sigma_1\sigma_2 - \frac{2}{3}\varepsilon_2\sigma_1^2$; $p = -\frac{2}{3}\varepsilon_1(\sigma_1 + \sigma_2)^2$; $q = \varepsilon_1\left(-\frac{1}{2}\sigma_2^2 - \frac{1}{3}\sigma_1\sigma_2 + \frac{5}{6}\sigma_1^2\right) - \frac{1}{2}\varepsilon_2\sigma_2^2$; $r = \frac{5}{6}\varepsilon_1\sigma_1\sigma_2 - \frac{1}{6}\varepsilon_1\sigma_1^2$; $t = \frac{8}{3}\varepsilon_1\sigma_2^2 + 2\varepsilon_2\sigma_2^2$; $u = \frac{8}{3}\varepsilon_2\sigma_1^2 + 2\varepsilon_1\sigma_1^2$; $v = \frac{1}{3}\varepsilon_2$; $w = \frac{1}{6}\varepsilon_2\sigma_1^2$; $y = \frac{1}{3}\varepsilon_1\sigma_1^2$ ($\sigma_1 = 1/r_{Ge-O(Ge)}$; $\sigma_2 = 1/r_{Ge-O^-}$; $\varepsilon_1 = 1/m_{Ge}$; $\varepsilon_2 = 1/m_O$).

The representation of the 3n − 4 nonzero normal vibrations of the chain under consideration is divided into irreducible representations of the factor group in the following way:

$$\Gamma = 3A_1 + 2B_1 + 2B_2 + A_2$$

The elimination of the redundant coordinates in the symmetry type A_1 is carried out by means of additional relationships between the mutually dependent coordinates. Written in internal coordinates, these relations have the following form:

$$\gamma_{12} + \gamma_{13} + \gamma_{14} + \gamma_{23} + \gamma_{24} + \gamma_{34} = 0$$
$$\alpha - \gamma_{12} = 0$$
$$\alpha + \sqrt{2}\,(q_1 + q_2)\,\sigma_1 = 0$$

The first of the three conditions constitutes the ordinary relation between the changes in the six angles with common vertices at the Ge atom, while the two other conditions reflect the requirement that the distances between translationally equivalent atoms of the chain (i.e., the unit-cell dimensions) should be constant. The matrices of kinematic interaction not containing dependent coordinates are shown below.

$$\begin{array}{c|c|c|c}
q^{A_1} & q^{A_1} & \gamma^{A_1} & \gamma^{A_2} \\
\frac{2}{3}(\varepsilon_1 + \varepsilon_2) & & & \\
\hline
-\frac{2}{3}\varepsilon_1 & \frac{2}{3}\varepsilon_1 + \varepsilon_2 & & \\
\hline
\frac{2}{3}[\varepsilon_1(\sigma_1 - \sigma_2) + \varepsilon_2\sigma_1] & \frac{2}{3}\varepsilon_1(\sigma_2 - \sigma_1) & \frac{2}{3}\varepsilon_1(\sigma_1 - \sigma_2)^2 + \varepsilon_2\left(\frac{2}{3}\sigma_1^2 + \frac{1}{2}\sigma_2^2\right) - \frac{4}{3}\varepsilon_1\sigma_1\sigma_2 & \\
\hline
 & & & \frac{2}{3}\varepsilon_2\sigma_2^2
\end{array}$$

$$\begin{array}{c|c|c|c}
q^{B_1} & \gamma^{B_1} & q^{B_2} & \gamma^{B_2} \\
\frac{4}{3}(\varepsilon_1 + \varepsilon_2) & & & \\
\hline
-2\varepsilon_1\sigma_2 - \frac{2}{3}\varepsilon_1\sigma_1 - \frac{2}{3}\varepsilon_2\sigma_1 & \frac{1}{3}\sigma_1^2(\varepsilon_1 + \varepsilon_2) + 2\varepsilon_1\sigma_1\sigma_2 + \frac{3}{2}\sigma_2^2(2\varepsilon_1 + \varepsilon_2) & & \\
\hline
 & & \frac{4}{3}\varepsilon_1 + \varepsilon_2 & \\
\hline
 & & \left(2\sigma_1 + \frac{2}{3}\sigma_2\right)\varepsilon_1 & 3\sigma_1^2(\varepsilon_1 + \varepsilon_2) + 2\varepsilon_1\sigma_1\sigma_2 + \sigma_2^2\left(\frac{1}{3}\varepsilon_1 + \frac{1}{2}\varepsilon_2\right)
\end{array}$$

We shall not present the potential energy matrices here, as they may be obtained in a completely analogous manner.

In relation to the spectra of the silicon–oxygen anions in silicates, a method of this kind was first used, as already indicated, by Stepanov and Prima; in addition to chains, systems with translational symmetry in two and three directions were also considered. In these cases translational symmetry is allowed for bypassing to coordinates of the form $qe^{i(n_1\theta_1+n_2\theta_2+n_3\theta_3)}$ and applying cyclical boundary conditions. We note that the foregoing example indicates the necessity (when eliminating the redundant coordinates) of allowing for certain additional conditions, which are not encountered when setting up the secular equations for molecules of finite dimensions. In the secular equations given in [87] these conditions were not taken into account, and the equations accordingly contain a number of zero roots. Thus, for the pyroxene chain, with eight atoms in the one-dimensional unit cell, the order of the secular equation in [87] is 24, yet there are only 20 nonzero fundamental frequencies. We also note that, in solving the mechanical problem for systems with translational symmetry, the most frequent error is incorrect or incomplete allowance for the symmetry operations of the factor group, due to the (by no means always permissible) identification of the latter with the set of operations of point symmetry for the finite number of atoms composing the unit cell. Thus in Saksena's paper [89] a chain of the pyroxene type was erroneously attributed to C_{2h} symmetry, which is satisfied by two tetrahedra (taken separately) in a single unit cell. The symmetry of the factor group was only partly taken into account in [87] when considering a plane hexagonal layer of tetrahedra: C_{2v} was taken instead of C_{6v}. As a result of this, some of the vibrations of symmetry types B_1 and B_2 or A_1 and A_2 were found to coincide in frequency in numerical calculations [86], as we should expect from the correlation between the irreducible representations of the C_{2v} and C_{6v} groups.

The foregoing limitations as to the applicability of particular methods of calculating the frequencies of isolated anions in order to interpret the spectra of silicates remain in force for anions with translational symmetry also. In subsequent sections, when considering the frequencies of the internal vibrations of the anions in specific structures, we shall keep to the semiempirical method by simultaneously using the symmetry properties of the sample, data obtained from approximate calculations of the vibrational frequencies of relatively simple anions, and, finally, analogies with the spectra of model organosilicon systems.

§ 3. Esters of Silicic Acids—"Models" of Silicate Anions

The difficulties arising when attempting to interpret the vibrational spectra of silicates has for long stimulated the synthesis of model systems. In this respect the greatest interest lies in the tetraalkoxysilanes $Si(OR)_4$ and the products of their hydrolytic condensation, the alkoxypolysiloxanes $(RO)_3Si[OSi(OR)_2]_nOR$ (n = 1, 2, 3, ...). These compounds may be considered as esters of orthosilicic acid and polysilicic acids. The presence of an SiO_4 tetrahedron or a chain of tetrahedra linked by common vertices as a necessary structural unit of the molecule leads us to expect a considerable similarity between the spectra of the esters of the silicic acids and the complex silicon–oxygen anions in silicates.

```
        R       R          R
        |       |          |
        O       O          O
        |       |          |
R—O—Si—O—Si—O— ... Si—O—R
        |       |          |
        O       O          O
        |       |          |
        R       R          R
```

The possibility of studying the spectra of these compounds in the gas and liquid phase, i.e., the possibility of studying both the infrared and the Raman spectra, and the relative simplicity of interpreting the vibrational spectrum of an individual molecule presented clear advantages in using these organosilicon models of silicates. The existence of a series of similar molecules with a successively increasing chain length was also a considerable advantage when studying the influence of the condensation of the silicon–oxygen tetrahedra on the vibrational spectrum. On the other hand, although when analyzing the spectra of the esters of the silicic acids there is no need to consider such factors as the interaction of a complex anion with the cation and the resonance of the vibrations of several anions in the unit cell of the crystal, it is not quite clear *a priori* to what extent the vibrations of the silicon–oxygen radical are "separated" from the vibrations of the C–O bonds and the vibrations of the alkyl radicals.

The first attempt at using the analogies in the structure of the esters of silicic acids and silicates in order to interpret the spectra of the latter was made by Signer and Weiler [103, 104], who even in 1932 studied the Raman spectrum of tetramethoxysilane $Si(OCH_3)_4$ and the products of its hydrolytic condensation, the "dimer" $(CH_3O)_3SiOSi(OCH_3)_3$, the "trimer" $(CH_3O)_3SiOSi(OCH_3)_2OSi(OCH_3)_3$, and a mixture of higher boiling-point fractions. However, errors in the interpretation of the spectrum of $Si(OCH_3)_4$ and the incompleteness of the experimental data led them to the conclusion that frequencies of the SiO_4 tetrahedron were relatively little changed in the condensation products. The spectra of the esters of orthosilicic acid were later studied by a number of authors [28, 29, 105-107]. Using spectra of the series $SiCl_4 \rightarrow ROSiCl_3 \rightarrow (RO)_2SiCl_2 \rightarrow (RO)_3SiCl \rightarrow Si(OR)_4$, various frequencies were attributed to the stretching vibrations of the SiO_4 tetrahedron in the $Si(OCH_3)_4$ molecule, including one at about 640 cm^{-1} (totally symmetric) and one at 840 cm^{-1} (triply degenerate); correspondingly in the $Si(OC_2H_5)_4$ spectrum vibrations occurred at 652 and 782 cm^{-1} [28, 107]. Attempts at studying changes in the spectrum in the course of condensation polymerization were never renewed except in the case of [108], in which the spectra of several ethoxypolysiloxanes were obtained. The calculation of the spectrum of tetramethoxysilane carried out by Iguchi [109, 110], and the more complete experimental data of [26, 111-114], aid in refining the interpretation of the spectra formed by the esters of orthosilicic acid and in estimating the force constants of the Si–O bonds.

Tetraalkoxysilanes. Let us devote some more detailed attention to the interpretation of the spectra and vibration frequencies of the silicon–oxygen tetrahedron in the $Si(OCH_3)_4$ molecule. The structure of this molecule, according to electron-diffraction studies of the vapor [10], corresponds to V_d symmetry with r_{SiO} = 1.64 Å, r_{CO} = 1.42 Å, $\angle$ SiOC=113°, and the central SiO_4 group retains its tetrahedral configuration. The results of dipole-moment measurements [118] suggest that the molecule retains its shape in the liquid phase also. In order to decide the question as to the suitability of the $Si(OCH_3)_4$ molecule as a spectroscopic model of the orthosilicate anion, we must discover the extent of the splitting of the degenerate vibrations of the SiO_4 groups resulting from the lower-than-T_d symmetry of the molecule; we must also determine to what extent the frequencies of the silicon–oxygen tetrahedron are displaced (both as a result of the change in the force constants and also as a result of interaction with the vibrations of the $O–CH_3$ groups) in comparison with the $[SiO_4]^{4-}$ ion.

The $T_d \rightarrow V_d$ transition suggests that the splitting of the irreducible representations (degenerate for the T_d group) takes place in accordance with the following scheme:

$$\begin{array}{l} T_d \rightarrow V_d \\ F_2 \rightarrow E \\ \quad\ \searrow B_2 \\ E \rightarrow B_1 \\ \quad\ \searrow A_1 \\ A_1 \rightarrow A_1 \end{array}$$

Thus the four normal vibrations of the SiO_4 tetrahedron ($2F_2$, E, A_1) decompose, on reducing the symmetry, into seven vibrations: 2E, $2B_2$, B_1, $2A_1$. The totally symmetric vibration naturally remains totally symmetric for the lower symmetry of the molecule. The extent of the splitting of the degenerate vibrations cannot be estimated from symmetry considerations; hence we must examine the identification of the experimentally observed frequencies. These data are presented in Table 8, which shows the identification of 21 degrees of freedom corresponding to the motion of the heavy atoms in the nine-mass system $Si(OC)_4$. The identification of the stretching frequencies of the Si–O and C–O bonds causes no difficulties: the two stretching vibrations of symmetry type A_1 are observed in the form of polarized Raman lines at about 640 cm^{-1} ($\nu_s SiO_4$) and 1100 cm^{-1} (ν_s C–O). In the latter the C–O stretching vibrations of other symmetry types are superimposed; this is undoubtedly due to the weakness of the interactions of the four equivalent C–O groups not directly linked to each other in the molecules. We may ascribe the two strong absorption bands (almost coinciding in frequency) at 838 and 825 cm^{-1} to stretching vibrations of the Si–O bond of the E and B_2 types of symmetry, corresponding to the triply degenerate stretch of the SiO_4 tetrahedron. In the Raman spectrum, the completely depolarized band at 844 cm^{-1} corresponds to these. As indicated by Table 8, the vibration frequencies of the CH_3 group are not very different from the frequencies of the corresponding vibrations in the spectra of other methoxysilanes. One notices the fact that the line at 1186 cm^{-1} corresponding to rocking of the CH_3 groups is partly polarized in the Raman spectrum. This is probably associated with the strong interaction of the νC–O and ρCH_3 vibrations.

The calculated frequencies of tetramethoxysilane shown on the right of Table 8 were obtained by Iguchi [110] for a nine-atom model (the CH_3 groups were taken as single atoms with mass 15). The force field of the molecule was described by a quadratic potential function with five force constants: two stretching constants Si–O (K_1) and C–O (K_2), two deformation constants SiOC (g) and OSiO (f), and an O...O interaction constant (h). Among Iguchi's force constants, $K_1 = 3.7$, $K_2 = 5.7$, $g = 0.30$, $f = 0.22$, $h = 0.06 \cdot 10^5$ dyn/cm, it is probable that only the force constant of the Si–O bond has a relatively high reliability, since the neglect of interactions between νC–O and ρCH_3 may lead to a considerable change in K_2; further, in the frequency range of the deformation vibrations the experimental data are incomplete and the interpretation of the spectrum is insufficiently reliable. Iguchi's value of K_1 coincides with that proposed by Siebert [17] for the force constant of Si–O with a multiplicity factor equal to 1.0. This also agrees with the nearly tetrahedral value of $\angle$SiOC. We note that Kriegsmann and Licht's [111]

Table 8. Frequencies in the Vibrational Spectrum of Tetramethoxysilane $Si(OCH_3)_4$

RS (liquid)			IRS (frequency in cm^{-1} and intensity)		Assignments and classification			
frequency cm^{-1}	intensity	degree of depolarization ρ	liquid	gas	calculated frequencies for the Si(OC)$_4$ group [110]	symmetry species of the vibrations: group[1] $V_d(Si[OC]_4)$	symmetry species of the vibrations: group $T_d(SiO_4)$	assignments
—	—	—	—	—	—	2E, B_2,	—	τCOSi and
221	0.5 d.	—	—	—	—	B_1, A_2, A_1	—	δCOSi
312	1	—	311 s.[2]	—	303, 279	B_1, A_1	E	δOSiO
412	1 v.d.	0.80	405 v.s.[2]	—	292, 400	B_2, E	F_2	
—	—	—	430 sho.[2]	<420	—	—	—	—
640	10 sh.	0.08	639 w.	—	665	A_1	A_1	$\nu_s SiO_4$
844	3 d.	0.80	825 s. 838 s.	832 s. 847 s.	825 899	B_2 E	F_2	$\nu_{as} SiO_4$
1082	5 [3]	0.20	1093 v.s.	1108 v.s.	1145	A_1	—	νCO
1111	4			1141 m.	1146 1142	E B_2	— —	
1186	4 d.	0.32	1190 s.	1197 s.	Rocking and internal deformation vibrations of CH_3			
1463 1476	6 2	0.80	1469 m.	1474 m.				

Note: 1. Selection rules: in the IRS symmetry types E and B_2 are active, in the RS types E, B_2, B_1, A_1.
2. According to [27].
3. For polarization measurements, with photoelectric recording of the spectrum, the lines 1082 and 1111 are not resolved.

Notation: d. = diffuse; v.d. = very diffuse; sh. = sharp; sho. = "shoulder" on the sloping part of a stronger band; s. = strong; w. = weak; v.s. = very strong; m. = medium.

force constants for the Si–O bonds in $Si(OCH_3)_4$ and $Si(OC_2H_5)_4$ calculated by means of a five-atom model, i.e., without allowing for the interaction of the νSi–O and νC–O vibrations, are much higher, 4.17 and $4.61 \cdot 10^5$ dyn/cm, respectively. This leads to an overestimation of the orders of the Si–O bonds (1.12 and 1.24) which once more confirms the ineffectiveness of using over-simplified models in discussing the nature of chemical bonds.

Let us now turn to a discussion of the changes in the vibrations of the SiO_4 tetrahedron on passing from the isolated $[SiO_4]^{4-}$ ion to the $Si(OCH_3)_4$ molecule. It follows from the foregoing that the splitting of the degenerate vibrations of the tetrahedron in the tetramethoxysilane molecule is relatively small and for the stretching vibration is no greater than 10 to 15 cm^{-1}. The frequency displacement is considerably greater. Thus, by comparison with the stretching frequencies of SiO_4 in the spectrum of the orthosilicate ion in solution (page 59), the frequency of the triply degenerate stretching vibration falls by 11%. The frequency of the totally symmetric vibration falls still more (17.5%). This is not hard to explain if we remember that the displacement of the oxygen atoms in the totally symmetric vibration is much greater, so that the interaction with vibrations of the C–O bonds is greater. The reduction in the frequency of the triply degenerate vibration is, it would appear, almost entirely a force constant effect, since the change in the order of the Si–O bonds from about 1.2 in the $[SiO_4]^{4-}$ ion to about 1.0 in the $Si(OCH_3)_4$ molecule by itself leads us to expect a reduction of some 10% in the vibrational frequencies.

In the $Si(OC_2H_5)_4$ spectrum, in the frequency range corresponding to bond stretching of the main framework, four C–C vibrations appear in the 940- to 980-cm^{-1} range; the frequencies of the Si–O and C–O vibrations in the spectrum of tetraethoxysilane differ comparatively little from the corresponding frequencies of tetramethoxysilane and may be identified as in [115]. However, the identification of the frequencies in the spectra of the other tetraalkoxysilanes $Si(OR)_4$ becomes rapidly more complicated as the dimensions of the aliphatic radical increase, although in all compounds of this type the strong absorption around 1100 to 1050 cm^{-1}, attributed to C–O vibrations, remains. In the Raman spectra of $Si(OC_3H_7\text{-}n)_4$ and $Si(OC_4H_9\text{-}n)_4$, a fairly strong, narrow line representing the symmetric stretch of the SiO_4 group may be identified; its frequency rises relative to that of tetramethoxysilane thus:

$Si(OCH_3)_4$	$Si(OC_2H_5)_4$	$Si(OC_3H_7\text{-}n)_4$	$Si(OC_4H_9\text{-}n)_4$
640 cm^{-1}	656 cm^{-1}	666 cm^{-1}	668 cm^{-1}

The rapid damping-out of this displacement as the size of the radical increases suggests that the rise in frequency is associated with a slight increase in the order of the Si–O bond arising from an increase in the positive inductive effect on the radicals by the O atom: O: Si $\equiv$ O $\leftarrow CH_2 \leftarrow CH_3$. However, the change in the order of the Si–O bond is small and hardly exceeds 8%, all the more so since the part played by the kinematic factors in the frequency rise observed has not been considered.

Vibrations of the Chains of Silicon–Oxygen Tetrahedra in Alkoxypolysiloxanes [115-117]. Having established the possibility of identifying the stretching vibrations of the silicon–oxygen tetrahedron in the spectra of the simplest tetraalkoxysilanes, we may now consider the changes of these vibrations in hydrolytic-condensation products, i.e., after the formation of groups of tetrahedra linked to each other by Si–O–Si "bridges." It is natural to start with the least complex cases, the hexaalkoxydisiloxanes, constituting models of the pyrosilicate anions Si_2O_7:

$$2Si(OR)_4 + H_2O \rightarrow (RO)_3SiOSi(OR)_3 + 2ROH.$$

The $Si(OCH_3)_4 \rightarrow (CH_3O)_3SiOSi(OCH_3)_3$ transition is accompanied by considerable changes in the vibrational spectrum in the range of frequencies corresponding to vibrations of the heavy

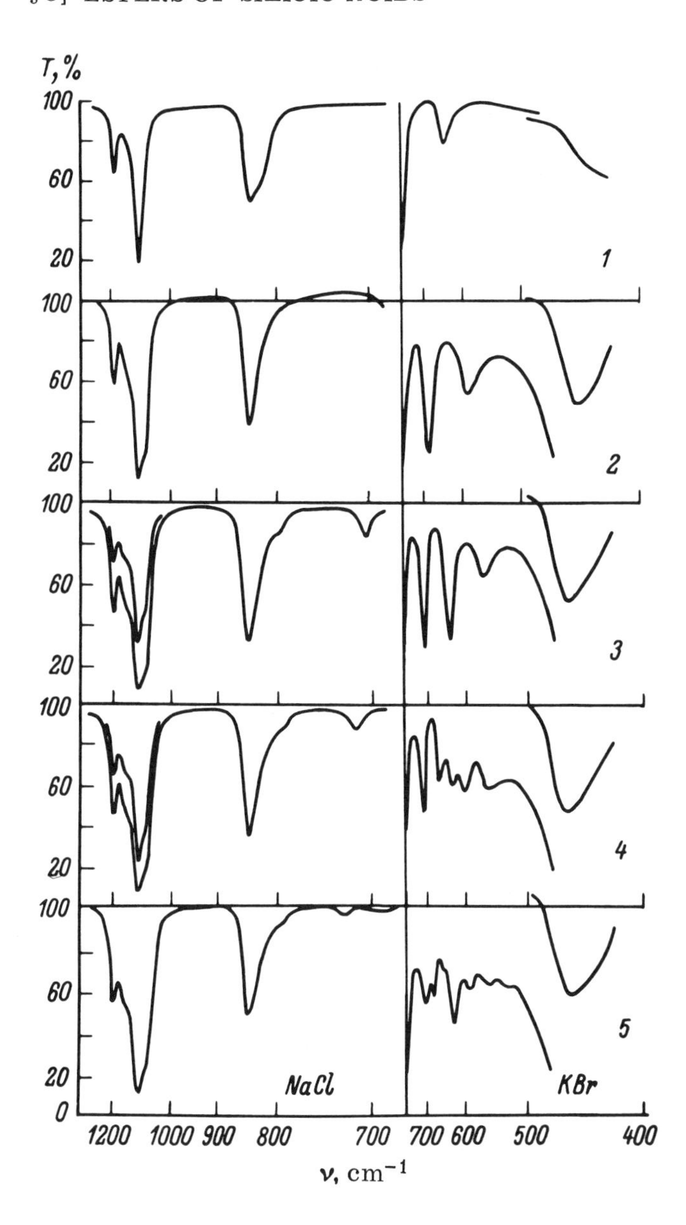

Fig. 20. Infrared spectra of the methoxypolysiloxanes.

atoms. These changes reveal themselves in the splitting or displacement of many frequencies and a change in the state of polarization of the lines in the Raman spectrum (Fig. 20, Table 9). Thus, instead of the symmetric stretch (pulsation) vibration of the SiO_4 tetrahedra observed in the infrared spectrum of $Si(OCH_3)_4$ with a very low intensity (and only in the liquid phase), in the $(CH_3O)_3SiOSI(OCH_3)_3$ spectrum two bands occur in the same region, the higher frequency band having a considerable intensity. Both frequencies are also present in the Raman spectrum, the lower frequency one being very intense and strongly polarized. Corresponding to the triply degenerate vibration of the SiO_4 tetrahedron, in the spectrum of the $Si(OCH_3)_4$ molecule there is a double absorption band 838 to 825 cm^{-1} and a wide depolarized Raman line at 844 cm^{-1}. In the spectrum of hexamethoxydisiloxane there is only one absorption band at 841 cm^{-1}, which corresponds to the 838-cm^{-1} line in the Raman spectrum. In addition to this, there is a sharp Raman line at 808 cm^{-1}. The lines at 808 and 838 cm^{-1} were not resolved in polarization measurements, and for the whole band the value of ρ fell from 0.8 to 0.5 on passing from the "monomer" to the "dimer," thus enabling the 808-cm^{-1} frequency to be assigned to the totally symmetric vibration. The band of C–O stretching vibrations is also complicated, exhibiting three maxima in the infrared spectrum.

TABLE 9. Frequencies in the Vibrational Spectra of the Methoxypolysiloxanes

$Si(OCH_3)_4$		$OSi_2(OCH_3)_6$				$O_2Si_3(OCH_3)_8$			$O_4Si_5(OCH_3)_{12}$			$O_6Si_7(OCH_3)_{16}$
		RS				RS			RS			
RS	IRS	frequency	intensity[1]	degree of depolarization	IRS	frequency	intensity[2]	IRS	frequency	intensity[2]	IRS	IRS
—	—	168	0.5	—	—	—	—	—	—	—	—	—
221	—	250—	1	—	—	254	1	—	210	0	—	—
—	—	—270		—	—	—	—	—	—	—	—	—
312	311	325	0.5	—	—	330	0	—	311	1	—	—
—	—	389	1	—	—	416	0	—	403—	0	—	—
412	405	445	1.5	—	442 s.	—	—	450 s.	—440		454 s.	453 s.
—	430	—	—	—	—	—	—	—	—	—	—	—
—	—	—	—	—	—	—	—	—	—	—	—	530 w.
—	—	—	—	—	—	—	—	—	—	—	557 v.w.	561 w.
—	—	577	10	0.05	588 w.	554	8	560 w.	—	—	599 w.	597 w.
640	639	—	—	—	—	639	0	629 m.	638	8	626 w.	627 m.
—	—	692	0.5	—	681 m.	702	3	701 m.	—	—	654 w.	651 v.w.
—	—	—	—	—	—	—	—	—	—	—	710 m.	681 w.
—	—	—	—	—	—	—	—	—	—	—	—	712 w.
—	—	808	7	0.50	—	817	5	804 w.	827	2	797 w.	797 w.
844	825	838	4		841 s.	840	5	843 s.	842	3	844 s.	843 s.
—	838	—	—	—	—	—	—	—	—	—	—	—
1082	—	—	—	—	1081 v.s.	—	—	1080 v.s.	—	—	1074 v.s.	1079 v.s.
—	1098	1090—	4	0.40	1098 v.s.	1094	3	1099 v.s.	1086—	2	1100 v.s.	1099 v.s.
1111	—	—1120			1133 m.	—	—	1138 m.	—1126		1140 m.	1151 m.
—	—	1163	0.5	—	—	1164	0	—	1164	0	—	—
1186	1190	1198	2	0.30	1197 s.	1196	1	1196 s.	1192	1	1196 s.	1196 s.
1463	1469	1461	9	0.80	1470 m.	1464	8	1470 m.	1457	6	1470 m.	1470 m.
1476		1471				—	—		1472			

Note: 1. Photoelectric measurements; 2. visual estimate.

In the Raman spectrum a band at 1090 to 1126 cm^{-1} corresponds to these vibrations; the structure of this has not been resolved.

As it was shown earlier that the splitting of the frequencies of the degenerate vibrations of the SiO_4 group in the $Si(OCH_3)_4$ spectrum was fairly small, it is easy to see that, in order to interpret the changes in these vibrations associated with the $Si(OCH_3)_4 \rightarrow (CH_3O)_3SiOSi(OCH_3)_3$ transition, the simple $SiO_4 \rightarrow Si_2O_7$ scheme may be used. It is not hard to convince oneself that, in the Si_2O_7 group, for each (nondegenerate) vibration of the SiO_4 tetrahedron, there are two vibrations corresponding in the first approximation to the in-phase and anti-phase vibrations of two coupled tetrahedra. Thus on allowing for the degeneracy we may establish a correspondence† between the nine vibrations of the isolated tetrahedron and 18 (of a total of 21) normal vibrations of the Si_2O_7 group:

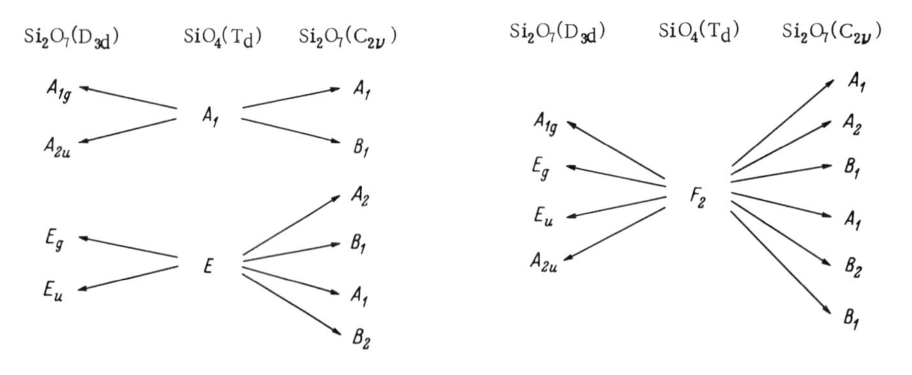

† We note that there are no unequivocal relations between the irreducible representations of the point groups D_{3d} and C_{2v}, as one is not a subgroup of the other.

The remaining three vibrations of the Si_2O_7 group corresponding to deformation of the SiOSi angle and internal rotations around the Si–O(Si) bonds lie in the low-frequency part of the spectrum.

The scheme given here enables us to derive a clear interpretation for the observed changes in the Si–O stretching vibrations. In fact, instead of one frequency for the totally symmetric vibration of the tetrahedra there are two, of which the lower-frequency one corresponds to the totally symmetric species for Si_2O_7. We note that both are active in the infrared and the Raman, clearly indicating the noncentrosymmetrical nature of the hexamethoxydisiloxane molecule, i.e., $\angle$SiOSi < 180°; thus the symmetry of the molecule is no higher than C_{2v}. The changes in the spectrum in the frequency range of the triply degenerate stretch of the tetrahedron may also be explained by the frequency-splitting scheme; in particular, the appearance of a polarized line in the Raman spectrum is due to the presence of components of the A_1 type of symmetry, while in the $SiO(CH_3)_4$ spectrum none of the components of this vibration are totally symmetric. We would direct attention to one further important conclusion: the splitting of the totally symmetric (pulsation) vibration of the tetrahedron is very great, the difference between the two components being over 110 cm^{-1}, while all six vibrations corresponding to the triply degenerate vibration of the SiO_4 group evidently fall in a narrow range, 808 to 841 cm^{-1}. This is not hard to explain, as in considering the spectrum of tetramethoxysilane we have already noted a considerably higher sensitivity of the pulsation vibration of the SiO_4 tetrahedron to interactions with vibrations of the groups adjoining the Si–O bond by way of the oxygen atom.

The C–O valence vibrations in the spectrum of hexamethoxydisiloxane should be represented by six frequencies relating (for a C_{2v} symmetry of the molecule) to the irreducible representations B_2, B_1(2), A_2, and A_1(2). The comparatively weak interaction existing between the vibrations of vibrators which are not directly coupled has the effect that these vibrations are concentrated in a narrow frequency range 1133-1081 cm^{-1}, and not all of them can be resolved in the spectrum.

Passing to the spectrum of the "trimer" $(CH_3O)_3SiOSi(OCH_3)_2OSi(OCH_3)_3$ leads to further changes (Fig. 20, Table 9), of which the most important is the appearance of three frequencies in the region of the totally symmetric vibration of the tetrahedron. Further increase in the length of the methoxypolysiloxane chain, followed right up to the "heptamer" $O_6Si_7(OCH_3)_{16}$, also shows itself in an increase in the number of bands in this part of the spectrum. The picture observed is typical of many instances of the formation of a chain of equivalent coupled vibrators. Similar changes were observed, for example, in the spectra of aliphatic hydrocarbons for the frequencies of the C–C vibrations, while among organosilicon compounds they were recently described for the spectra of methylpolysiloxanes [27]. In the latter a successive increase was observed in the number of frequencies of the ν_s SiOSi vibrations in accordance with the number of siloxane links in the molecule.

Thus a study of the spectra of the methoxypolysiloxanes shows that the formation of a chain of two or more silicon–oxygen tetrahedra linked by common vertices leads to the splitting of the vibrations of the tetrahedron; this is particularly great in the case of the symmetric stretching vibration. The number of components observed in this region enables us to determine the number of coupled vibrators (silicon–oxygen tetrahedra) in a chain of finite length. In the formation of chains comprising a large number of tetrahedra, the observed spectrum shows no new details and gradually becomes diffuse: the frequencies of the coupled vibrators form a "branch" in which, for a considerable length of the chain, only the limiting vibrations have a substantial intensity.

The changes in the spectrum in the frequency range of the C–O and Si–O stretching vibrations on passing from tetraethoxysilane to the ethoxypolysiloxanes are completely analogous with those described for the methoxy derivatives. Thus the transition $Si(OC_2H_5)_4 \rightarrow (C_2H_5O)_3SiOSi(OC_2H_5)_3$ leads to severe splitting of the pulsation vibrations of the SiO_4 tetrahedra, while the much less severe splitting of the other stretching vibrations (with the appearance of components belonging to the totally symmetric irreducible representation) in general only shows up experimentally by virtue of a change in the state of polarization of a band lying at about 800 cm^{-1} in the Raman spectrum: the value of ρ falls from 0.62 to 0.31. A further transition to octaethoxytrisiloxane leads to the appearance of three vibrations corresponding to pulsations of the silicon–oxygen tetrahedra; the strongest and most polarized (ρ = 0.15) in the Raman spectrum is that of lowest frequency.

Attention should be drawn to the fact that the splitting of the pulsation vibration of the tetrahedron in the spectra of the ethoxypolysiloxanes is smaller than in the methoxypolysiloxanes. Thus for $(CH_3O)_3SiOSi(OCH_3)_3$ and $(C_2H_5O)_3SiOSi(OC_2H_5)_3$ the frequency difference of the two components, referred to the frequency of the corresponding orthosilicic ester, is 16.6 and 11.3%. This difference can hardly be explained by interactions with other vibrations; it evidently constitutes an indication that there is a difference in the configuration of the molecules. The main factor determining the value of the splitting in the totally symmetric vibration of the tetrahedron when two tetrahedra are joined by a common oxygen atom is clearly the SiOSi angle: the nearer SiOSi is to 180°, the stronger is the kinematic coupling between the vibrations of the two tetrahedra. We may therefore suppose that in hexamethoxydisiloxane the ∠SiOSi is greater than in hexaethoxydisiloxane. This (at first unexpected) conclusion cannot be derived from steric considerations but may be explained on considering possible differences in the orders of the $Si-O(CH_3)$ and $Si-O(C_2H_5)$ bonds. It was noted earlier that, as a result of the great electropositivity of the C_2H_5 group, some increase might take place in the multiplicity factor of the $Si-O(C_2H_5)$ bond as compared with that of $Si-O(CH_3)$ (approximately unity). It then seems probable that the intensification of the $p\pi-d\pi$ interactions in the Si–O(C) bonds on passing from the hexamethoxy to the hexaethoxydisiloxane is accompanied by a weakening of these interactions in the Si–O–Si bridge. A reduction in the multiplicity factor of the Si–O–Si bonds, i.e., a reduction in the order of the π bonds, leads to a fall in the equilibrium value of ∠SiOSi. This kind of "competition" between Si–O bonds involving the same silicon atoms has already been mentioned when considering other organosilicon compounds and is also equally probable for silicate anions, for example, the Si–O(Si) and $Si-O^-$ bonds of the pyrosilicate anion, models of which are provided by the molecules of the hexaalkoxydisiloxanes.

Limited Nature of the Analogies between the Spectra of the Alkoxypolysiloxanes and Silicate Anions. This limitation is due to a whole series of factors, some of which have already been mentioned. Thus before passing on to a consideration of the spectra of silicates we may mention the most important differences which will have to be taken into account when using the results of our study of the esters of silicic acids in order to interpret the frequencies of the vibrations of silicate anions. We noted earlier that, by comparison with the spectrum of the orthosilicate anion $[SiO_4]^{4-}$, the stretching frequencies of the silicon–oxygen tetrahedron in $Si(OCH_3)_4$ are too low. Moreover, the frequency of the totally symmetric vibration is reduced more than the others, so that the difference between ν_{as} and ν_s is much greater than for the $[SiO_4]^{4-}$ ion. Then on passing from the spectrum of this ion to the simplest of the anions containing condensed tetrahedra $[Si_2O_7]^{6-}$, it may prove that one of the two components formed in the splitting of the ν_s SiO_4 vibration will fall close to frequencies (including some of the same symmetry type) derived from the splitting of the ν_{as} SiO_4 vibration. This is shown schematically in Fig. 21, from which it follows in particular that among the stretching vibrations of the Si_2O_7 group, only the low-frequency component of ν_s SiO_4, corresponding to the vibration designated as ν_s SiOSi in the spectra of the disiloxanes, may turn out to be comparatively isolated in frequency. Equally, on passing to the trisilicate group Si_3O_{10}, it may be possible to identify only the two lower frequencies of the three components of the ν_s vibration of the SiO_4 tetrahedron. These two frequencies can be considered as the ν_s SiOSi vibrations of the two Si–O–Si bridges existing in this group.

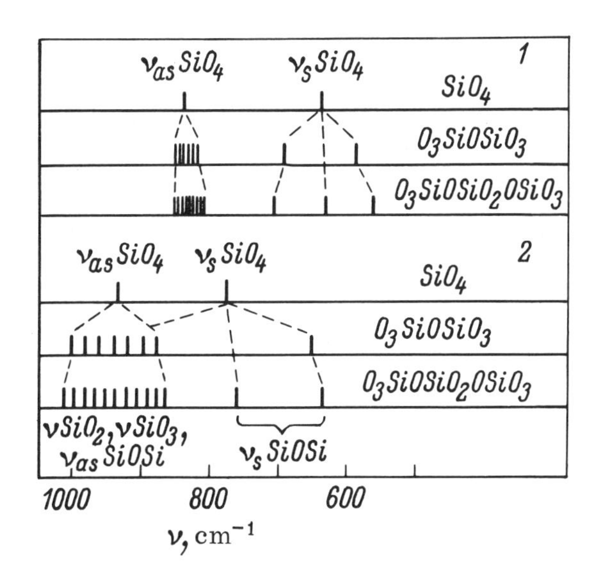

Fig. 21. Possible differences in the frequencies of the SiO_4, Si_2O_7, Si_3O_{10}, etc. groups forming the frameworks of the alkoxypolysiloxane molecules (1) or the complex anions in silicates (2).

It follows from this discussion that, in order to give an approximate description of the vibrations of a group of coupled silicon–oxygen tetrahedra, it is more appropriate to classify these, not by reference to the "genetical" link with the vibrations of the isolated tetrahedron (as many authors do), but by reference to their preferential association with the vibrations of the Si–O–Si bridges and the terminal Si–O$^-$ bonds (or $Si(O^-)_3$, $Si(O^-)_2$ groups). This method of characterizing the vibrations of individual structural elements of a complex anion is in keeping with the comparatively large mass of the Si atom and also the differences in the force constants due to the differences in the orders of the Si–O$^-$ and Si–O(Si) bonds. A calculation of the forms of vibration of several silicate anions confirms the validity of this kind of classification.

§4. Ortho- and Pyrosilicates

Interpretation of Orthosilicate Spectra. The interpretation of these spectra in the frequency range of the stretching vibrations of the SiO_4 tetrahedron encounters no serious difficulties: the triply degenerate $\nu_{as} SiO_4$ vibration belongs to the F_2 species of the T_d group and is always active in the infrared spectrum. In the spectra of the majority of orthosilicates the corresponding band decomposes into several components embracing a fairly wide frequency range (850 to 1000 cm^{-1}). The main reasons for the multiplet nature of the $\nu_{as} SiO_4$ band include the removal of the degeneracy for a lower site symmetry of the SiO_4 group in the lattice, and resonance interactions between translationally nonequivalent tetrahedra. The extent of the splitting due to the latter cause increases considerably when the bonds between the complex anion and the cations are partly covalent. Fermi resonance with overtones and combinations of the deformation vibrations may further complicate the spectrum. The totally symmetric vibration of type A_1 of the SiO_4 anion is inactive in the infrared spectrum, but is often observed in the spectra of the orthosilicates, which have a reduced site symmetry. The intensity of this vibration in the infrared spectrum is usually low, and its identification is not always unambiguous. In comparing the spectra of isomorphous orthosilicates, the intensity of the $\nu_s SiO_4$ band may be used as a characteristic of the internal field in which the anion is situated.

Despite many investigations of the spectra of the orthosilicates [65-68, 119-121], at the present time there is still no clarity in the choice of "ideal" values for the frequencies of the

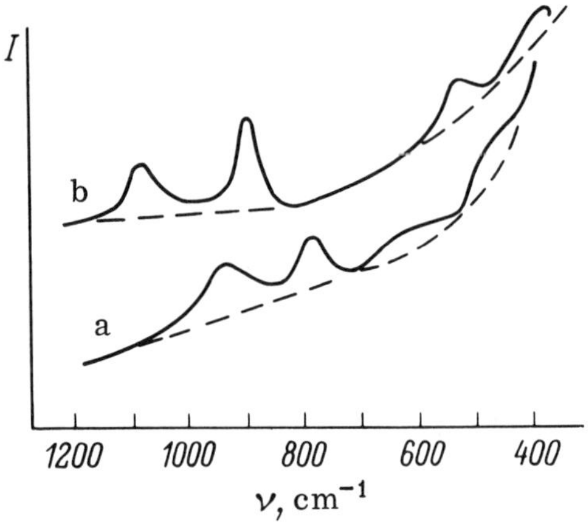

Fig. 22. Raman spectra of the SiO_4 and PO_4 ions. a) In an aqueous solution of sodium silicate; b) in an aqueous solution of monosodium phosphate.

SiO_4 tetrahedron such as might be used for estimating force constants.† Particularly great difficulties are associated with determining the frequencies of the deformation vibrations of the tetrahedron (δ_{as} SiO_4 type F_2 and δ_s SiO_4 type E), the more so in view of the fact that, because of their very hygroscopic properties, no study has been made of the alkali orthosilicates, in the spectra of which one might expect a relatively weak influence of the lattice vibrations on the internal vibrations of the anion. In the following table we compare the frequencies and force constants of the SiO_4 ion, obtained by Siebert [17] by averaging the data over the spectra of a number of orthosilicates, and the results of Tarte [120], who used the spectrum of Ba_2SiO_4, which contains a large Ba^{2+} cation, having a relatively low "specific charge" e/r:

	Frequencies, cm^{-1}				Force constants ($\cdot 10^5$ dyn/cm)			
	$\nu_1(A_1)$	$\nu_2(E)$	$\nu_3(F_2)$	$\nu_4(F_2)$	K_{SiO}	k_{OSiO}	$h_{SiO,SiO}$	$l_{OSiO,OSiO}$
Siebert	775	275	935	460	4.44	0.62	0.41	0.19
Tarte	826	260	910	500	4.2	0.8	0.7	0.3

The discrepancies are considerable even in the range of frequencies of the stretching vibrations, particularly the "pulsation" vibration ν_1, the most sensitive to kinematic interaction with the vibrations of the oxygen–cation bonds. The reduction in the characteristic frequency of this vibration on passing to the GeO_4 tetrahedron enhances the interaction; hence Tarte's force constant for the Ge–O bond obtained from the Ba_2GeO_4 spectrum was even a little higher than that for the Si–O bond.

Another source of information regarding the frequencies of the vibrations of isolated SiO_4 and GeO_4 tetrahedra is provided by the spectra of aqueous solutions of the alkali silicates and germanates. For a high concentration of alkali the amount of ortho ions in the solution is quite high. Figure 22 shows the Raman spectrum of an aqueous solution of sodium silicate [122]: the frequencies of stretching vibrations (935 to 777 cm^{-1}) are close to the values obtained by Siebert. We should remember, however, that these frequencies correspond to the vibrations of partly hydrated tetrahedra ($[HSiO_4]^{3-}$, $[H_2SiO_4]^{2-}$, and partly $[H_3SiO_4]^{-}$) and differ

† It should be remembered that the determination of the frequencies of the four fundamental vibrations of the SiO_4 tetrahedron still fails to solve the problem of finding the five coefficients in the expansion of the potential energy of the system. For this reason, a set of four force constants, allowing for the interaction of homogeneous coordinates only, was used in [17, 120]. However, even with such a set, the calculation of the constants from only four experimentally determinable parameters fails to secure adequate accuracy. In the structurally analogous organic molecule, CH_4, the determination of the force constants is based on 27 frequencies by using isotopically substituted molecules.

from the frequencies of the $[SiO_4]^{4-}$ ion both on account of the kinematic interaction of the vibrations of the tetrahedron with the vibrations of the hydroxyls and also as a result of changes in force constants: the order of the Si−O(H) bond is lower than that of the $Si-O^-$, and the latter may increase on increasing the number of Si−O(H) bonds associated with the same silicon atom. Additional difficulties are associated with the presence in solution of tetrahedra condensed by way of Si−O−Si bridges. In particular, in the lower-frequency region there are weak bands at 448 and 607 cm^{-1}, of which the latter probably corresponds to the vibrations of these bridges. A similar result is obtained when studying the spectra of alkaline solutions of germanium dioxide [123]. The Raman spectra contains bands at 235, 345, 667, and 765 cm^{-1}, ascribed to the vibrations of hydrated tetrahedra, chiefly $[H_2GeO_4]^{2-}$, and a band at 529 cm^{-1} associated with vibrations of the Ge−O−Ge bonds of condensed tetrahedra. The presence of absorption in the range 950 to 970 cm^{-1} in the infrared spectrum, in addition to the bands with frequencies close to those just mentioned, is also an indication of the existence of highly condensed groups of tetrahedra (vibrations of the ν_{as} GeOGe type). Absorption in the range 650 to 800 cm^{-1} in the spectra of the crystalline orthogermanates corresponds to stretching vibrations of the GeO_4 tetrahedron; the frequency of the ν_s GeO_4 vibration, which is independent of the mass of the central atom, is equal to or even greater than that of the ν_{as} GeO_4 vibration [120].

For PO_4 tetrahedra, the characteristic frequencies of which are higher, the problem of identifying the vibrations of the "isolated" tetrahedron is also slightly simplified by the existence of data relating to the spectra of the alkali orthophosphates and the spectra of PO_4 tetrahedra in aqueous solutions. Hence the set of frequencies of the PO_4 group used by Siebert for estimating the force constants and orders of the bonds has a relatively higher degree of reliability: $\delta_s PO_4 = 420$ cm^{-1}, $\delta_{as} PO_4 = 562$ cm^{-1}, $\nu_s PO_4 = 937$ cm^{-1}, $\nu_{as} PO_4 = 1022$ cm^{-1} (from the Raman spectrum of K_3PO_4 in solution).

Subsequently it is important to remember that, in calculating the frequencies of complex anions in structures with condensed tetrahedra, we must pay attention to the possibility of the deformation force constants obtained from the vibration frequencies of the ortho ions in crystals being slightly too high.

Spectra of the Pyrosilicates. The Transition $SiO_4 \rightarrow Si_2O_7$. The material presented in the earlier sections enables us to approach the problem of elucidating the characteristic features of the pyrosilicate spectra empirically, without initially appealing to a calculation of the frequencies of the Si_2O_7 group. This is quite important, as there are no compounds containing alkali cations among the spectroscopically studied pyrosilicates. The establishment of spectroscopic criteria for the presence of Si_2O_7 groups in a structure is of interest not only as a first stage in passing from the spectra of the orthosilicates to the spectra of crystals with highly condensed anions, but also in itself. The diorthosilicate groups Si_2O_7, until quite recently regarded as rare, have now been found in many natural and synthetic compounds, and the development of the crystal chemistry of silicates in recent years has been closely bound up with views as to the part played by these groups as a characteristic structural element in crystals with large cations [124]. In spite of the relatively simple structure of the Si_2O_7 group, it should show all the main features of condensed silicate anions associated with the existence and mutual influence of bonds of the $Si-O^-$ and Si−O(Si) type. The existence of structurally similar ions and X_2O_7 molecules with X = Ge, P, S, Cl also stimulates interest in the nature of the chemical bonds in the pyrosilicate anion.

The transition from the spectra of orthosilicates to those of pyrosilicates is best illustrated by choosing compounds containing the same cations, and by avoiding those in which frequencies corresponding to the vibrations of the complex anions relative to the cation sublattice might interfere. For this purpose, of course, it is desirable to choose cations with only slightly covalent bonds connecting them to the oxygens and a comparatively large mass.

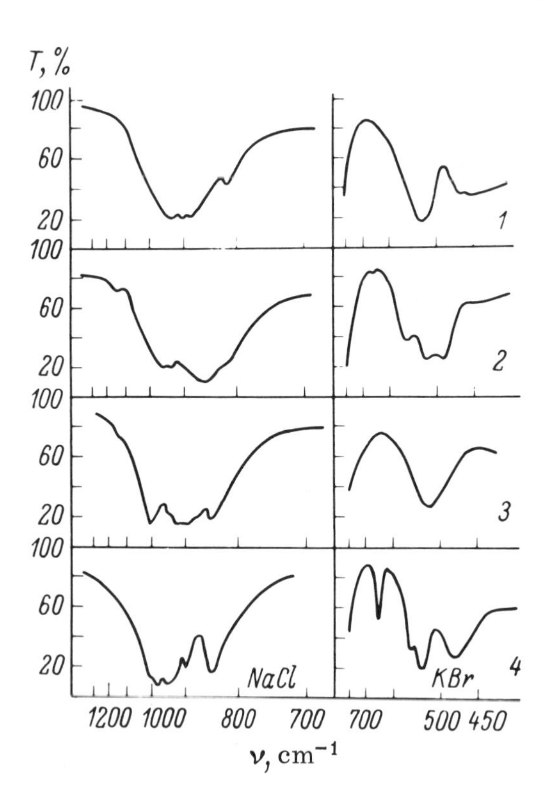

Fig. 23. Infrared spectra of calcium ortho- and pyrosilicates. 1) $Ca_3(SiO_4)O$; 2) γ-Ca_2SiO_4; 3) β-Ca_2SiO_4; 4) $Ca_3Si_2O_7$. Samples prepared by pressing in KBr: sample diameter, 25 mm; total weight, 2 g; content of sample material, 5 mg.

Let us consider several specific examples [125]. Figure 23 shows the infrared spectra of the calcium orthosilicates and calcium pyrosilicate, rankinite $Ca_3Si_2O_7$. The spectra of the calcium orthosilicates and oxyorthosilicates containing "isolated" silicon–oxygen tetrahedra exhibit intense absorption bands of Si–O stretching in the frequency range 1000 to 800 cm^{-1}. In the lower-frequency part of the spectrum absorption occurs near 500 cm^{-1}; this may be associated with the internal deformation vibrations of the complex anion, while the slight rise in the frequencies of these vibrations as compared with the free $[SiO_4]^{4-}$ ion is evidently explained by interaction with the lattice vibrations,† i.e., the Ca . . . O bonds. All these compounds except $Ca_3Si_2O_7$ have been quite fully studied by x rays [126-130]; from symmetry considerations no restrictions are imposed on the selection rules for the internal vibrations of the SiO_4 ions; all the vibrations are optically active, and the degeneracy is completely removed. This explains the existence of no less than three components of the $\nu_{as}\,SiO_4$ vibration, embracing a range of 933 to 883 cm^{-1} in the spectrum of $Ca_3(SiO_4)O$, 952 to 855 cm^{-1} in γ-Ca_2SiO_4, and 996 to 845 cm^{-1} in β-Ca_2SiO_4. To the lower-frequency $\nu_s\,SiO_4$ vibration we may ascribe the weaker 812 cm^{-1} band in the spectrum of $Ca_3(SiO_4)O$ and probably the 818 cm^{-1} in that of γ-Ca_2SiO_4. The splitting of the triply degenerate deformation vibration $\delta_{as}\,SiO_4$ is only clearly observed in the spectrum of γ-Ca_2SiO_4.

The spectrum of rankinite $Ca_3Si_2O_7$, the formula of which suggests the existence of complex $[Si_2O_7]^{6-}$ anions, is primarily distinguished by the appearance of a medium-intensity band at about 652 cm^{-1}, i.e., in the part of the spectrum free from absorption bands for calcium silicates with isolated SiO_4 tetrahedra. In the range 1000 to 840 cm^{-1} there is a complex group of five or six bands, and there are three bands between 560 and 480 cm^{-1}. These data are in good agreement with the expected changes in the spectrum associated with the $SiO_4 \rightarrow Si_2O_7$ transi-

† The actual vibrational frequencies of Ca. . . O evidently lie at lower frequencies; this follows from comparing the spectra of the calcium silicates with the spectra of the isostructural silicates of the divalent rare earths (see p. 253).

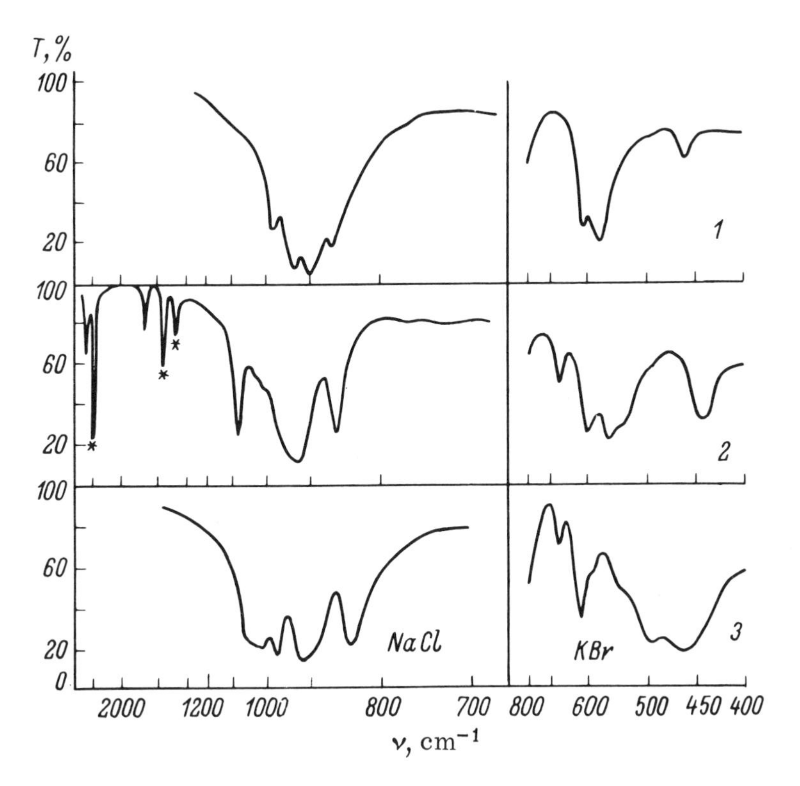

Fig. 24. Infrared spectra of zinc ortho- and pyrosilicates. 1) Zn_2SiO_4 willemite (pressed in KBr, 5 mg); 2) $Zn_4(OH)_2$ $(Si_2O_7) \cdot H_2O$ hemimorphite (suspension in paraffin oil, absorption bands of the oil indicated by asterisks); 3) $Ca_2ZnSi_2O_7$ hardystonite (pressed in KBr, 5 mg).

tion based on the esters of silicic acids if we take the 652-cm^{-1} band as the ν_s SiOSi vibration, i.e., as the lower-frequency of the two components formed by the splitting of the pulsation vibration of the tetrahedron. The second component, as was suggested before, apparently falls in a range containing a large number of frequencies "genetically" associated with the ν_{as} SiO_4 vibration, and its identification is difficult. For the same reason (partial superposition of bands), the number of maxima observed in the range 1000 to 840 cm^{-1} is smaller than expected.

Another comparison between the spectra of the orthosilicates and pyrosilicates can be made for the zinc silicates (Fig. 24).

The structures of willemite Zn_2SiO_4 [131], hardystonite $Ca_2ZnSi_2O_7$ [132, 133], and hemimorphite $Zn_4(OH)_2(Si_2O_7) \cdot H_2O$ [134, 157] have been studied by x-ray diffraction and the existence of SiO_4 and Si_2O_7 anions has been established reliably. The presence of Zn^{2+} cations occupying positions with tetrahedral coordination relative to oxygen in all these crystals suggests that there will be a substantial proportion of covalence in the bonds joining the complex anions to the cations. Hence the frequencies of the Zn–O vibrations may be higher, so that their interaction with the internal vibrations of the complex anions will be much stronger. Nevertheless, the willemite spectrum contains no absorption bands in the frequency range between 610 cm^{-1}, i.e., above the frequencies of the deformation vibrations of the tetrahedron and the Zn–O vibrations, and 850 to 1000 cm^{-1}, where the frequencies of the Si–O stretching vibrations lie. The transition to the pyrosilicate spectra leads to changes in the frequency range of both the stretching and deformation

Table 10. Classification of the Vibrations of the Si_2O_7 Anion

Description of vibration	Symmetry type C_{2v}	Symmetry type D_{3d}	Description of vibration	Symmetry type C_{2v}	Symmetry type D_{3d}
ν_{as}SiOSi	B_1	A_{2u}	ρSiO_3	$\{B_2, A_1\}$	E_g
$\nu_{as}(\delta_{as})SiO_3$	$\{B_2, A_1\}$	E_g	$\rho' SiO_3$	$\{B_1, A_2\}$	E_u
$\nu'_{as}(\delta'_{as})SiO_3$	$\{B_1, A_2\}$	E_u	δSiOSi	A_1	E_u
$\nu_s(\delta_s)SiO_3$	A_1	A_{1g}	τSiO_3	$\{B_2, A_2\}$	A_{1u}
$\nu'_s(\delta'_s)SiO_3$	B_1	A_{2u}			
ν_sSiOSi	A_1	A_{1g}			

Note: ν = stretch, δ = deformation, ρ = rock, τ = torsional; the prime indicates that the vibrations of the two SiO_3 groups are opposite in phase.

vibrations of Si–O; however, in the range 610 to 850 cm^{-1} there is only one new band of comparatively low intensity at about 660 to 680 cm^{-1}. As in the rankinite spectrum, this frequency may be ascribed to the ν_sSiOSi.

Thus the changes in the vibration frequencies of the complex anion associated with the $SiO_4 \rightarrow Si_2O_7$ transition demonstrate the possibility of identifying the ν_s SiOSi vibration band if the frequencies of the lattice vibrations are fairly low and if these do not perturb the deformation vibrations of the anion too much. Subsequently we shall consider the spectra of a whole series of pyrosilicates satisfying this requirement; in all cases in which the selection rules allow the occurrence of the ν_s SiOSi frequency in the infrared spectrum, a single band of medium or low intensity is found experimentally in the range 730 to 600 cm^{-1}. On passing to the spectra of pyrosilicates with cations capable of forming covalent bonds with oxygen we find a rise in the M–O vibration frequencies, and this impedes or makes impossible the identification of the ν_s SiOSi vibration frequency. Thus, whereas in the spectrum of epidote (Moenke [21]), the structural formula of which may be written in the form $Ca_2Al_2FeO(SiO_4)(Si_2O_7)OH$ [135, 136], the ν_s SiOSi vibration of the Si_2O_7 ion can still reliably be associated with the band at 650 cm^{-1}, in zoisite $Ca_2Al_3O(SiO_4)(Si_2O_7)OH$ five bands occur in the range 730 to 600 cm^{-1}; hence the presence of Si_2O_7 groups cannot be established simply on the basis of the external form of the spectrum.† This also applies to pyrosilicates containing aluminum as cation, such as lawsonite, gehlenite, and melilite. The latter two minerals are structurally similar to hardystonite; the presence of aluminum cations in tetrahedral coordination linking the Si_2O_7 groups into a single system evidently produces such strong coupling between their vibrations that, even for the stretching vibrations, the concept of an isolated Si_2O_7 ion is unsuitable. The partial replacement of Si by Al in the diortho groups themselves may further complicate the spectrum.

Spectra and Configurations of the Si_2O_7 Groups. Possibility of Estimating the Force Constants of the Bonds. It is particularly interesting to study the possibility of drawing any conclusions regarding the form of the Si_2O_7 groups and the force constants of bonds of the Si–O and Si–O(Si) types in these groups. The first problem reduces mainly to that of determining the SiOSi angle, since for $\angle$SiOSi = 180° the only stable configuration is the one with D_{3d} symmetry, while for $\angle$SiOSi < 180° changes in the spectrum on passing from the highest symmetry C_{2v} to C_s or C_2 are insignificant. The spectra of rankinite,

† The interpretation of the spectra of these crystals may also be complicated by the presence of OH groups and oxy ions O^{2-} connected to the cations by bonds having a greater proportion of covalence than those in the silicate anion.

Table 11. Frequencies of the Vibrations Calculated for Various Configurations of the Si_2O_7 Groups

	C_{2v}				D_{3d}	
Symmetry type	∠SiOSi=120° frequencies, cm^{-1}		∠SiOSi=145° frequencies, cm^{-1}		∠SiOSi=180° frequencies, cm^{-1}	Symmetry type
	I	II	I	II	I	
B_1	1040	986	1104	1075	1139	A_{2u}
A_1	907	925	908	914	906	E_g
B_2	907	925	907	913		
A_2	906	924	906	912	906	E_u
B_1	907	923	907	913		
A_1	842	838	828	827	820	A_{1g}
B_1	776	786	784	788	783	A_{2u}
A_1	682	660	603	592	539	A_{1g}
B_2	415	415	415	418		
A_1	398	398	406	406		
B_1	388	389	388	388	380	A_{2u}
B_1	355	355	355	355	362	E_g
A_2	352	356	352	356	362	E_u
A_1	352	352	350	350	329	E_g
B_2	350	350	350	349	329	E_u
A_2	259	254	259	255	224	A_{1g}
B_1	254	253	255	255		
A_1	243	240	235	233		

Note: I) For all configurations the force constants of the $Si-O^-$ and Si–O(Si) bonds are the same and equal to $7.7 \cdot 10^6$ cm^{-2}; II) for ∠SiOSi = 120° the force constant of the $Si-O^-$ bond equals $8.0 \cdot 10^6$ cm^{-2} and that of the Si–O(Si) equals $6.8 \cdot 10^6$ cm^{-2}; for ∠SiOSi = 145° the constants for the $Si-O^-$ and Si–O(Si) bonds are 7.8 and $7.2 \cdot 10^6$ cm^{-2} respectively. The other force constants are taken the same: angular O^-SiO^- and O^-SiO = $1.5 \cdot 10^6$ cm^{-2}; interaction of Si–O bonds with general Si = $0.4 \cdot 10^6$ cm^{-2} and with general O = $0.4 \cdot 10^6$ cm^{-2}; A-B = a–b = $0.2 \cdot 10^6$ cm^{-2}. Here, as always, A and a are the interactions of the Si–O(Si) or $Si-O^-$ bond with the OSiO angle of which it forms a side, while B and b are those with the opposite OSiO angle. The remaining interactions are equated to zero.

hardystonite, and hemimorphite studied indicate a bent form of the Si–O–Si bridge, since for a centrosymmetric configuration and a straight group the ν_s SiOSi vibration is inactive in the infrared spectrum (Table 10). However, allowance for possible changes in the selection rules on account of the lower symmetry of the internal field and resonance interactions between the vibrations of the Si_2O_7 groups often make any conclusions regarding the form of these groups less definite. In the rankinite crystal $Ca_3Si_2O_7$ belonging to the $P2_1/b-C_{2h}^5$ space group [137], there are four Si_2O_7 ions in the unit cell. The four sets of special positions with site symmetry C_i and multiplicity factor 2 make it possible to place four centrosymmetric Si_2O_7 groups in two of them; then only irreducible representations of the C_{2h} factor group inactive in the infrared spectrum will correspond to the ν_s SiOSi vibrations. On the other hand, a comparatively high intensity of the ν_s SiOSi band may be considered as evidence in favor of the assumption that a four-fold set of general positions is populated by bent Si_2O_7 groups. We shall show that the analysis of frequency values also agrees with the latter view.

In order to elucidate the way in which the frequencies of the Si_2O_7 group depend on its shape and the ratio of the orders of the Si–O and Si–O(Si) bonds, it is useful to calculate the frequencies of the normal vibrations for several models. Let us consider two configurations of a bent† Si_2O_7 group with symmetry C_{2v} and a SiOSi angle equal to 120 and 145°, and also a straightened group with symmetry D_{3d}. Since any calculation of the frequencies of an isolated Si_2O_7 ion bears an approximate character in view of our neglect of interactions with the lattice vibrations, it is convenient to allow only for the most important elements of the force constant matrix. An approximate estimate of the values of these elements may be based on data relating to the spectra of the siloxanes and alkoxysilanes. Table 11 shows the results of a calculation with two versions of the force constants of the $Si-O^-$ and Si–O(Si) bonds. An angle of SiOSi = 180° can only be achieved by considerably increasing the order of the Si–O(Si) bond and probably at the same time reducing the orders of the $Si-O^-$ bonds in the Si_2O_7 group. Hence for the centrosymmetrical model we may take $K_{Si-O(Si)} = K_{Si-O^-}$ which agrees with the similar distances $r_{Si-O(Si)}$ and r_{Si-O^-} in the structures of silicates with ∠SiOSi = 180° (see p. 288). The frequencies of the vibrations of the Si_2O_7 group for this case are given in the columns headed I. On the other hand, for ∠SiOSi < 180°, as a rule, $r_{Si-O(Si)} > r_{Si-O^-}$ [9, 134], and we may suppose that this relationship is reflected in the force constants. Hence in the columns headed II the frequencies of the bent Si_2O_7 groups are calculated on the assumption that $K_{Si-O^-} > K_{Si-O(Si)}$ for ∠SiOSi ≠ 180° and that $K_{Si-O(Si)} \simeq 6.6 \cdot 10^6$ cm^{-2} for the tetrahedral SiOSi angle.

The frequencies given in Table 11 agree satisfactorily with the experimental spectra of the pyrosilicates in the frequency range of the stretching vibrations of the anion, although for the frequency of the ν_s SiOSi vibration and particularly for the deformation vibrations the calculated values are too low, this being partly a consequence of neglecting interactions with the lattice vibrations. One reason for the difference between the experimental and calculated stretching frequencies may also be the assumption (made when choosing the force constants in version II) that the sum of the force constants of the four Si–O bonds at each Si atom was constant. Actually this sum may very well increase on increasing the basicity of the cations, i.e., on passing from the pyrosilicates with ∠SiOSi = 180° to the pyrosilicates with bent Si_2O_7 groups. However, further variation of the force constants is unjustified in view of the limited applicability of the model of the isolated anion. It should also be noted that calculation of the form of the vibrations of the Si_2O_7 groups confirms the views expressed when analyzing the spectra of the model organosilicon compounds. The ν_s SiOSi vibration involves little change in the lengths of the end $Si-O^-$ bonds, while the contribution made by changes in these distances to the ν_{as} SiOSi vibration is considerably greater and it is less characteristic in form.

The results of our calculation of the frequencies of the Si_2O_7 groups suggest that the greatest changes in the spectrum in relation to the value of ∠SiOSi will be associated with the positions of the ν_s SiOSi and ν_{as} SiOSi vibration bands. As an example of crystals with severely bent Si_2O_7 ions we may consider ilvaite: $CaFe_2^{\cdot\cdot}Fe^{\cdot\cdot\cdot}O(Si_2O_7)OH$. According to x-ray structural data, the SiOSi angle in this mineral is 122° [138]. The infrared spectrum of ilvaite differs from the spectra of pyrosilicates with larger SiOSi angles (considered earlier) in respect of the higher frequency and intensity of the ν_s SiOSi vibration band (Fig. 25):

ν_{as}SiOSi, ν_{as} and $\nu'_{as}SiO_3$, ν_s and ν'_sSiO_3
1038, 945, 901, 818, 754
ν_sSiOSi δSi_2O_7 (and νM—O)
701 570, 536, 494, 452 cm^{-1}

---

† Increasing the SiOSi angle to 180° in the model with symmetry C_{2v} leads to an improbable configuration with symmetry D_{3h}. Hence for values of the angle SiOSi close to 180° a structure with symmetry C_s is more probable for the Si_2O_7 ion.

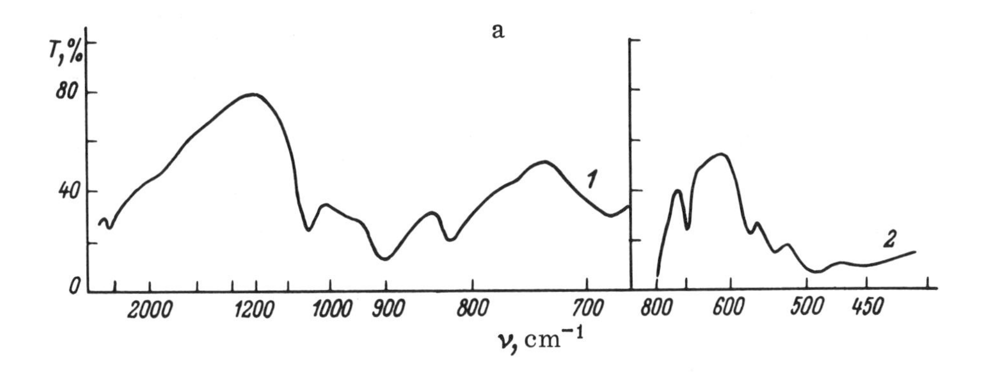

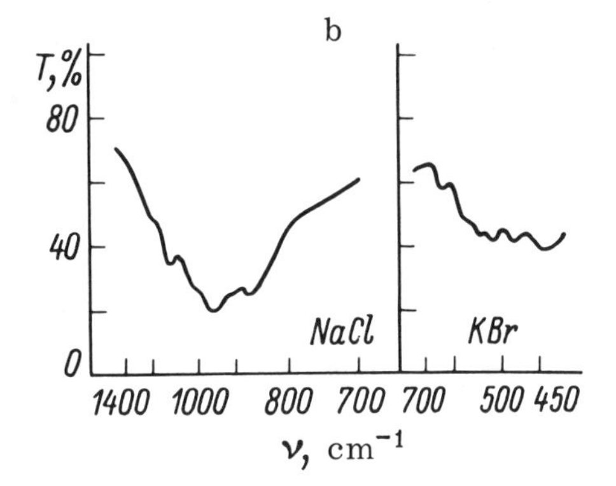

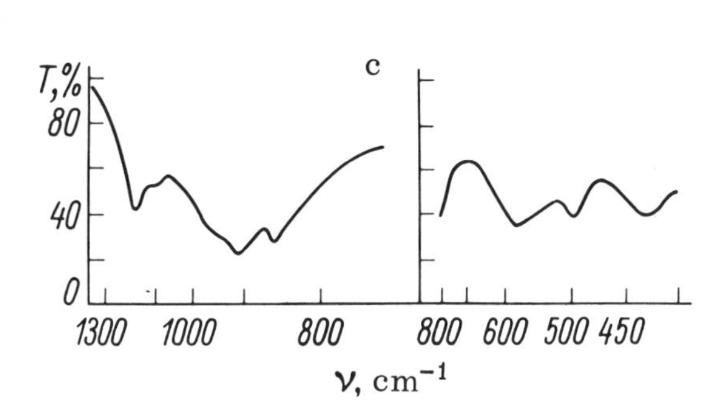

Fig. 25. Infrared spectrum of pyrosilicates distinguished by different configurations of the Si_2O_7 groups. a) Ilvaite $CaFe_2^{\cdot\cdot}Fe^{\cdot\cdot\cdot}O(Si_2O_7)OH$ [pressed in KBr: 1) 5 mg, 2) 10 mg]; b) seidozerite $Na_8(Zr_3Ti)Ti_2Mn_2O_4(Si_2O_7)_4F_4$ (pressed in KBr, 5 mg); c) thortveitite $Sc_2Si_2O_7$ (pressed in KBr, 5 mg).

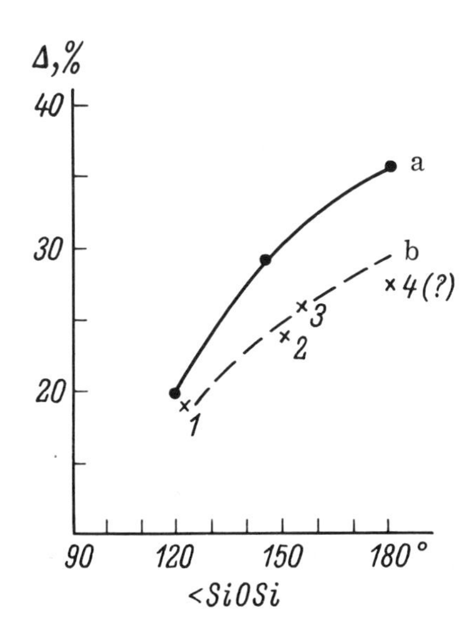

Fig. 26. Dependence of the splitting of the stretching vibrations of the Si–O–Si bridge ($\Delta = [\nu_{as} - \nu_s SiOSi]/[\nu_{as} + \nu_s SiOSi]$) on the value of ∠SiOSi in the Si_2O_7 group. a) Calculated data; b) experimental data; 1) ilvaite, 2) calamine, 3) epidote, 4) seidozerite.

In order to obtain a spectroscopic estimate of the SiOSi angle in pyrosilicates with an unknown geometry of the Si_2O_7 group, we may attempt to use the relation between the SiOSi angle and the splitting of the pulsation vibration of the SiO_4 tetrahedron. The latter may conveniently be characterized by the quantity $\Delta = (\nu_{as} - \nu_s \text{SiOSi})/(\nu_{as} + \nu_s \text{SiOSi})$ [139]. The dependence of this quantity on $\angle$SiOSi according to calculation is shown in Fig. 26a. On plotting the analogous relationship from experimental data, identification of the ν_sSiOSi band causes no difficulty, while for the ν_{as}SiOSi band we may take the highest frequency of the Si–O stretching bands in the spectrum of the corresponding pyrosilicate. This relationship, plotted from the few available data, is shown in Fig. 26b; in character it resembles the calculated curve, but is displaced in the direction of smaller Δ. The main reason for the discrepancy is probably the considerable difference in the influence of lattice vibrations on ν_{as} and ν_sSiOSi. Interaction with the vibrations of the cation–oxygen bonds should raise the ν_sSiOSi frequency more than ν_{as}SiOSi so that the value of Δ will fall. For large SiOSi angles this effect is particularly significant both as a result of the relatively low frequency ν_sSiOSi and also as a result of the higher vibration frequencies of the cation–oxygen bonds, which in these structures usually possess a considerable degree of covalence. The difference in the slopes of the calculated and experimental slopes is probably an indication of a stronger dependence of the force constant of the Si–O(Si) bond on the value of $\angle$SiOSi than that taken in the calculation. The accuracy of estimating the angle SiOSi by the spectroscopic method may be considerably improved as experimental data accumulate, but it can hardly become better than $\pm 10°$.

It follows from Fig. 26 that the configuration of the Si_2O_7 groups in seidozerite established by x-ray diffraction in [140], with an angle SiOSi = 180°, is at variance with spectroscopic data. The unit cell formula of seidozerite may be written $Na_8(Zr_3Ti)Ti_2Mn_2O_4(Si_2O_7)_4F_4$. In a crystal belonging to the $P2/c-C_{2h}^4$ space group, the four Si_2O_7 groups occupy a set corresponding to site symmetry C_1. Hence the centrosymmetric properties of these groups do not follow from the lattice symmetry. On the other hand, even for a centrosymmetric structure of the Si_2O_7 groups, the selection rules for the factor group of the crystal allow the appearance of a ν_sSiOSi vibration band in the infrared spectrum. In the spectrum of seidozerite [139] shown in Fig. 25, the ν_{as}SiOSi and ν_sSiOSi bands are identified without any difficulty, while the value of Δ obtained from their frequencies (1115 and 633 cm^{-1}) suggests that the most probable value of $\angle$SiOSi in this crystal is about 170 to 160°. This conclusion found confirmation in the results of a recent investigation into the structure of calcium seidozerite which established a bent configuration of the Si_2O_7 groups [141, 142].

In crystals of pyrosilicates containing Si_2O_7 groups, the centrosymmetric properties of which are due to the symmetry of the lattice, the detection of diortho groups by the spectroscopical method involves great difficulties, since allowance for the Davydov splitting leads to no change in the selection rules. The most reliable evidence as to the presence of Si_2O_7 ions may in these cases be obtained from Raman spectra. Indirect indications as to the existence of centrosymmetric Si_2O_7 groups may be secured by considering the number of bands observed in the range of the Si–O stretching and also considering the ν_{as}SiOSi vibration frequencies. At the present time the existence of Si_2O_7 groups with $\angle$SiOSi = 180° has been fairly reliably established in two structures by x-ray diffraction. Figure 25 shows the spectrum of thortveitite $Sc_2Si_2O_7$, in the structure of which the centrosymmetric properties of the Si_2O_7 groups were confirmed by a special study [143]. Thortveitite crystals belong to the $B2/m-C_{2h}^3$ space group and contain only one Si_2O_7 anion in the primitive cell; this has the full symmetry of the C_{2h} factor group. This excludes Davydov splitting, at least for the unexcited state of the lattice. The selection rules for the internal vibrations are not difficult to obtain by expanding the representations of the symmetry group of the isolated ion D_{3d} with respect to the representations of the C_{2h} group. The loss of the three-fold axis should lead to the removal of degeneracy but retention of the mutual exclusion rule. It is then easy to relate the frequencies of the four strong absorption bands in the range of stretching vibrations†:

Frequencies, cm^{-1}	1169	975	907	852
Symmetry types C_{2h}:	B_u	B_u	A_u	B_u
Symmetry types D_{3d}:	A_{2u} (ν_{as}SiOSi)	E_u ($\nu'_{as}SiO_3$)		A_{2u} ($\nu'_s SiO_3$)

The three bands at lower frequencies (578, 498, and 434 cm^{-1}) probably correspond to the vibrations $\delta'_s SiO_3(A_{2u})$, $\rho' SiO_3$ and $\delta'_{as} SiO_3$ (both E_u), or to a vibration of the A_{2u} type and one of the E_u vibrations, split as a result of the removal of the degeneracy. On studying the spectrum of thortveitite in the lower-frequency range (down to 300 cm^{-1}), five bands were found with a frequency of under 600 cm^{-1} [120], which indicates the splitting of both the degenerate vibrations. The frequencies of these vibrations differ considerably from those of the free anion owing to interaction with lattice vibrations. We note that, in the absence of data relating to the Raman spectrum, the only indication as to the existence of Si–O–Si bridges in the crystal is the ν_{as} SiOSi band, the frequency of which in this case is much higher than the upper limit of the frequency range for stretching of isolated SiO_4 groups in silicates. Subsequently it will be shown that, even in silicon–oxygen anions more complicated than Si_2O_7, the appearance of absorption bands in the range 1150 to 1200 cm^{-1} is associated with the existence of straightened Si–O–Si bridges with an angle SiOSi = 180° (or nearly so).

Another pyrosilicate containing centrosymmetric Si_2O_7 groups is $K_2Pb_2(Si_2O_7)$. The unit cell of a crystal belonging to the $P\bar{3}-C^1_{3i}$ space group contains only one Si_2O_7 group [144]; hence there are no grounds for expecting the removal of degeneracy or any disruption of the selection rules. The infrared spectra of $K_2Pb_2(Si_2O_7)$ and the isostructural crystal $K_2Pb_2(Ge_2O_7)$ contain six bands each in the range 300 to 1200 cm^{-1} [145], which exactly corresponds to the number of internal vibrations of the anion with active types of symmetry A_{2u} and E_u (if we reject the low-frequency vibration δXOX). The vibration ν_s XOX is not observed, as in the spectrum of thortveitite. The identification of the frequencies in [145] is based on the satisfactory agreement between the experimental values and the results of a calculation carried out with a simple model of valence forces (Table 12).

† The weaker band at 1100 cm^{-1} does not belong to the set of fundamental vibrations; this follows from the temperature dependence of its intensity (see p. 286).

Table 12. Frequencies of the Pyrosilicate and Pyrogermanate

Frequency, cm^{-1} experimental Si_2O_7	experimental Ge_2O_7	calculated Si_2O_7	calculated Ge_2O_7	Symmetry of the vibrations
1015	905	1010	900	A_{2u}
915	745	925	740	E_u
770	715	760	710	A_{2u}
570	475	540	480	E_u
480	395	420	360	A_{2u}
425	320	380	320	E_u

Table 13. Calculated Frequencies of the Ge_2O_7 Groups with Symmetry C_{2v}

∠GeOGe=145° B_1	A_1	B_2	A_2	∠GeOGe=120° B_1	A_1	B_2	A_2
928	764			840	775		
764	701			774	714		
686	474	741	784	691	569	731	774
309	325	352	307	310	316	352	307
308	306	304	211	306	306	304	211
210	199			209	203		

In contrast to the spectrum of thortveitite, the frequency ν_{as}XOX is low, like that of the other stretching vibrations. It should also be mentioned that the ratio of the force constants used in the calculation $K_{X-O^-}/K_{X-O(X)} = 1.2$ contradicts the results of the x-ray structural study of $K_2Pb_2(Ge_2O_7)$ [146], since the length of the Ge–O(Ge) bond is less than that of the Ge–O⁻ bond by an amount exceeding any possible error in their determination: $r_{Ge-O(Ge)} = 1.69 \pm 0.01$ Å, and $r_{Ge-O^-} = 1.79 \pm 0.06$ Å. Hence the agreement between the experimental and calculated frequencies is to a certain extent illusory, all the more so in view of the fact that the Raman spectra of these compounds have not been studied. The X_2O_7 groups in these crystals are united into layers by the Pb–O bonds, the length of which (2.20 Å in $K_2Pb_2Si_2O_7$ [144] and 2.13 Å in $K_2Pb_2Ge_2O_7$ [146]) is much smaller than the sum of the ionic radii, which indicates a considerable covalent character, as does the trigonal coordination of the lead by the oxygen. It follows from this that the model of the isolated anion is inapplicable even for calculating the stretching vibrations of Si–O(Ge–O), and their low frequencies may be associated not only with a change in the potential-energy constants of the anion but also with the considerable contribution of the O–Pb bonds to those vibrations.†

Since the absence of x-ray structural and spectroscopic data prevents the establishment of empirical correlations between the spectrum and the configuration of the Ge_2O_7 group in the pyrogermanates, we confine ourselves to calculating the frequencies of these groups for a GeOGe angle equal to 120 and 145° (Table 13).

In order to carry out the calculation we use rather arbitrary force constants: $K_{Ge-O^-} = 7.1 \cdot 10^6$ cm^{-2} and $K_{Ge-O(Ge)} = 6.6 \cdot 10^6$ cm^{-2} with ∠GeOGe = 145°; with ∠GeOGe = 120° their values are respectively 7.3 and $6.2 \cdot 10^6$ cm^{-2}; the angular coefficient K_{OGeO} is the same in both cases, $1.5 \cdot 10^6$ cm^{-2}. The calculation shows that Δ is quite sensitive to changes in angle; however, in the experimental relationship the influence of lattice vibrations on the frequencies ν_s and ν_{as}GeOGe will be still more noticeable than in the case of silicates.

Pyrophosphates. These have been spectroscopically studied in more detail than pyrosilicates: the infrared spectra of the pyrophosphates have been studied for a whole series of compounds, including those with alkali cations [148-152]; information regarding the Raman spectrum of the $[P_2O_7]^{4-}$ ion has been obtained thanks to the relative stability of this ion in aqueous solution and in the melt [152-154]. This enables us not only to identify the stretching vibrations of the P_2O_7 group but also approximately to estimate the force constants of the P–O(P) and P–O⁻ bonds. Only one of the alkali pyrophosphate crystals, $Na_4P_2O_7 \cdot 10H_2O$, has been solved by x-ray diffraction to a reasonably high degree of accuracy [155, 156]: the differences in the lengths of the P–O(P) and P–O⁻ bonds reaches 0.1 Å (1.61 ± 0.015 Å and 1.51 ± 0.015 Å, respectively), while the angle POP equals 133.5°. According to later results [157] the angle POP is 130.2°. There is no doubt, however, that a transition to compounds with less basic cations capable of forming partly covalent bonds with oxygen will lead to a reduction of the difference in the distances $r_{P-O(P)}$ and r_{P-O^-} and to an increase in the angle POP, up to values close to 180° (see §3 of Chap. VI). It is thus natural to expect an increase in the order of the P–O(P) bond in structures with an increased POP angle.

In identifying the stretching vibrations of the P_2O_7 anion, the least difficulties are associated with the recognition of the ν_sPOP and $\nu_s PO_3$ frequencies appearing in the Raman spectrum in the form of strong polarized ($\rho < 0.3$) lines (Fig. 27). The vibration $\nu_{as}POP(B_1)$ is ob-

† These considerations are probably also applicable to the recently studied barysilite $MnPb_8(Si_2O_7)_3$ [147]: the use of curve b in Fig. 26 leads to an anomalously small angle SiOSi (about 115°) owing to the reduced (to 970 cm^{-1}) ν_{as}SiOSi frequency. The estimate of ∠SiOSi in [147] based on correlation between the angle and ν_s SiOSi (702 cm^{-1}) gives approximately 130°, in agreement with x-ray structural data; however, the accuracy of the latter is very low.

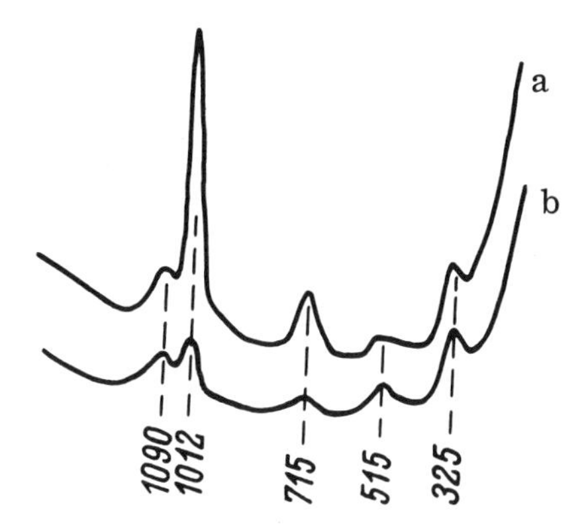

Fig. 27. Raman spectrum of an aqueous solution of $K_4P_2O_7$ in polarized light. Excitation from the Hg 4358 Å line; a) "strong," b) "weak" component.

served in the form of a strong band in the infrared spectrum at about 900 cm^{-1}. In order to explain the frequency of this vibration, lower than the ν_{as} SiOSi in the pyrosilicates, we must allow for the fact that, for all the similarity between the masses of Si and P and the configurations of the Si_2O_7 and P_2O_7 groups, the differences between the spectra of the pyrosilicates and pyrophosphates are determined mainly by the differences in the force constants. Comparison of the calculated forms of the vibrations of the P_2O_7 and Si_2O_7 ions leads to the conclusion that there is a change in the relative contribution of the elongation of the $X-O^-$ and X–O(X) bonds to the two strongly interacting vibrations of symmetry type B_1, ν_{as} XOX and ν_s' XO_3. As a result of the considerable increase in the force constants of the $X-O^-$ bond, the changes in the lengths of the X–O(X) bonds in the spectrum of the pyrophosphate are chiefly localized in the lower-frequency member of these vibrations, while the changes in the $X-O^-$ distance make the greatest contribution to the other vibration situated close to 1100 cm^{-1}, i.e., somewhat lower than the ν_{as} PO_3 frequencies at 1130 to 1150 cm^{-1}. In the spectrum of the pyrosilicate ion, the higher-frequency vibration is mainly associated with the elongation of the Si–O–Si bridge. The increase in the differences between the force constants of the end and bridge X–O bonds in the X_2O_7 groups, where X are elements of groups VI and VII of the periodic table, leads to a separation of the frequencies and forms of the XO_3 and XOX vibrations, more pronounced than in pyrophosphates.

It is interesting to consider the possibility of a spectroscopic estimate of the POP angle in the P_2O_7 group. However, the ν_s POP frequency in the spectra of the majority of the pyrophosphates never extends beyond the limits of the range 720 ± 20 cm^{-1}. On increasing the ∠POP the ν_s POP frequency should diminish; however, the rise in the force constant of the P–O(P) bond usually accompanying a rise in ∠POP may compensate the change in this frequency. Moreover the two factors (kinematic and dynamic) act in the same direction on ν_{as} POP, leading to a rise in frequency from 900 cm^{-1} for alkali pyrophosphates to about 980 cm^{-1} in the pyrophosphates of magnesium and zirconium. The use of the quantity $\Delta = [\nu_{as} - \nu_s \mathrm{POP}]/[\nu_{as} + \nu_s \mathrm{POP}]$ in order to judge the magnitude of the angle POP shows that the latter falls gradually in the order $Li_4P_2O_7$, $Na_4P_2O_7$, $K_4P_2O_7$ in accordance with the values Δ = 12.0, 11.85, and 10.55%. Analogous changes also occur in the series of alkaline-earth pyrophosphates from $Mg_2P_2O_7$ with Δ = 14.05% to $Ba_2P_2O_7$ with Δ = 11.9%.

In addition to the normal pyrophosphates, spectroscopic investigations have been made into the $H_2P_2O_7$ ions in solution [153, 158] and acid salts $M_2H_2P_2O_7$ [148], in the spectra of which there is a rise in the frequencies of some of the vibrations of the PO_3 end groups, undoubtedly associated with an increase in the force constants of the $P-O^-$ bonds at the expense of the more nearly single P–O(H) bonds. Interpretations have also been given for the spectra of other ions analogous to P_2O_7, namely $[As_2O_7]^{4-}$ and $[AsPO_7]^{4-}$ [152] and also $[V_2O_7]^{4-}$ [159]. Many elements of Groups VI and VII such as S, Se, Cr, and Cl form compounds containing the $[X_2O_7]^{2-}$ anion or X_2O_7 molecule, respectively. The structure of the Cl_2O_7 molecule in vapor has been established by electron diffraction [160]; the structure of the crystal $K_2S_2O_7$ has been solved by x-ray diffraction [161]. The infrared and Raman spectra of crystalline normal and acid pyrosulfates $M_2S_2O_7$ and MHS_2O_7 and also the Raman spectrum of the pyrosulfate ion in the melt were studied in [163-166]. So were the spectra of pyrochromates [168], pyroselenates [162], and pyromolybdates and pyrotungstates [159], although the possibility of the existence of X_2O_7 ions with X = W or Mo raises some doubts. Thus, in $Na_2Mo_2O_7$ crystals, instead of Mo_2O_7 groups, structural analysis [167] establishes a lattice of MoO_4 tetrahedra and MoO_6 octahedra. The Cl_2O_7 spectrum has been studied both for the gas and for the liquid and solid phases [170].

As there were until recently contradictions in the assignment of the S_2O_7 and Cl_2O_7 frequencies, it is interesting to calculate the vibrational frequencies of X_2O_7 groups with X = P, S, Cl; this should also enable us to discuss the chemical structure of these groups [170a]. We take a geometrical model with $\angle XOX = 120°$ and symmetry C_{2v}; then the end O atoms lying in the plane of the XOX bridge are oriented in the same direction from the X...X line as the O bridge atom. The model is similar in shape to the S_2O_7 ion in $K_2S_2O_7$ and the Cl_2O_7 molecule, although actually $\angle SOS = 124°$, $\angle ClOCl = 118°$, the XO_3 groups are turned around the X–O(X) bonds, the angles OXO in the XO_3 groups are greater, and the angles OXO involving the bridge O are smaller than tetrahedral. The PO_3 groups are turned still further and the angle POP differs still more from 120° for the P_2O_7 ion in $Na_4P_2O_7 \cdot 10H_2O$. However, in the anhydrous $Na_4P_2O_7$ crystal, the spectrum of which was compared with the results of calcula-

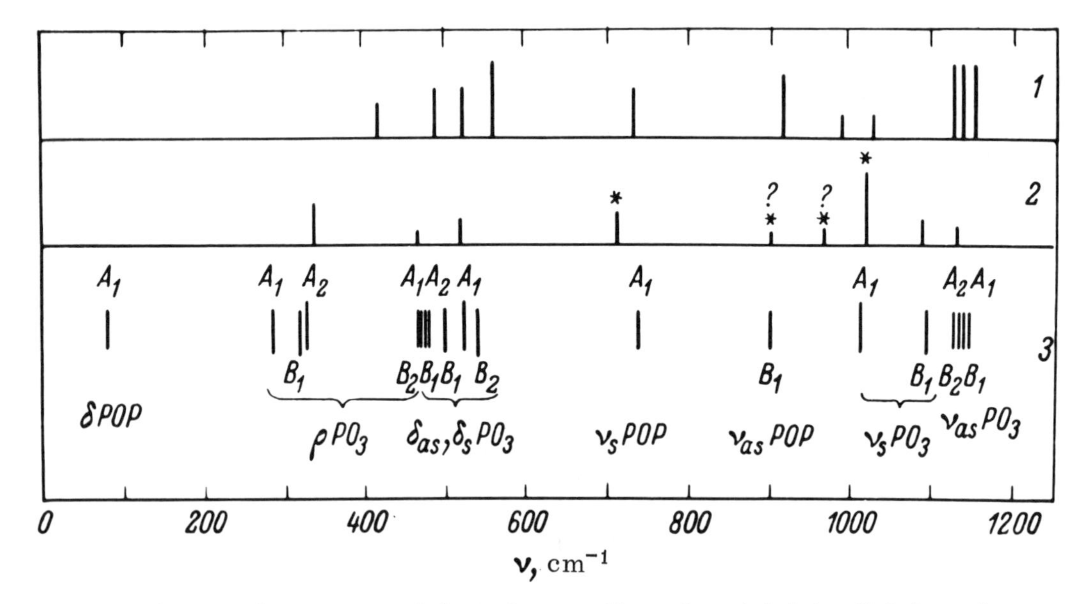

Fig. 28. Vibrational spectrum of the P_2O_7 ion. Experimental data: 1) infrared spectrum of crystalline $Na_4P_2O_7$ [152]; 2) Raman spectrum of an aqueous solution of $K_4P_2O_7$ [153] (polarized lines indicated by an asterisk); the query (?) indicates polarized lines 900 and 966 cm^{-1} probably belonging to the hydrated ortho ions and absent from the spectrum of the freshly prepared solution at low temperatures; see Fig. 27 [154]; 3) identification of the frequencies and results of a calculation of the P_2O_7 ion for $K_{PO^-} = 11.8 \cdot 10^6$ cm^{-2}, $K_{PO(P)} = 7.3$, $H_{PO,OP} = -0.5$, $h_{PO^-,PO^-} = 0.4$, $h_{PO^-,PO(P)} = 0.2$, $K_{\delta O\text{-}PO^-} = 2.6$, $K_{\delta O\text{-}PO(P)} = 2.3$, $K_{\delta POP} = 0.5$. The lengths of the $P-O^-$ and P–O(P) bonds are taken as 1.51 and 1.61 Å. For symmetry C_2, which more exactly describes the structure of the P_2O_7 ion, the representation A_2 of group C_{2v} transforms into the totally symmetric representation.

tion, the deviations from C_{2v} symmetry may be smaller than in the crystal hydrate, while the POP angle is probably smaller than 130° owing to the lower order of the bridge bonds. Owing to the absence of sufficient experimental data and the limited applicability of the isolated-anion model to the spectra of the crystals, we confine ourselves to considering the diagonal force constants and the mutual interaction constants of the bonds. A calculation of the partial derivatives of the frequencies with respect to all the force constants will enable us to estimate the probable error in determining the force constants of the bonds on using this simplified set.

The results of the calculation shown schematically in Figs. 28 to 30 solve the problem as to the identification of the stretching frequencies fairly reliably and enable us to estimate the orders of the X–O bonds. In Fig. 31a (allowing for possible error) we compare the force constants of the end and bridge bonds K_{XO^-} and $K_{XO(X)}$ with those of the single X–O bonds of order 1.0 considered by Siebert, K'_{XO}. Owing to the absence of data relating to the alkali pyrosilicates $M_6Si_2O_7$ (required for comparison with the spectra of the alkali pyrophosphates and pyrosulfates), the probable frequencies of the pyrosilicates were estimated (on the assumption that ∠SiOSi ≃ 120° for M = Li, Na, K) by extrapolation from the known spectra of ilvaite and $La_2Si_2O_7$, in which ∠SiOSi is smallish, and the spectrum of $Ca_3Si_2O_7$, in which the influence of lattice vibrations on the frequency of the anion is less substantial. In Fig. 31b we show the changes in the orders of the end and bridge bonds of the X_2O_7 groups N_{XO^-} and $N_{XO(X)}$ calculated by the Siebert method and the formal order of the end bond $N^*_{XO^-}$, obtained on the assumption that the sum of the orders of the bonds belonging to atom X equals the number of its group in the periodic system, while $N_{XO(X)} = 1$. We see from this figure that the amount by which the order of the X–O bond exceeds the formally expected value falls on passing from X = Si to X = Cl. The changes in the sum of the orders of the bonds of the XO_4 tetrahedron in the X_2O_7 groups $\left(\sum_{[X_2O_7]} N_{XO_4}\right)$ agrees with the well-known empirical rules limiting the number of electrons in the outer shell of atoms of the third period to twelve (probably occupying one s, three p, and two d orbits of X): $\sum_{[X_2O_7]} N_{XO_4}$ is no greater than 6, and on passing from X = S to X = Cl, not only does it not increase, but on the contrary it falls, which is reflected in the paradoxically

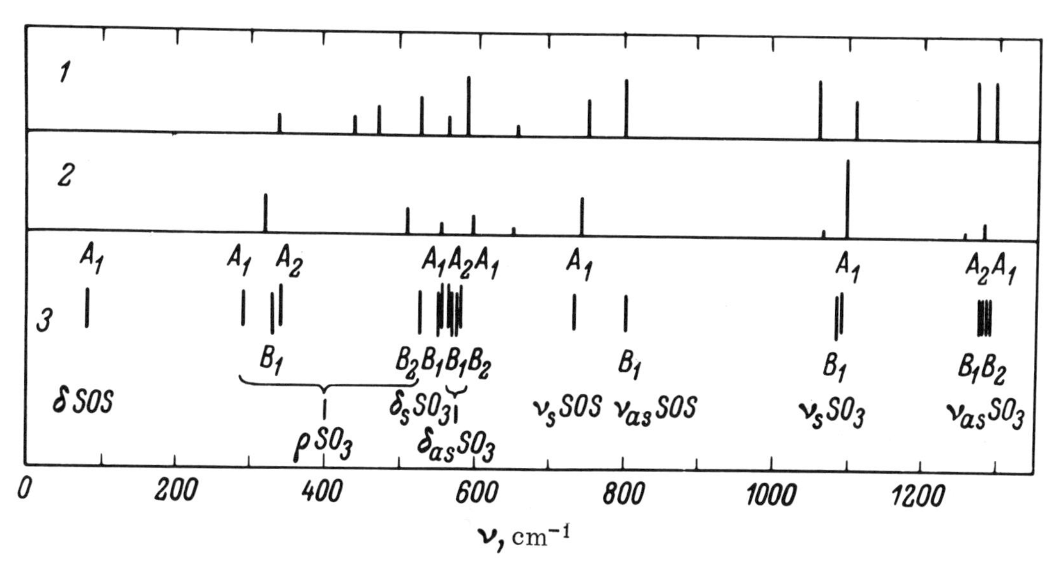

Fig. 29. Vibrational spectrum of the S_2O_7 ion. Experimental data [165]: 1) infrared spectrum of crystalline $K_2S_2O_7$; 2) Raman spectrum; 3) calculated frequencies of the S_2O_7 ion for $K_{SO^-} = 14.8 \cdot 10^6$ cm^{-2}, $K_{SO(S)} = 5.2$, $H_{SO,OS} = 0.5$, $h_{SO^-,SO^-} = 0.1$, $h_{SO^-,SO(S)} = 0.2$; $K_{\delta O^-SO^-} = 3.4$, $K_{\delta O^-SO(S)} = 2.4$, $K_{\delta SOS} = 0.5$. Geometrical parameters $r_{SO^-} = 1.44$ Å, $r_{SO(S)} = 1.64$ Å.

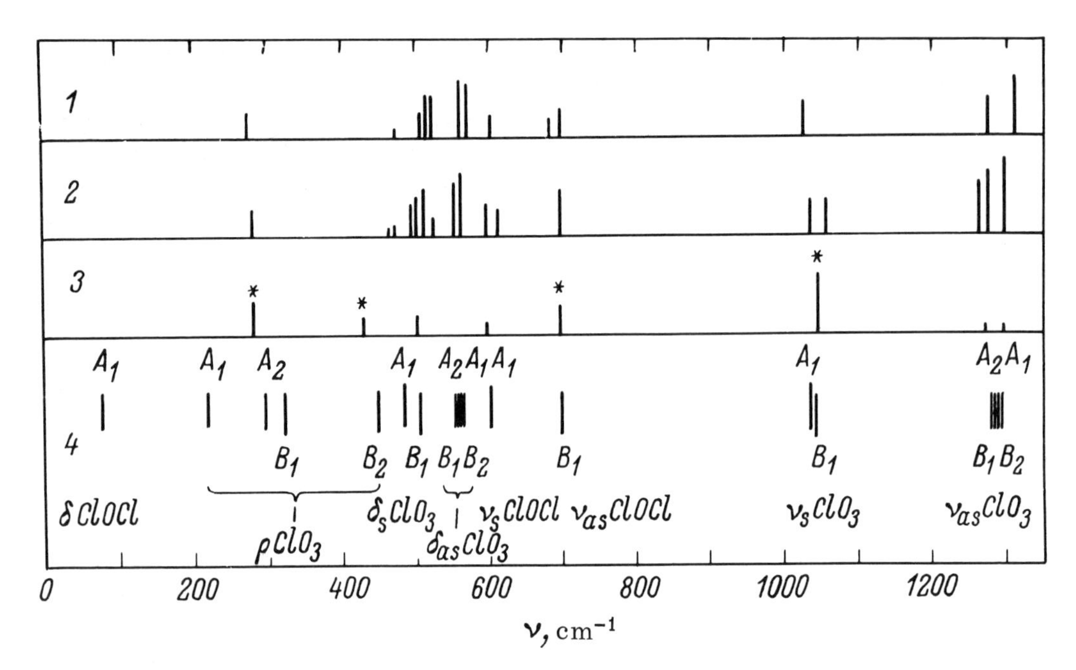

Fig. 30. Vibrational spectrum of the Cl_2O_7 molecule. Experimental data [170]: 1) infrared spectrum of Cl_2O_7 vapor; 2) infrared spectrum of the crystal; 3) Raman spectrum of a solution in CCl_4 (polarized lines indicated by an asterisk); 4) calculated frequencies of the Cl_2O_7 molecules for $K_{Cl=O} = 15.3 \cdot 10^6$ cm^{-2}, $K_{Cl-O} = 2.9$, $H_{Cl-O,Cl-O} = -0.5$, $h_{Cl=O,Cl=O} = -0.4$, $h_{Cl=O,Cl-O} = 0.2$, $K_{\delta O=Cl-O} = 1.8$, $K_{\delta O=Cl=O} = 3.2$, $K_{\delta ClOCl} = 0.5$. The constants $h_{Cl=O,Cl=O}$ and $h_{Cl=O,Cl-O}$ are determined from the $HClO_4$ spectrum calculated for the five-atom model (C_{3v}). The smaller Cl–O(H) distance in this molecule correlates with the higher constant $K_{Cl-O} = 5.5 \cdot 10^6$ cm^{-2}. Other constants: $K_{Cl=O} = 15.0$, $K_{\delta O=Cl=O} = 3.0$, $K_{\delta O=Cl-O} = 2.2$; (HO)ClO_3 frequencies (calculated): 1274, 582, 301 cm^{-1} for E and 1047, 738, 560 cm^{-1} for A_1; frequencies (experimental): 1326 and 1263, 1050, 725, 579, 560, 318, and 307 cm^{-1} [170]. Geometrical parameters of Cl_2O_7: $r_{Cl=O} = 1.41$ Å, $r_{Cl-O} = 1.71$ Å.

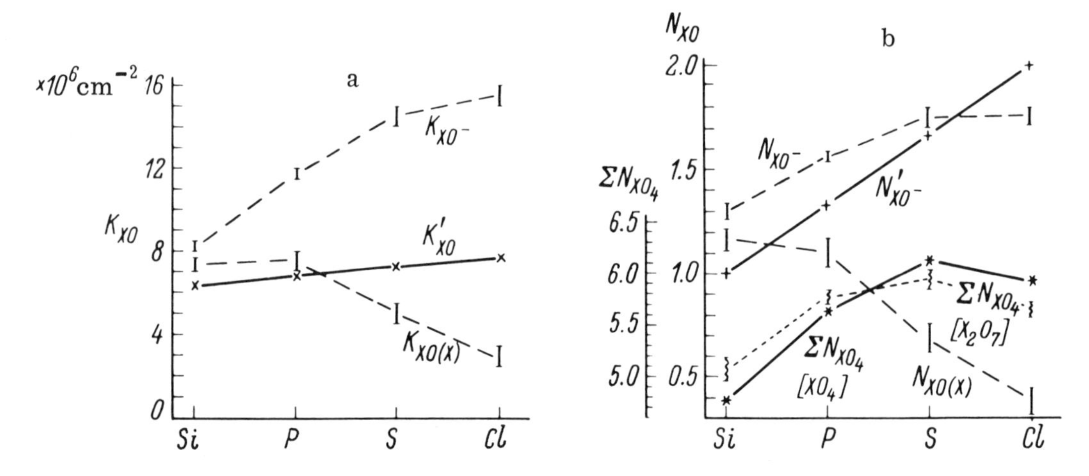

Fig. 31. Force constants (a) and orders of the bonds (b) in the X_2O_7 groups.

low value of $K_{ClO(Cl)}$. The sums of the orders of the bonds in the ions $[XO_4]^{(8-n)-}$ $\left(\sum_{[XO_4]} N_{XO_4}\right)$, taken from Siebert's data [17] and shown in Fig. 31b, vary analogously.

The difference in the values of $\sum_{[X_2O_7]} N_{XO_4}$ and $\sum_{[XO_4]} N_{XO_4}$ on the assumption of constant energy of the electrostatic bonds may be regarded as a measure of the energetic contribution favoring the process

$$2\{M^{+}_{8-n}[XO_4]^{(8-n)-}\} \rightarrow M^{+}_{2}O^{2-} + M_{14-2n}[X_2O_7]^{(14-2n)-}$$

by which we mean the decomposition of the ortho salt with the formation of two phases (pyro compound and metal oxide), or one phase (by the formation of a structure with pyro ions and additional O^{2-} anions). The latter version is evidently more favorable for partly covalent M...O bonds. In this respect Fig. 31b indicates a relatively high suitability of structures with pyro groups for silicates, approximately equal suitability of the ortho and pyro structures for X = P, and a preference for the ortho structures in the case of X = S or Cl.†

Complex anions in the form of chains of finite size with a number of tetrahedra greater than two have only recently been found in silicates (Si_3O_{10}); however, these are not infrequently encountered in phosphates and sulfates and have been studied by x-ray diffraction [171-174]. In the spectra of tri- and tetraphosphates (ions P_3O_{10} and P_4O_{13}), bands of vibrations of the PO_2 groups are observed at higher frequencies (up to 1250 cm^{-1}) than those of the "end" PO_3 groups in these or in pyrophosphate anions. In the frequency ranges of the vibrations ν_s and ν_{as}POP the number of bands increases to two or three, corresponding to the number of P–O–P bridges in the complex anion [148-149].

§ 5. Cyclic Silicate Anions

For an Si:O ratio of 1:3, cyclic $(SiO_3)_n^{2n-}$ silicate anions, like chains, are associated with metasilicate structures. Here we shall consider the spectra of silicates containing anions of this kind, since these may be related to structures of the "bridge" type with groups comprising a finite number of condensed silicon–oxygen tetrahedra. We must nevertheless remember that each of the silicon atoms in the cyclic anion forms only two bonds of the $Si-O^-$ and two of the Si–O(Si) type, in contrast to the ortho- and pyrosilicates. It is thus not impossible that under favorable conditions the order of the $Si-O^-$ bonds may prove to be higher than that associated with the cases considered earlier.

Even in the first papers devoted to the infrared spectra of silicates Matossi and Kruger [66] directed attention to the very strong band at about 750 cm^{-1} characteristic of silicates with ring anions, i.e., in the region usually free from strong bands. These observations were not, however, correctly interpreted, as the band at 750 cm^{-1} was ascribed to the activation of the totally symmetric vibration of the silicon–oxygen tetrahedron in the infrared spectrum. Later investigations of Bokii and Plyusnina [175, 176] confirmed the results of Matossi and Kruger; they established that the frequency of the "ring band" in the spectra of silicates with six-fold rings $(Si_6O_{18})^{12-}$ was somewhat higher than in the case of silicates containing rings of three tetrahedra. With only one exception [81], no attempt was made to interpret the spectra of silicates with cyclic anions, until that given in [177]. The secular equations were derived by Prima [88] for plane rings with symmetry D_{nh}, using internal coordinates. Saksena et al. [90] considered the same problem for the rings Si_3O_9, Si_4O_{12}, and Si_6O_{18} in Cartesian coordinates and carried out numerical calculations with the aid of a simplified model of the force field, described

† We note that the use of data relating to the force constants for making a direct estimate and comparison of the orders of the π bonds in the systems under consideration is impeded by the change in the strength of the σ bonds X–O on passing from X = Si to X = Cl [169].

Table 14. Classification of the Vibrations of the Cyclic Ion X_3O_9 with Symmetry D_{3h}

Description of the vibration	Symmetry type	Description of the vibration	Symmetry type
$\nu_{as}O^-XO^-$	E'', A_2''	γO^-XO^-	E', A_2'
$\nu_{as}XOX$	E', A_2'	ρO^-XO^-	E'', A_2''
$\nu_s O^-XO^-$	E', A_1'	δO^-XO^-	E', A_1'
$\nu_s XOX$	E', A_1'	τO^-XO^-	E'', A_1''
δXOX	E', A_1'	$\delta_\perp XOX$	E'', A_2''

Note: In [180, 181] δXOX is denoted ν_{ring}.

by only three force constants: $K_{Si-O^-} = 5 \cdot 10^5$, $K_{Si-O(Si)} = 4 \cdot 10^5$, and $K_\delta = K_{OSiO}/r^2_{SiO} = 0.7 \cdot 10^5$ dyn/cm.

The $[Si_3O_9]^{6-}$ and $[Ge_3O_9]^{6-}$ Rings. Let us consider the possibility of interpreting the frequencies of the internal vibrations of the simplest anions X_3O_9 by supposing, as before, that the frequencies of the symmetrical vibrations of the X–O–X bonds are to some extent "separated" from other vibrations of the system. It is then reasonable, considering the relatively large mass of the atom X and the probable differences in the force constants of the $X-O^-$ and X–O(X) bonds, to describe the vibrations of the complex ion approximately as the vibrations of O^--X-O^- and X–O–X groupings. A classification of the vibrations of the Si_3O_9 ion of this kind is presented in Table 14. In approximately dividing the vibrations into those of the stretching and deformation types we must remember that the formation of a closed ring of Si–O–Si bonds, strictly speaking, leads to a reduction by one in the number of stretching vibrations as compared with an open chain of the same composition. Thus the totally symmetric vibration of the ring ν_s SiOSi (type A_1' with symmetry D_{3h}) is to some extent a deformation vibration. On the other hand, a planar deformation vibration of SiOSi of the same type of symmetry should mix strongly with this vibration† which leads to an anomalously high frequency for δSiOSi. The difference between these two vibrations may be illustrated by the following scheme for the directions of the displacements of the Si and O atoms

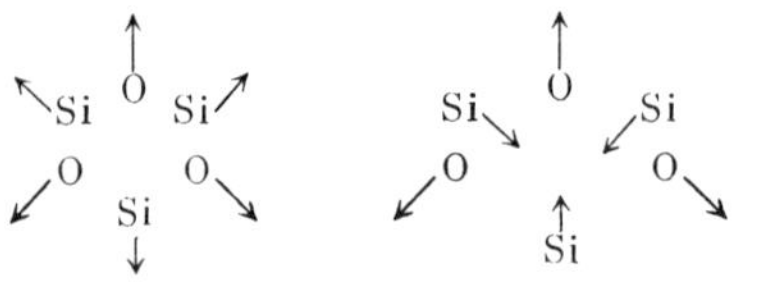

although quantitative calculation of the forms of the vibrations would hardly have any point in the absence of adequate data regarding the internal force field of the anion, and without allowing for the interaction with the lattice vibrations. The complete absence of data relating to the Raman spectra makes it impossible to obtain any information regarding the frequencies of these vibrations.

In order to identify the frequencies in the spectra of the rings under consideration we must also remember that, while the OSiO angles are close to tetrahedral, the values of the SiOSi angles in planar rings are no greater than 130°, while a deviation of the rings from the

† At the same time an out-of-plane deformation vibration $\delta_\perp$SiOSi reduces, in essence, to an internal rotation around the Si–O(Si) bonds in the ring, and its frequency (as may easily be shown by calculation) is no greater than a few tens of reciprocal centimeters.

plane configuration can only reduce the value of ∠SiOSi. As we shall show for a number of examples in later chapters, existing data indicate that the plane configuration of the Si_3O_9 cycles is preserved even in those structures in which this configuration is not a necessary consequence of the symmetry of the crystal but is due, as it would appear, to the energetically unfavorable effect of further reducing the angle SiOSi. The low value of the angle SiOSi (as compared with that of other silicate structures) suggests that there will be a reduction in the ν_{as} SiOSi and a rise in the ν_s SiOSi frequencies. We may further suppose that the first of these effects will be more pronounced, since the dynamic factors should exert their influence in the same direction: a reduction in ∠SiOSi worsens the conditions for the formation of the $p\pi-d\pi$ bonds in the Si–O–Si bridges.

The identification of the vibration frequencies of the cyclic anion is made easier if a three-fold symmetry axis (or one of higher order) is a symmetry element of the crystal. The moments of the transitions can only be oriented parallel or perpendicular to this axis, and hence we may distinguish simple vibrations from degenerate vibrations by comparing the spectra (absorption or reflection) of appropriately oriented single-crystal plates, without having to use polarized radiation. Subsequently we shall use data relating to the reflection spectra of oriented single crystals of ring silicates taken from [66, 176].

The foregoing considerations enable us to give an interpretation for the spectra of silicates and germanates with $[X_3O_9]^{6-}$ anions possessing intrinsic D_{3h} symmetry as in Table 15. According to x-ray structural analysis, benitoite crystals $BaTi(Si_3O_9)$ belong to the $P\bar{6}c2-D_{3h}^2$ space group and contain two plane Si_3O_9 rings in the unit cell [178], occupying special positions with symmetry C_{3h}. Consideration of the selection rules for an isolated ion lead us to expect the appearance of only four bands in the infrared spectrum in the stretching region, of which three are polarized perpendicularly to the optic axis and the other one parallel. Allowance for the site symmetry of the Si_3O_9 group in the crystal will not alter this result, but on allowing for resonance interactions between the vibrations of two such groups in the cell, i.e., analyzing the spectrum by means of the factor group of the space group, splitting of the degenerate vibrations becomes possible. The three reflection bands observed in the reflection spectrum of a plate having its plane perpendicular to the optic axis in the Si–O stretching region [66] may evidently be ascribed to three degenerate vibrations. No ν_{as} O^-SiO^- vibration with electric vector along the optic axis is observed in the spectrum for this orientation of the cut; the absence of any additional bands in the absorption spectrum of powder [175] probably indicates the superposition of the band of this vibration on one of the bands of type E′. The spectrum of catapleiite shown in Fig. 32 may be interpreted in a similar way. According to x-ray structural data [179], catapleiite $Na_2Zr(Si_3O_9)\cdot 2H_2O$ contains two $[Si_3O_9]^{6-}$ ions in the unit cell and belongs to one of the space groups D_{6h}^4, D_{3h}^4, or C_{6v}^4. It follows from data relating to the reflection spectrum of catapleiite [176] that only one band (the one with the highest frequency at about 1041 cm^{-1}) becomes stronger when the cut is oriented parallel to the optic axis (in Fig. 32b this band may be seen as a "shoulder" on the sloping part of the 1007-cm^{-1} band). Thus the band at 1041 cm^{-1} may be unambiguously ascribed to the ν_{as} $O^-SiO^-(A_2'')$ vibration, and this confirms the view that the ν_{as} SiOSi vibration band is not the highest-frequency band of all those belonging to the cyclic Si_3O_9 anion. At the same time the frequency ν_s SiOSi(E′) is much higher than the corresponding frequency in the spectra of pyrosilicates. The two other stretching vibrations of the same type of symmetry differ comparatively little in frequency, and the ascription of the higher-frequency one to ν_{as} SiOSi and the other to ν_s O^-SiO^- is to a certain extent arbitrary. It is nevertheless characteristic that the latter is slightly split, while the two other stretching vibrations of the E′ type of symmetry show no splitting at all. Since for the hexagonal symmetry of the crystal this splitting may be associated not with the removal of degeneracy but simply with interaction between the ions (Davydov effect), it is natural that the splitting of the vibration chiefly associated with the $Si-O^-$ side bonds should prove to be most noticeable. It is possible that it is for

Table 15. Interpretation of the Spectra of Silicates and Germanates with X_3O_9 Ions (Symmetry D_{3h})

Assignments	Symmetry types of the vibrations and selection rules			IRS (frequency, cm^{-1})				Factor group of space group (symmetry types and selection rules) D_{6h}
	intrinsic symmetry D_{3h}	site symmetry C_{3h}	factor group of space group D_{3h}	benitoite $BaTi(Si_3O_9)$, reflection spectrum (cut perp. to optic axis) [66]	$SrGeO_3$ (II), absorption spectrum (powder)	$Na_2Zr(Si_3O_9)\cdot 2H_2O$ absorption spectrum (powder)	$Na_2Zr(Si_3O_9)\cdot 2H_2O$ polarization of bands in the reflection spectrum [176]	
ν_{as} O^-XO^-	E'' ina; a, dp	E'' ina; a, dp	E'' ina; a,dp E'' ina; a,dp	—	—	—	—	E_{1g} ina;a,dp E_{2u} ina; ina
	A_2'' a ∥ ; ina	A'' a ∥ ; ina	A_1'' ina; ina A_2'' a ∥ ; ina	—	854 v.s.	1041 s.	∥	A_{2u} a ∥ ; ina B_{1g} ina;ina
ν_{as} XOX	E' a ⊥; a, dp	E' a ⊥; a, dp	E' a ⊥; a,dp E' a ⊥; a,dp	1065	829 s.	1007 v.s.	?	E_{1u} a ⊥ ; ina E_{2g} ina; a,dp
	A_2' ina; ina	A' ina; a, p	A_1' ina;a,p A_2' ina; ina	—	—	—	—	A_{2g} ina; ina B_{1u} ina; ina
ν_s O^-XO^-	E' a ⊥; a, dp	E' a ⊥; a,dp	E' a ⊥; a,dp E' a ⊥; a,dp	930	763 v.s.	948 v.s. 935 v.s.	⊥	E_{1u} a ⊥; ina E_{2g} ina; a,p
	A_1' ina; a,p	A' ina; a,p	A_1' ina;a,p A_2' ina;ina	—	—	—	—	A_{1g} ina;a,p B_{2u} ina; ina
ν_s XOX	E' a ⊥; a,dp	E' a ⊥; a, dp	E' a ⊥; a, dp E' a ⊥; a, dp	767	500 v.s.	740 s.	⊥	E_{1u} a ⊥; ina E_{2g} ina; a,dp
	A_1' ina; a, p	A' ina; a, p	A_1' ina; a, p A_2' ina; ina	—	—	—	—	A_{1g} ina; a,p B_{2u} ina; ina
X–O deformation (and M–O stretching) vibrations				496 461 385	— — —	544 w. 492 v.w. 437 m.	— — —	

Note: In the transition $C_{3h} \rightarrow D_{3h}$ each of the pair of complex-conjugate representations forming separately degenerate representations of types E' and E'' transform independently into the corresponding degenerate representations of the D_{3h} group.

Apart from those shown in the table, the spectrum of catapleiite contains the bands 3300 and 1604 cm^{-1} (νH_2O and δH_2O).

According to [176], the band at 1007 cm^{-1} in the reflection spectrum of catapleiite is approximately equally strong for orientations cut parallel and perpendicular to the optic axis. We may suppose that in the parallel case the great intensity of this band is associated with the superposition of the 1041 cm^{-1} band, which is strongest for a cut parallel to the optic axis.

Notation: a = active; ina = inactive, ⊥ and ∥ represent orientations of the dipole moment relative to the C_3 axis; p = polarized; dp = depolarized.

the same reason (the greater influence of the cation) that the width of the ν_{as} O^-SiO^- band is increased in the spectra of many silicates containing Si_3O_9 ions.

We may also attempt to use spectroscopic data in order to make a more reasonable choice between the space groups D_{6h}^4, D_{3h}^4, and C_{6v}^4 for catapleiite. If in fact we reject the last of these three groups, which contains special positions of multiplicity factor 2 with symmetry C_{3v}, considerably reducing the strictness of the selection rules, the choice between the other

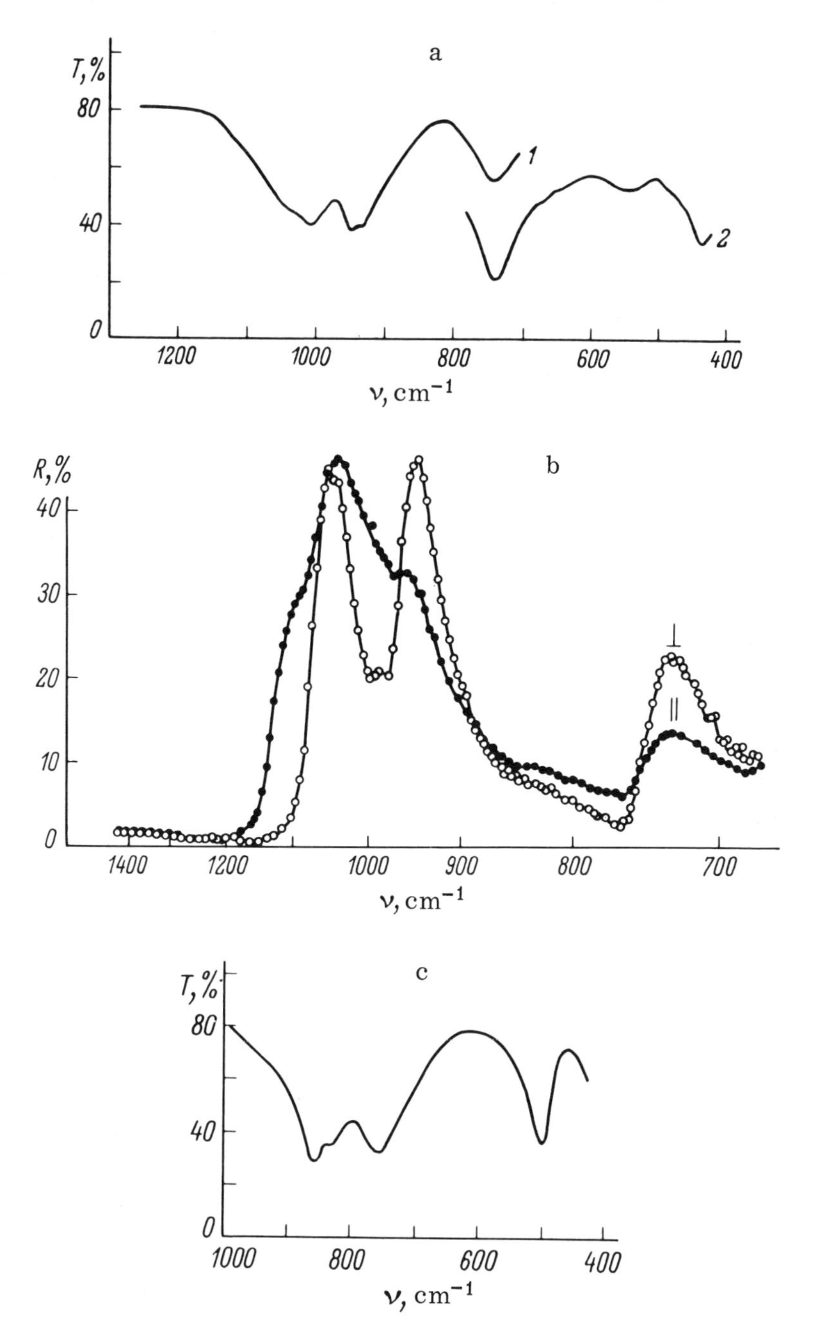

Fig. 32. Infrared spectra of silicates and germanates with rings comprising three tetrahedra. a) Absorption spectrum of catapleiite $Na_2Zr(Si_3O_9) \cdot 2H_2O$ (pressed in KBr): 1) 2 mg, 2) 5 mg; b) reflection spectrum of a catapleiite single crystal [176] (|| and ⊥ orientations of the electric vector of the incident radiation relative to the optic axis of the crystal); c) absorption spectrum of the high-temperature form of strontium metagermanate $Sr_3(Ge_3O_9)$ (pressed in KBr, 5 mg).

two groups follows immediately from the observed doublet structure of the $\nu_s O^-SiO^-(E')$ band. As indicated in Table 14, this is only possible for an anion site symmetry C_{3h} and a crystal symmetry $D_{3h}^4-P\bar{6}2c$, while in a crystal with symmetry D_{6h}^4 (site symmetry D_{3h}) only one of the Davydov components is active in the infrared spectrum. It should be noted that in this respect also the C_{6v}^4 group is improbable, since for the site group C_{3v} (the only one here possible) representations E_1 and E_2 of the factor group correspond to each degenerate irreducible representation of this group, and of these only one is active in the infrared spectrum.

The identification of the frequencies of the deformation vibrations, of which only a proportion fall into the part of the spectrum under consideration, is considerably more complicated. Saksena et al. [90] relate the highest frequency of the deformation vibrations of the ring to symmetry type A_2'' on the basis of frequency calculations with a simplified force field: $\nu(E') = 1082, 904, 766, 470, 404, 203$ cm^{-1}; $\nu(A_2'') = 1043, 567$ cm^{-1}. The absence of data relating to the orientation of the electric vectors of the vibrations in the frequency range below 700 cm^{-1} prevents a direct check of this proposition, while the approximate character of the calculation in [90], leading to certain contradictions with respect to experimental data even in the frequency range of the stretching vibrations, makes the values of the deformation frequencies particularly unreliable. If, in accordance with calculation, we relate the frequency 544 cm^{-1} in the catapleiite spectrum to the vibration $\rho O^-SiO^-(A_2'')$, then the vibration δSiOSi of type E′ corresponds to the band at 437 cm^{-1}, which agrees with the frequency of the analogous vibration of the trisiloxane ring of the molecule $[OSi(CH_3)_2]_3$ [180] (also see p. 267). We note, however, that in the spectrum of the phosphate ring $[P_3O_9]^{3-}$ some authors ascribe considerably lower frequencies to the δPOP vibrations [181].

The high-temperature form of strontium metagermanate $SrGeO_3$ contains (according to x-ray structural data) planar $[Ge_3O_9]^{6-}$ rings [182, 183]. The infrared spectrum of this compound (Fig. 32) in the frequency range above 400 cm^{-1} contains altogether four bands, which evidently correspond to four active stretching vibrations of the anion. The characteristic "ring band" (ν_s GeOGe of type E′) is situated at 500 cm^{-1} for the germanate, while the identification of the three high-frequency vibrations is less clear. The subsequent discussion of their assignment, and of assignments in the spectra of Si_3O_9 rings is given in Chap. IV using the spectra of solid solutions between silicates and germanates with X_3O_9. The obvious absence of any violations of the selection rules for D_{3h} symmetry (or C_{3h}) indicates a higher site symmetry than C_3, which arises from the fact that the crystal belongs to the $R\bar{3}-C_{3i}^2$ space group. We may suppose that the apparent absence of infringements of the selection rules is associated with the existence of a higher pseudosymmetry. In other words, for very slight deviations of the symmetry of the internal field from higher symmetry and comparatively weak and nondirectional (electrostatic) interactions between the ions, the observed vibrational spectrum may correspond to an approximate symmetry of the complex ion higher than its point symmetry in the crystal lattice.

The $[Si_4O_{12}]^{8-}$ and $[Si_6O_{18}]^{12-}$ Rings. These have been less studied spectroscopically than rings of three tetrahedra. The majority of rings of this type known from x-ray structural investigations exist in the crystals of silicates containing such elements as Al and Be. The number of materials having spectra capable of being considered qualitatively without allowing for lattice vibrations is therefore extremely small. In contrast to silicates with Si_3O_9 rings, for which a deviation from the planar configuration is improbable, in the Si_4O_{12} ring (the tetrahedral OSiO angles being preserved) the SiOSi angles must reach 160° in order to achieve a planar structure. Thus in many cases it is probable that the Si_4O_{12} and Si_6O_{18} rings will be nonplanar, with a reduction in symmetry from D_{nh} to $D_{(n/2)d}$ or C_{nv}, or an even greater reduction in symmetry. Thus in kainosite $Ca_2(Y, Ln)_2(Si_4O_{12})CO_3 \cdot H_2O$ the only symmetry element of the nonplanar Si_4O_{12} ring is a mirror plane [184].

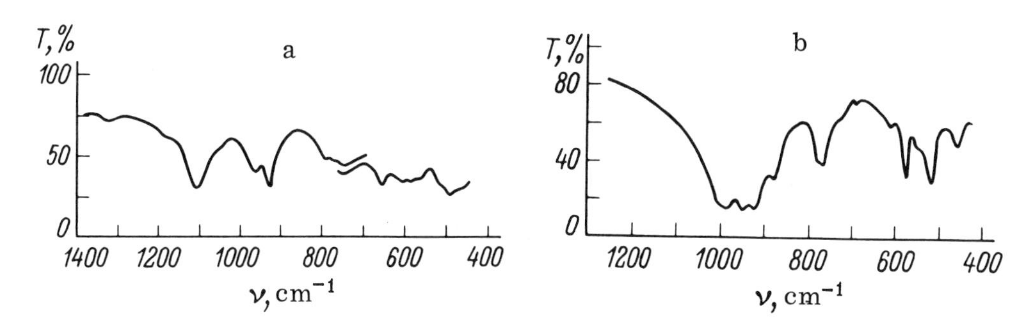

Fig. 33. Infrared spectra of some silicates with cyclic anions. a) Baotite $Ba_4(Ti,Nb)_6ClO_{16}(Si_4O_{12})$ (pressed in KBr, 8 mg); b) dioptase $Cu_6(Si_6O_{18})\cdot 6H_2O$ (pressed in KBr, 5 mg).

Si–O stretching vibrations in the Si_4O_{12} and Si_6O_{18} rings may as earlier be classified as ν_{as} and ν_s SiOSi and ν_{as} and ν_s O^-SiO^-. To each of these types of vibrations correspond two nondegenerate and one or two (for Si_6O_{18}) degenerate irreducible representations. However, for symmetry D_{nh} of the planar rings, each of the four types of vibrations contains only one representation active in the infrared spectrum. Hence the number of bands in the spectrum is not increased relative to the spectra of the Si_3O_9 rings, and any judgment as to the number of tetrahedra in the planar ring is only possible on the basis of the band frequencies. An increase in the SiOSi angle should in fact lead to a rise in the ν_{as} SiOSi frequency both on account of kinematic factors and possibly also dynamic factors (i.e., rise in the order of the Si–O–Si bonds). It is probable moreover that there will be a certain reduction in the order of the $Si{-}O^-$ bonds. Kinematic effects should decrease the frequency of the ν_s SiOSi vibrations of the planar rings as the dimensions of the ring increase: for the value of force constants used by Saksena et al. [90] the frequency of the degenerate ν_s SiOSi vibration is 766, 684, and 602 cm^{-1}, respectively, for the rings Si_3O_9, Si_4O_{12}, and Si_6O_{18}.

The only silicate with an Si_4O_{12} anion that has been studied spectroscopically [185] is baotite $Ba_4(Ti, Nb)_8ClO_{16}(Si_4O_{12})$. According to x-ray structural data [186] a planar configuration of the rings is probable, although their site symmetry in the lattice of a crystal with space group $I4_1/a{-}C_{4h}^6$ is only S_4. In the infrared spectrum of baotite (Fig. 33a), in the frequency range above 800 cm^{-1} there are only three strong absorption bands, of which (in accordance with what has already been said) that of highest frequency is probably associated with the ν_{as} SiOSi vibration and the two others with the vibrations ν_{as} and ν_s O^-SiO^-. Except for a weak band at 1040 cm^{-1}, in this part of the spectrum there are no signs of an increase in the number of bands (for symmetry S_4, among the vibrations of the types indicated six are potentially active in the infrared spectrum). In the lower-frequency range the baotite spectrum contains a large number of bands (see Table 16), which cannot be explained by vibrations of the ring. In the baotite lattice, in addition to Si atoms, there are also other atoms capable of forming substantially covalent bonds with oxygen: Ti(Nb) and Cl. In view of the large distances between the Cl atoms and the closest of the O atoms in the present case it is hard to imagine the presence of covalent Cl–O bonds, while this is very probable for titanium–oxygen octahedra. In the majority of silicates the frequencies of the titanium–oxygen octahedra are no greater than 500 to 600 cm^{-1}, including the $BaTi(Si_3O_9)$ silicate just considered (see also p. 165). One of the reasons for the comparatively low Ti–O frequencies in these cases is interaction with the vibrations of Si–O bonds formed by the same oxygen atoms. In baotite, as in titanates, this factor is excluded, and the formation of Ti⟨O,O⟩Ti bonds in the condensation of the octahedra by way of common edges leads

Table 16. Frequencies of the Absorption Maxima in the IRS of Baotite

Frequency, cm^{-1}, and intensity	Assignment	Frequency, cm^{-1}, and intensity	Assignment
(1325 v.w.)	—	745 s.	ν_s SiOSi, νTi—O
(1170 v.w.)	—	656 s.	
1106 v.s.	ν_{as} SiOSi(E_u)	603 s.	
(1040 v.w.)	—	588 s.	
962 s.	ν_{as} O^-SiO^-(A_{2u}); ν_s O^-SiO^-(E_u)	568 s.	
926 v.s.		517 s.	Si–O deformation and M–O stretching vibrations
789 m.	ν_s SiOSi, νTi—O	491 v.s.	
770 w.		470 v.s.	

to a rise in some of the vibrational frequencies. The partial replacement of titanium by niobium acts in the same direction. Hence the existence of a large number of bands in the frequency range below 800 cm^{-1} may be associated with the Ti(Nb)–O vibrations. Thus with respect to the lattice vibrations baotite should be considered as a titanosilicate rather than a silicate of titanium.

Evidently still more mixing of the vibrations occurs in spectra of the series of silicates with planar Si_6O_{18} anions, both as a result of the presence of B, Be, and Al cations and as a result of the partial replacement of Si by Al in the ring. It is probable that the ascription of a considerably higher frequency [66, 175, 176] than would follow from calculation to the so-called "ring band" (ν_s SiOSi) is associated with the impossibility of clearly distinguishing the frequencies of the complex anion, or at least recognizing a severe displacement of these frequencies, in the spectra of these silicates. However, since the influence of interaction with the lattice vibrations affects the higher frequencies of the complex anion less, we may suppose that the appearance of a very high frequency band at 1180 to 1200 cm^{-1} is largely associated with special features in the kinematics (and dynamics) of the actual anion. In the reflection spectra of beryl $Al_2Be_3(Si_6O_{18})$ and vorobyevite $CsAl_2Be_3[(Si, Al)_6O_{18}]$, this band is considerably stronger when the electric vector is oriented in the plane normal to the optic axis, which justifies us in assigning it to vibrations of type E_{1v}(ν_{as} SiOSi). Thus in contrast to rings composed of three tetrahedra the frequency ν_{as} O^-SiO^- is not the highest of the frequencies of the internal vibrations of the six-member ring; as mentioned earlier, this may be explained as being due to the rise in the frequency ν_{as} SiOSi when the angle ∠SiOSi approaches 180°.

Worthy of more detailed attention is the spectrum of dioptase $Cu_6(Si_6O_{18}) \cdot 6H_2O$, in which there are no grounds for expecting such substantial changes in the internal vibrations of the anion under the influence of lattice vibrations. In the infrared spectrum of dioptase (Fig. 33b) we notice the absence of bands with frequencies above 1000 cm^{-1} and the appearance of a fairly strong doublet at 773 to 780 cm^{-1}. Both these features of the spectrum may be satisfactorily explained by a nonplanar structure of the Si_6O_{18} ring with fairly small SiOSi angles [187, 188]: the frequency ν_{as} SiOSi in this kind of ring should fall considerably in comparison with a planar ring of the same composition, while the frequency ν_s SiOSi should rise. The intrinsic (and site) symmetry of the $[Si_6O_{18}]^{12-}$ ion in the dioptase crystal is $C_{3i} \equiv S_6$. The number of stretching vibrations allowed by the selection rules in the infrared spectrum is in this case doubled as compared with a planar ring, four vibrations having a transition moment in the direction of the S_6 axis and four others perpendicular to the latter (Table 17). The use of data relating to the reflection spectrum of dioptase single crystals [176] enables us to include the frequencies 992, 952, and 923 cm^{-1} and at least one component of the doublet 780 to 773 cm^{-1} among the number of vibrations with a vector normal to the optic axis. Polarized along the optic axis we

Table 17. Stretching Vibrations of the Si_6O_{18} Ions

Assignment	Symmetry types of the vibrations and selection rules		IRS of dioptase $Cu_6(Si_6O_{18})\cdot 6H_2O$	
	D_{6h}	S_6	absorption spectrum of the polycrystalline sample	polarization of bands (with respect to the S_6 axis) in the reflection spectrum [176]
ν_{as}SiOSi	B_{1u} ina; ina	A_u a $\|$; ina	1003 v.s.	?
	E_{1u} a $\perp$; ina	E_u a $\perp$; ina	992 v.s.	$\perp$
	A_{2g} ina; ina	A_g ina; a, p	—	—
	E_{2g} ina; a, dp	E_g ina; a, dp	—	—
$\nu_{as}O^-SiO^-$	A_{2u} a $\|$; ina	A_u a $\|$; ina	—	—
	E_{2u} ina; ina	E_u a $\perp$; ina	952 v.s.	$\perp$
	B_{1g} ina; ina	A_g ina; a, p	—	—
	E_{1g} ina; a, dp	E_g ina; a, dp	—	—
$\nu_s O^-SiO^-$	E_{1u} a $\perp$; ina	E_u a $\perp$; ina	923 v.s.	$\perp$
	B_{2u} ina; ina	A_u a $\|$; ina	881 s.	$\|$
	E_{2g} ina; a, dp	E_g ina; a, dp	(827 v.w.)	—
	A_{1g} ina; a, p	A_g ina; a, p	—	—
ν_s SiOSi	E_{1u} a $\perp$; ina	E_u a $\perp$; ina	780 } s.	$\perp$?
	B_{2u} ina; ina	A_u a $\|$; ina	773 }	$\|$
	E_{2g} ina; a, dp	E_g ina; a, dp	(734 w.)	—
	A_{1g} ina; a, p	A_g ina; a, p	(692 w.)	—
	—	—	(612 w.)	—
Deformation vibrations of Si_6O_{18}			576 s.	—
			558 w.	—
			518 s.	—
			458 m.	—

Note: In parentheses are the frequencies of very weak maxima, the interpretation of which is not clear. Frequencies νH_2O and δH_2O are not shown in the table.

have a vibration with frequency 881 cm^{-1} and, judging from the considerable displacement of the reflection maximum near 780 cm^{-1} for different orientations of the crystal, one of the components of the 780- to 773-cm^{-1} doublet. The band at 1003 cm^{-1} in the reflection spectrum is not resolved; however, the change in the shape of the neighboring band at 992 cm^{-1} suggests that the frequency of 1003 cm^{-1} corresponds to a third vibration with a vector parallel to S_6. Thus we can identify seven of the expected eight stretching vibrations; their assignment is given in Table 17; however, without checking by calculation it is hard to decide whether we should not reverse the assignments of frequencies $\nu_{as} O^-SiO^-$ and ν_{as} SiOSi of type E_u.

Metaphosphates with Ring Anions. These have been frequently studied by x-ray diffraction and by the methods of infrared spectroscopy, Raman spectra, and nuclear magnetic resonance. It is therefore interesting to compare the spectra of the metaphosphates with the spectra of the corresponding silicates, the more so in view of the fact that the cyclic metaphosphates may be produced and studied in aqueous solution as well as in crystal form. The differences in the spectra of the metasilicate and metaphosphate rings are determined not so much by the slight difference in the masses of Si and P as by the tendency of the phosphates to form structures with smaller valence angles at the oxygen atoms in the bridge bonds P–O–P. The considerable differences in the order of the bonds of types $P–O^-$ and P–O(P), greater than for corresponding bonds in silicates (already mentioned in our discussion of the spectra of pyrophosphates), become still more substantial in rings: for the two bonds of type P–O(P) and two $P–O^-$ formed by each phosphorus atom, the order of the latter may exceed 1.5. For the structures of cyclic phosphates solved by x-ray diffraction [189-194], an estimate of the orders of the π bonds on the assumption of a linear dependence on interatomic distances [9] gives about 0.65 for the $P–O^-$ and 0.25 to 0.35 for P–O(P) bonds. The existence of a certain double-bonded character in the rings suggests that there may be some alternation of the bonds in the ring [195], which corresponds to a reduction in symmetry, while the comparatively small values of the POP angles raise the possibility of a nonplanar configuration even for the trimetaphosphate ring. The symmetry of this ring may clearly lose not only the plane σ_h normal to the C_3 axis but also this axis itself under the influence of the two factors mentioned. This leads to the possibility of there being different conformations of the ring, and with this we may associate, for example, the existence of three crystal forms of anhydrous sodium trimetaphosphate [196]. For a nonplanar P_3O_9 ring there may be two fundamentally differing conformations: one with symmetry C_{3v} (the cis form or "boat" shape), in which all the P atoms lie on one side of the plane formed by the bridge O atoms, and the other with symmetry C_s (trans form or "armchair" shape), in which two P atoms lie on one side and one on the other side of the plane of oxygen atoms. For the tetrametaphosphate ring the number of possible conformations is much larger; the most important of these are the configurations with symmetry C_{4v} (boat), C_{2h} (armchair, two neighboring P atoms above the plane of the O bridge atoms and two others below it), and D_{2d}, or S_4, the P atoms arranged alternately above and below the plane of the O atoms. Other less symmetrical configurations may be derived from the three given by eliminating certain symmetry elements.

Of the conformations of the metaphosphate rings indicated, only some have been identified by x-ray diffraction in the structures of crystals. Some arguments in favor of the existence of various conformations of tri- and tetrametaphosphate rings in solutions are advanced in Van Wazer's book [196]. Here we shall only consider data relating to spectroscopic investigations in order to estimate the possibility of using these methods for discussing the structure of rings.

The infrared spectrum of anhydrous sodium trimetaphosphate has been studied many times [148, 197, 198]; so has the Raman spectrum of the trimetaphosphate anion in aqueous solution [181, 199, 200]. Figure 34 represents a schematic diagram of the infrared spectrum of crys-

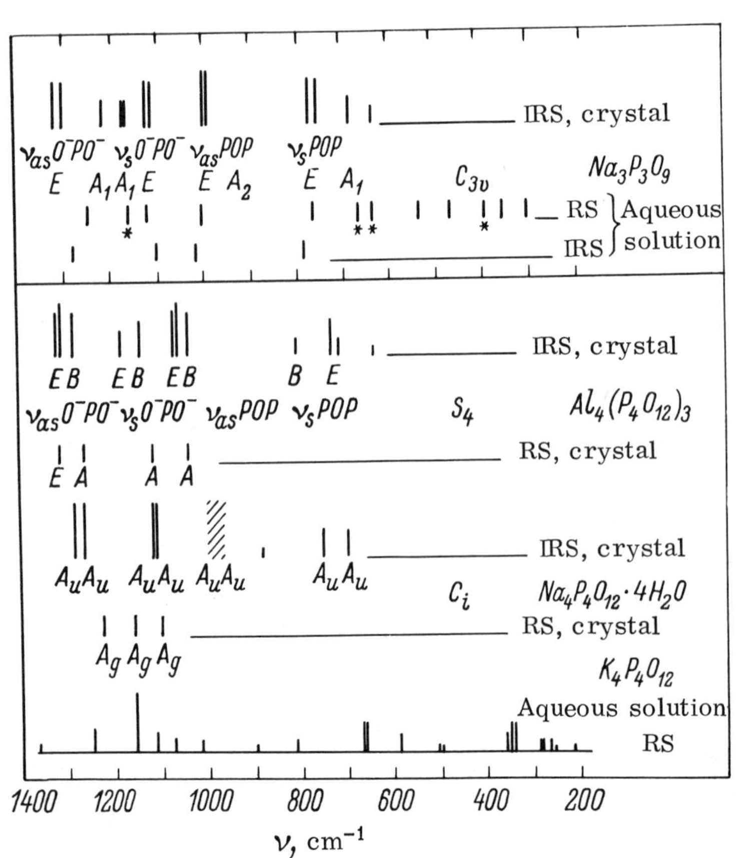

Fig. 34. Frequencies in the infrared and Raman spectra of cyclic metaphosphate anions. For $Na_3P_3O_9$ with a P_3O_9 ring devoid of a three-fold axis, C_{3v}-group symmetry types are arbitrarily shown. The polarized lines in the Raman spectrum are indicated by an asterisk. In the spectra of the tetrametaphosphates the symmetry types of the vibration are shown for the intrinsic symmetry of the anion.

talline $Na_3P_3O_9$, together with the assignment [198, 200] of the stretching frequencies of the ring, based on the assumption of C_{3v} symmetry. As in the spectra of the cyclic silicate anions, an approximate classification of the vibrations with respect to the motions of the O^-PO^- and POP groups (ring bonds) is appropriate. The frequencies of the vibrations of the O^-PO^- groups considerably exceed the corresponding frequencies in the spectra of the silicates, and also the vibration frequencies of the PO_3 groups in pyrophosphates; this is clearly associated with an increase in the multiplicity factor of the $P-O^-$ bonds in the metaphosphates. All the degenerate vibrations of the ring in the spectrum of the crystal are split; this may be due both to a lower-than-C_{3v} site symmetry of the anion in the crystal and also to its lower intrinsic symmetry. The latter cause seems the more likely, since the splitting of the degenerate vibrations is comparatively large. An x-ray diffraction study of the structure of $Na_3P_3O_9$ (orthorhombic form) confirms this assumption: the crystals belong to the Pmcn–D_{2h}^{16} space group. The four P_3O_9 rings are arranged in a set with symmetry C_s but their configuration is close to the more symmetrical C_{3v}, while the conformation corresponds to the "boat" type [191]. The average value of ∠POP in the ring is about 127°, which agrees with the lengths of the ring bonds, 0.13 Å longer than the $P-O^-$. Passing from the site group to the factor group of the crystal suggests type A′ anion–vibration splitting, which is only observed experimentally for one vibration, $\nu_s\, O^-PO^-$.

Steger studied the infrared and Raman spectra of a solution of $Na_3P_3O_9$ in H_2O and D_2O [200] and came to the conclusion that the selection rules were satisfied for D_{3h} symmetry (Table 18), i.e., that the P_3O_9 ion had a plane configuration in aqueous solution.

Data relating to the polarization of the Raman lines supports the identification of some of the stretching vibrations, including those in the spectrum of the crystal. No reasonably reliable explanation has been found for the presence of two polarized lines in the region of 665 to 635 cm^{-1}: whereas one of these probably corresponds to the totally symmetric ring stretch (ν_s POP of type A_1'), the second may be ascribed to either the vibration δ_sPOP of the same type of symmetry [198] or else to the vibration δO^-PO^- [200], or even to Fermi resonance with one of the overtones of the lower-frequency vibrations.

The spectra of the tetrametaphosphates differ little from the spectra of the trimetaphosphates [148]. The differences between them are of the same order as between the spectra of crystals of different tetrametaphosphates. X-ray-diffraction solutions of the structures of crystalline tetrametaphosphates have been carried out for $Al_4(P_4O_{12})_3$ [189], $(NH_4)_4P_4O_{12}$ [190] and two forms of

Table 18. Stretching Vibration Frequencies of the P_3O_9 Ion in Solution

Vibration	Symmetry types	Frequencies, cm^{-1}	
		IRS	RS
$\nu_{as}\, O^-PO^-$	A_2''	1277	—
	E''	—	1243
$\nu_s\, O^-PO^-$	A_1'	—	1158
	E'	1098	1119
ν_{as}POP	A_2'	—	—
	E'	1015	1000
ν_sPOP	E'	783	763
	A_1'	—	634

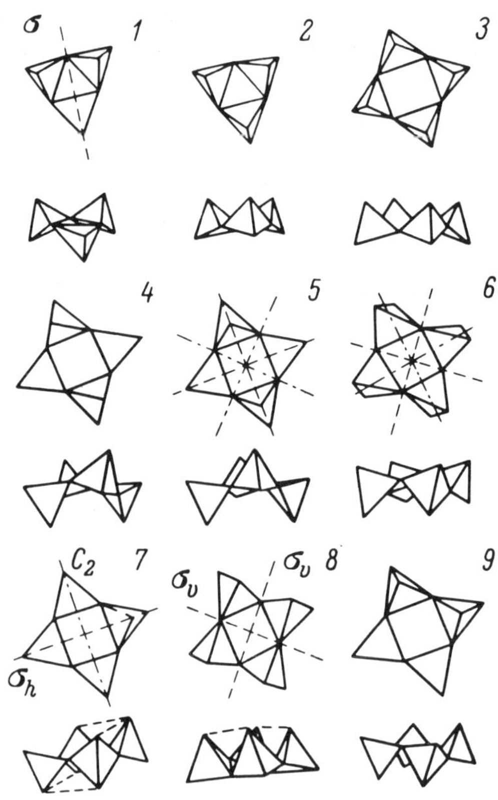

Fig. 35. Some of the possible configurations of the tri- and tetrametaphosphate anions. 1) C_s, $NH_4MgP_3O_9$ (orthorhombic)*; 2) C_{3v}, $Na_3P_3O_9$ (orthorhombic)*; 3) C_{4v}*; 4) S_4, $Al_4(P_4O_{12})_3$*; 5) D_{2d} (V_d−I), $Cu_2P_4O_{12}$**; 6) D_{2d} (V_d−II), $Mg_2P_4O_{12}$**; 7) C_{2h}, $(NH_4)_4P_4O_{12}$*; 8) C_{2v}, $Na_4P_4O_{12}$**; 9) C_i, $Na_4P_4O_{12} \cdot 4H_2O$*.
*Indicates structures established by x-ray diffraction method.
**Structures proposed on the basis of spectroscopic research.

$Na_4P_4O_{12} \cdot 4H_2O$ [193, 194].† On the basis of Steger's study of the RS [201] and IRS [202] of crystalline tetrametaphosphates, attempts have been made to predict the configuration of the ring in a number of crystals the structure of which has never been solved by x-ray diffraction. It is interesting to note in particular that "armchair" conformation with intrinsic symmetry C_i (or C_{2h}) was predicted for $Na_4P_4O_{12} \cdot 4H_2O$ long before its x-ray investigation, simply on the basis of data relating to the selection rules for the stretching vibrations of the ring. Figure 35 gives a schematic picture of the different configurations of the P_3O_9 and P_4O_{12} rings, and indicates in what crystals they have been established by x-ray structural or spectroscopic data [202].‡

† The earlier view that there was a connection between the existence of triclinic and monoclinic forms of $Na_4P_4O_{12} \cdot 4H_2O$ and the different conformations of the P_4O_{12} ring have not been justified: the symmetry of the rings in both cases is C_i, and their configurations differ little, although in the monoclinic form the ring has an obvious C_{2h} pseudosymmetry.

‡ According to the results of E. V. Poletaev [202a, 202b], who studied the infrared and Raman spectra of a number of tetrametaphosphates, in $Mg_2P_4O_{12}$ it is more probable that there is a centrosymmetric configuration of the anion, similar to that established by x-ray diffraction in $(NH_4)_4P_4O_{12}$ (ion symmetry C_{2h}).

Let us give some more detailed consideration to the assignment of the frequencies of the P–O stretching vibrations in the spectra of two tetrametaphosphates distinguished both by the shape of the ring and by the nature of the cation. Aluminum tetrametaphosphate contains P_4O_{12} rings possessing intrinsic (and site) symmetry S_4 in a crystal belonging to the $I\bar{4}3d-T_d^6$ space group. For symmetry S_4 each of the four quasi-characteristic groups of stretching vibrations (ν_{as} and $\nu_s\,O^-PO^-$ and ν_{as} and ν_s POP) gives two vibrations of symmetry types E and B, active in the infrared spectrum. In the Raman spectrum a third vibration of type A is active and also a vibration of type E. The primitive cell of the crystal contains six such rings, and in order to determine the selection rules for the Davydov components of the internal vibrations of the anions we may use the scheme of correlations between the irreducible representations of the site and factor group (the symmetry species printed in bold type are active in the infrared spectrum).

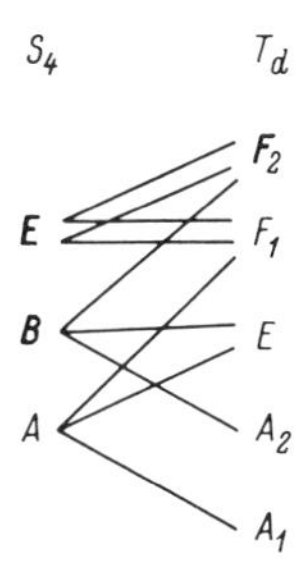

It is clear from this that no representation active in the infrared spectrum corresponds to a vibration of type A, one corresponds to a vibration of type B, and two vibrations of type F_2, not necessarily coinciding in frequency, correspond to a vibration of type E. The presence of Al^{3+} cations, capable of a partly covalent interaction with the anions, makes the coupling between the latter fairly strong and directional, so that there may be considerable splitting of the degenerate vibration. After these preliminary comments there is no difficulty in identifying the frequencies in the infrared spectrum shown in Fig. 34 (from experimental data of [148, 202]): all the vibrations of type E except one are in fact slightly split. Of the four vibrations observed in the Raman spectrum, three belong to type A and one to type E.

In the spectrum of the monoclinic form of $Na_4P_4O_{12} \cdot 4H_2O$ all the frequencies of the P–O stretching vibrations are slightly lower than in $Al_4(P_4O_{12})_3$, where they are probably raised as a result of interaction with the comparatively high frequencies of the lattice vibrations. The P_4O_{12} rings with symmetry C_i are situated on inversion centers (set with multiplicity factor 2) of a crystal with space symmetry $P2_1/a-C_{2h}^5$ [193]. Hence each of the internal vibrations of the anions gives two vibrations of the crystal.

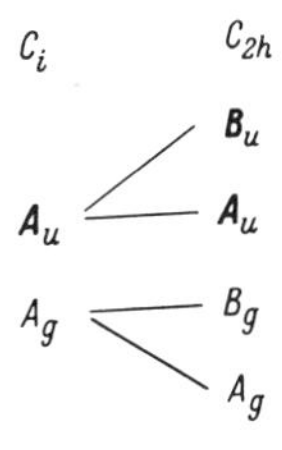

However, in the infrared spectrum, in each of the frequency ranges for ν_{as} and $\nu_s\,O^-PO^-$, ν_{as} and ν_s POP, there are only two bands corresponding to two A_u vibrations expected for the intrinsic symmetry of the anion (Fig. 34): each of the four types of stretching vibrations of the anion include two vibrations of type A_u and two of A_g for symmetry C_i. Both frequencies of type A_g are observed in the Raman spectrum for the $\nu_s\,O^-PO^-$ vibrations, but of the $\nu_{as}\,O^-PO^-$ vibrations only one is recorded, and of the ν_{as} POP vibrations none at all, these being usually of low intensity in the Raman spectrum. We note that the absence of Davydov splitting of the internal vibrations of the anion in the spectrum of the crystal, associated with the relative weakness of the interactions between the ions for alkali cations, is typical both for $Na_4P_4O_{12} \cdot 4H_2O$ and also for the majority of the frequencies of $Na_3P_3O_9$ considered earlier. Hence the appearance of doublets in the infrared spectrum in these cases is a direct indication of the removal of the degeneracy as a result of the reduction in the intrinsic symmetry of the anion, i.e., it may serve as a method of conformation analysis (see also p. 270).

The assignment of the frequencies of the deformation vibrations of the tetrametaphosphate anion, and equally the totally symmetric ("pulsation") vibration of the ring, still remains unsolved. The structure of the P_4O_{12} anion in aqueous solution has been studied by means of Raman spectra in [203]. It is suggested that the anion has symmetry C_{2h} and is similar to the P_4O_{12} ring in the $(NH_4)_4P_4O_{12}$ crystal.

References

1. C. Eaborn, Organosilicon Compounds, Butterworths, London (1960).
2. W. Noll, Chemie und Technologie der Silicone, Verl. Chemie, Weinheim (1960).
3. E. A. V. Ebsworth, Volatile Silicone Compounds, McMillan Co., New York (1963).
4. L. H. Sommer, Stereochemistry, Mechanism, and Silicon, McGraw-Hill Co., New York (1965).
5. V. Schomaker and D. P. Stevenson, J. Amer. Chem. Soc., 63:37 (1941).
6. W. Moffitt, Proc. Roy. Soc., A200:409 (1950).
7. G. Dzhaffe, Usp. Khim., 26:1060 (1957).
8. D. P. Craig, A. Maccoll, R. S. Nyholm, L. E. Orgel, and L. E. Sutton, J. Chem. Soc. (L.), 332 (1954).
9. D. W. J. Cruickshank, J. Chem. Soc. (L.), 5486 (1961).
10. K. Yamasaki, A. Kotera, M. Yokoi, and Y. Ueda, J. Chem. Phys., 18:1414 (1950).
11. M. Yoki, Bull. Chem. Soc. Japan, 30:100, 106 (1957).
12. A. Almenningen, O. Bastiansen, V. Ewing, K. Hedberg, and M. Traetteberg, Acta Chem. Scand., 17:2455 (1963).
13. E. H. Aggarwal and S. H. Bauer, J. Chem. Phys., 18:42 (1950).
14. R. M. Badger, J. Chem. Phys., 2:128 (1934): 3:710 (1935).
15. H. Siebert, Z. Anorgan. und Allgemein. Chem., 273:170 (1953).
16. J. Goubeau, Angew. Chemie, 69:77 (1957).
17. H. Siebert, Z. Anorgan. und Allgemein. Chem., 275:225 (1954).
18. E. A. Robinson, Canad. J. Chem., 41:3021 (1963).
19. R. J. Gillespie and E. A. Robinson, Canad. J. Chem., 42:2496 (1964).
20. M. A. Porai-Koshits and S. P. Ionov, Zh. Strukt. Khim., 5:474 (1964).
21. E. L. Wagner, J. Chem. Phys., 37:751 (1962).
22. D. W. J. Cruickshank and E. A. Robinson, Spectrochim. Acta, 22:555 (1966).
23. H. Kriegsmann, Z. Elektrochem., 61:1088 (1957).
24. D. W. Scott, J. F. Messerly, S. S. Todd, G. B. Guthrie, I. A. Hossenlopp, R. T. Moore, Ann Osborn, W. T. Berg, and J. P. McCullough, J. Phys. Chem., 65:1320 (1961).
25. A. N. Lazarev, Kr. Poiker, and T. F. Tenisheva, Dokl. Akad. Nauk SSSR 175:1322 (1967).
25a. A. N. Lazarev, Kr. Poiker, and L. L. Shchukovskaya, Izv. Akad. Nauk SSSR, Neorg. Mat., 3:2021 (1967).
25b. A. N. Lazarev, Kr. Poiker, and E. V. Kukharskaya, Izv. Akad. Nauk SSSR, Neorg. Mat., 3:2029 (1967).
25c. Kr. Poiker and A. N. Lazarev, Izv. Akad. Nauk SSSR, Neorg. Mat., 4:1513 (1968).
25d. Kr. Poiker and A. N. Lazarev, Izv. Akad. Nauk SSSR, Neorg. Mat., 4:1716 (1968).
25e. A. N. Lazarev and Kr. Poiker, Izv. Akad. Nauk SSSR, Neorg. Mat., 4:1720 (1968).
25f. T. F. Tenisheva and A. N. Lazarev, Izv. Akad. Nauk SSSR, Neorg. Mat., 4:1952 (1968).
26. R. Forneris and E. Funk, Z. Elektrochem., 62:1130 (1958).

27. A. L. Smith, Spectrochim. Acta, 19:849 (1963).
28. H. Murata, J. Chem. Phys., 20:347 (1952).
29. J. Goubeau and H. Behr, Z. Anorgan. und Allgemein. Chem., 272:2 (1953).
30. A. A. V. Stuart, C. La Lau, and H. Breederveld, Recueil Trav. Chim., 74:747 (1956).
31. I. F. Kovalev, Opt. i Spekt., 6:594 (1959).
32. I. F. Kovalev, in: Physical Problems of Spectroscopy, Vol. 1, Izd. AN SSSR, Moscow (1962), p. 360.
33. I. F. Kovalev, Opt. i Spekt., 12:550 (1962).
34. P. P. Shorygin and E. M. Popov, Zh. Fiz. Khim., 38:1429 (1964).
35. M. I. Batuev, M. F. Shostakovskii, V. I. Belyaev, A. D. Matveeva, and E. V. Dubrova, Dokl. Akad. Nauk SSSR, 95:531 (1954).
36. V. I. Kasatochkin, M. F. Shostakovskii, O. I. Zil'berbrand, and D. A. Kochkin, Izv. Akad. Nauk SSSR, Ser. Fiz., 18:726 (1954).
37. W. Tatlock and E. G. Rochow, J. Organ. Chem., 17:1555 (1952).
38. J. F. Hyde, J. Amer. Chem. Soc., 75:2166 (1953).
39. E. Richards and H. W. Thompson, J. Chem. Soc. (L.), 124 (1949).
40. M. Kakudo, N. Kasai, and T. Watase, J. Chem. Phys., 21:1894 (1953).
41. R. West and R. H. Baney, J. Amer. Chem. Soc., 81:6145 (1959).
42. Ya. I. Ryskin and M. G. Voronkov, in: Chemistry and Practical Application of Organosilicon Compounds, No. 3, Izd. TsBTI, Leningrad (1958).
43. Ya. I. Ryskin, Opt. i Spekt., 4:532 (1958).
44. Ya. I. Ryskin, M. G. Voronkov, and Z. I. Shabarova, Izv. Akad. Nauk SSSR, Otd. Khim. Nauk, 1019 (1959).
45. Ya. I. Ryskin, Opt. Spektr., 7:278 (1959).
46. L. Allred, E. G. Rochow, and F. G. A. Stone, J. Inorg. and Nucl. Chem., 2:416 (1956).
47. K. Licht and H. Kriegsmann, Z. Anorgan. und Allgemein. Chem., 323:190 (1963).
48. A. N. Lazarev, in: Structural Transformations in Glasses at High Temperatures, Izd. Nauka, Moscow–Leningrad (1965), p. 233.
49. C. M. Huggins, J. Phys. Chem., 65:1881 (1961).
50. R. West, L. S. Whatley, and K. J. Lake, J. Amer. Chem. Soc., 83:761 (1961).
51. R. C. Lord, D. W. Robinson, and W. C. Schumb, J. Amer. Chem. Soc., 78:1327 (1956).
52. R. A. Wegener and B. Post, Amer. Cryst. Assoc. Program and Abstr., Milwaukee (1958), Paper A-2 (cited by G. S. Smith and L. E. Alexander, Acta Crystallogr., 16:1015 (1963)).
53. A. N. Lazarev, M. G. Voronkov, and T. F. Tenisheva, Opt. i Spektr., 5:365 (1958).
54. T. F. Tenisheva, A. N. Lazarev, and M. G. Voronkov, in: Chemistry and Practical Application of Organosilicon Compounds, No. 3, Izd. TsBTI, Leningrad (1958).
54a. J. E. Griffiths and D. F. Sturman, Spectrochim. Acta, 25A:1415 (1969).
55. D. C. McKean, Spectrochim. Acta, 13:38 (1958).
56. H. Kriegsmann, Z. Anorgan. und Allgemein. Chem., 299:78 (1959).
57. H. Kriegsmann and K. Engelgardt, Z. Anorgan. und Allgemein. Chem., 310:320 (1961).
58. A. Almenningen, K. Hedberg, and R. Seip, Acta Chem. Scand., 17:2264 (1963).
59. H. R. Linton and E. R. Nixon, J. Chem. Phys., 29:921 (1958).
60. M. I. Batuev, V. A. Ponomarenko, A. D. Matveeva, and A. D. Petrov, Dokl. Akad. Nauk SSSR, 95:705 (1954).
61. I. P. Vaganova and I. F. Kovalev, Opt. i Spektr., 17:960 (1964).
62. J. Goubeau and W. D. Hiersemann, Z. Anorgan. und Allgem. Chem., 290:292 (1957).
63. T. D. Goldfarb and S. Sei, J. Amer. Chem. Soc., 86:1679 (1964).
64. A. N. Lazarev, T. F. Tenisheva, and V. P. Davydova, Dokl. Akad. Nauk SSSR, 158:648 (1964).
65. C. Schaefer, F. Matossi, and K. Wirtz, Z. Phys., 89:210 (1934).
66. F. Matossi and H. Kruger, Z. Phys., 99:1 (1936).
67. F. Matossi and O. Bronder, Z. Phys., 111:1 (1938).

68. F. Matossi, Helv. Phys. Acta, 11:469 (1938).
69. Ya. I. Gerlovin, Dokl. Akad. Nauk SSSR, 38: 136 (1943).
70. Ya. I. Gerlovin, Dokl. Akad Nauk SSSR, 38:186 (1943).
71. V. A. Florinskaya and R. S. Pechenkina, Dokl. Akad. Nauk SSSR, 89:57 (1953).
72. V. A. Florinskaya, Dokl. Akad. Nauk SSSR, 89:261 (1953).
73. V. A. Florinskaya and R. S. Pechenkina, in: Structure of Glass, Izd. AN SSSR, Moscow-Leningrad (1955), p. 80.
74. B. D. Saksena, Proc. Indian Acad. Sci., 12A:93 (1940).
75. B. D. Saksena, Proc. Indian Acad. Sci., 16A:270 (1942).
76. B. D. Saksena, Proc. Indian Acad. Sci., 19A:357 (1944).
77. B. D. Saksena, Proc. Indian Acad. Sci., 22A:379 (1945).
78. B. D. Saksena, Proc. Indian Acad. Sci., 30A:308 (1949).
79. B. D. Saksena, Proc. Phys. Soc., 72:9 (1958).
80. J. Barriol, J. Phys. et Radium, 7:209 (1946).
81. F. Matossi, J. Chem. Phys., 17:679 (1949).
82. M. V. Vol'kenshtein, M. A. El'yashevich, and B. I. Stepanov, Vibrations of Molecules, Vols. 1 and 2, GITTL, Moscow-Leningrad (1949).
83. E. B. Wilson, J. C. Decius, and P. C. Cross, Theory of the Vibrational Spectra of Molecules (Molecular Vibrations: The Theory of Infrared and Raman Spectra, McGraw-Hill, N.Y. (1955)).
84. L. S. Mayants, Theory and Calculation of the Vibrations of Molecules, Izd. AN SSSR, Moscow (1960).
85. B. I. Stepanov and A. M. Prima, Opt. i Spektr., 4:734 (1958).
86. B. I. Stepanov and A. M. Prima, Opt. i Spektr., 5:15 (1958).
87. A. M. Prima, Trans. Inst. Physics and Mathematics of the Academy of Sciences of the Belorussian SSR, No. 2, Izd. AN BSSR, Minsk (1957), p. 124.
88. A. M. Prima, Opt. i Spektr., 9:452 (1960).
89. B. D. Saksena, Trans. Faraday Soc., 57:242 (1961).
90. B. D. Saksena, K. S. Agarwal, and G. S. Jauchri, Trans. Faraday Soc., 59:276 (1963).
91. O. Theimer, Monatsh. Chem., 81:424 (1950).
92. A. S. Davydov, Theory of the Absorption of Light in Molecular Crystals, Izd. AN Ukr. SSR, Kiev (1951).
93. J. G. Kirkwood, J. Chem. Phys., 7:506 (1939).
94. S. E. Whitcomb, H. H. Nielsen, and L. H. Thomas, J. Chem. Phys., 8:143 (1940).
95. B. I. Stepanov, Zh. Fiz. Khim., 14:474 (1940).
96. A. I. Ansel'm and N. N. Porfir'eva, Zh. Éksp. Teor. Fiz., 19:438 (1949).
97. T. Shimanouchi and S. Mizushima, J. Chem. Phys., 17:1102 (1949).
98. L. Kellner, Proc. Phys. Soc. (L), 64A:521 (1951).
99. L. I. Vidro and B. I. Stepanov, Dokl. Akad. Nauk SSSR, 82:557 (1952).
100. B. I. Stepanov and O. P. Girin, Trudy Gos. Opt. Inst., 23(139):3 (1953).
101. P. Higgs, Proc. Roy. Soc. (L.), 220A:472 (1953).
102. T. Miyazawa, J. Chem. Phys., 35:693 (1961).
103. R. Signer and J. Weiler, Helv. Chim. Acta, 16:115 (1932).
104. J. Weiler, Z. Phys., 80:617 (1933).
105. B. Thosar and R. Bapat, Z. Phys., 109:472 (1938).
106. J. Duchesne, J. Chem. Phys., 16:1009 (1948).
107. H. Murata, J. Chem. Phys., 20:1184 (1952).
108. R. Okawara, Bull. Chem. Soc. Japan, 31:154 (1958).
109. K. Iguchi, J. Chem. Phys., 22:1937 (1954).
110. K. Iguchi, J. Phys. et Radium, 16:401 (1955).
111. H. Kriegsmann and K. Licht, Z. Elektrochem., 62:1163 (1958).
112. A. N. Lazarev and M. G. Voronkov, Opt. i Spektr., 4:180 (1958).

113. A. N. Lazarev, T. P. Tulub, and Ya. Ṡ. Bobovich, Opt. i Spektr., 4:417 (1958).
114. A. N. Lazarev, Opt. i Spektr., 8:511 (1960).
115. A. N. Lazarev, Opt. i Spektr., 4:805 (1958).
116. A. N. Lazarev and M. G. Voronkov, in: Chemistry and Practical Application of Organosilicon Compounds, No. 3, Izd. TsBTI, Leningrad (1958), p. 52.
117. A. N. Lazarev and M. G. Voronkov, Opt. i Spektr., 8:614 (1960).
118. B. A. Arbuzov and T. G. Shavsha, Dokl. Akad. Nauk SSSR, 68:859 (1949); 79:599 (1951).
119. P. J. Launer, Amer. Mineralogist, 37:764 (1952).
120. P. Tarte, Étude Expérimentale et Interprétation du Spectre Infrarouge des Silicates et des Germanates. Application à des Problèmes Structuraux Relatifs à l'État Solide, Acad. Royale de Belgique, Mémoires, Vol. 35, Nos. 4a, 4b, Brussels (1965).
121. H. Moenke, Mineralspectren, Academie-Verlag, Berlin (1962).
122. D. Fortnum and J. O. Edwards, J. Inorg. and Nucl. Chem., 2:264 (1956).
123. C. E. Walrafen, J. Chem. Phys., 42:485 (1965).
124. N. V. Belov, Crystal Chemistry of Silicates with Large Cations, Izd. AN SSSR, Moscow (1961).
125. A. N. Lazarev, Opt. i Spektr., 9:195 (1960).
126. C. M. Midgley, Acta Crystallogr., 5:307 (1952).
127. D. W. J. Cruickshank, Acta Crystallogr., 17:685 (1964).
128. D. K. Smith, A. Majumdar, and F. Ordway, Acta Crystallogr., 18:787 (1965).
129. J. W. Jeffery, Acta Crystallogr., 5:26 (1952).
130. L. S. Dent Glasser and F. F. Glasser, Acta Crystallogr., 18:453, 455 (1965).
131. W. L. Bragg and W. H. Zachariasen, Z. Kristallogr., 72:518 (1930).
132. B. E. Warren, Z. Kristallogr., 74:131 (1930).
133. B. E. Warren and O. R. Trautz, Z. Kristallogr., 75:525 (1930).
134. G. A. Barclay and E. G. Cox, Z. Kristallogr., 113:23 (1960).
135. N. V. Belov and I. M. Rumanova, Dokl. Akad. Nauk SSSR, 89:853 (1963).
136. T. Ito, N. Morimoto, and N. Sadanaga, Acta Crystallogr., 7:53 (1954).
137. K. M. Moody, Mineral. Mag., 30:79 (1953).
138. N. V. Belov and V. I. Mokeeva, Dokl. Akad. Nauk SSSR, 81:581 (1951).
139. A. N. Lazarev, Kristallografiya, 6:125 (1961).
140. V. I. Simonov and N. V. Belov, Kristallografiya, 4:163 (1959).
141. S. M. Skshat and V. I. Simonov, Kristallografiya, 10:591 (1965).
142. N. V. Belov, Mineralog. Sb. L'vov, Gos. Univ., 19:131 (1965).
143. D. W. J. Cruickshank, H. Linton, and G. A. Barclay, Acta Crystallogr., 15:491 (1962).
144. I. Naray-Szabo, Phys. Chem. of Glasses, 4:38A (1963).
145. J. Lajzerowicz, Bull. Soc. Franc. Miner. Crist., 87:520 (1964).
146. G. Bassi and J. Lajzerowicz, Bull. Soc. Franc. Miner. Crist., 88:332 (1965).
147. J. Lajzerowicz, Acta Crystallogr., 20:357 (1966).
148. D. E. C. Corbridge and E. J. Lowe, J. Chem Soc. (L.), 493 (1954).
149. W. Bues and H.-W. Gehrke, Z. Anorgan. und Allgemein. Chem., 288:291 (1956).
150. A. Mutschin and K. Maennchen, Z. Analyt. Chem., 160:81 (1958).
151. A. Simon and H. Richter, Z. Anorgan. und Allgemein.Chem., 301:154 (1959).
152. W. Bues, K. Buhler, and P. Kuhnle, Z. Anorgan. und Allgemein. Chem., 301:154 (1959).
153. T. J. Hanwick and P. O. Hoffmann, J. Chem. Phys., 19:708 (1951).
154. A. N. Lazarev and V. S. Aksel'rod, Opt. i Spektr., 9:326 (1960).
155. D. M. McArthur and C. A. Beevers, Acta Crystallogr., 10:428 (1957).
156. D. W. J. Cruickshank, Acta Crystallogr., 17:672 (1964).
157. W. S. MacDonald, D. W. J. Cruickshank, B. Beagley, and T. G. Gevit, Seventh International Conf. on Crystallography, Summaries of Contributions [Russian translation], Izd. Nauka, Moscow (1966), p. 61.

158. E. Steger and C. Fischer-Bartelek, Z. Anorgan. und Allgemein. Chem., 38:15 (1965).
159. T. Dupuis and M. Viltange, Mikrokhim. et Ichnoanal. Acta, 232 (1963).
160. B. Beagley, Trans. Faraday Soc., 61:1821 (1965).
161. H. Lynton and M. R. Truter, J. Chem. Soc. (L.), 5112 (1960).
162. R. Partzold, H. Amoulong, and A. Ruzicka, Z. Anorgan. und Allgemein. Chem., 336:278 (1965).
163. H. Gerding, J. W. Steeman, and L. J. Revallier, Réc. Trav. Chim. de Pays-Bas, 69:944 (1950).
164. P. A. Giguere and R. Savoie, Canad. J. Chem., 38:2467 (1960).
165. A. Simon and H. Wagner, Z. Anorgan. und Allgemein. Chem., 311:102 (1962).
166. A. Walrafen, D. E. Irish, and T. F. Young, J. Chem. Phys., 37:662 (1962).
167. M. Seleborg, Acta Chem. Scand., 21:499 (1967).
168. H. Stammreich, D. Bassim, O. Sala, and H. Siebert, Spectrochim. Acta, 13:192 (1958).
169. D. S. Urch, J. Inorgan. and Nucl. Chem., 25:771 (1963).
170. R. Savoie and P. A. Giguere, Canad. J. Chem., 40:495, 991 (1962).
170a. A. N. Lazarev and T. Tenisheva, Dokl. Akad. Nauk SSSR, 177:849 (1967).
171. D. R. Davies and D. E. C. Corbridge, Acta Crystallogr., 11:315 (1958).
172. D. W. J. Cruickshank, Acta Crystallogr., 17:674 (1964).
173. R. Westrik and C. H. MacGillavry, Acta Crystallogr., 7:764 (1954).
174. D. W. J. Cruickshank, Acta Crystallogr., 17:684 (1964).
175. G. B. Bokii and I. I. Plyusnina, Nauchn. Dokl. Vyssh. Shkoly (Geologo-Geogr. Nauki), No. 3, 116 (1958).
176. I. I. Plyusnina and G. B. Bokii, Kristallografiya, 3:752 (1958).
177. A. N. Lazarev, Opt. i Spektr., 12:60 (1962).
178. W. H. Zachariasen, Z. Kristallogr., 74:139 (1930).
179. B. Brunowsky, Acta Physicochim. URSS, 5:863 (1936).
180. H. Kriegsmann, Z. Anorgan. und Allgemein. Chem., 298:223 (1959).
181. A. Simon and E. Steger, Z. Anorgan. und Allgemein. Chem., 277:209 (1954).
182. F. Liebau, Acta Crystallogr., 10:790 (1957).
183. W. Hilmer, Naturwiss., 45:238 (1958).
184. T. F. Volodina, I. M. Rumanova, and N. V. Belov, Dokl. Akad. Nauk SSSR, 149:173 (1963).
185. A. N. Lazarev and T. F. Tenisheva, Zh. Strukt. Khim., 4:703 (1963).
186. V. I. Simonov, Kristallografiya, 5:544 (1960).
187. N. V. Belov, V. P. Butuzov, and N. I. Golovastnikov, Dokl. Akad. Nauk SSSR, 87:953 (1952).
188. H. G. Heide, K. Boll-Dornberger, E. Thilo, and E. M. Thilo, Acta Crystallogr., 8:425 (1955).
189. L. Pauling and J. Sherman, Z. Kristallogr., 18:226 (1965).
190. C. Romers, J. A. A. Ketelaar, and C. H. MacGillavry, Acta Crystallogr., 4:411 (1951).
191. H. M. Ondik, Acta Crystallogr., 18:226 (1965).
192. E. D. Eanes and H. M. Ondik, Acta Crystallogr., 15:1280 (1962).
193. H. M. Ondik, S. Block, and C. H. MacGillavry, Acta Crystallogr., 14:555 (1961).
194. H. M. Ondik, Acta Crystallogr., 17:1139 (1964).
195. D. W. J. Cruickshank, Acta Crystallogr., 17:675 (1964).
196. J. R. Van Wazer, Phosphorus and Its Compounds, Vol. I., Chemistry, Interscience, New York (1958).
197. J. Lecomte, A. Boulle, and M. Domine-Berges, Bull. Soc. Chim. Franc., 15:764 (1948).
198. W. Bues and H.-W. Gehrke, Z. Anorgan. und Allgemein. Chem., 288:307 (1956).
199. H. J. Hofmann and K. R. Andress, Naturwiss., 41:94 (1954).
200. E. Steger, Z. Anorgan. und Allgemein. Chem., 296:305 (1958).
201. E. Steger and A. Simon, Z. Anorgan. und Allgemein. Chem., 294:1 (1958).
202. E. Steger, Z. Anorgan. und Allgemein. Chem., 294:146 (1958).
202a. É. V. Poletaev, Izv. Akad. Nauk Kazakh. SSR, Ser. Khim., No. 1, 42 (1968).
202b. É. V. Poletaev, Izv. Akad. Nauk Kazakh. SSR, Ser. Khim., No. 5, 1 (1968).
203. E. Steger, Z. Anorgan. und Allgemein. Chem., 291:76 (1957).

CHAPTER III

RELATIONS BETWEEN THE VIBRATIONAL SPECTRUM AND STRUCTURE OF THE COMPLEX ANIONS IN SILICATES. HIGHLY-CONDENSED STRUCTURES

In this chapter we shall consider the spectra of highly condensed silicate anions possessing spatial periodicity of the structure in one or two dimensions. These include the silicon–oxygen chains of metasilicates, complex anions in the form of ribbons of several interlinked chains or in the form of tubes, i.e., rolled or twisted ribbons, and finally layer structures possessing translational symmetry in two directions. In view of the difficulty of consistently applying the methods of calculating vibrational frequencies, the problem of the qualitative interpretation of the internal vibrations of complex anions is solved by considering the symmetry properties of the systems in question. Here attention is devoted principally to elucidating the possibilities of using spectroscopic methods for establishing the type of structure of the complex silicate anion.

§1. Single Silicate Chains $[(SiO_3)_n]_\infty$

It is typical of structures of this type that, as in the case of silicates with simple ring anions, there should be a ratio of Si:O = 1:3; each of the silicon atoms forms two bonds of the Si–O(Si) type, used for constructing the chains, and two side bonds of the $Si{-}O^-(M^+)$ type. The differences between the lengths of the bonds of these two types evidently depend on the chemical nature of the cations, the coordination of the O atoms in the $Si{-}O^-$ and Si–O(Si) bonds with respect to the cations, and the geometry of the chains (values of the angle SiOSi), and may reach approximately 0.1 Å; this indicates considerable differences in the orders of the bonds and the values of the force constants. The number of x-ray diffraction investigations into the structures of chain metasilicates carried out with an accuracy sufficient to permit the detailed discussion of the multiplicities of the Si–O bonds on the basis of interatomic distances is as yet quite small [1-3]. At the same time, even at the present moment a larger number of structural types of metasilicate chains have been identified [4-6], these being distinguished both by the number of silicon–oxygen tetrahedra in the identity period and also by their mutual orientation. Figure 36 shows the most characteristic types of metasilicate and metagermanate chains. The number of tetrahedra n in the identity period of the chain $[(XO_3)_n]_\infty$ varies from unity in copper metagermanate $CuGeO_3$ and two in pyroxenes and alkali metasilicates to three in wollastonite, $CaSiO_3$, and bustamite, $CaMn(SiO_3)_2$, and, finally, to five in rhodonite and seven in pyroxmangite. This list has recently been supplemented by the structure of batisite with n = 4 [7] and stokesite $Ca_2Sn_2(Si_6O_{18}) \cdot 4H_2O$ with spiral chains containing six SiO_4 tetrahedra in one turn [8], and also $PbSiO_3$ with n = 12 [8a]. The values of the SiOSi and GeOGe angles in the chains (not always established too reliably) vary over a wide range: from normal tetrahedral values in $CuGeO_3$ up to 160° or 170° and even 180° in $PbSiO_3$. However, in the majority of cases the valence angles at the oxygen remain within a narrower range: 130 to 150°.

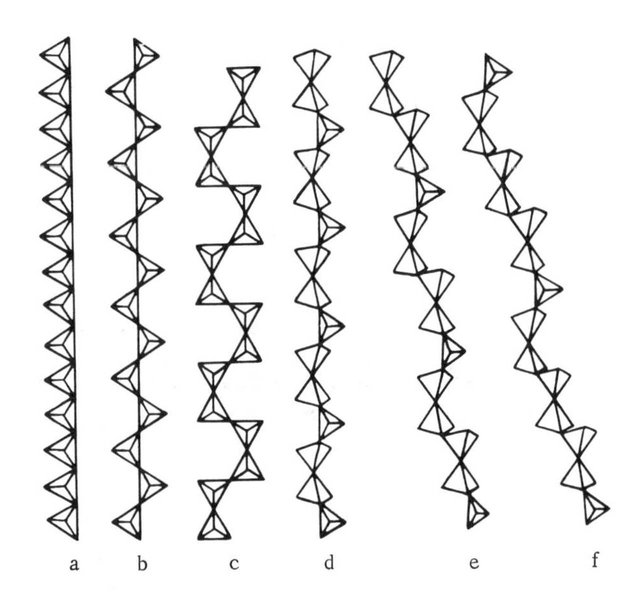

Fig. 36. Configurations of the chains of silicon– and germanium–oxygen anions in metasilicates and metagermanates. a) copper metagermanate; b) pyroxene; c) batisite; d) wollastonite; e) rhodonite; f) pyroxmangite.

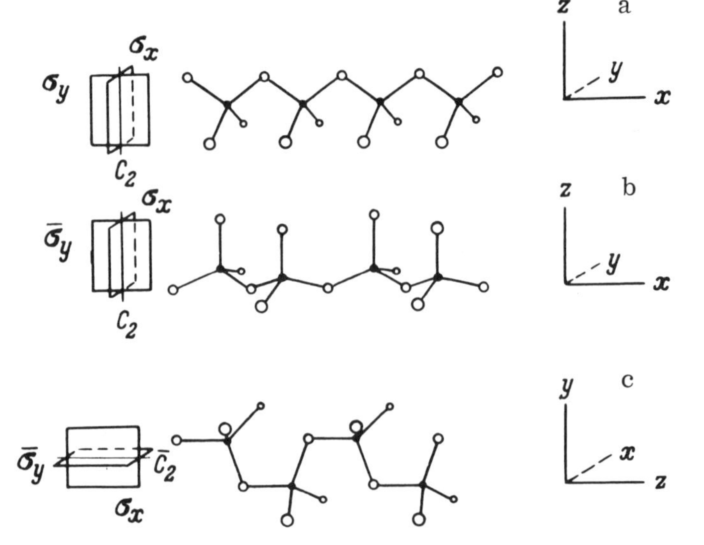

Fig. 37. Symmetry elements of the $(XO_3)_\infty$ and $(X_2O_6)_{2\infty}$ chains. a) $(XO_3)_\infty$, structure of the $CuGeO_3$ type; b) $(X_2O_6)_\infty$, pyroxene chain; c) $(X_2O_6)_\infty$ chains of the metasilicates and metagermanates of Li and Na. C_n and σ_i are symmetry axes and planes; $\bar{C}_n$ and $\bar{\sigma}_i$ are screw axes and glide planes.

Spectra of Metasilicates and Metagermanates with Chains Containing One or Two Tetrahedra in the Identity Period [10]. Copper metagermanate is the only representative of structures with chains of the $[XO_3]_\infty$ type [9]. This is evidently associated with the energetically unfavorable aspect of the formation of a metasilicate chain with ∠SiOSi = 110°. Since chains of the $[X_2O_6]_\infty$ type may also be represented by the structure of alkali metagermanates, which are convenient for spectroscopic investigation, it is natural to begin by considering the spectra of the metagermanate chains.

The chain of germanium–oxygen tetrahedra in $CuGeO_3$, with one tetrahedron in the identity period, has the following symmetry elements, apart from the unitary element and a pure translation along the axis of the chain: a symmetry plane σ_x perpendicular to the chain axis, a symmetry plane σ_y containing the Ge atoms and the bridge O atoms, and a two-fold symmetry axis C_2 normal to the chain axis (Fig. 37). Thus the factor group of the one-dimensional space group is isomorphic with the point group C_{2v}, which helps in determining the symmetry properties and selection rules for the eight nonzero vibrations of the chain. The four stretching vibrations belong to symmetry types A_1 (2), B_1, and B_2 (Table 19). Of these four limiting vibrations of the infinite chain, for the sake of a qualitative description three may be compared with the three degenerate vibrations of the GeO_4 tetrahedron and one with its totally symmetric (pulsation-type) vibration. Subsequently, as in the case of cyclic anions, the vibrations of the chains will be classified by reference to the motions of the O^-GeO^- and GeOGe groups. Both methods of description, the relationships between which are given in Table 20, constitute a coarse approximation, and fail to reflect the true forms of the vibrations completely. The latter can only be estimated on analyzing the vibrations by the method of normal coordinates.

In the infrared spectrum of $CuGeO_3$ (Fig. 38), in the frequency range above 400 cm^{-1} there are altogether only five bands with a simple shape. The foregoing considerations enable us to attribute three vibrations in the range 870 to 700 cm^{-1} to the vibrations ν_{as} GeOGe, ν_{as}, and ν_s O^-GeO^-, and the weaker band at 630 cm^{-1} to the symmetric vibration of GeOGe in the chain. The comparatively high frequency of the latter may be explained by the anomalously small GeOGe angle. For the same reason, and also on account of interaction with the Cu–O vibrations, one of the deformation vibrations of the chain, appearing in the form of an intense, wide band at 535 cm^{-1}, falls into this part of the spectrum.

The spectra of the alkali metagermanates contain a far greater number of bands in the same frequency range, despite the undoubtedly weaker interaction of the internal vibrations of the anion with the lattice vibrations. According to the results of x-ray structural investiga-

TABLE 19. Symmetry Types and Number of Stretching Vibrations of the $(XO_3)_\infty$ and $(X_2O_6)_\infty$ Chains

Factor group of chain C_{2v}	Characters			Selection rules		Orientation of dipole moment (to chain axis)		No. of tetrahedra (X atoms) in identity period	
								1	2
symmetry types of the vibrations	C_2	σ_x	σ_y	IRS	Raman spectrum (RS)	axis C_2 ⊥ to chain axis	axis C_2 ∥ to chain axis	no. of stretching vibrations	
A_1	+1	+1	+1	a	a, p	⊥	∥	2	3
A_2	+1	−1	−1	ina	a, dp	—	—	—	1
B_1	−1	+1	−1	a	a, dp	⊥	⊥	1	3
B_2	−1	−1	+1	a	a, dp	∥	⊥	1	1

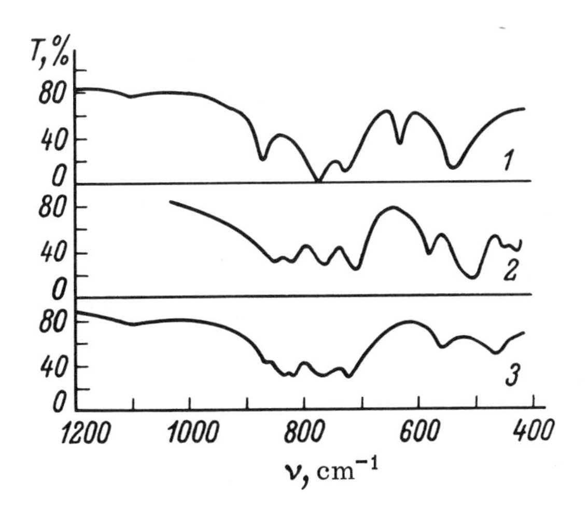

Fig. 38. Infrared spectra of crystalline metagermanates. 1) $CuGeO_3$ (pressed in KBr, 7 mg); 2) Li_2GeO_3; 3) Na_2GeO_3 (pressed in KBr, 5 mg).

TABLE 20. Stretching Vibrations of the XO_4 Groups and the Chains Formed

Structure	Symmetry group (factor group)	Symmetry types and description of vibrations							
Chain $(XO_3)_\infty$	C_{2v}		B_2	B_1	A_1	A_1			
			as_{XOX}	as_{OXO}	s_{OXO}	s_{XOX}			
Tetrahedron XO_4	T_d			F_2		A_1			
Chain $(X_2O_6)_\infty$ (structure in pyroxene; C_2 axis ⊥ to axis of chain)	C_{2v}	B_2	B_1	B_1	A_1	A_1	B_1	A_2	A_1
		as'_{XOX}	as_{XOX}	as'_{OXO}	as_{OXO}	s_{OXO}	s'_{OXO}	s'_{XOX}	s_{XOX}
Chain $(X_2O_6)_\infty$ (structure in Na_2GeO_3 and Na_2SiO_3; C_2 axis ‖ to axis of chain)	C_{2v}	B_1	A_1	A_2	B_2	A_1	B_1	B_1	A_1

Note: Prime denotes vibrations opposite in phase in neighboring XOX or OXO groups.

tions [11, 12], in the Li_2GeO_3 and Na_2GeO_3 crystals the chain anions contain two tetrahedra in the identity period. The configuration of these chains is shown in Fig. 37: the Ge–O–Ge bonds lie in one plane constituting a symmetry plane σ_x for the whole chain; the axis of the chain is a screw axis (two-fold), while the plane passing through it perpendicular to σ_x is a glide plane $\bar{\sigma}_y$. Since a translation along the chain axis reduces to the identity transformation on passing from the one-dimensional space group to its factor group, the factor group of the chain under consideration is isomorphic with C_{2v}. Of the eight stretching vibrations of such a chain, seven refer to the irreducible representations of the C_{2v} group active in the infrared spectrum (Table 19). We see from Table 20 that five of these vibrations are probably grouped in a region of relatively high frequencies, while the two symmetric GeOGe stretching vibrations (ν_s and ν'_s GeOGe) should, by analogy with copper metagermanate, fall in a lower-frequency part of the spectrum.† The number of bands in the spectra of Li_2GeO_3 and Na_2GeO_3 observed in the frequency ranges 850 to 700 cm^{-1} and 600 to 450 cm^{-1} agrees exactly with these principles. The frequencies of the ν_s and ν'_s GeOGe vibrations (lower than in the spectrum of $CuGeO_3$) are a natural consequence of the increase in the GeOGe angles in the chains of the alkali metagermanates. An additional cause of the reduction in frequencies may be the lower frequency of the lattice vibrations, interacting more weakly with vibrations of the chain. This assumption is in agreement with the fact that there is a reduction of 30 to 40 cm^{-1} in the frequencies of ν_s and ν'_s GeOGe on passing from Li_2GeO_3 to Na_2GeO_3, i.e., on reducing the frequencies of the lattice vibrations and preserving the configuration of the complex anion almost unaltered.

The assignment of the frequencies in the spectra of the metagermanates is given in Table 21. Individual assignments of the frequencies of each of the five vibrations in the range 850 to

† Table 20 also shows the "genetical" relationship between the vibrations of the chain and the vibrations of an isolated tetrahedron.

TABLE 21. Frequencies of the Absorption Maxima in the Infrared Spectra of the Metagermanates

$CuGeO_3$ chain $(GeO_3)_\infty$	Assignment		Li_2GeO_3	Na_2GeO_3
			chain $(Ge_2O_6)_\infty$	
868 v.s.	B_2 ν_{as}GeOGe	B_1, A_1	857 v.s.	866 s.
780 v.s.	B_1 ν_{as}O⁻GeO⁻	B_2, A_2	848 v.s.	835 s.
724 v.s.	A_1 ν_sO⁻GeO⁻	B_1, A_1	809 v.s.	820 v.s.
			761 v.s.	770 v.s.
			707 v.s.	721 s.
630 m.	A_1 ν_s GeOGe	B_1, A_1	582 s. 506 v.s.	557 s. 464 v.s.
535 v.s.	Deformation GeO		447 m. 427 m.	— —

Note: The stretching frequencies of $CuGeO_3$ agree satisfactorily with those calculated from the secular equations which were derived in Chap. II, § 2. $K_{Ge-O^-} = 7.5$, $K_{Ge-O(Ge)} = 6.5$ and $K_{OGeO} = 1.5 \cdot 10^6$ cm^{-2}; B_2 = 832, 252 cm^{-1}; B_1 = 785, 350 cm^{-1}; A_1 = 752, 616, 142 cm^{-1}; A_2 = 221 cm^{-1}. The low values of the frequencies of the deformation vibrations are largely associated with the limited applicability of the isolated-anion model.

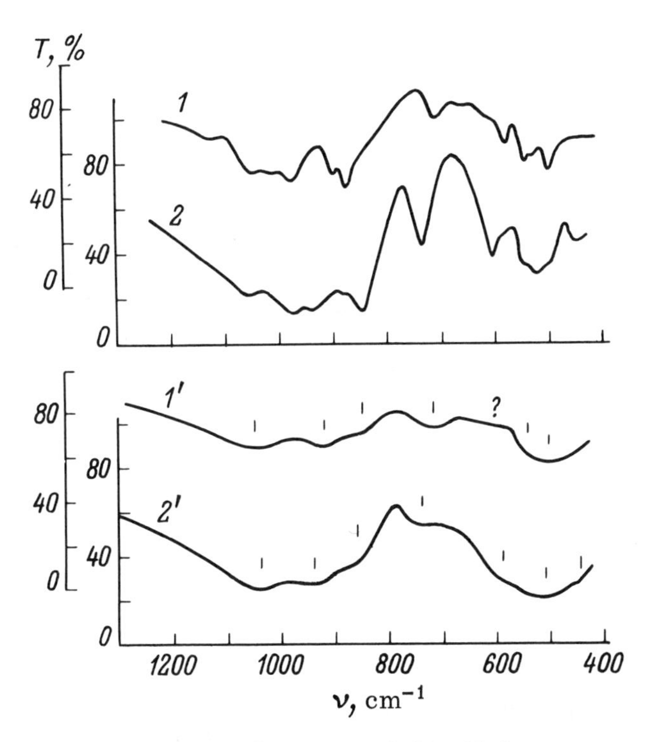

Fig. 39. Infrared spectra of the alkali metasilicates. 1) Crystalline Na_2SiO_3; 2) crystalline Li_2SiO_3; 1') and 2') glasses of the same compositions. Pressed in KBr, 5 mg.

700 cm^{-1} is hardly possible without calculation. Of the two vibrations in the range 600 to 450 cm^{-1} the lower frequency is ascribed to a vibration of type B_1 and the higher to A_1. This is based on the expected greater intensity of the bands of vibrations of type B_1 in the infrared spectrum and on data relating to the Raman spectrum of potassium germanate glass, in composition very similar to the metagermanate (48 mole % K_2O, 52% GeO_2), in which frequencies of 840, 790 (on the background of the band at 860 to 780), 505, and 370 cm^{-1} were observed. If we suppose a similarity between the structure of the metagermanate chains in this glass and in the germanates of Li and Na, we may ascribe the frequencies of 840, 790, and 505 cm^{-1} to the three stretching vibrations of the chain with symmetry type A_1. The latter satisfactorily agrees with the values of the corresponding frequencies in Li_2GeO_3 and Na_2GeO_3, 582 and 557 cm^{-1}, if we remember the fall in frequencies on passing from Li to Na and assume a further fall on passing to potassium germanate.

According to the results of x-ray structural analysis [13-15], there are two planar chains in lithium and sodium metasilicate crystals, similar in structure to the chains of the alkali metagermanates, but distinguished by slightly larger values of the SiOSi angles. Figure 39 shows the infrared spectra of Li_2SiO_3 and Na_2SiO_3; the assignment of the frequencies is analogous to that described earlier for the alkali metagermanates, with the one difference that, of the two bands of vibrations ν_s and ν_s' SiOSi in the range 740 to 580 cm^{-1}, the band of lower frequency is ascribed to a vibration of type A_1. This agrees with data relating to the Raman spectrum of vitreous sodium metasilicate, containing a polarized band at about 600 cm^{-1} [16]. Calculation of the frequencies of the ν_sXOX vibrations of a chain of the pyroxene type belonging to symmetry species A_1 and A_2 by means of the secular equations of [27] leads to analogous changes in frequencies on passing from X = Ge to X = Si:

$$\nu_s \mathrm{GeOGe}\,(A_1) > \nu_s' \mathrm{GeOGe}\,(A_2), \quad \text{but} \quad \nu_s \mathrm{SiOSi}\,(A_1) < \nu_s' \mathrm{SiOSi}\,(A_2)$$

independently of the chosen values of the force constants. Hence it is not surprising that there is a similar change in the relation between the frequencies ν_sXOX of types A_1 and B_1 in the spectra of the chains of the metagermanates and metasilicates. We note that these crystals, belonging to the space group $Cmc2_1 - C_{2v}^{12}$, only contain special positions with symmetry C_s, while the one-dimensional site group of an infinite chain (p. 26), which also includes a screw axis and glide plane, corresponds to a higher symmetry. The factor group of this group is isomorphic with C_{2v}, in agreement with the existence of just one Si_2O_6 "molecule," the elementary link of the chain in the primitive (three-dimensional) cell. It follows from the foregoing that in the crystals of the alkali metasilicates and metagermanates there are no grounds for considering that there will be any disruption of the selection rules derived by considering the symmetry of the isolated chain; in particular, the vibration of type A_2 remains inactive in the infrared spectrum. As in the spectra of the metagermanates, the change from Li_2SiO_3 to Na_2SiO_3 is accompanied by a fall in the frequencies ν_s and ν_s'SiOSi (Table 22). The smaller absolute value of this displacement is associated with the higher (relative to ν_sGeOGe) frequencies of the vibrations of the metasilicate chain, which interact more weakly with the lattice vibrations. The preservation of the two maxima in the spectra of vitreous Na_2SiO_3 and Li_2SiO_3 in the frequency range ν_s SiOSi indicates the existence of several distorted $(Si_2O_6)_\infty$ chains in these glasses, the configuration of these being similar to those observed in crystals (Fig. 39).

A large group of natural and synthetic silicates containing chains of the $[Si_2O_6]_\infty$ type in their structure is formed by the pyroxenes. The general chemical formula of this group of minerals is

$$W_{1-p}(X, Y)_{1+p} Z_2 O_6,$$

where W = Ca, Na; X = Mg, Fe^{2+}, Mn, Ni, Li; Y = Al, Fe^{3+}, Cr, Ti, and Z = Si, Al [17].

TABLE 22. Frequencies in the Infrared Spectrum of Metasilicates with $[(SiO_3)_2]_\infty$

Li_2SiO_3	Na_2SiO_3	Assignments	$NaFe^{\cdots}Si_2O_6$ aegirine	$CaFe^{\cdot\cdot}Si_2O_6$ hedenbergite
—	(1117 w.)	—	(1414 w.)	—
1055 s.	1043 s.		1059 v.s.	1089 v.s.
973 v.s.	1007 s.	ν'_{as} and ν_{as} SiOSi	1004 v.s.	1056 v.s.
942 v.s.	976 s.	ν'_{as} and ν_{as} O^-SiO^-	950 v.s.	959 v.s.
879 s.	898 s.	ν'_s and ν_s O^-SiO^-	897 s.	912 v.s.
847 v.s.	873 s.		864 s.	860 v.s.
736 m.	716 m.		725 w.	663 m.
—	(667 v.w.)	ν'_s and ν_s SiOSi	—	—
605 m.	583 m.		639 m.	624 m.
(587 v.w.)	—	—	—	(616?)
545 m.	545 s.	Si–O deformation	560 s.	—
521 s.	534 s.	(and M–O	545 s.	518 v.s.
500 s.	—	stretching in	507 s.	492 v.s.
447 m.	452 s.	pyroxenes)	467 s.	466 v.s.

Note: In the aegirine spectrum the band at 1414 cm^{-1} is due to a trace of carbonate ion.

For crystallographic reasons it is usual to separate the orthorhombic pyroxenes, for which $1 - p \simeq 0$ and the number of Y atoms is also small, while the X atoms are mainly represented by Mg and Fe^{2+}, and the monoclinic pyroxenes, in which there may be both divalent cations (diopside, hedenbergite) and exclusively mono- and trivalent cations (spodumene, aegirine), i.e., p may vary from 0 to 1 in the same way as the content of X and Y atoms. The degree of replacement of Si by Al in the complex anion is insignificant for the majority of the pyroxenes, which is a favorable circumstance in any attempt at analyzing the vibrations of the silicon–oxygen chains. On the other hand, the presence of a large number of cations forming strong and partly covalent bonds with the oxygen, Al, Fe^{3+}, and also Mg, leads us to expect quite profound changes in the spectra of the complex anions in pyroxenes as compared with the spectrum expected of the isolated anion.

An idealized scheme for a chain of the pyroxene type is shown in Fig. 37b. The elements of symmetry are: a two-fold axis perpendicular to the chain axis, a symmetry plane also perpendicular to the chain axis, and a glide plane parallel to the chain axis. The whole set of these elements forms a one-dimensional space group, the factor group of which is isomorphic with C_{2v}. The distribution of the stretching vibrations of the chain with respect to the irreducible representations of this group is shown in Table 18. For our own purposes the most interesting fact is that ν_s and $\nu_s^!$ SiOSi are in the present case of symmetry types A_1 and A_2, and hence for an isolated chain only the first of them is active in the infrared spectrum. Actually, as indicated in Fig. 40, which shows the absorption spectra of two monoclinic pyroxenes, in the range 600 to 750 cm^{-1} there are two bands of medium intensity; the stronger, low-frequency band evidently corresponds to the in-phase vibrations of the Si–O–Si bridges of type A_1 and the other to a vibration of type A_2. In the frequency range 1100 to 800 cm^{-1} both minerals give five bands, but it is hard to exclude the possibility of there being a partial or complete coincidence between vibrations differing little in frequency. In the spectra of the orthorhombic pyroxenes [18], the number of bands resolved in the region 1100 to 800 cm^{-1} is not much greater, while the bands in the range of frequencies ν_sSiOSi give a whole series of discrete maxima, which are grouped in the regions 760 to 720 and 690 to 630 cm^{-1}.

Thus it is desirable to consider the influence of the symmetry of the crystals and the resonance interactions between the ions on the selection rules and multiplet characteristics of

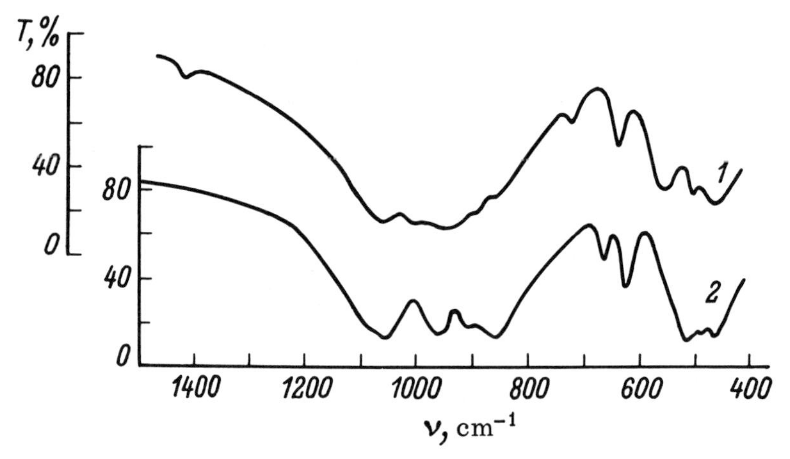

Fig. 40. Infrared spectra of monoclinic pyroxenes. 1) Aegirine $NaFe^{\cdot\cdot\cdot}Si_2O_6$; 2) hedenbergite $CaFe^{\cdot\cdot\cdot}Si_2O_6$ (pressed in KBr, 5 mg).

the internal vibrations. The majority of the monoclinic pyroxenes belong to the $C2/c-C_{2h}^6$ space group and contain four Si_2O_6 links in the unit cell. The primitive (noncentered) cell of this kind of lattice is half this size. All the silicon and oxygen atoms lie in a general position, but the glide plane is a symmetry element of both the chain and the whole crystal [19]. Hence the factor group of the one-dimensional site group is isomorphic with C_s.

It is useful to verify the correctness of this kind of conclusion by means of the relation between the orders of the factor group and the site group of the crystal and the number of molecules in the primitive cell. In the present case for a crystal factor group of the fourth order and two molecules in the cell the order of the site group should in fact be equal to two.

Some of the monoclinic pyroxenes (clinoenstatite) are slightly different in structure; they belong to the $P2_1/c-C_{2h}^5$ space group and contain four sections of Si_2O_6 chains in the unit cell. Even from this fact, without having recourse to the x-ray solution of the structure [21], we may conclude that there is a trivial (C_1) site symmetry of the chains in these crystals. Let us now write down the scheme of correlation between the irreducible representations of the intrinsic, site, and space groups, remembering of course the corresponding factor groups:

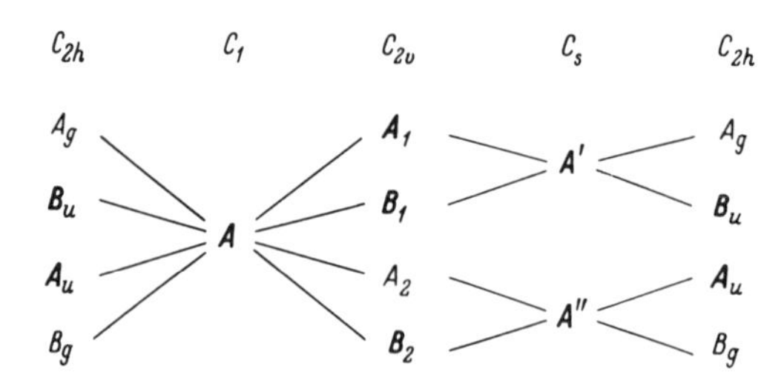

It will be seen from this scheme that the site symmetry of the chain in the crystal requires that all its internal vibrations be active. For crystals with symmetry $C2/c$ the interaction between the vibrations of the chains leads to the appearance of only one Davydov component active in the infrared spectrum, while for symmetry $P2_1/c$ splitting of the internal vibrations of the chains may be observed in the infrared spectrum.

The crystals of the orthorhombic pyroxenes belong to the space group $Pbca-D_{2h}^{15}$, and there are eight Si_2O_6 sections of the pyroxene chains in the unit cell [22]. None of the symmetry elements of the isolated chain is a symmetry element of the crystal; the site group is C_1, which also follows from Z = 8 for a factor group of the eighth order. Thus the scheme of correlations between the representations of the factor groups of intrinsic, site, and space symmetries now appears as follows:

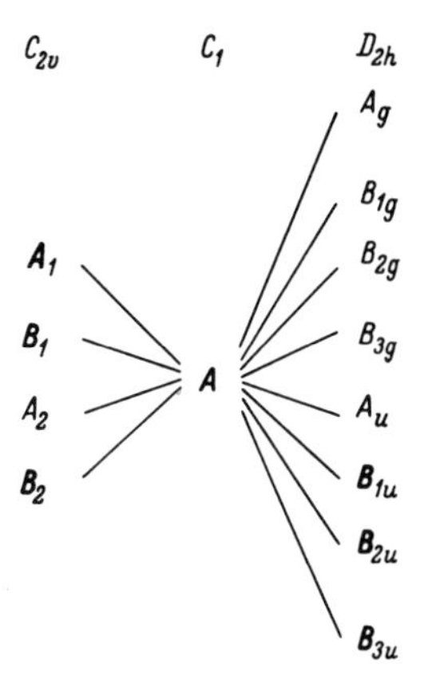

Three crystal vibrations active in the infrared spectrum correspond to each of the internal vibrations of the anion.

It follows from the foregoing that in the infrared spectra of the pyroxenes we may expect the appearance of two, four, or six bands in the frequency range of the symmetrical Si–O–Si stretching vibration, depending on whether the crystal belongs to the space group C2/c, $P2_1/c$, or Pbca respectively. We see from Table 23, which correlates existing data regarding frequencies in the ν_s SiOSi region, that these principles are in complete agreement with the experimen-

TABLE 23. Frequencies of the Absorption Maxima in the Spectra of the Pyroxenes in the Region of the Symmetrical Si–O–Si Stretching Vibration

Mineral	Formula	Space group	Frequency of vibrations ν_sSiOSi	Refs.
Hedenbergite	$CaFe^{\cdot\cdot}(Si_2O_6)$	$C2/c - C_{2h}^6$	663, 624	[10]
Diopside	$CaMg(Si_2O_6)$		672, 635	[18]
Augite	$Ca(Mg, Fe^{\cdot\cdot}, Al)[(Si, Al)_2O_6]$		676, 637	[18]
Aegirine	$NaFe^{\cdot\cdot\cdot}(Si_2O_6)$		725, 639	[10]
Spodumene	$LiAl(Si_2O_6)$		. . ., 642	[18]
Jadeite[1]	$NaAl(Si_2O_6)$		748, 718 (w.), 615	[18]
Clinoenstatite[2]	$Mg_2(Si_2O_6)$	$P2_1/c - C_{2h}^5$	740, 726; 680, 640	—
Enstatite[2]	$Mg_2(Si_2O_6)$	$Pbca - D_{2h}^{15}$	766, 745, 728; 690, 641,. . .	—
Hypersthene	$(Fe, Mg)_2(Si_2O_6)$		760 (doub.), 725; 692, 650, 635	[18]
Bronzite	$(Mg, Fe)_2(Si_2O_6)$		745, 722; 695, 650	[18]

Note: 1) Results of V. A. Kolesova [23], 744 and 663 cm^{-1}. 2) Results of V. A. Kolesova.

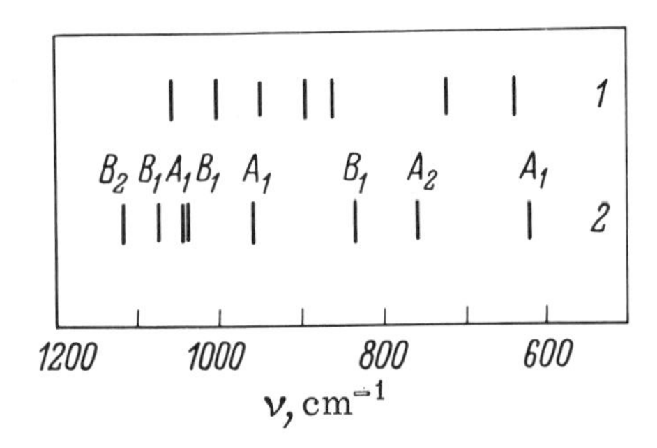

Fig. 41. Stretching vibrations of the pyroxene chain. 1) Experimental frequencies of the infrared spectrum of aegirine; 2) calculated frequencies [28] for $K_{Si-O(Si)} = 7.51 \cdot 10^6\ cm^{-2}$. $K_{Si-O^-} = 1.25\ K_{Si-O(Si)}$, $K_{\delta OSiO} = {}^1/_{13}\ K_{Si-O(Si)}$, $K_{\delta SiOSi} = 0.3\ K_{\delta OSiO}$.

tal results. We note that the difference in frequencies between the Davydov components corresponding to one and the same internal vibration of the chain may reach 40 to 50 cm^{-1}, which is almost equal to the difference in the frequencies of the vibrations ν_s and ν_s'SiOSi. In the range of the other, higher-frequency stretching vibrations, the number of observed maxima is no greater than six or seven, even in the case of orthorhombic pyroxenes, where the appearance of 18 bands is theoretically possible. This is probably associated not only with their superposition but also with the smaller distance between the Davydov components of these vibrations.

Attempts at calculating the frequencies of the pyroxene chains have been undertaken repeatedly, although Matossi's paper [24] contains an error in the geometrical configuration of the chain used for the calculation, while Saksena [25] made an error in the symmetry of the chain, in particular ignoring the presence of translational symmetry. Secular equations for the pyroxene chain with symmetry C_{2v} were obtained by Stepanov and Prima [26, 27]. The same authors made some numerical calculations of the stretching frequencies of the chain, characterizing the force field by four force constants, two bond stretching (Si–O (Si) and Si–O⁻) and two angle deformation (∠OSiO and ∠SiOSi) [28]. For a ratio of $K_{Si-O^-} = 1.25\,K_{Si-O(Si)}$, which appears physically justified,† and deformation coefficients 10 to 20 times smaller than the stretching coefficients, the sequence of values of the calculated frequencies agrees qualitatively with experiment. In particular, the two lowest frequencies relate, as one might expect, to the symmetry types A_2 and A_1, while the frequencies of the vibrations ν_{as} SiOSi and particularly ν'_{as}SiOSi lie at the high-frequency end of the spectrum (Fig. 41). Any additional variation of the force field hardly has any meaning, since correspondence between the calculated and experimental frequencies, particularly in the region of frequencies ν_s and ν_s'SiOSi, achieved without considering the interaction with the cations, remains illusory.

Metasilicate Chains $[(SiO_3)_n]_\infty$ with n = 3 to 7. At the present time these are represented by a comparatively small number of x-ray structural solutions. The number of such substances studied spectroscopically [29] is also very small. The most studied is the structure of the group of calcium and manganese silicates including wollastonite β-$CaSiO_3$[30, 31] and bustamite (Ca, Mn) SiO_3 [32] with infinite chains containing three silicon–oxygen tetrahedra in the identity period (Fig. 36). Both structures have recently been refined [33, 34], and slight differences in the form of the chains have been found. In crystals of rhodonite (Mn, Ca) SiO_3 and pyroxmangite (Mn, Fe) SiO_3 the configuration of the chains, also shown in Fig. 36, has some similarity with wollastonite chains; however, the identity period contains respectively five and seven silicon–oxygen tetrahedra [35-37]. Since the model of the one-dimensional crystal was suitable for a qualitative interpretation of the spectra of silicates with two tetrahedra in the identity period, we may expect that for chains with larger identity periods an analogous identi-

†In the pyroxenes the lengths of the Si–O bridge are evidently about 0.05 to 0.02 Å greater than the length of the Si–O⁻ side bonds [1, 20]; in Na_2SiO_3 the average lengths of the Si–O⁻ and Si–O (Si) bonds differ by 0.08 Å [13a].

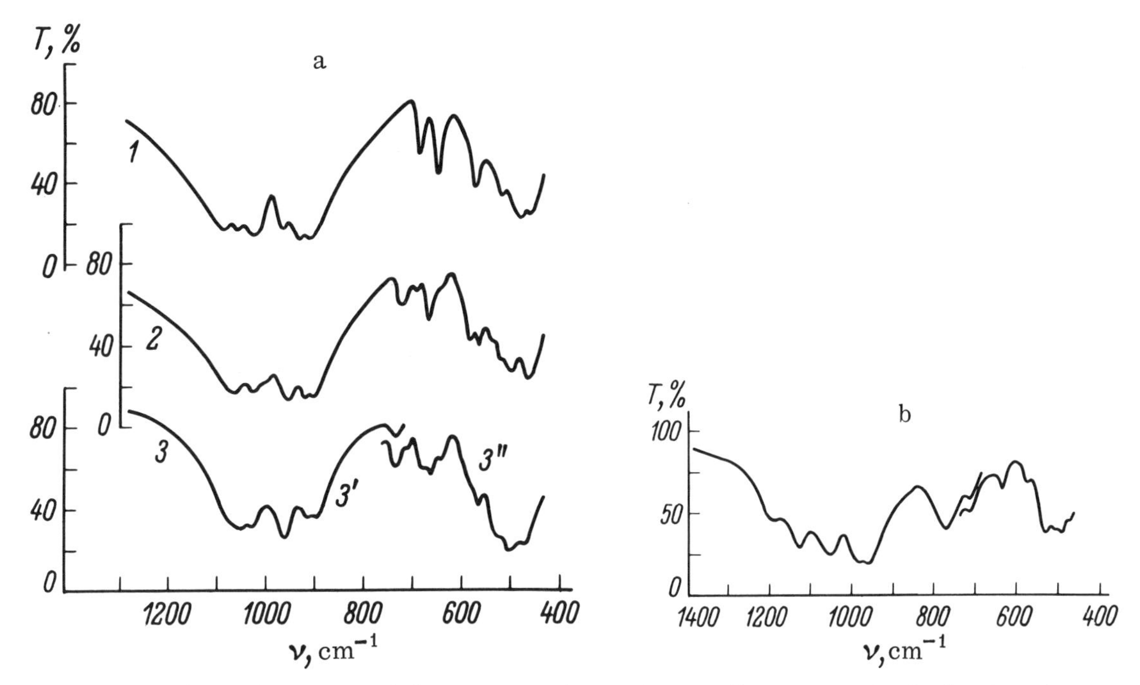

Fig. 42. Infrared spectra of the pyroxenides. a) β-wollastonite (1), rhodonite (2), pyroxmangite (3) [pressed in KBr: 1 and 2) 5 mg, 3') 2.5 mg, 3") 5 mg]; b) batisite (pressed in KBr, 5 mg).

fication of the frequencies should be possible. It is, in particular, extremely interesting to examine the possibility of using the number of maxima observed in the range of frequencies of the ν_sSiOSi vibrations as a criterion for estimating the dimensions of the identity period of the chain. We may expect that in silicates with the cations Ca^{2+}, Mn^{2+}, Fe^{2+} the influence of the interionic interactions on the internal vibrations of the anion will be less significant than in pyroxenes with the cations Mg^{2+}, Al^{3+}, Fe^{3+}.

Figure 42a shows the infrared spectra of wollastonite, rhodonite, and pyroxmangite. Owing to the low symmetry of the chains there are no grounds for expecting that any of the internal vibrations of the chains will be absent in the infrared spectrum. Three bands of medium intensity, 680, 642, and 566 cm^{-1}, observed in the spectrum of wollastonite in the interval between the regions of strong absorption below 500 cm^{-1} and from 900 to 1100 cm^{-1} may naturally be ascribed to the three "symmetrical" stretching vibrations of the three Si–O–Si bridges forming the elementary link in the chain. In the interval 900 to 1100 cm^{-1} there are only six strong bands instead of the nine expected (for an isolated chain). This once again demonstrates the difficulty of identifying the frequencies and separating the characteristic bands in this part of the spectrum if the symmetry of the complex anion is low. We note, by the way, that in the spectrum of bustamite (Table 24) the number of bands in this range of frequencies almost reaches the expected number. In the region of vibrations ν_sSiOSi there are three bands, as in wollastonite, but the lowest-frequency band is weak and only seen in the form of a diffuse shoulder on the side of the stronger bands of the deformation vibrations. It is not impossible that this may be connected with an increase in one of the SiOSi angles in the structure of bustamite (and a reduction in the rigidity of this angle, see Ch. VI, § 3).

Passing to the spectra of silicates with larger identity periods of the chains hardly affects the number of bands observed in the frequency range of the vibrations ν_{as}SiOSi, ν_{as}, and $\nu_s O^-SiO^-$. However, the number of maxima in the range ν_sSiOSi increases sharply: in the rhodonite spectrum there are six; and in that of pyroxmangite, seven. Evidently the appearance of

TABLE 24. Frequencies of the Absorption Maxima in the Spectra of Silicates with Anions $[(SiO_3)_n]_\infty$ for n = 3 to 7

Compound					Assignments
wollastonite β-$CaSiO_3$	bustamite $(Ca, Mn)SiO_3$	batisite $Na_2BaTi_2O_2Si_4O_{12}$	rhodonite $(Mn, Ca)SiO_3$	pyroxmangite $(Mn, Fe)SiO_3$	
structure of chain anion					
$[(SiO_3)_3]_\infty$		$[(SiO_3)_4]_\infty$	$[(SiO_3)_5]_\infty$	$[(SiO_3)_7]_\infty$	
		1189 s.			$\nu_{as}SiOSi$, $\nu_{as}O^-SiO^-$, $\nu_s O^-SiO^-$
1087 v.s.	1086 p.,s.	1124 v.s.	1080 p.,s.	1077 p., s.	
1056 v.s.	1062 v.s.	1050 v.s.	1063 v.s.	1050 v.s.	
1019 v.s.	979 v.s.	998 p.	1023 v.s.	1027 v.s.	
964 v.s.	951 v.s.	982 v.s.	998 s.	959 v.s.	
925 v.s.	911 v.s.	961 v.s.	951 v.s.	946 p?	
904 v.s.	872 v.s.		916 v.s.	908 v.s.	
	850 p.,s		900 v.s.	890 s.	
	803 m.				
			720 m.	732 m.	$\nu_s SiOSi$
		768 s.	690 w.	710 p.,w.	
680 m.	660 m.	713 m.	661 m.	676 m.	
642 m.	619 m.	634 m.	638 p., w.	658 m.	
566 m.	560 p?	573 w.	577 m.	637 m.	
			559 m.	579 p., m.	
				561 m.	
508 m.	525 p.	531 s.	532 m.	523 s.	δSiO, $\nu M-O$
471 v.s.	510 v.s.	509 s.	514 m.	496 v.s.	
452 v.s.	465 v.s.	492 s.	492 v.s.	468 v.s.	
	450 p.,m.	469 m.	458 v.s.		

one extra maximum in the rhodonite spectrum is associated with interionic interactions or a Fermi resonance.

Figure 42b shows the spectrum of yet another chain metasilicate, batisite [38]. In the structure of batisite the existence of $[Si_4O_{12}]_\infty$ chains has been established [7], the four tetrahedra forming the identity period of the chain being linked in pairs, so that the chain consists of Si_2O_7 groups, with two such groups in one identity period (Fig. 36). Correspondingly the SiOSi angles in these groups and the Si−O−Si bonds connecting these groups into a chain are substantially different. As we should expect, the batisite spectrum shows four bands in the ν_sSiOSi vibration frequency range. The number of bands in the higher-frequency range is small; one notices the presence of a band with a frequency of about 1190 cm^{-1}, which indicates the existence of SiOSi angles close to 180° in the complex anions. This is contradicted, however, by the presence of all four ν_sSiOSi bands in the infrared spectrum, unless one regards their appearance as a result of the infringement of the mutual exclusion rule in the noncentrosymmetric batisite crystal (space group $Ima2-C_{2v}^{22}$). A "compromise" assumption of an SiOSi angle of about 170 to 160° agrees closely with the value of 165° obtained from x-ray data.

The examples presented indicate the possibility of using spectroscopic data for judging the dimensions of the identity period in the chain of silicon−oxygen tetrahedra. The main limitations imposed on the applicability of the method arise from the same examples.

1. An increase in the covalent character in the bonds of the cations with the anions, particularly when the mass of the cation is reduced, leads to an increase in the interaction of the vibrations, and for a suitable crystal symmetry this produces a considerable splitting in the bands of the internal vibrations of the anion. Among the stretching vibrations of a complex anion, vibrations of the ν_sSiOSi type are particularly sensitive to interionic interaction. Hence it is desirable to have information regarding the lattice parameters and space group of the crystal so as to be able to establish the expected multiplet nature of the bands of internal vibrations of the anion.

2. As the number of tetrahedra in the identity period of the chain increases, the separation of the bands of ν_sSiOSi vibrations diminishes, and the demarcation of the effects due to the interaction of vibrations within and between chains becomes correspondingly more complex.

These difficulties are particularly substantial when one attempts to make a spectroscopic analysis of the structure of the chains of the metagermanates, as their stretching frequencies are lower than those of silicates. Thus in the spectrum of $CaGeO_3$, isostructural with wollastonite, the lowest of the three ν_sGeOGe frequencies lies at about 450 cm^{-1} (p. 137). In the spectrum of $CdGeO_3$ studied in [39], there are also three ν_sGeOGe bands between 560 and 490 cm^{-1}, an agreement expected from the similarity between the ionic radii of Ca^{2+} and Cd^{2+}, 0.99 and 0.97 Å (Arens). However, in the spectra of $MnGeO_3$ and $CoGeO_3$, which in view of the smaller radii of the cations (0.80 and 0.72 Å) probably form chains of the pyroxene type,† the separation of the ν_sGeOGe frequencies from the frequencies of the deformation vibrations is less clear: apart from the two ν_sGeOGe bands in the range 500 to 560 cm^{-1} there are bands at about 450 to 470 cm^{-1}, i.e., in the same range of frequencies as that containing one of the ν_sGeOGe bands in the $CaGeO_3$ spectrum. The identification of the latter as deformation vibrations of the anion can only be based on their comparatively high intensity.

We also note that the low symmetry of the majority of the silicate chains usually enables us to avoid difficulties associated with the inactivity of some of their ν_sSiOSi vibrations in the infrared spectrum and with the possibility of a reduction in the number of bands observed on account of degeneracy. Some examples of chains with degenerate vibrations will be given later.

Metaphosphates with Infinite Chains of Phosphorus – Oxygen Tetrahedra. These constitute a widespread class of compounds; their structure has been repeatedly studied by x-ray diffraction and shows a great variety of possible chain configurations. Some of the forms of chains reliably established by x-ray structural analysis of the crystals are shown in Fig. 43. Some of these configurations (2, 3, 4) are encountered in silicate structures also. Since almost all the structures of chain metaphosphates solved by x-ray diffraction only contain alkali cations, while the accuracy of determining the interatomic distances is low, it proved difficult [42] to establish a clear correspondence between the lengths of the $P-O^-$ bonds and the basicity of the cations. The $P-O^-$ bonds were in all cases shorter than the P–O(P) by 0.10 to 0.15 Å. The latter also evidently possess a certain proportion of double-bondedness, since the POP angles lie in the range 130 to 140°, i.e., substantially exceeding the tetrahedral values.

The infrared spectra of the alkali metaphosphates with infinite chains of phosphorus–oxygen tetrahedra have been repeatedly studied (see, for example, [50]), but only in [51] was an attempt made at interpreting the spectra and establishing a link between the spectrum and the structure of the chain anion. However, the conclusions drawn in [51] as to the existence of chains of the $[(PO_3)_2]_\infty$ type both in the potassium Kurrol salt (KPO_3) and in the sodium Kurrol salt of the B type are not entirely in accordance with x-ray data. It is therefore interesting to discover possible causes of the discrepancy and also to compare the spectra of the chain metaphosphates with the structurally similar silicates. Figure 44 gives a schematic picture of the infrared spectra of the two phosphates in which, by analogy with the cyclical metaphosphates, it is easy to identify vibration frequencies ν_{as} and $\nu_s O^-PO^-$, ν_{as}POP and ν_sPOP. As already noted, the better "separability" (as compared with the corresponding silicates) of the vibrations of the O^-PO^- and POP groups is associated not with differences in kinematics, which are insignificant,‡

†The space groups of orthorhombic $MnGeO_3$–Pbca–D_{2h}^{15} and monoclinic $CoGeO_3$–C2/c–C_{2h}^{6} established by x-ray diffraction [40, 41] agree with this view.

‡The small difference in the masses of Si and P cannot have any great significance, but the slightly smaller average values of the POP angles (as compared with SiOSi in the corresponding structures) help in separating the POP and O^-PO^- frequencies.

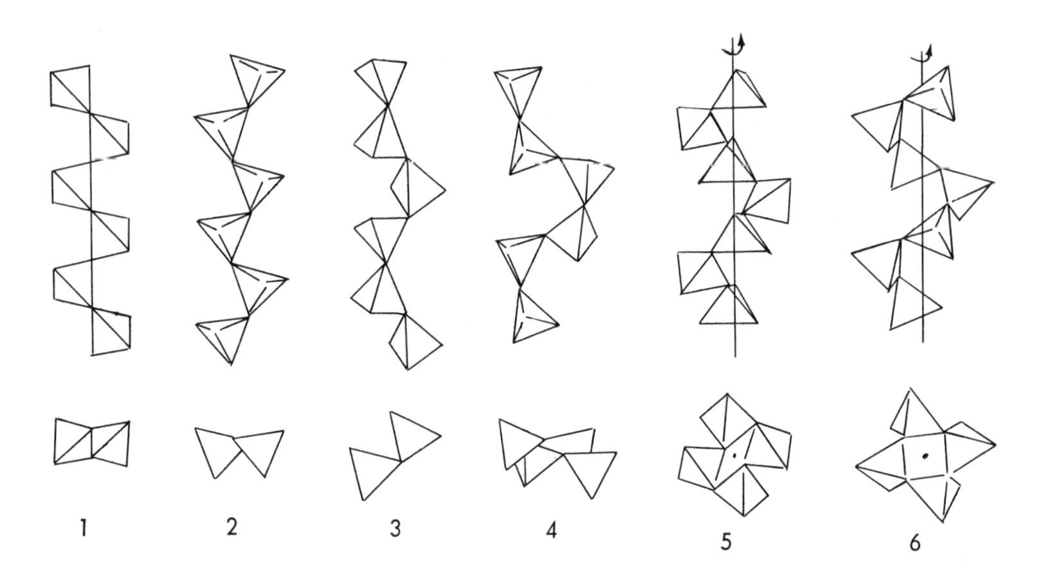

Fig. 43. Configuration of the infinite chains of phosphorus–oxygen tetrahedra in crystals of metaphosphates according to x-ray data [42]. 1) $RbPO_3$ [43, 44] ($CsPO_3$); 2) KPO_3 [45] (low-temperature form of $LiPO_3$); 3) $NaPO_3$, Madrell's salt, high-temperature form [30] and $Na_2H(PO_3)_3$ [46]; 4) $Pb(PO_3)_2$ [42] ($Ca(PO_3)_2$); 5) $NaPO_3$, Kurrol salt A [47] and $AgPO_3$ [48]; 6) $NaPO_3$, Kurrol salt B [49]. The compounds for which the formulas are given in brackets are plausibly related to the corresponding structural type on the basis of indirect considerations.

but chiefly with the greater differences in the force constants of the $P-O^-$ and $P-O(P)$ bonds. These differences, formally defined by a ratio of 3/2:1 in the orders of the bonds, may actually be still greater, despite the fact that the order of the $P-O(P)$ bond is probably slightly greater than unity. In the presence of the cations of alkali metals, the increase in the effective negative charge on the "side" oxygen atoms leads us to expect a considerable intensification of the π bonds involving the participation of the d orbitals of the P atom.

In the spectra of both phosphates shown in Fig. 44, the region of each of the four listed quasi-characteristic types of vibrations contains two absorption maxima. In KPO_3 crystals, which contain slightly distorted chains of the pyroxene type with two tetrahedra in the identity period [45], this is not unexpected. The appearance of the two ν_sPOP vibrations in the infrared spectrum is explained by the low (C_1) symmetry of the chain. The KPO_3 crystals belong to the

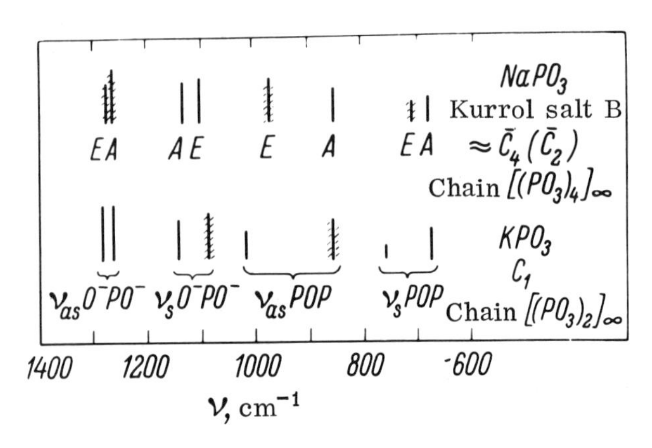

Fig. 44. Identification of the frequencies of P–O stretching in the spectra of two chain metaphosphates.

$P2_1/a-C_{2h}^5$ space group and contain four elementary sections of the metaphosphate chains in the unit cell (site symmetry C_1). Thus the multiplet structure of the internal vibrations of the anion expected as a result of interionic interactions is exactly the same as in the clinoenstatite considered earlier. However, in the KPO_3 spectrum not one of the internal vibrations is appreciably split, and we must ascribe this fact to the far weaker coupling between the vibrations of the complex anions, effected by purely electrostatic interactions between the K^+ ion and the anions, whereas in clinoenstatite the carriers of the interionic interactions are the partly covalent Mg–O bonds. Only a more careful study of the spectrum of a KPO_3 single crystal in polarized light revealed the splitting of some of the stretching vibrations of the chain into narrow doublets [51a].

In the $NaPO_3$ (Kurrol salt B) crystal, belonging to the same space group as KPO_3, the unit cell contains two sections of the spiral chains with four tetrahedra in the identity period [49]. Correspondingly the order of the site group of the chain equals two; a screw axis (two-fold) $\bar{C}_2$ is a symmetry element of both the isolated chain and the whole crystal. Then the spectrum of a chain of the $[(PO_3)_4]_\infty$ type should have four vibrations active in absorption in each of the frequency ranges $\nu_s O^-PO^-$, $\nu_{as}POP$ and $\nu_s POP$ (allowance for interaction between the ions in lattice does not in this case lead to further increase in the number of frequencies active in the infrared spectrum). However, in each of these four frequency ranges there are only two bands; this may only be explained by turning to the configuration of the metaphosphate chain in this crystal illustrated in Fig. 43. In fact, the arrangement of the four tetrahedra forming one turn in the spiral chain deviates little from the higher-symmetry configuration possessing a four-fold screw axis along the axis of the chain. Then to each of the four types of stretching vibrations of the chain there correspond two simple (B and A) and one doubly degenerate (E) irreducible representation of the factor group of the chain, isomorphic with the point group C_4. The frequencies observed in the infrared spectrum are referred in Fig. 44 to the active representations A and E. It is characteristic that almost all the bands of the quasi-degenerate vibrations are slightly broadened.

In conclusion we note that the spectrum of $NaPO_3$-II, the high-temperature form of Madrell's salt, which contains wollastonite-like $[(PO_3)_3]_\infty$ chains [30], exhibits (as we should expect) three frequencies (742, 719, and 702 cm^{-1}) in the region of the $\nu_s POP$ vibrations. In the higher-frequency region, instead of the nine bands expected there are only seven maxima, although one of these is very broad.

§ 2. Silicates with Ribbon Anions

By complex silicate ribbon anions [52] we shall understand structures possessing translational symmetry in only one direction, like single chains, but, in contrast to the latter, capable of being considered as the result of the "fitting together" of several chains by way of transverse Si–O–Si bridges. Apart from normal ribbons, structures of this type may naturally include complex anions in the form of the skeletal tubes found in narsarsukite [53, 54] and vlasovite [55]. In the crystal chemistry of silicates ribbon anions are customarily classified according to the structure of the single chains forming the ribbon by condensation [5, 6]. Thus $(Si_4O_{11})_\infty$ ribbons in amphiboles may be considered as the result of the condensation of two pyroxene chains. Xonotlite ribbons $(Si_6O_{17})_\infty$ may analogously be considered as the product of the fitting together of chains of the wollastonite type; a different mutual orientation (and a different number of transverse Si–O–Si bridges) will lead to the formation of ribbons of another type, for example, the ribbons in epididymite [56] and elpidite [57], also associated with the wollastonite "motif." The ratio of the silicon and oxygen contents in the ribbon anions may vary according to their structure, over the range $1/3 < Si/O \leq 2/5$.

The spectra of silicates with anions in the form of infinite ribbons may also be considered on the basis of the model of a "one-dimensional" crystal on the assumption of the relative weak-

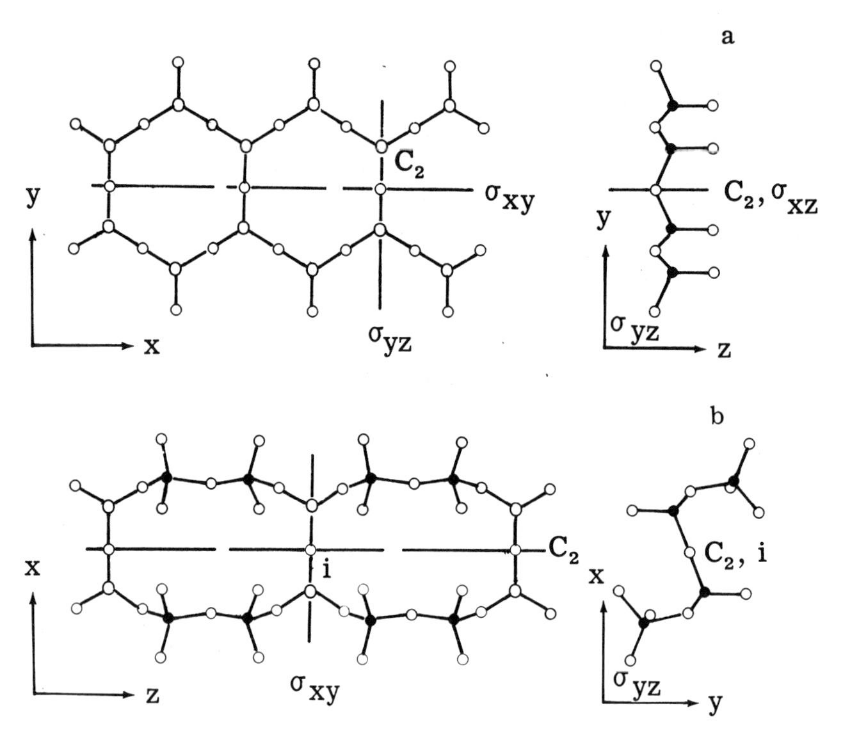

Fig. 45. Idealized arrangement of ribbons. a) amphiboles; b) xonolite. Symmetry elements: C_n and σ_i , axes and planes; i, inversion center.

ness of interionic interactions. The latter compels us to limit consideration to the higher frequency stretching vibrations of the ribbon. By analogy with the spectra of chain and cyclic metasilicates we may expect that the symmetric vibrations of the Si–O–Si bonds will also be quasi-characteristic in the present case. The number of such bonds in a ribbon the identity period of which comprises p silicon–oxygen tetrahedra is p + r, where r is the number of transverse bridges. These transverse Si–O–Si bridges are formed by silicon atoms participating in three bonds of the Si–O (Si) type and only one $Si-O^-$. Hence chemically they are probably somewhat different from the Si–O (Si) bonds in simple chains, and this may be reflected both in the kinematics of the system (magnitude of ∠ SiOSi) and in the values of the force constants. In the most highly-condensed ribbons in the limiting case p = r, all the Si atoms may each form three Si–O (Si) bonds, and chemically such anions are similar to the layer silicate anions considered later. In considering the spectra of silicate ribbons we must also remember that the presence of one or several closed rings of Si–O–Si bonds in the identity period of the ribbon together with other favorable conditions, for example, a large ∠ SiOSi, may lead to a displacement of one or several symmetrical SiOSi stretching vibrations towards the frequencies of the deformation vibrations of the ribbon.

Amphiboles. These constitute the most widespread class of ribbon silicates. The silicon–oxygen ribbons of amphiboles are formed of two pyroxene chains symmetrical with respect to a plane parallel to the axes of the chains. Other symmetry elements of the amphibole ribbons are planes perpendicular to the ribbon axis in which the transverse Si–O–Si bonds lie, and a C_2 axis at the intersection of two planes (Fig. 45). The combination of these symmetry elements, together with the periodicity of the structure of the ribbon, forms a group the factor group of which is isomorphic with C_{2v}. The identity period of the $(Si_4O_{11})_\infty$ ribbon contains four silicon–oxygen tetrahedra, five Si–O–Si bonds, two $Si\genfrac{}{}{0pt}{}{O^-}{O^-}$ groups, and two $Si-O^-$. The general chemical formula of the amphiboles may be written in the form

$$X_{2-3}Y_5Z_8O_{22}(OH, F)_2 \ [58],$$

where X is a di- or monovalent cation with an ionic radius intermediate between the radii of Mg^{2+} and K^+, Y is a di- or tervalent cation with a radius in the range $r_{Al^{3+}} - r_{Mn^{3+}}$; Z is predominantly Si, perhaps partially replaced by Al and $Fe^{\cdots}$. In accordance with the strict relationship observed between the structures of amphiboles and pyroxenes [59], amphibole crystals may, depending on the nature of the cations, belong either to the monoclinic or orthorhombic systems. The structure of both classes of amphiboles has been studied by x-ray diffraction.

Figure 46 shows the infrared spectra of two monoclinic amphiboles, tremolite $Ca_2Mg_5 \cdot (Si_4O_{11})_2(OH)_2$ and actinolite $Ca_2(Mg, Fe)_5(Si_4O_{11})_2(OH)_2$, and also one of the orthorhombic amphiboles, anthophyllite $(Mg, Fe)_7(Si_4O_{11})_2(OH)_2$. The spectra of all three minerals contain a complex group of not less than seven or eight strong overlapping bands between 900 and 1100 cm^{-1}, and another group of intense bands below 550 cm^{-1}. In the frequency range 780 to 640 cm^{-1} there is a group of comparatively narrow, less intense bands, amounting to five maxima in the spectra of the monoclinic amphiboles and six in the spectrum of anthophyllite. These bands may clearly be compared with the five ν_sSiOSi vibrations in the spectrum of the isolated amphibole ribbon. An approximate description of the stretching vibrations of the $(Si_4O_{11})_\infty$ ribbon and their classification with respect to the irreducible representations of the C_{2v} group are given in Table 25, from which it follows in particular that of the five ν_sSiOSi vibrations one is inactive in the infrared spectrum. Hence it is desirable to return to x-ray structural data regarding the spatial symmetry and structure of these crystals.

Tremolite and actinolite belong to the space $C2/m - C_{2h}^3$ and contain four unit sections of Si_4O_{11} ribbons in the unit cell [60, 61]. The noncentered, primitive cell is half as large. Among

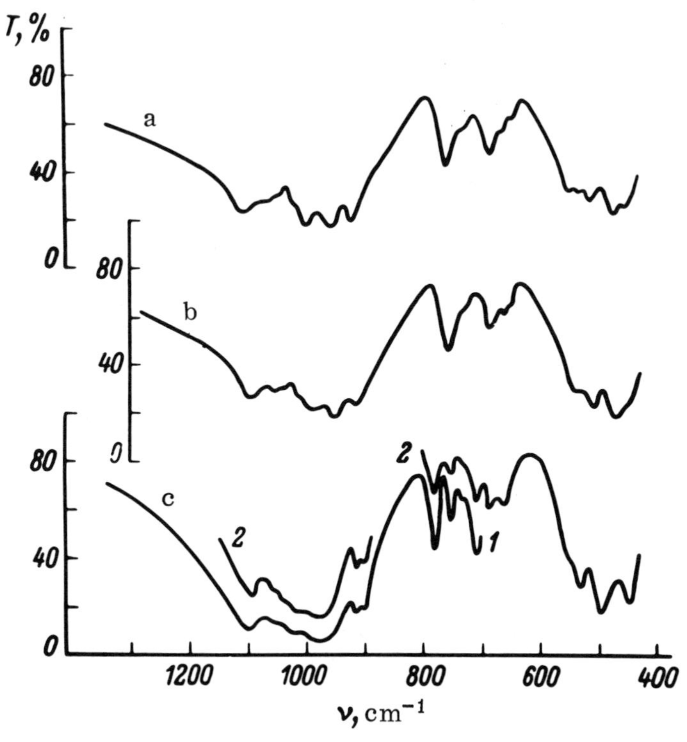

Fig. 46. Infrared spectra of amphiboles. a) tremolite (pressed in KBr, 5 mg); b) actinolite (5 mg); c) anthophyllite: 1) 10 mg, 2) 5 mg.

TABLE 25. Frequencies of the Absorption Maxima in the Spectra of the Amphiboles

Description of vibrations of the $(Si_4O_{11})_\infty$ ribbon	Symmetry types of the C_{2v} group	Frequencies, cm^{-1}		
		tremolite	actinolite	anthophyllite
ν_{as}SiOSi	B_2			
	B_2			
	A_1	1107 v.s.	1094 v.s.	1097 v.s., double?
	A_2	1074 s. double?	1052 s.	1059 s.
	B_1	1048 s.	1037 s.	1047 s.
		1020 s.	1012 s.	1016 v.s.
$\nu_{as}O^-SiO^-$	B_2	998 v.s.	985 v.s. double?	976 v.s., double?
	A_1	952 v.s.	948 v.s.	916 s.
$\nu_s O^-SiO^-$	B_2	920 v.s.	914 v.s.	903 s.
	A_1			
νSiO^-	B_2			
	A_1			
ν_sSiOSi	B_1	758 m.	755 m.	781 m.
	A_2	730 w.	727 w.	756 w.
	B_2	682 m.	682 m.	735 v.w.
	A_1	662 w.	658 w.	710 m.
	A_1	645 w.	646 w.	688 m.
				662 m.
Deformation vibrations of the ribbon (and M–O stretching vibrations)		545 s.	540 s.	553 s.
		530 s.	—	532 s.
		510 v.s.	507 v.s.	494 v.s.
		466 v.s.	476, 466 v.s.	—
		451 v.s.	453 v.s.	448 v.s.

the symmetry elements of the chain only the plane parallel to the axis of the chain is at the same time a symmetry element of the crystal. The relation between the irreducible representations of the factor groups of the one-dimensional group of the chain C_{2v}, its site group C_s in the crystal, and the three-dimensional group of the crystal C_{2h} is similar to that considered earlier for the monoclinic pyroxenes and leads to the conclusion that all the internal vibrations of the chains are active in the infrared spectrum of the crystal and that for each of these vibrations there is only one component active in the infrared spectrum. These conclusions entirely agree with the observed spectrum (it is not hard to explain the rather smaller than expected number of bands in the high-frequency region by partial superposition). The unit cell of anthophyllite, which belongs to the D_{2h}^{16}–Pnma space group, contains eight Si_4O_{11} groups, which might be considered as an indication of a site symmetry C_1 of the chain, since the order of the factor group of the space group is also equal to eight. Actually it is known from the results of x-ray analysis [62] that the silicate ribbons in the anthophyllite crystal possess a site symmetry C_s and form two sets of four Si_4O_{11} groups each, populating the same set of special positions (C_s) with nonfixed parameters x and z. The scheme of correlations of the irreducible representations of the factor groups C_{2v}, C_s, and D_{2h} for each of these sets has the form

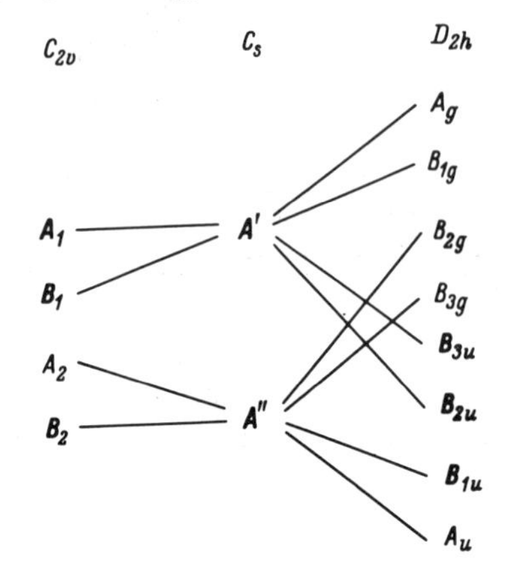

It follows from this that each of the sets (on allowing for splitting as a result of resonance interactions of the vibrations) will give the overall system eight components active in the infrared spectrum and lying in the ν_sSiOSi frequency range. Vibrations of types A_2 and B_2 for the isolated chain each give one, while the B_1 vibration and the two vibrations of the type A_1 each give two active components. The greater proportion of these splittings are evidently insignificant, since in the anthophyllite spectrum experiment only reveals one band more (in the ν_sSiOSi frequency range) than in the spectra of the monoclinic amphiboles.

Ribbons of More Complex Structure. According to x-ray structural data [63], xonotlite $Ca_6(Si_6O_{17})(OH)_2$ contains ribbons formed from two wollastonite chains. A characteristic feature of these chains is the transverse bridges, straightened to $\angle$ SiOSi = 180°. An idealized scheme of the construction of the xonotlite ribbon is shown in Fig. 45b, from which we see that the ribbon is symmetrical with respect to the two-fold axis directed along the ribbon axis, and the plane perpendicular to this axis, i.e., the factor group of the ribbon is isomorphic with C_{2h}, while the O atoms of the transverse Si–O–Si bridges lie an inversion centers. Considering that from the seven Si–O–Si bonds in the identity period of the strip it is possible to form three sets of equivalent bonds of this type, of which one is invariant with respect to all symmetry operations of the C_{2h} group and consists only of one Si–O–Si bond, the second is invariant to the operation σ_{xy} and contains two Si–O–Si bonds, while each of the four bonds of the third set is invariant only to the identity operation, it is not hard to convince oneself that the ν_sSiOSi vibrations are distributed in the following manner with respect to the irreducible representations of the C_{2h} group: A_g; A_g and B_u; A_g, B_g, A_u, and B_u. This indicates that only three of the ν_sSiOSi frequencies can appear in the infrared spectrum. However, of the symmetry elements of the ribbon only the two-fold axis is at the same time a symmetry element of the crystal, which belongs to the space group $P2/a-C_{2h}^4$ (Z = 2). Thus, writing the scheme of correlations in the form

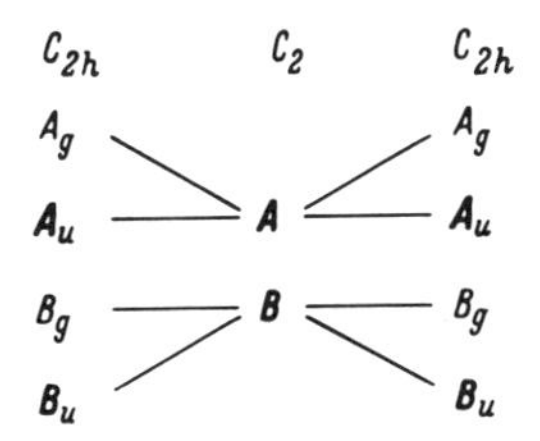

we arrive at the conclusion that one active representation of the factor group of the crystal corresponds to each of the seven normal vibrations of the ν_sSiOSi type. We note that not only in this case but also whenever the center of symmetry of the crystal does not coincide with the center of symmetry of the molecule or ion, the mutual exclusion rule is removed.

The infrared spectrum of xonotlite (Fig. 47) in the ν_sSiOSi frequency range shows four strong (539, 608, 630, and 670 cm^{-1}) and two weaker (710 and 739 cm^{-1}) bands. The latter may probably be ascribed to two of the ν_sSiOSi vibrations inactive in the spectrum of the isolated (and centrosymmetric) chain. Of the four stronger bands two, around 610 and 630 cm^{-1}, are associated with vibrations of the OH group in the xonotlite lattice, which follows from the study of the deuteroxonotlite spectrum in [63a]. Hence only two of the expected three fairly strong bands corresponding to the ν_sSiOSi vibrations active even in the spectrum of the isolated ribbon can be identified in the frequency range above 500 cm^{-1} (two further bands corresponding to vibrations inactive for the isolated chain may have a low intensity). If we exclude an improbable chance degeneracy, we have to assume the displacement of at least one of the ν_sSiOSi frequencies into the region of deformation vibrations of the complex anion. It should be noted that, apart from the small number of reasonably strong bands in the ν_sSiOSi frequency range, there are also other arguments indicating the insignificance of the deviation of the ribbon configuration from the "ideal" (centrosymmetric) configuration. The presence of an Si–O–Si bond with

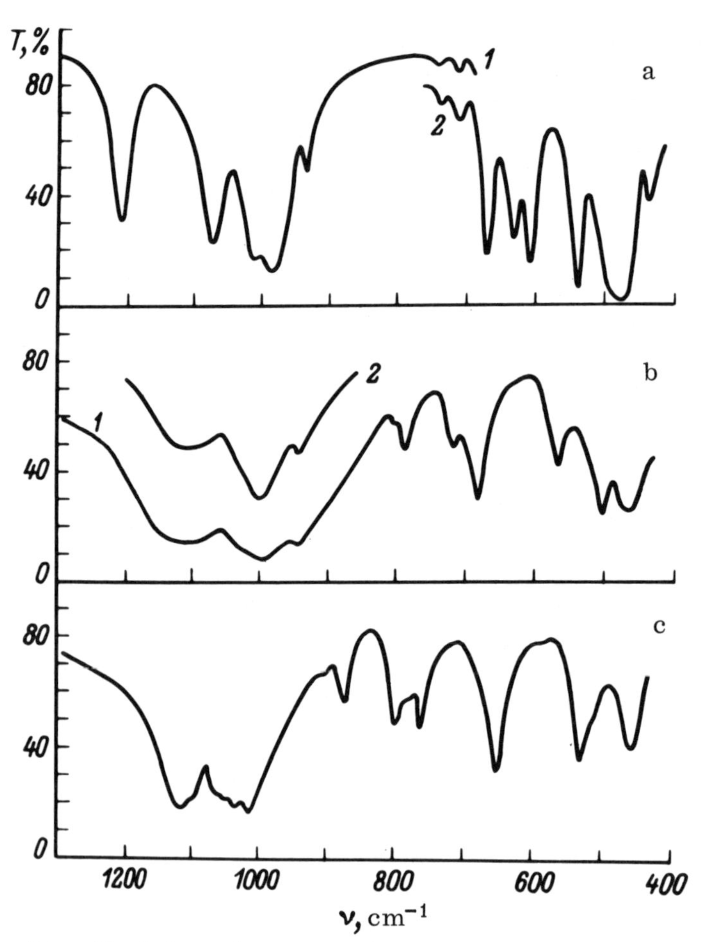

Fig. 47. Infrared spectra of silicates with ribbon anions of complex structure. a) xonotlite [pressed in KBr: 1) 2.5 mg, 2) 8 mg]; b) vlasovite [1) 5 mg, 2) 2.5 mg]; c) narsarsukite [2.5 mg].

∠SiOSi = 180° finds direct reflection in the high-frequency part of the spectrum: the spectrum of xonotlite contains a single narrow band† at 1200 cm^{-1}, i.e., considerably higher than the ordinary values of ν_{as} SiOSi, but close to the corresponding frequencies of thortveitite and batisite, which also have large SiOSi angles. The remaining six ν_{as} SiOSi vibrations, two vibrations of the $Si-O^-$ group, and eight vibrations of the $Si\genfrac{}{}{0pt}{}{<O^-}{<O^-}$ groups are represented in all by four bands observed between 1100 and 900 cm^{-1}.

In crystals of vlasovite $Na_4Zr_2(Si_8O_{22})$ x-ray examination [55] showed ribbons of a new type, which may be regarded as the result of the condensation not of chains but of rings of four silicon–oxygen tetrahedra (Fig. 48). Such a ribbon contains two rings in the identity period, the centers of the rings being inversion centers of the ribbon, while one ring transforms into another by the operation of a glide plane passing along the ribbon axis, or by rotation around a two-fold axis normal to the ribbon axis. Since the oxygen atoms of the Si–O–Si bonds joining the rings into the ribbon lie on the two-fold axes, the SiOSi angles in these bonds are equal to

†The ascription of this band to vibrations of the complex anion in xonotlite is proved by Ya. I. Ryskin, who observed it in the spectrum of deuteroxonotlite.

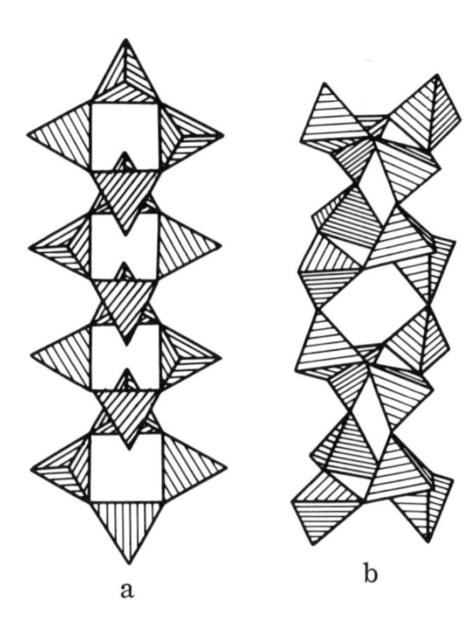

Fig. 48. Ribbons of silicon–oxygen tetrahedra: a) in vlasovite crystals; b) narsarsukite.

180° (the SiOSi angles in the rings lie at about 145°). All the symmetry elements indicated, composing a one-dimensional group the factor group of which is isomorphic with C_{2h}, are also symmetry elements of the crystal (space group $C2/c - C_{2h}^6$), the primitive cell of which contains one Si_8O_{22} grouping. Correspondingly the selection rules and number of vibrations for the isolated chain and for the whole crystal coincide. The ten Si–O–Si bonds form three sets of symmetrically equivalent groups: two each with four Si–O–Si bonds "in the general position" and one of two Si–O–Si bonds with intrinsic symmetry C_2. Then the ten symmetric-stretching SiOSi vibrations are distributed over the irreducible representations of the C_{2h} group in the following way: $2(A_g, B_g, A_u, B_u) + (A_g, A_u)$, which gives five vibrations active in the infrared spectrum. To these five vibrations correspond the five bands observed in the spectrum of vlasovite (Fig. 47) between 800 and 550 cm^{-1}: 798 (p), 786, 717, 682, 567 cm^{-1}. In the higher-frequency region five maxima in all are resolved, i.e., far fewer than might be expected from symmetry considerations. The highest frequency in the spectrum is not greater than 1140 cm^{-1}, although, as in the structure of xonotlite, the vlasovite ribbon contains SiOSi angles equal to 180°.

Figure 47 also shows the spectrum of narsarsukite $Na_2TiO(Si_4O_{10})$, a mineral containing silicon–oxygen chains in the form of tubes created by the condensation of rings of four tetrahedra (Fig. 48). These anions may also be considered as "double-threaded" helixes (their axis in the crystal coincides with the 4_2 axis), obtained by the condensation of two simple helical chains. Narsarsukite crystals belong to the $I4/m - C_{4h}^5$ space group and contain eight Si atoms in the primitive cell. Correspondingly the grouping Si_8O_{20} forming the identity period of the ribbon possesses the full symmetry of the factor group of the space group of the crystal. It is thus not hard to show that of the twelve ν_sSiOSi vibrations of the narsarsukite ribbon only three correspond to active representations of the C_{4h} group (two in E_u and one in A_u). However, in the infrared spectrum of narsarsukite, instead of the expected two, numerous bands are observed in the frequency range characteristic of the ν_sSiOSi vibrations. Considering the high convergence obtained when refining the structure of narsarsukite by the method of least squares [54], one can hardly assume a lower $(I\bar{4} - S_4^2)$ symmetry of the crystal. A reason for the discrepancies observed, as in the case of baotite, is probably that a considerable proportion of covalence in the Ti–O bonds raises the frequencies of the Ti–O vibrations. This primarily relates to the bonds between titanium and the "excess" oxygen atoms, not taking part in the formation of silicon–oxygen ribbons. Analogous conclusions were reached in [54] on consideration of the interatomic distances in the structure of narsarsukite by way of Pauling's electrostatic valence rule.

§3. Silicates with Layer Anions

Layer-like structures consisting of one or several sorts of monatomic cations and complex anions possessing translational symmetry in two dimensions are characteristic of a large number of natural and synthetic silicates. Layer structures with pronounced effects on the physical properties characterize, in particular, such practically important minerals as talc and mica. Together with two-dimensional silicate anions, many of these minerals contain $[OH]^-$ ions, lying like the cations between the layers. Several silicon–oxygen layers alternating with cations and hydroxyls often form electrically neutral "packets" connected to each other by Van der Waals interactions only.

In the majority of known structures with silicate anions in the form of simple layers, the ratio of the silicon and oxygen contents in the layer corresponds to the formula Si_2O_5, i.e., each silicon atom forms three bonds of the Si–O (Si) type and only one $Si-O^-$ (... M^+). We may expect that under otherwise equal conditions (same electropositivity of the atom M) the $Si-O^-$ bonds will possess a rather higher multiplicity than the corresponding bonds in chain or orthosilicate anions. The quantity and accuracy of existing data regarding the interatomic distance in silicates are insufficient to give a reliable verification of this proposition. The average bond lengths in the alkali meta- and disilicates suggest the possibility of a shortening of the Si–O (Si) bonds also, and not only the $Si-O^-$ (the accuracy of available data for the lithium silicates is considerably lower than for the sodium silicates):

	Na_2SiO_3 (chain) [13]	β-$Na_2Si_2O_5$ (layer) [64]	α-$Li_2Si_2O_5$ (layer) [65]
r_{Si-O^-}, Å	1.57	1.51	1.56
$r_{Si-O(Si)}$, Å	1.68	1.64	1.65
Δ, Å	0.11	0.13	0.09

Comparison of the data in the first two columns with those of the last indicates that the difference in the lengths of the $Si-O^-$ and Si–O (Si) bonds between the groups

$$\begin{matrix}(Si)O \\ (Si)O\end{matrix}\!\!>\!Si\!<\!\!\begin{matrix}O^- \\ O^-\end{matrix} \quad \text{and} \quad \begin{matrix}(Si)O \\ (Si)O \\ (Si)O\end{matrix}\!\!\Big>\!Si-O^-$$

does not exceed in magnitude the changes in length caused by differences in the electronegativity of the cations for the same type of structure of the complex anion. On analyzing the spectra we may suppose with fair certainty that the ratio of the bond orders $N_{Si-O^-}/N_{Si-O(Si)}$ is no greater than 1.5. This ratio is clearly much too high for structures of the mica type, where the presence of cations capable of forming partly covalent bonds with oxygen may lead to a substantial leveling of the orders of the $Si-O^-$ and Si–O (Si) bonds. As we shall show later, estimates of the ratio of the Si–O bond orders in layer-like anions on the basis of spectroscopic data are only possible in certain very simple cases, and even then they are not very accurate.

The classification of known configurations of the layers of silicon–oxygen tetrahedra is usually carried out in accordance with the "genetic" relationship between the structure of the layer and the disposition of the tetrahedra in the chains which may be regarded as generating the layer by condensation. Preserving this classification, we may now consider existing data regarding the relationship between the spectra and structure of layer-like anions. Methods of allowing for the symmetry properties of layer-like structures when determining the number of active vibrations in the spectrum have already been described in Chap. I.

It should be noted that silicates with a layer structure constitute one of the classes of silicate minerals most intensively studied by spectroscopic methods. Special attention has been paid to the infrared spectra of argillaceous minerals, primarily as a method of identification and analysis [66-75]. Less developed have been investigations associated with attempts at using spectroscopic methods for studying the structure of layer-like silicates. First place among such investigations relate to research into the vibrations of hydroxyl groups with the aim of classifying their degree of coupling, establishing the orientation of the hydroxyls relative to the crystallographic axes, and so on [76-82] (see also § 2, Chapter IV). An attempt at analyzing the vibrations of complex layer anions in talc was first made by Farmer [83], and later the vibrations of the deformed talc-like lattice in the spectra of micas were also considered [84, 85]. Calculations of the vibrational frequencies of the layer of silicon–oxygen tetrahedra were only carried out for the simplest case of the hexagonal lattice of talc. The secular equations were obtained for this model by Prima and Stepanov [26-28], using internal coordinates, and later by Vedder [85]. A calculation of the vibrations of a layer of the talc type was also made by Saksena [25] using Cartesian coordinates. The investigations of Stubican and Roy [86-88] are particularly important in the interpretation of the spectra of layer silicates. These authors made no attempt at comparing the bands in the spectra of the layer silicates with the vibrations of any simplified models, but made a detailed study of the changes in frequency and to some extent the intensities and shapes of the bands associated with isomorphous substitutions both in the complex anion (Si by Ge, Si by Al) and the cation positions ((Mg^{2+} by Ni^{2+} or Fe^{2+} and Al^{3+} by Ga^{3+} and Fe^{3+}), as well as the substitution of OD for OH. The results thus obtained will be used shortly in discussing the spectra of talc and micas. The majority of spectroscopic investigations into layer silicates relate to the infrared spectra of finely crystalline samples, and only in a few cases (for studying dichroism) have oriented polycrystalline films [83, 84] or the reflection spectra of single crystals [89, 90] been used. Investigations involving polarized radiation have usually merely embraced the range of vibrations of the hydroxyls, and only occasionally the bands of vibrations of the silicon–oxygen anion [85]. There are hardly any data regarding the Raman spectra of layer silicates.

Talc and Micas [89]. Talc $Mg_3(Si_4O_{10})(OH)_2$ is one of the most typical and spectroscopically well-studied silicates with layer silicon–oxygen anions. Figure 49 shows the structure of talc crystals derived from x-ray diffraction [91]; two layers of plane hexagonal lattices of tetrahedra face each other with their apical (unshared) oxygen atoms, one layer being displaced relative to the other, but related to it by a center of symmetry; the inversion center coincides with the Mg^{2+} ion, shown as a broken line. In the same layer as these ions are the remaining Mg^{2+} ions, also surrounded octahedrally by the oxygen atoms of the silicon–oxygen tetrahedra and $(OH)^-$ ions, the latter lying in spaces formed by the free vertices of the tetrahedra in the six-fold rings of the silicon–oxygen layers. In the plan we only show one of the two silicon–oxygen layers forming the packet and connected to each other through inversion centers at the points occupied by Mg^{2+}. In the similar structure of pyrophyllite $Al_2(Si_4O_{10})(OH)_2$ these cation positions are vacant, and the two others are occupied by Al^{3+} ions. The unit cell of the crystal, belonging to the $C2/c-C_{2h}^6$ (Z = 4) space group, contains two of the "bricks" thus described, connected to each other by van der Waals forces only; the total number of Si atoms in the cell is thus 16, and the smallest translationally-nonidentical (noncentered) cell contains eight Si atoms. The layer of silicon–oxygen tetrahedra considered as isolated from the remaining structure in talc may be obtained by the condensation of chains of the pyroxene type and possesses hexagonal symmetry; the factor-group of the two-dimensional space group is isomorphic with C_{6v}. The primitive cell of such a layer contains only two tetrahedra; correspondingly Si atoms occupy special positions with symmetry C_{3v} in the two-dimensional space lattice (p. 32), while the formula of the layer may be written in the form $(Si_2O_5)_{\infty,\infty}$.

It is desirable first of all to consider the changes in the spectrum on passing from one layer to the "brick," since the very weak interactions between the bricks can hardly have much influence on the vibrational spectrum. The factor-group of the two-dimensional space group of such a brick is isomorphic with C_{2h}; it forms inversion centers, siting 1/3 of all the Mg^{2+}

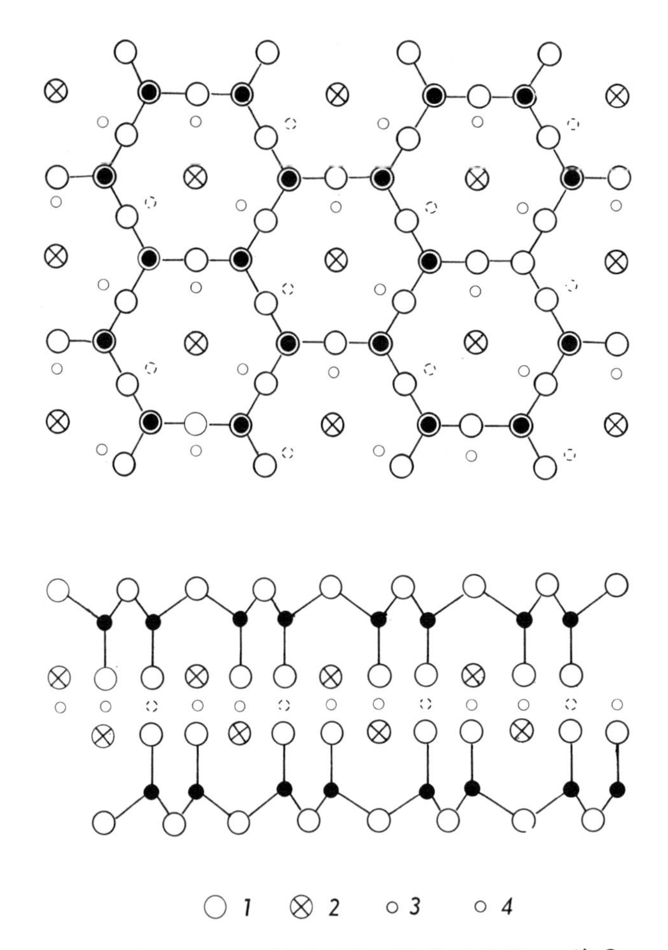

Fig. 49. Structure of talc $Mg_3(Si_4O_{10})(OH)_2$. 1) Oxygen; 2) hydroxyl; 3) silicon; 4) magnesium or aluminum.

ions, planes (normal to the plane of the layer) also containing these Mg^{2+} ions, the $(OH)^-$ ions, and 1/3 of all the bridge oxygen atoms, and finally two-fold axes passing normally to the plane of symmetry through the Mg^{2+} ions of both sorts. The $(Si_2O_5)_{\infty,\infty}$ layers are only invariant with respect to one of these elements (symmetry planes), and among the symmetry elements of the brick only the inversion centers are symmetry elements of the three-dimensional lattice. Hence the scheme of correlations between the factor-groups of the $(Si_2O_5)_{\infty,\infty}$ layer, its site symmetry in the brick, the total symmetry of the brick $Mg_3(Si_4O_{10})(OH)_2$ itself, its site symmetry in the crystal, and the space group of the crystal has the following form:

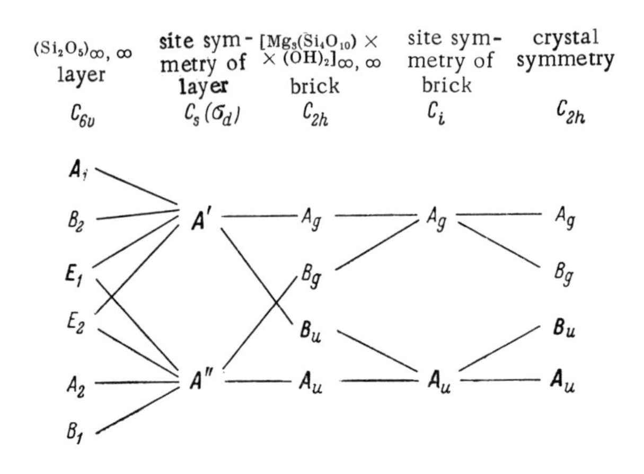

It follows from this that for each nondegenerate vibration of the layer there are two vibrations of the "brick," of which one is active in the infrared spectrum and the other in the Raman spectrum; the frequencies may differ considerably. In the spectrum of the three-dimensional crystal, for each vibration of the layer there are four branches; two of them, active in the infrared spectrum, are of types A_u and B_u and similar in frequency, save that one has its electric vector along the two-fold axis and the other in a plane normal to this axis. Hence the conclusion drawn from the symmetry of the isolated layer that the dipole moments are oriented either strictly in the plane of the layer or at right angles to it cannot be fulfilled absolutely exactly.

Let us now try to identify the frequencies in the infrared spectrum of talc (Fig. 50). Since the primitive cell of the silicon–oxygen layer contains only two tetrahedra, i.e., two bonds of the $Si-O^-$ and three of the $Si-O-Si$ types, in the vibrational spectrum of the "two-dimensional" crystal there will be eight $Si-O$ stretching vibrations: $3\nu_{as}$ SiOSi, 2ν SiO^- and $3\nu_s$ SiOSi. In view of the high intrinsic symmetry of the layer, some of the vibrations are degenerate. We see from Table 26 that in the infrared spectrum only three stretching vibrations are active, of which two relate to symmetry type A_1 and one to E_1. It is natural to compare these three vibrations with the three strong bands observed in the talc spectrum between 1100 and 600 cm^{-1}. The lowest-frequency vibration ν_sSiOSi (A_1) clearly corresponds to the band at 672 cm^{-1}, and the vibrations ν_{as} SiOSi (E_1) and νSiO^- (A_1) to the bands at 1043 and 1018 cm^{-1}. In the reflection spectrum of a plate of talc cut along the cleavage plane, i.e., parallel to the silicon–oxygen layers, for an angle of incidence deviating by 17° from the normal, there are also two high-frequency bands, of which the higher has the greater intensity. In the absorption spectrum of a sample obtained by pressing powder mixed with potassium bromide, in which any preferred orientations of individual single crystals are clearly improbable, the ratio of intensities of these bands is reversed. Hence the vibration of type E_1 for which the dipole moment in the plane of the layer changes may be associated with the frequency 1043 cm^{-1} and the vibration of type A_1 with 1018 cm^{-1}, in agreement with the proposed frequency ratio of the stretching vibrations: ν_{as} SiOSi > νSiO^- > ν_sSiOSi.

In Farmer's paper [83] the relative intensities of the bands in the spectrum of a talc sample obtained by pressing with KBr were compared with those in the spectra of samples prepared by depositing a suspension in isopropyl alcohol. The relative intensity of the high-frequency component of the doublet diminished in the spectrum of the sample obtained by sedimentation, and on the assumption that the talc plates were preferentially oriented along the plane of the substrate this was taken as an indication that this band belonged to a vibration of type A_1. A probable explanation for the contradiction between these results and the reflection spectrum of talc, according to Farmer [84], lay in the dependence of the spectrum on the interaction between the lattice vibrations and the field of dielectric polarization induced by these vibrations

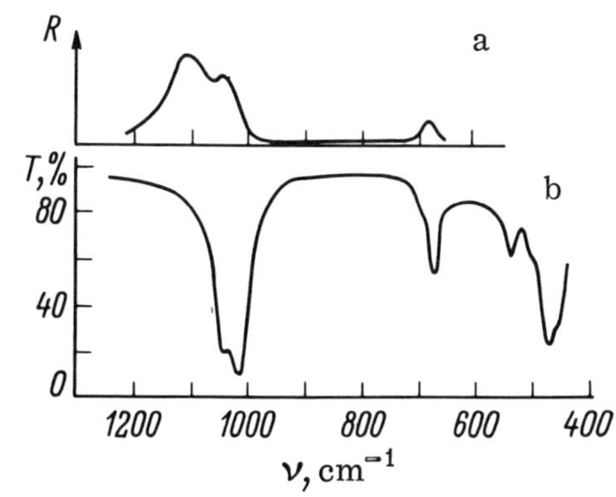

Fig. 50. Infrared spectrum of talc. a) Reflection spectrum of a plate cut along the cleavage plane (incident angle 17°); b) absorption spectrum (pressed in KBr, 1 mg).

TABLE 26. Interpretation of the IR Spectra of Talc

IR frequencies (cm^{-1}) and intensity		Ascription of frequencies	
Absorption of polycrystalline aggregate	Reflection of single crystal (plane of sample along cleavage planes)	Symmetry types (and selection rules) of vibrations of the $(Si_2O_5)_{\infty,\infty}$ layer with symmetry C_{6v}	Description of vibrations
1043 v.s. —	1110 v.s. —	E_1 $a\perp$; a, dp B_1 ina; ina	ν_{as}SiOSi
1018 v.s. —	1043 v.s. —	A_1 $a\parallel$; a, p B_1 ina; ina	νSiO$^-$
— 672 s.*	— 685 m.	E_2 ina; a, dp A_1 $a\parallel$; a, p	ν_sSiOSi
536 m. 501 m. 469 v.s. 456 v.s.	Not studied	Deformation vibrations of the layer modified by lattice vibrations	

Note: The selection rules are first given for the IRS with an indication of the orientation of the dipole moment relative to the C_6 axis, then for the Raman spectrum with an indication of the state of polarization of the line.

In the absorption spectrum of a single crystal 0.02 mm thick, bands are found at 943 and 779 cm^{-1} corresponding to the vibrations νSiO$^-$ (B_1) and ν_sSiOSi (E_2).

*The band at 672 cm^{-1} has been identified as a δOH vibration [104]. The A_1 vibration is a "shoulder" at 690 cm^{-1}, also present in the reflection spectrum: see also footnote, p. 122.

in the crystal. If the talc crystal is regarded as uniaxial, then, for dimensions larger than the wavelength of the lattice vibrations, the frequency of the maximum in the absorption spectrum lies between the frequencies ν_t and ν_l of the transverse and longitudinal vibrations of the unbounded (infinite) crystal. For the vibration with dipole moment along the optic axis the dependence of the frequency on the angle θ between the optic axis and the wave vector of the lattice vibrations takes the form

$$\nu_0^2 = \nu_t^2 + (\nu_l^2 - \nu_t^2)\cos^2\theta$$

Correspondingly, for the vibration with a dipole moment perpendicular to the optic axis and lying in the plane defined by this axis and the wave vector

$$\nu_0^2 = \nu_t^2 + (\nu_l^2 - \nu_t^2)\sin^2\theta$$

while, for the vibration with a dipole moment normal both to the optic axis and to the wave vector, $\nu_\theta = \nu_t$ for any θ. However, in the spectrum of a plane, uniaxial crystal with a thickness much smaller than the wavelength of the lattice vibrations and an optic axis normal to the plane of the plate, vibrations perpendicular to the plane will be observed with a frequency ν_l, and vibrations in the plane of the plate with a frequency ν_t, independent of the orientation of the infrared radiation exciting these vibrations. From this point of view, in the absorption spectrum of finely crystalline talc the νSiO$^-$ (A_1) vibration appears with a frequency ν_l and the ν_{as}SiOSi (E_1) with a frequency ν_t, to which Farmer respectively ascribed the 1045 and 1018 cm^{-1} bands. Then in the reflection spectrum from the surface of a single crystal perpendicular to the optic axis the main part of the intensity of the maximum associated (neglecting extra-axial beams and the deviation of the incident angle from the normal) with the E_1-type vibration lies between the frequencies ν_l and ν_t for this vibration, from which it follows that $\nu_l \simeq 1120$ cm^{-1}. Thus in accordance with the foregoing proposition ν_{as}SiOSi(E_1) > νSiO$^-$(A_1).

Arguments in favor of the proposed identification of the ν_{as}SiOSi(E_1) and νSiO$^-$(A_1) vibration frequencies may also be obtained by calculating the frequencies of the stretching vibrations of the layer. In Fig. 51 we compare the frequencies of the absorption maxima in the infrared

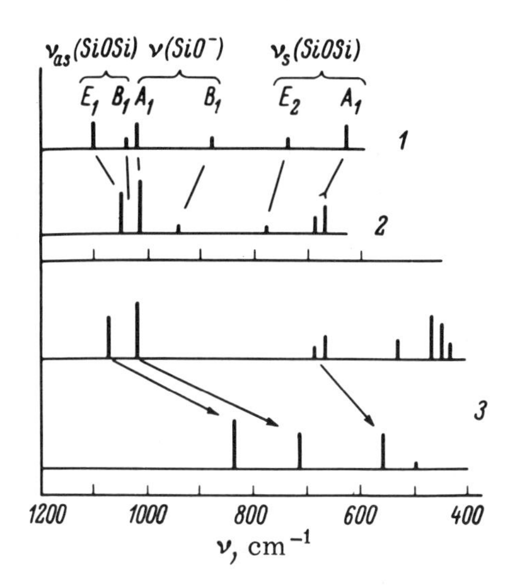

Fig. 51. Diagram to illustrate the interpretation of the talc spectrum. 1) Calculated frequencies [28] of the $(Si_2O_5)_{\infty,\infty}$ lattice (C_{6v}); 2) infrared spectrum of talc; 3) change in latter on putting Ge for Si [87].

spectrum of talc with the results of a calculation of the frequencies of the isolated layer carried out by Stepanov and Prima [28] for the following values of the force constants:

$$K_{Si-O(Si)} = 7.51 \cdot 10^6\ cm^{-2},\ K_{Si-O^-} = 1.25\ K_{Si-O(Si)},$$
$$K_{\delta OSiO} = \tfrac{1}{13} K_{Si-O(Si)},\ K_{\delta SiOSi} = 0.3\ K_{\delta OSiO}.$$

An analogous calculation carried out by Saksena [25] using only three force constants: $K_{Si-O(Si)} = 4.0$, $K_{Si-O^-} = 5.0$, and $K'_{\delta OSiO} = K_{\delta OSiO}/r^2_{Si-O} = 0.7 \cdot 10^5$ dyn/cm, also leads to ν_{as} SiOSi$(E_1) > \nu_{as}$ $SiO^-(A_1)$ (1128 and 1007 cm^{-1} respectively). Since we are concerned with the two highest-frequency vibrations of the complex anion, to a first approximation we may confine consideration simply to the two stretching force constants and use the simplified secular equations of the vibrations of the layer taken from [26] in order to consider the relationship between the frequencies of these two vibrations and the ratio of the orders of the bonds, and hence of the force constants K_{Si-O^-} and $K_{Si-O(Si)}$. A calculation of the frequencies for different values of the ratio $K_{Si-O^-}/K_{Si-O(Si)}$ shows that $\nu\,SiO^-(A_1) < \nu_{as}$ SiOSi(E_1), if $K_{Si-O^-}/K_{Si-O(Si)}$ is no greater than 1.5. The most satisfactory agreement between the calculated and observed frequencies is achieved for values of $K_{Si-O^-}/K_{Si-O(Si)}$ of 1.2 to 1.3. This not only confirms the validity of the ascription of the higher-frequency vibration in the talc spectrum to type E_1 but also shows that in the silicon–oxygen tetrahedron it is impossible for a π-bond to be formed between the Si and O, even if approximately equal in strength to the σ-bonds, despite the fact that, in the present case, the existence of only one bond of the $Si-O^-$ type with three other bonds of the Si–O(Si) type tends to suggest the concentration of the donor–acceptor $p\pi-d\pi$ interaction primarily in this bond. It is not impossible, of course, that the replacement of magnesium by an alkali cation may substantially raise the $K_{Si-O^-}/K_{Si-O(Si)}$ ratio in comparison with the value of 1.2 to 1.3 characteristic of talc.

An additional confirmation of the assignment of the bands of the three stretching vibrations of the layer in the talc spectrum may be obtained by comparing the spectrum with the germanium analog of talc $Mg_3(Ge_4O_{10})(OH)_2$ studied by Stubican and Roy [87]. The three strong bands in the spectrum of Ge talc (Fig. 51), according to the interpretation of the spectrum of Si talc, should be ascribed to the vibrations ν_{as} GeOGe(E_1), $\nu\,GeO^-(A_1)$, and ν_s GeOGe(A_1): 837, 717, and 561 cm^{-1}. We can show that the ratio of the frequencies of the two first bands is in agreement with the values of the corresponding frequencies in the spectrum of the silicon analog. The frequencies of the ν_s SiOSi and ν_s GeOGe vibrations depend considerably on the deformation force constants and come out too low for any values of the stretching force constants alone. If in order to determine the vibration frequencies in Ge talc we use the simplified secular equations, then the frequency 840 cm^{-1} for the vibration of type E_1 is obtained for $K_{Ge-O(Ge)}$ = $3.2 \cdot 10^5$ dyn/cm. Taking (as in the silicon analog) K_{Ge-O^-} = 1.2 $K_{Ge-O(Ge)}$, for the frequency of the Ge–O$^-$ vibration of type A_1 we obtain 727 cm^{-1}, in good agreement with experiment. This firstly confirms the validity of the assignment of the frequencies of the stretching vibrations in the talc spectrum, and secondly indicates that the degree of nonequivalence between the Ge–O$^-$ and Ge–O–Ge bonds is about the same as in the corresponding bonds in the silicates, or slightly less so.

As already noted, allowance for the interaction between the vibrations of the layers in the talc crystal removes the ban from the vibrations inactive in the infrared spectrum of the isolated layer. In the absorption spectrum of comparatively thick (20 μ) talc plates there are bands at 779 and 943 cm^{-1}, which may presumably be ascribed to the ν_s SiOSi (E_2) and νSiO$^-$ (B_1) vibrations. According to an approximate calculation the frequencies should lie at about 750 and 900 cm^{-1} (Fig. 51). Thus we are able to identify altogether five out of the six (allowing for degeneracy) stretching vibrations of the layer of silicon–oxygen tetrahedra. The sixth vibration, ν_{as} SiOSi of type B_1, according to calculation falls between the frequencies of ν_{as} SiOSi (E_1) and νSiO$^-$ (A_1), and, although it is not forbidden in the spectrum of the crystal, it cannot be observed on the background of the far stronger neighboring bands. The bands observed in the talc spectrum at 530 to 450 cm^{-1} mainly relate to the deformation vibrations of the silicon–oxygen layer. Farmer's attempt, based on an analogy with the spectra of MgO and Mg $(OH)_2$, to compare the absorption at about 450 cm^{-1} with the frequencies of the Mg–O vibrations fails to agree with the strong displacement of these bands on passing to the spectrum of Ge talc.

Let us now discuss possible reasons for the doublet nature of the ν_sSiOSi (A_1) vibrations. Apart from the absorption maximum at 672 cm^{-1} the spectrum shows a "shoulder" at about 690 cm^{-1}; the doublet structure of the band is confirmed by the reflection spectrum as well. Farmer, comparing the spectra of samples obtained by pressing with KBr and by sedimentation from a suspension, observed differences in the relative intensity of the two components of this band and ascribed them to vibrations of types A_1 and E_1 [83]. Saksena's calculated vibration frequencies of the layer agree with the proposition as to the existence of A_1 and E_1 vibrations of similar frequency in the region under consideration: A_1 – 1007, 650 cm^{-1}; E_1 – 1128, 646, 344 cm^{-1}; B_1 – 1009, 859, 381 cm^{-1}; E_2 – 768, 547, 93 cm^{-1}. On the other hand, an analysis of the spectra of a considerable number of silicates indicates the improbability of deformation vibrations of the silicon–oxygen anion appearing above 650 to 600 cm^{-1}. We may purely formally relate the frequencies at 672 and 690 cm^{-1} to the two components of types A_u and B_u into which the ν_sSiOSi (A_1) vibration decomposes in the spectrum of the crystal. However, this splitting, as indicated earlier, is due to the van der Waals interaction of the layers and can hardly be observed experimentally. A more probable cause of the double nature of the band may be the disruption of the selection rules (activation of the A_g vibration of the talc "brick") by the lower symmetry of the real crystals.† In this case the splitting of the ν_sSiOSi (A_1) vibration is associated with the interaction of the $(Si_2O_5)_{\infty,\infty}$ layers through the cations as illustrated in Fig. 52, which compares the spectrum of talc with that of synthetic talc in which the Mg^{2+} ions are replaced by Ni^{2+} and that of pyrophyllite.

As the proportion of covalence in the bonds of the cation increases (and hence the interaction between the vibrations of the layers intensifies) in the order Mg–Ni–Al, the stretching frequencies of the complex anion rise. At the same time the splitting of the ν_s SiOSi (A_1) vibrations increases, while the number of components remains the same. It should be noted that the considerable changes in the frequencies of the complex anion in the spectrum of pyrophyllite,

†Results from the inelastic scattering of neutrons in layer-like silicates [A. W. Naumann et al., in: Clays and Clay Minerals, S. W. Bailey, ed., Pergamon (1966), p. 367] enable us to associate one of the components of the doublet with the δMgOH vibrations in talc.

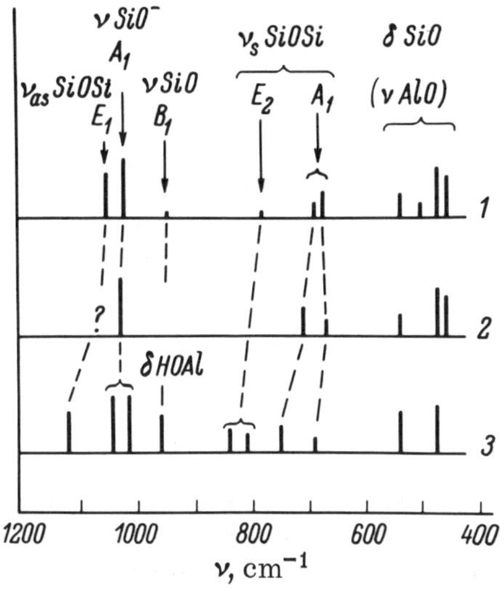

Fig. 52. Comparison of the infrared spectra of: 1) talc $Mg_3(Si_4O_{10})(OH)_2$; 2) nickel talc $Ni_3(Si_4O_{10})(OH)_2$; and 3) pyrophyllite $Al_2(Si_4O_{10})(OH)_2$.

particularly the splitting of the νSiO^- (A_1) vibration into two components with differently oriented electric vectors [83], and the presence of two [87] or three [84] fairly strong bands in the $\nu_s SiOSi$ (E_2) vibration range, are probably associated not only with the change in the nature of the cations but also with a change in the symmetry of the field created by them. The structures of talc and pyrophyllite are similar, but for every three Mg^{2+} ions in talc the pyrophyllite contains two Al^{3+} cations; hence the cation positions shown by broken lines in Fig. 49 remain free in pyrophyllite, which corresponds to a further deviation of the silicon–oxygen layer from ideal hexagonal symmetry.†

It is interesting to note that the frequency of the $\nu_s SiOSi$ (A_1) vibration of the silicon–oxygen layer in talc shows a clear dependence on the substitutions in the anion part of the structure. Stubican and Roy [86], studying the spectra of saponites, which may be considered as the results of replacing some of the Si atoms in talc by Al (with compensation for the loss of the electroneutrality of the talc bricks by the introduction of Na^+ ions between them: $Mg_3[(Al_xSi_{4-x})\cdot O_{10}](OH)_2\cdot Na_x$), established that an increase in aluminum content led to a reduction in the frequency of this band, roughly in accordance with a linear law. At the same time there is a decrease in the absorption intensity at the maximum of the band and an increase in its width; this is evidently associated with the random character of the distribution of Al atoms in the structure of the anion. For a substantial Al content the $\nu_s SiOSi$ band is greatly blurred and becomes less obvious.

Spectra similar in general features to the simplest spectra considered here characterize many minerals with layer-like anions such as micas, clay minerals, and so on. Thus, for example, the spectrum of antigorite $Mg_3(Si_2O_5)(OH)_4$, a 1:1 trioctahedral mineral, contains three bands in the Si–O stretching range [86], close in frequency to the three active vibrations of the layer in the talc spectrum: $\nu_{as} SiOSi$ (E_1), νSiO^- (A_1), and $\nu_s SiOSi$ (A_1). These bands were also observed in the reflection spectrum [92] at 1048, 993, and 673 cm^{-1}, and (as in talc) of the

†According to the accepted terminology, talc and pyrophyllite are layer silicates of the 2:1 type, i.e., containing one layer of cations to two silicon–oxygen layers, but talc furthermore belongs to the trioctahedral type (all three possible positions of the cations are occupied) while pyrophyllite belongs to the dioctahedral. Minerals of 1:1 type are distinguished from those of 2:1 type by the fact that one of the two silicon–oxygen layers is replaced by a layer of hydroxyls. Kaolinite $Al_2(Si_2O_5)(OH)_4$ is an example of a dioctahedral 1:1 mineral; chrysotile $Mg_3(Si_2O_5)(OH)_4$, a trioctahedral 1:1 mineral.

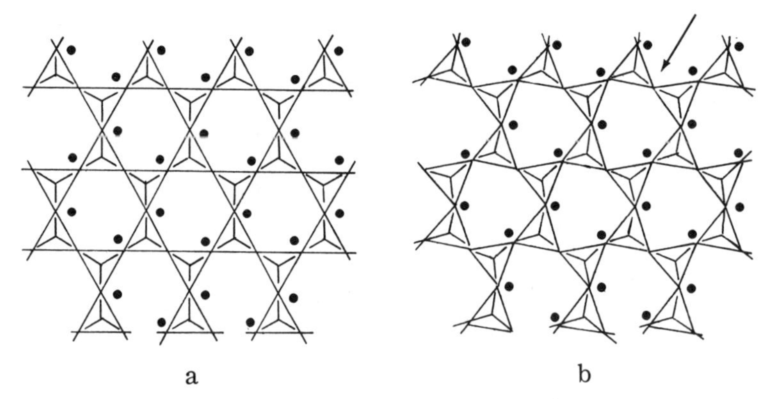

Fig. 53. Configuration of the silicon–oxygen layer in dioctahedral minerals. a) Idealized; b) real, retaining only one symmetry plane, normal to the layer (indicated by an arrow) of the elements of the C_{6v} group.

two bands 1048 and 993 cm^{-1} the high-frequency one was stronger in the reflection spectrum, but the low-frequency one in the absorption spectrum.

The spectrum of the most important 1:1 dioctahedral mineral, kaolinite $Al_2(Si_2O_5)(OH)_4$ [18, 84, 86, 87], and the analogous dickite, nacrite, and halloysite differ little from the pyrophyllite spectrum shown in Fig. 52. The number of strong bands in the frequency range of Si–O stretching is greater than in the talc spectrum; this is attributed [84, 85] to the substantial deviations in the form of the layer from hexagonal symmetry. As shown in Fig. 53, the real configuration of the layer in these minerals corresponds to symmetry C_s. Farmer and Russell [84] sought to identify the vibrational frequencies of the complex anion on the basis of a classification in terms of the representations of the C_{2v} group, which in the opinion of those authors was suitable for describing the symmetry of the layer, and also data relating to differences in the intensities of the absorption bands for a disordered or partly ordered orientation of the crystallites in the sample. The proposed interpretation of the spectrum, however, leads to the conclusion that the $\nu SiO^-(A_1)$ vibration frequency is higher than the $\nu_{as}SiOSi(E_1)$, which appears unlikely on considering the ratios of the orders of the bonds and the force constants of $Si-O^-$ and Si–O(Si). Possible reasons for this contradiction were discussed earlier.

Among the 2:1 dioctahedral minerals the spectrum of muscovite $KAl_2(AlSi_3O_{10})(OH)_2$ [85], together with pyrophyllite, has been studied in great detail, both in absorption and reflection. The spectra showed considerable differences not only in relation to whether the electric vector were perpendicular or parallel to the cleavage plane, but also in relation to the orientation of the electric vector in the cleavage plane along the a or b axes; this confirms the view that the model of the isolated $(Si_2O_5)_{\infty,\infty}$ layer has only a limited application. Vedder's proposed interpretation of the Si–O vibration bands is based on calculating the frequencies of a two-dimensional lattice of the talc type (C_{6v}) with three force constants, and agrees satisfactorily with the assignment of the Si–O vibration bands in the talc spectrum considered here. The only important difference lies in the assignment of the strong band† at 520 cm^{-1} to the $\nu_s SiOSi(A_1)$ vibration. A 30% fall in the frequency of the internal vibration as compared with that of talc is hard to explain. It is probably more reasonable to associate the $\nu_s SiOSi(A_1)$ vibration with the weak bands at 750 to 700 cm^{-1}; their low intensity and blurred contour may be ascribed to the

†The agreement of this frequency with the results of the calculation carries little weight, since the model of the isolated anion as a rule leads to too low values for the lowest of the stretching frequencies although satisfactory at higher frequencies.

influence of the partial replacement of Si by Al in the complex anion, in the same way as was observed in the spectra of the saponites.

Layer Silicates Formed from $(Si_2O_6)_\infty$ Chains [90]. These are represented by a fairly large number of synthetic and natural alkali and alkaline-earth silicates. A common feature in these structures is the existence of layers consisting of six-fold rings of silicon–oxygen tetrahedra, although not even approximately possessing the hexagonal symmetry characteristic of talc. Whereas the layers in talc may be considered as the result of the condensation of pyroxene chains with symmetry C_{2v}, in these silicates the layers may be derived from chains of alkali metasilicates or barium metasilicate [64, 93, 94]. A characteristic of such chains is the existence of two-fold screw axes along the axis of the chain. In the crystals of the alkali metasilicates the chains also possess two planes: a glide plane, and a symmetry plane in which Si atoms and bridge O atoms lie. However, on condensation, as indicated by Liebau [65, 96, 97], the configuration of the chains is somewhat distorted, and some of their symmetry elements are lost.

Figure 54 gives a schematic representation of the structure of the silicon–oxygen layer in sanbornite, a low-temperature form of $BaSi_2O_5$ [95]. The smallest translationally-nonidentical cell of this layer (shown by a broken line) contains four $Si-O^-$ and six $Si-O-Si$ bonds. Hence the vibrational spectrum of the $(Si_4O_{10})_{\infty,\infty}$ layer should contain six ν_s SiOSi vibrations, which probably lie in the range 550 to 800 cm^{-1}; in view of the low symmetry of the layer none of the vibrations are degenerate. Thus we may expect a considerable complication of the spectrum as compared with that of talc. The "two-dimensional" lattice of the layer in sanbornite possesses the following symmetry elements: screw axes $\bar{C}_2$ along the direction of the metasilicate chains, reflection planes σ linking these chains into the layer, and a glide plane $\bar{\sigma}$ with a diagonal translation. By setting up the multiplication table for these elements it is not difficult to see that the factor-group of the two-dimensional space group formed by them is isomorphic with respect to C_{2v}. Table 27 gives the distribution of the stretching vibrations of the layer with respect to the symmetry types of this group. A transformation to the factor group of the three-dimensional space group Pmcn-D_{2h}^{16} (the cell contains sections of two layers) will not alter the number of frequencies expected in the infrared and Raman spectrum.

The infrared spectrum of sanbornite is given in Fig. 55 and the frequencies of the absorption maxima and their proposed assignments in Table 27. Of the six ν_sSiOSi vibrations, one is of type A_2 inactive in the infrared spectrum, but of the five vibrations allowed by the selection rules only four are observed, corresponding to the absorption bands between 800 and 620 cm^{-1}. We may suppose that yet another ν_sSiOSi vibration, probably of type A_1, lies in the lower frequency range and may be masked by the side of the very strong band at 530 to 540 cm^{-1}. In view of its high intensity the latter can hardly be ascribed to the ν_sSiOSi vibration. This view is all the more likely in view of the fact that the lowest-frequency ν_sSiOSi vibration is the "symmetrical" vibration of the six-fold ring of tetrahedra in the structure of the layer, partly bearing the character of a deformation vibration, and should be low in frequency and possess

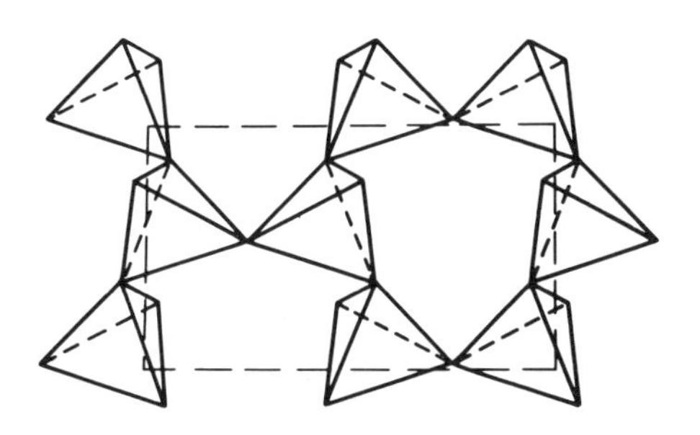

Fig. 54. Structure of the silicon–oxygen layer in sanbornite $Ba_2Si_4O_{10}$ [95].

TABLE 27. Spectra of Silicates with $(Si_4O_{10})_{\infty,\infty}$ Layers

Sanbornite $BaSi_2O_5$			Sodium disilicate $Na_2Si_2O_5$		
Frequencies, cm^{-1}, and intensities in the IRS of the polycrystalline aggregate	description of the vibrations of the layer, symmetry type, and selection rules for the C_{2v} group		crystalline α-$Na_2Si_2O_5$: Freq., cm^{-1}, intensities, in polycrystalline aggregate IRS	vitreous $Na_2Si_2O_5$: frequencies, intensities, and polarization of the RS lines	classification of the vibrations
			1185 s.	—	
		$2B_2$ *a*	—	1132 (2)	
(1197 v.w.	ν_{as}SiOSi	B_1 *a*	1105 s.	1105 (10, p.)	
~1138 v.s.		$2A_2$ *ina*	—	1071 (1)	
1089 v.s.		A_1 *a*	1054 s.	—	ν_{as}SiOSi и νSiO^-
~1060 ?			1032 w.	—	
1038 m.		B_2 *a*	1010 v.s.	—	
998 v.s.	νSiO^-	B_1 *a*	966 v.s.	—	
978 v.s.		A_2 *ina*	—	942 (3, p.)	
		A_1 *a*	(867 v.w.)	—	
800 m.			—	780 (3, dp.)	
759 m.		B_2 *a*	768 m.	—	
694 m.	ν_sSiOSi	$2B_1$ *a*	738 m.	—	ν_sSiOSi
621 w.		A_2 *ina*	~700 p.	—	
?		$2A_1$ *a*	602 m.	600 (2)	
			—	565 (7, p.)	
540 s.			519 v.s.	—	
534 s.	δSiO		481 v.s.	—	δSiO
483 v.s.					
463 v.s.					

Note: In brackets we indicate the frequencies of weak maxima not belonging to the fundamental vibrations of the complex anion. The intensities in the RS of the vitreous $Na_2Si_2O_5$ are given in accordance with visual estimates [99], and the state of polarization of the lines is taken from [100].

a negligible intensity in the infrared spectrum. In view of the absence of data relating to the Raman spectrum of sanbornite, we give the frequencies obtained in the spectrum of a glass of composition $BaSi_2O_5$: 1115–m., 1035–m., 920–v.w., 715–?, 640–m., 595–w., 518–v.w., 170–w. The structure of the glass is probably more similar to that of the high-temperature form of $BaSi_2O_5$, but this differs little from the low-temperature form. Accordingly, the presence of the band at 595 cm^{-1} may presumably be associated with the ν_s SiOSi vibration remaining unobserved in the infrared spectrum. In the ν_{as}SiOSi and νSiO^- frequency range there are six maxima instead of seven expected from the selection rules. These bands are grouped in two regions of strong absorption, possibly in keeping with their assignment to vibrations of one type or the other.

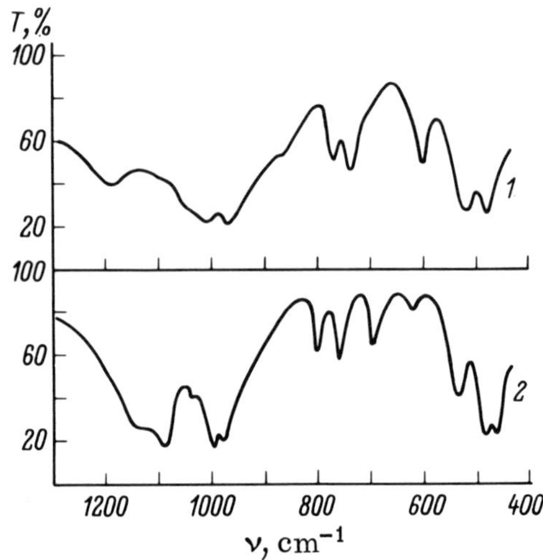

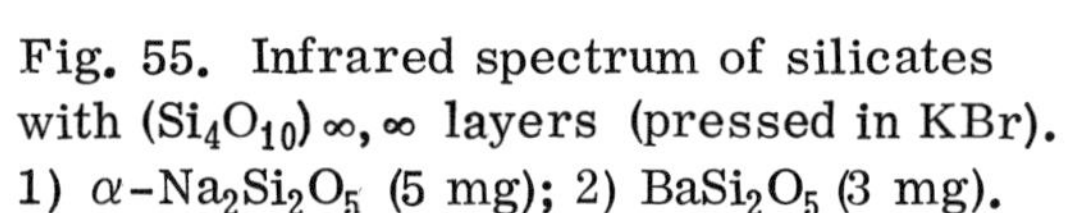

Fig. 55. Infrared spectrum of silicates with $(Si_4O_{10})_{\infty,\infty}$ layers (pressed in KBr). 1) α-$Na_2Si_2O_5$ (5 mg); 2) $BaSi_2O_5$ (3 mg).

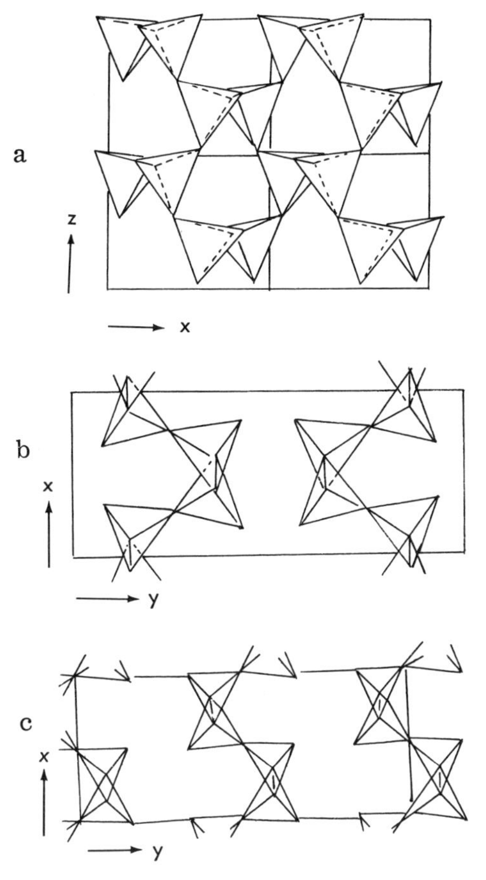

Fig. 56. Structure of the silicon–oxygen layer (schematic). a) and b) $Li_2Si_2O_5$ [65]; c) α-$Na_2Si_2O_5$ [96].

The silicon–oxygen layers in crystals of alkali disilicates $M_2Si_2O_5$ possess a similar type of structure. The structure of the complex anions in $Li_2Si_2O_5$ [65] and α-$Na_2Si_2O_5$ [96] is shown in Fig. 56. For α-$Na_2Si_2O_5$ only one projection is given, since the parameters along the z axis were not determined by Liebau [65]. The unit cell of the $(Si_4O_{10})_{\infty,\infty}$ layer, as in sanborite, contains six Si–O–Si bonds, which leads us to expect the appearance of six ν_sSiOSi frequencies in the spectrum. The infrared spectrum of α-$Na_2Si_2O_5$ is given in Fig. 55 (1) and contains altogether three bands in the ν_sSiOSi frequency range, or four if one counts the weak "shoulder" on the 738 cm^{-1} band at about 700 cm^{-1}. We note, however, that the absorption around 700 and 867 cm^{-1} in the α-$Na_2Si_2O_5$ spectrum may be due to a trace of sodium metasilicate.

The $Li_2Si_2O_5$ spectrum is very similar, but most of the bands are a little higher in frequency [98]. Table 27 also shows some data relating to the Raman spectrum of glass of composition $Na_2Si_2O_5$. We may suppose that the relative arrangement of the tetrahedra in the glass is similar to the structure of α-$Na_2Si_2O_5$ crystals, since the crystallization of the glass at high temperatures leads to the formation of precisely this form of sodium disilicate [14, 96]. The use of Raman spectra may facilitate the identification of the ν_sSiOSi vibration frequencies, particularly in the region bordering with the frequencies of the deformation vibrations of the layers. Actually the Raman spectrum of the glass shows a very strong line at 565 cm^{-1} [99], almost completely polarized [100], which compels us to ascribe it to the totally symmetric stretching vibration of the anion. Another ν_sSiOSi vibration not observed in the absorption spectrum may be associated with the depolarized band at 780 or 600 cm^{-1} in the Raman spectrum of the glass.

The unexpectedly small number of ν_sSiOSi vibration bands in the infrared spectrum of sodium disilicate and the small number of frequency coincidences with the Raman spectrum suggests a high symmetry of the complex anion. However, the symmetry established by x-ray diffraction is not very high, not over C_{2v} in α-$Na_2Si_2O_5$ and C_s in $Li_2Si_2O_5$, which suggests that almost all the frequencies of the fundamental vibrations should appear in the infrared spectrum. These characteristics of the spectra may be associated with the existence of a higher pseudosymmetry, i.e., approximate symmetry of the complex anions. According to Liebau's data, the α-$Na_2Si_2O_5$ crystals belong to the $P2_1nb$-C_{2v}^9 space group and the $Li_2Si_2O_5$ to the $Cc-C_s^4$, but they may approximately be described within the framework of the $Pcnb-D_{2h}^{14}$ and $Ccc2-C_{2v}^{13}$ groups. The essence of this pseudosymmetry is that, in two of the six Si–O–Si bonds contained in the unit cell of the layer, the "bridge" O atoms are only slightly displaced from special positions, a center of symmetry for α-$Na_2Si_2O_5$ and a two-fold axis for $Li_2Si_2O_5$. If the O atoms were situated exactly at the center of symmetry, the SiOSi in α-$Na_2Si_2O_5$ would be equal to 180°. Actually, according to Liebau, these angles are less than 180°, but are much greater than the SiOSi angles in the other four bonds. Hence the two ν_sSiOSi vibrations chiefly associated with the motions of the "straightened" Si–O–Si bridges may correspond to low-intensity bands in the infrared spectrum, which will lead to the simplication of the observed spectrum. The existence of Si–O–Si bonds with anamolously large SiOSi angles in the structures under consideration is also a probable cause of the appearance of bands with frequencies around 1200 cm^{-1} in the ν_{as}SiOSi region.

Analogous considerations yield a qualitative interpretation of the spectrum of petalite $LiAl(Si_4O_{10})$. According to Moenke [18] the spectrum of this silicate shows two groups of strong bands: 1210, 1160, 1080, 1018 cm^{-1} and 558, 547, 530, 472, 430, 415 cm^{-1}, and in the ν_sSiOSi range of frequencies four narrow bands of medium intensity at 779, 757, 733, and 706 cm^{-1}. An x-ray structural study of petalite [97] established the existence of layers of silicon–oxygen tetrahedra very similar in structure to the layers in alkali disilicates. The unit cell contains two such layers, and the two-dimensional cell of each of them has the composition Si_4O_{10}. The petalite crystal belongs to the $Pa-C_s^2$ space group; hence all six ν_sSiOSi vibrations of the isolated layer are allowed by the selection rules in the infrared spectrum, and each may in principle have a doublet structure as a result of interionic interactions. The smaller number of bands observed experimentally is evidently associated with the obvious pseudosymmetry of the crystal, which enables it to be described within the $P2/a-C_{2h}^4$ space group. In this case two of the six bridge O atoms lie in a special position, inversion centers, and hence two of the six ν_sSiOSi vibrations are forbidden in the infrared spectrum (this ban is not removed on allowing for interionic interactions). Structurally, this pseudosymmetry finds expression in the considerably differing values of the angles SiOSi; the four O atoms in the general position make an SiOSi angle of about 150°, while the two O atoms nearly falling in special positions have $\angle$SiOSi = 165° [97]. As in the earlier-considered cases, the presence of Si–O–Si bonds with angles close to 180° produces the appearance of ν_{as}SiOSi frequencies around 1200 cm^{-1}.

Recent more reliable data regarding the symmetry and structure of the α-$Na_2Si_2O_5$ crystal [64a] refine the foregoing considerations regarding the assignment of ν_sSiOSi frequencies. The space group of the crystal is Pcnb, but the O atoms in the "straightened" Si–O–Si bridges lie not on inversion centers but on two-fold axes ($\angle$SiOSi = 160°, the four other SiOSi angles are 138.9°). The configuration of the layer is very similar to that shown in Fig. 56 for $Li_2Si_2O_5$; the cell (three-dimensional) of the crystal contains elementary sections of two such layers. The factor group of the two-dimensional space group of the layer is isomorphic with C_{2v} (in accordance with the doubled order of the factor group of the crystal D_{2h}). Of the six translationally-nonequivalent Si–O–Si groups of the layer two occupy a double set with symmetry C_2

and give ν_sSiOSi vibrations of types A_1 and A_2, while the remaining four have symmetry C_1, which gives four ν_sSiOSi vibrations of types A_1, A_2, B_1, B_2. Thus only four ν_sSiOSi vibrations are allowed by the selection rules in the infrared spectrum (this is valid also for the crystal as a whole, since one layer transforms into the other by an inversion, and the transition $C_{2v} - D_{2h}$ gives one active component each for B_1, B_2, and A_1 but none for A_2; in the Raman spectrum each vibration of the layer gives one allowed component each). Of these four vibrations three clearly correspond to the infrared bands 768, 738, and 602 cm^{-1}(A_1?), and the fourth (type A_1) corresponds to the polarized band at 565 cm^{-1} in the Raman spectrum of the glass. Of the two ν_sSiOSi vibrations of type A_2, one possibly lies at 780 cm^{-1} (depolarized band in the glass spectrum), while the identification of the other remains obscure.

Two-Dimensional Lattices Based on $(Si_3O_9)_\infty$ Chains [90]. Silicon–oxygen layers with structures more complex than those hitherto considered can be derived by the condensation of chains of the wollastonite type with three tetrahedra in the identity period. The number of structures of this kind studied by x-ray diffraction is very small.

In apophyllite crystals $KCa_4(Si_8O_{20})F \cdot 8H_2O$ belonging to the $P4/mnc-D_{4h}^6$ space group the unit cell [101] contains sections of two layers of composition Si_8O_{20} (Fig. 57a). The basic structure comprises interlinked four-fold rings of tetrahedra, there being two forms of these rings: In some the free vertices of the tetrahedra are turned up and the others down. Such layers may be regarded as the result of the condensation of slightly deformed chains of the wollastonite type with the assistance of two side Si_2O_7 [102]. The symmetry elements of the layer are four-fold axes passing through the centers of the rings and two sets of mutually perpendicular axes in a plane parallel to the layer: a) simple two-fold axes (at an angle of 45° to the x and y axes) passing through the O atoms joining the four-fold rings into the layer, and b) two-fold screw axes (2_1) parallel to the x and y axes. The factor group of the two-dimensional space group formed by these elements is isomorphic with D_4, which agrees with the doubled order of the factor group of the three-dimensional lattice, containing two layers transforming into each other by a symmetry plane in its unit cell. It should be noted that here once again we encounter a case in which the site group of the complex anion in the crystal does not coincide with any one of the site groups allowable for the space group in question. This is only possible in the presence of translational symmetry of the complex anion in one or two directions; of course in this case also the site group is a subgroup of the space group, as they share the C_4, C_2, and $\bar{C}_2$ axes.

The unit cell of the $(Si_8O_{20})_{\infty,\infty}$ lattice consists of eight tetrahedra and contains twelve Si–O–Si and eight Si–O⁻ bonds. However, the number of bands in the spectrum should not be very high owing to the high symmetry of the anion; some of the vibrations are degenerate and

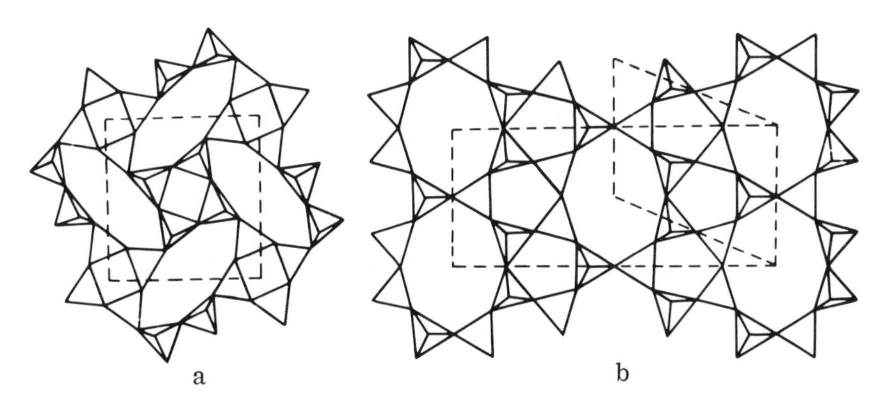

Fig. 57. Structure of silicon–oxygen layer. a) Apophyllite $(Si_6O_{20})_{\infty,\infty}$; b) okenite $(Si_6O_{15})_{\infty,\infty}$.

TABLE 28. Infrared Spectrum of Apophyllite and Okenite

Apophyllite $KCa_4(Si_8O_{20})F \cdot 8H_2O$			Okenite $Ca_3(Si_6O_{15}) \cdot 2H_2O \cdot 4H_2O$	
absorption (polycrystalline sample), frequencies, cm^{-1}, and intensities	reflection (single crystal); frequencies, cm^{-1}, and orientation to C_4 axis	description of vibrations of a $(Si_8O_{20})_{\infty,\infty}$ layer and selection rules for symmetry D_4, symmetry species of vibrations	IRS (absorption) of polycrystalline sample, frequencies, cm^{-1}, and intensities	description of vibrations of a $(Si_6O_{15})_{\infty,\infty}$ layer and selection rules for C_{2h} symmetry
1127 s. 1094 s.	— —	ν_{as} SiOSi: $3E$ $a\perp$, $2B_2$ *ina*, $2A_2$ $a\parallel$, B_1 *ina*, A_1 *ina*	1150 s. 1110 v.s. 1092 v.s. 1073 v.s. 1046 v.s.	ν_{as} SiOSi: $3B_u$ $a\perp$, $3A_u$ $a\parallel$, $2B_g$ *ina*, A_g *ina*
1047 v.s. 1015 v.s.	— —	ν SiO⁻: $2E$ $a\perp$, B_2 *ina*, A_2 $a\parallel$, B_1 *ina*, A_1 *ina*	1017 v.s. 985 v.s. 938 s. ~880 w.	ν SiO⁻: $2B_u$ $a\perp$, A_u $a\parallel$, B_g *ina*, $2A_g$ *ina*
790 s. 762 s. 624 s. 601 s.	792 ⊥ 766 ⊥ 627 ‖ 604 ⊥	ν_s SiOSi: $3E$ $a\perp$, B_2 *ina*, A_2 $a\parallel$, $2B_1$ *ina*, $2A_1$ *ina*	802 m. 773 m. 744 m. 725 p.,w. 662 m. 644 p.,m. 604 m.	ν_s SiOSi: $2B_u$ $a\perp$, A_u $a\parallel$, $2B_g$ *ina*, $4A_g$ *ina*
553 s. 536 s. 503 v.s. 478 v.s.	— — — —	δ SiO and ν *MO*	545 m. 515 s. 493 p.,s. 475 v.s. 453 v.s.	δ SiO and ν *MO*

Note: Orientation of electric vector given with respect to the C_4 axis of the D_4 group and the C_2 axis of the C_{2h} group.

some are forbidden in the infrared spectrum by the selection rules. For a classification of the stretching vibrations of the layer with respect to the irreducible representations of the D_4 group (Table 28) it is sufficient to remember that none of the eight $Si-O^-$ lies on a symmetry element, nor do eight of the Si–O–Si bonds (in both four-membered rings), while the O atoms of the other four Si–O–Si bonds are situated on two-fold axes. In the ν_sSiOSi frequency range we may expect the appearance of four bands corresponding to three vibrations of type E and one of type A_2. These vibrations may clearly be associated with the four bands observed in the spectrum of apophyllite between 800 and 600 cm^{-1} (Fig. 58). The reflection spectrum of a large

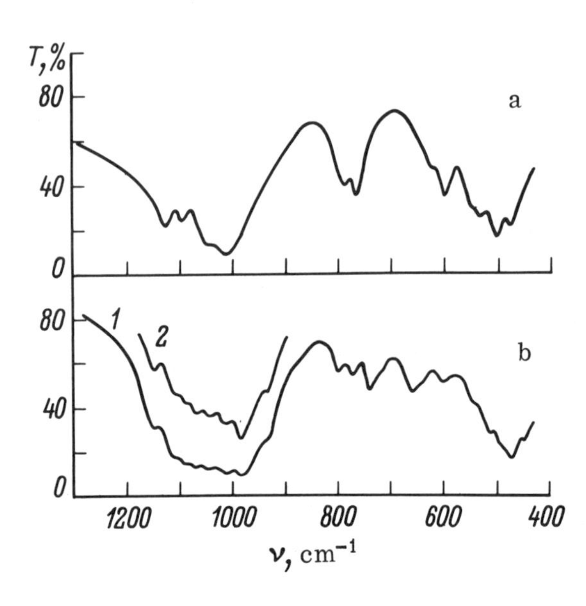

Fig. 58. Infrared absorption spectra of silicates with layers of complex structure. a) Apophyllite (pressed in KBr, 5 mg); b) okenite [1) 3.5 mg; 2) 1.5 mg].

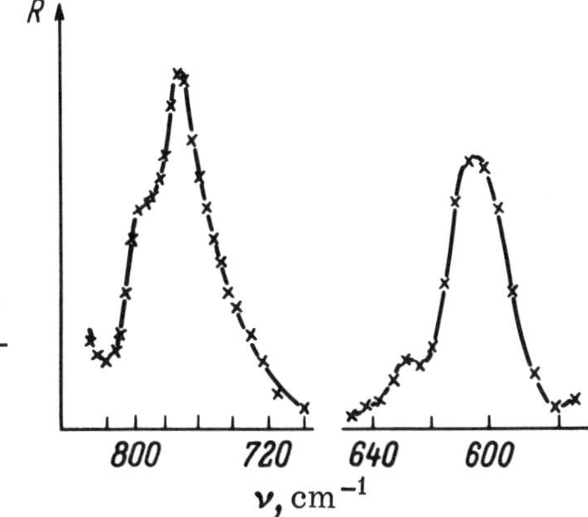

Fig. 59. Reflection spectrum of an apophyllite single crystal. Plane of the sample normal to the optical axis, incidence angle 8°.

sheet of apophyllite obtained from a face perpendicular to the optic axis (nearly normal incidence) showed a clear reduction in the relative intensity of one of the four bands, around 625 cm^{-1} (Fig. 59). This suggests ascribing it to vibrations of symmetry type A_2 and the other three to vibrations of type E. The number of maxima resolved in the ν_{as}SiOSi and νSiO$^-$ frequency range was half that expected. This was evidently due to frequency superposition.†

Figure 57b shows a different configuration of a silicon–oxygen layer found, according to Mamedov and Belov [103], in the structure of okenite $3(CaO \cdot 2SiO_2 \cdot 2H_2O) = Ca_3(Si_6O_{15}) \cdot 2H_2O \cdot 4H_2O$ (with three "molecules" of this composition in the cell). The structure of the two-dimensional lattice in okenite was derived by these authors as the result of the condensation of ribbons of the xonotlite type, i.e., ultimately wollastonite chains. According to the scheme of Fig. 57b, the proposed lattice of okenite possesses the following symmetry elements: inversion centers, parallel simple (C_2) and screw ($\bar{C}_2$) two-fold axes (in the plane of the layer), and symmetry planes σ perpendicular to these layers. The orthogonal (centered) cell of the layer contains twelve tetrahedra, and the primitive (oblique-angled) cell includes six tetrahedra, or in other words nine Si–O–Si and six Si–O$^-$ bonds. Table 28 gives a classification of the stretching vibrations of the $(Si_6O_{15})_{\infty,\infty}$ layer with respect to the symmetry species of the factor group of the two-dimensional group of the layer, which is isomorphic with respect to C_{2h}. The selection rules indicate the possibility of only three ν_sSiOSi vibrations appearing in the infrared spectrum (at any rate for the isolated layer). This is not supported by experiment (Fig. 58), in which at least six or seven maxima are resolved between 800 and 600 cm^{-1}. In addition to this, in the ν_{as}SiOSi vibration range there are no bands with frequencies above 1150 cm^{-1}, i.e., in this region also the spectrum does not support the presence of Si–O–Si bridges with angles close to 180° in okenite. These data lead to the conclusion that there is a lower symmetry of the layer in okenite; however, they give no grounds for a complete reexamination of the structure. In fact the number of bands in the ν_sSiOSi region approaches that expected if the symmetry of the layer is, for example, C_2 or C_s. A refusal to accept a layer structure of the anion would lead to the supposition that SiOH groups were present in order to accommodate a change in the Si:O ratio in the anion for the same empirical formula of the mineral. The spectrum of okenite in the frequency range of the OH stretching vibrations shows no strong bonds such as are ordinarily characteristic of SiOH groups; however, two bands at 3600 and 3400 cm^{-1} are in excellent agreement with the presence of water molecules of two types in the scheme of Mame-

†Allowance for interaction between the vibrations of the two layers in the present case cannot introduce anything new into the spectrum of the crystal; on making the change $D_4 \rightarrow D_{4h}$, for each of the vibrations active in the infrared spectrum there is only one active component with unaltered orientation of the electric vector, while the inactive vibrations remain inactive.

dov and Belov: zeolitic and constitutional. (A band of δH_2O vibrations is observed at 1650 cm^{-1}.) Thus spectroscopic data do not contradict the proposed structural formula of okenite, but indicate a lower symmetry of the complex anion.

References

1. D. W. J. Cruickshank, J. Chem. Soc. (L.), 5486 (1961).
2. J. V. Smith and S. W. Bailey, Acta Crystallogr., 16:801 (1963).
3. C. T. Prewitt and D. R. Peacor, Amer. Mineralogist, 49:1527 (1964).
4. N. V. Belov, Crystal Chemistry of Silicates with Large Cations, Izd. AN SSSR, Moscow (1961).
5. F. Liebau, Naturwiss., 49:481 (1962).
6. N. V. Belov, Kristallografiya, 8:587 (1963).
7. A. V. Nikitin and N. V. Belov, Dokl. Akad. Nauk SSSR, 146:1401 (1962).
8. A. Vorma, Mineral. Mag., 33:615 (1963).
8a. M. L. Boucher and D. R. Peacor, Z. Krist., 126:98 (1968).
9. Y. Ginetti, Bull. Soc. Chem. Belg., 63:209 (1954).
10. A. N. Lazarev and T. F. Tenisheva, Opt. i Spektr., 10:79 (1961).
11. Y. Ginetti, Bull. Soc. Chem. Belg., 63:460 (1954).
12. H. Hahn and U. Theune, Naturwiss., 44:33 (1957).
13. A. Grund and M. Pizy, Acta Crystallogr., 5:837 (1952).
13a. W. S. McDonald and D. W. J. Cruickshank, Acta. Crystallogr., 22:37 (1967).
13b. B. A. Maksimov, Yu. A. Kharitonov, V. V. Ilyukhin, and N. V. Belov, Dokl. Akad. Nauk SSSR, 178:1309 (1968).
14. G. Donnay and J. D. H. Donnay, Amer. Mineralogist, 38:163 (1953).
15. H. Seemann, Acta Crystallogr., 9:251 (1956).
16. Ya. S. Bobovich and T. P. Tulub, Usp. Fiz. Nauk, 66:3 (1958).
17. W. A. Deer, R. A. Howie, and J. Zussman, Rock-Forming Minerals, Vol. 2, Longmans, London (1963).
18. H. Moenke, Mineralspektren, Acad. Verl., Berlin (1962).
19. B. E. Warren and W. L. Bragg, Z. Kristallogr., 69:168 (1928).
20. C. T. Prewitt and C. W. Burnham, Amer. Mineralogist, 51:956 (1966).
21. N. Morimoto, D. E. Appleman, and H. T. Evans, Carnegie Inst. Washington, Ann. Rep. Dir. Geophys. Lab., 193 (1958-9).
22. B. E. Warren and D. I. Modell, Z. Kristallogr., 75:1 (1930).
23. V. A. Kolesova, Opt. i Spektr., 6:38 (1959).
24. F. Matossi, J. Chem. Phys., 17:679 (1949).
25. B. D. Saksena, Trans. Faraday Soc., 57:242 (1961).
26. B. I. Stepanov and A. M. Prima, Opt. i Spektr., 4:734 (1958).
27. A. M. Prima, Tr. Inst. Fiz. i Matem. Akad. Nauk Belorussian SSR, No. 2 (1957).
28. B. I. Stepanov and A. M. Prima, Opt. i Spektr., 5:15 (1958).
29. A. N. Lazarev and T. F. Tenisheva, Opt. i Spektr., 11:584 (1961).
30. K. Dornberger-Schiff, F. Liebau, and E. Thilo, Acta Crystallogr., 8:752 (1955).
31. Kh. S. Mamedov and N. V. Belov, Dokl. Akad. Nauk SSSR, 107:463 (1956).
32. F. Liebau, M. Sprung, and E. Thilo, Z. Anorgan. und Allgemein. Chem., 297:213 (1958).
33. D. R. Peacor and M. J. Buerger, Z. Kristallogr., 117:331 (1962).
34. D. R. Peacor and C. T. Prewitt, Amer. Mineralogist, 48:588 (1963).
35. Kh. S. Mamedov, Dokl. Akad. Nauk AzerbSSR, 14:445 (1958).
36. F. Liebau, W. Hilmer, and G. Lindemann, Acta Crystallogr., 12:182 (1959).
37. F. Liebau, Acta Crystallogr., 12:177 (1959).
38. A. N. Lazarev and T. F. Tenisheva, Zh. Strukt. Khim., 4:703 (1963).
39. J.-P. Labbé, Thesis, Univ. Paris (1965).

40. A. E. Ringwood and M. Seabrook, J. Geophys. Res., 68:4601 (1963).
41. P. Royen and W. Forweg, Z. Anorgan. und Allgemein. Chem., 326:113 (1963).
42. K.-H. Jost, Acta Crystallogr., 17:1539 (1964).
43. D. E. C. Corbridge, Acta Crystallogr., 9:308 (1956).
44. D. W. J. Cruickshank, Acta. Crystallogr., 17:681 (1964).
45. K.-H. Jost, Acta Crystallogr., 16:623 (1963).
46. K-.H. Jost, Acta Crystallogr., 15:951 (1962).
47. K.-H. Jost, Acta Crystallogr., 14:844 (1961).
48. K.-H. Jost, Acta Crystallogr., 14:779 (1961).
49. K. H. Jost, Acta Crystallogr., 16:640 (1963).
50. D. E. C. Corbridge and E. J. Lowe, J. Chem. Soc. (L.), 493 (1954).
51. W. Bues and H.-W. Gehrke, Z. Anorgan. und Allgemein. Chem., 288:307 (1956).
51a. V. Stahlberg and E. Steger, Z. Anorg. und Allgemein. Chem., 349:50 (1967).
52. A. N. Lazarev and T. F. Tenisheva, Opt. i Spektr., 12:215 (1962).
53. Yu. A. Pyatenko and Z. V. Pudovkina, Kristallografiya, 5:563 (1960).
54. D. R. Peacor and M. J. Buerger, Amer. Mineralogist, 47:539 (1962).
55. A. A. Voronkov and Yu. A. Pyatenko, Kristallografiya, 6:937 (1961).
56. N. V. Belov, Mineralog. Sb. L'vov. Geolog. Obshch., No. 11, p. 3 (1957).
57. N. N. Neronova and N. V. Belov, Dokl. Akad. Nauk SSSR, 150:642 (1963).
58. E. J. W. Whittaker, Acta Crystallogr., 13:291 (1960).
59. E. J. W. Whittaker, Acta Crystallogr., 13:741 (1960).
60. B. E. Warren, Z. Kristallogr., 72:42 (1930).
61. J. Zussman, Acta Crystallogr., 8:301 (1955).
62. B. E. Warren and D. I. Modell, Z. Kristallogr., 75:161 (1930).
63. Kh. S. Mamedov and N. V. Belov, Zap. Vses. Mineralog. Obshch., 85:13 (1956).
63a. Ya. I. Ryskin, G. P. Stavitskaya, and N. A. Mitropol'skii, Izv. Akad. Nauk SSSR, Neorg. Mat., 5:575 (1969).
64. A. Grund, Bull. Soc. Franc. Mineral. Crist., 77:775 (1954).
64a. A. K. Pant and D. W. J. Cruickshank, Acta Crystallogr., 24b:13 (1968).
64b. A. K. Pant, Acta Crystallogr., 24b:1077 (1968).
65. F. Liebau, Acta Crystallogr., 14:389 (1961).
66. W. D. Keller and E. E. Pickett, Amer. Mineralogist, 34:355 (1949).
67. W. D. Keller and E. E. Pickett, Amer. J. Sci., 248:264 (1950).
68. J. M. Hunt, M. P. Wisherd, and L. C. Bonham, Anal. Chem., 22:1478 (1960).
69. H. H. Adler, P. E. Kerr, E. E. Bray, N. P. Stephens, J. M. Hunt, W. D. Keller, and E. E. Pickett, Infrared Spectra of Reference Clay Minerals, Amer. Petrol. Inst. Project, 49 (1951).
70. J. M. Hunt and D. S. Turner, Anal. Chem., 25:1169 (1953).
71. W. M. Tuddenham and R. J. P. Lyon, Anal. Chem., 31:377 (1959).
72. G. W. Brindley and J. Zussman, Amer. Mineralogist, 44:185 (1959).
73. O. N. Setkina and N. M. Gopshtein, Tr. Leningr. Tekhnol. Inst. im. Lensoveta, No. 37, p. 79 (1957).
74. O. N. Setkina, Zap. Vses. Mineralog. Obshch., 88:39 (1959).
75. J. J. Fripiat, Bull. Groupe Francais Argiles, 12:25 (1960).
76. G. B. B. M. Sutherland, Nuovo Cimento, Suppl. II, Ser. 10, 635 (1955).
77. J. M. Serratosa and W. F. Bradley, J. Phys. Chem., 62:1164 (1958).
78. J. M. Serratosa and W. F. Bradley, Nature, 181:111 (1958).
79. J. M. Serratosa, A. Hidalgo, and J. M. Vinas, Nature, 195:486 (1962).
80. B. D. Saksena, Trans. Faraday Soc., 60:1715 (1964).
81. H. W. Van der Marel and J. H. L. Zwiers, Silicates Industr., 24:359 (1959).
82. W. Vedder and R. S. McDonald, J. Chem. Phys., 38:1583 (1963).
83. V. C. Farmer, Mineral. Mag., 31:829 (1958).

84. V. C. Farmer and J. D. Russell, Spectrochim. Acta, 20:1149 (1964); 22:389 (1966).
85. W. Vedder, Amer. Mineralogist, 49:736 (1964).
86. V. Stubican and R. Roy, Amer. Mineralogist, 46:32 (1961).
87. V. Stubican and R. Roy, Z. Kristallogr., 115:200 (1961).
88. V. Stubican and R. Roy, J. Amer. Ceram. Soc., 44:625 (1961).
89. A. N. Lazarev, in: Optics and Spectroscopy, Vol. 2, p. 286 (1963).
90. A. N. Lazarev and T. F. Tenisheva, in: Optics and Spectroscopy [in Russian], Vol. 2, p. 292 (1963).
91. W. A. Deer, R. A. Howie, and J. Zussman, Rock-Forming Minerals, Vol. 3, John Wiley and Sons, New York (1962).
92. F. Matossi and H. Krüger, Z. Phys., 99:1 (1936).
93. F. Liebau, Neues Jahrb. Mineral., Abhandl., 94:1209 (1960).
94. A. N. Lazarev, T. F. Tenisheva, and R. G. Grebenshchikov, Dokl. Akad. Nauk SSSR, 140:811 (1961).
95. R. M. Douglass, Amer. Mineralogist, 43:517 (1958).
96. F. Liebau, Acta Crystallogr., 14:395 (1961).
97. F. Liebau, Acta Crystallogr., 14:399 (1961).
98. A. M. Kalinina, V. N. Filipovich, V. A. Kolesova, and I. A. Bondar', in: The Glassy (Vitreous) State, No. 1: Catalyzed Crystallization of Glass, Izd. AN SSSR, Moscow-Leningrad (1963), p. 53.
99. E. F. Gross and V. A. Kolesova, Zh. Fiz. Khim., 26:1673 (1952).
100. Ya. S. Bobovich and T. P. Tulub, Opt. i Spektr., 2:174 (1957).
101. W. H. Taylor and St. Naray-Szabó, Z. Kristallogr., 77:146 (1931).
102. N. V. Belov, Mineralog. Sb. L'vov. Geolog. Obshch., No. 13, p. 23 (1959).
103. Kh. S. Mamedov and N. V. Belov, Dokl. Akad. Nauk SSSR, 121:720 (1958).
104. J. D. Russell, V. C. Farmer, and B. Velde, Mineralog. Mag., 292:869 (1970).

CHAPTER IV

USE OF SPECTROSCOPY IN THE CRYSTAL-CHEMICAL STUDY OF SILICATES

The methods systematically set out in the preceding chapters for qualitatively interpreting the spectra of complex anions in silicates and correlating the type of anion structure with the vibrational spectrum may also be used in various crystal-chemical investigations. In this chapter we shall give a number of examples of the use of spectroscopy in establishing the structure of the complex anion, particularly in those cases in which the impossibility of obtaining reasonably large single crystals, the low symmetry of the crystals, or other circumstances impede the x-ray solution of the structure. In view of this the examples given are confined to a consideration of the most accessible data regarding the study of polycrystalline samples. Some of the examples illustrate the possibility of using the spectra of powder samples together with x-ray diffraction data regarding the lattice constants and space group of the crystal. In many cases this combination of two methods gives reasonably full information regarding crystal structure in a very simple manner.

The next section of this chapter is devoted to the use of spectroscopic methods in studying hydrated silicates, phosphates, and other structures, the possibility of spectroscopically establishing the position of the proton in such structures, the identification of the so-called acid silicates, and the study of the hydrogen bonds in these minerals and their effect on the structure of the complex anions.

The third section considers the possibility of spectroscopically establishing the oxygen coordination of various (chiefly amphoteric) cations. This question has not been considered heretofore, although it is directly related to the problem of determining the limits of applicability of the method of analyzing the vibrations of complex anions on the assumption that these are independent of the lattice vibrations. Finally the spectra of some defect structures are considered (structures with isomorphous substituents), and the possibility of using data relating to silicate–germanate solid solutions to increase the reliability of frequency calculations in the vibrational spectra and estimates of force constants is discussed.

§1. Determination of the Structure of the Complex Anion in Silicates

The Spectroscopic Study of Alkaline-Earth Metasilicates and Metagermanates. This constitutes a good illustration [1] of the prospects of the method where one has to study a fairly large number of structures in order to elucidate various crystal-chemical laws. A comparison of a silicate and germanate structure (the compositions of these corresponding to the same formula) constitutes an extension of the method to crystal-chemical research, since it enables us to establish criteria determining the relative stability of different structures. In what follows we shall mainly consider the part played by the geometrical factor, i.e., the part played by the relationship between the ionic radius of the cation and the dimensions of the complex anion or its elementary unit, the XO_4 tetrahedron. The separation of these effects from other factors affecting the structure of the crystal requires the satisfaction of a number of conditions when selecting subjects for study. The compounds to be compared must not only have the same composition and contain cations with the same formal charge, but the structure of the electron shells of the cations must also be the same. Then the effect of changes in the nature of the chemical bonds between the cation and the complex anion on the structure of the latter may be determined (to a certain extent) by allowing for changes in the basicity of the cation.

The metasilicates and metagermanates of the alkaline-earth metals MXO_3 satisfy the foregoing conditions and enable us to vary the ionic radii of the cations over a fairly wide range, from 0.66 to 1.34 Å ($Mg^{2+} - Ba^{2+}$). These compounds have been synthesized and described for the majority of the alkaline-earth metals (apart from radium and beryllium), although at the time of carrying out the spectroscopic investigations a complete structural solution had only been obtained for some of them, while others still remain without any x-ray solution of the structure. For the majority of compounds of this type only the lattice constants have been determined; this has served as a basis for ascribing them provisionally to certain known structural types and for attempting a preliminary classification of these structures by reference to cation size [2]. In some cases these determinations have been based simply on the indexing of Debye photographs, since the production of even small single crystals has involved great difficulty. The compounds of the series in question are convenient for spectroscopic investigation, since for every one (except magnesium metagermanate) the requirement as to the "separability" of at least the stretching vibrations of the complex anion from the lattice vibrations is excellently satisfied.

Since the interpretation of the spectrum of magnesium metagermanate is unreliable without allowing for the lattice vibrations, and since the spectra of the pyroxenes (including $MgSiO_3$) and wollastonite were given earlier, we shall start by considering the metagermanates with the wollastonite structure. Figure 60 shows the spectrum of calcium metagermanate $CaGeO_3$ obtained by sintering stoichiometric amounts of the oxides [3]. The considerable number of bands in the $CaGeO_3$ spectrum compels us to choose the chain, rather than the ring structure that is also possible for an anion with a Ge:O ratio of 1:3, as being the less symmetrical and not containing degenerate vibrations. The lattice parameters of this crystal [2] suggest a similarity to the structure of β wollastonite; however, the spectral characteristics indicate that this analogy is not quite exact. The initially unidentitied [3] band at 950 cm^{-1} ascribed to an impurity lay at higher frequencies than the values of ν_{as} GeOGe typical for metagermanates, but the fact that this band really did belong to the $CaGeO_3$ spectrum was confirmed by subsequent analysis. The appearance of this band and the doubling of the lattice constants along the a axis mentioned in [2] (relative to β-$CaSiO_3$) suggests that $CaGeO_3$ contains ribbons of the xonotlite type with $\angle$ GeOGe equal or close to 180° (then the formula may be written as $6CaGeO_3 = Ca_6[Ge_6O_{17}]O$, or for a slightly different stoichiometry $5CaO \cdot 6GeO_2 = Ca_5[Ge_6O_{17}]$). Subsequent x-ray data [11, 13], however, fail to show the doubling of a, and our own investigation into the spectrum of $PbGeO_3$, an analog of $PbSiO_3$ with one of the SiOSi angles in the chain equal to 180°, indicated the possibility that, under certain geometrical conditions, unusually high ν_{as} GeOGe frequencies might occur in the spectra of the metagermanates. X-ray diffraction revealed the existence of a continuous series of β-$CaSiO_3$–$CaGeO_3$ solid solutions, but with a nonrandom distribution of the Si and Ge (§4 of this chapter). The spectra of the solid solutions in this system gave no indications of the presence of xonotlite ribbons formed by SiO_4 tetrahedra, or SiO_4 and GeO_4 together.

The identification of the frequencies in the $CaGeO_3$ spectrum may be carried out on assuming the presence of $[Ge_3O_9]_\infty$ chains:

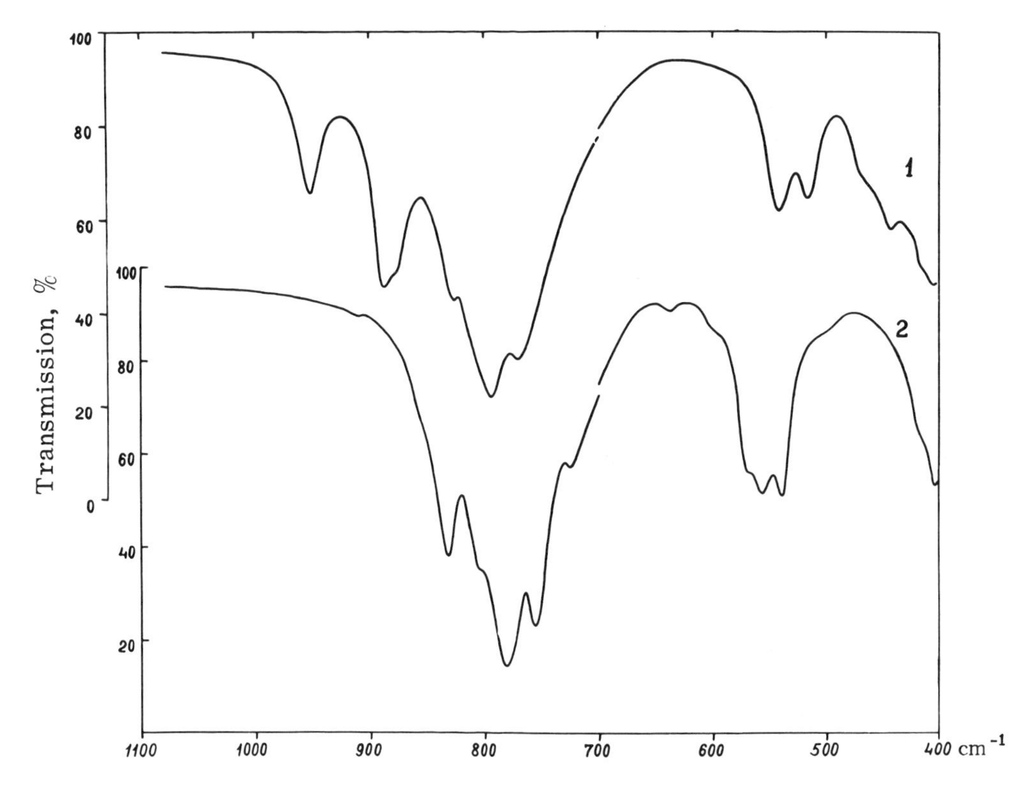

Fig. 60. Infrared spectra of $CaGeO_3$ and low-temperature form of $SrGeO_3$; 1) $CaGeO_3$; pressed in KBr, 3 mg; 2) $SrGeO_3$ (low-temperature form), 4 mg.

951 m.	826 s.	ν_{as}GeOGe	542 s.		443 s.		
805 } s.	794 v.s.	$\nu_{as}O^-GeO^-$	518 s.	ν_sGeOGe	416 p.	δGeO
887 } s.	770 v.s.	$\nu_s O^-GeO^-$	465 p.		405 v.s.	

Here the form of the chains probably differs from the wollastonite type in respect of a marked increase in one of the GeOGe angles (in this sense $CaGeO_3$ is closer to bustamite, but a complete analogy is hardly possible, since in bustamite there are cations of two sorts in ordered positions, Ca and Mn); these differences are in keeping with the nonrandom distribution of Si and Ge in the β-$CaSiO_3$–$CaGeO_3$ solid solutions.

Stavitskaya and Ryskin [4], by dehydrating strontium dihydrogermanate at 550°C, obtained a product evidently constituting strontium metagermanate $SrH_2GeO_4 \rightarrow SrGeO_3 + H_2O$. The spectrum of this compound suggests a similarity to the structure of wollastonite. However, the more discrete spectrum (Fig. 60) obtained recently from samples synthesized by R. G. Grebenshchikov and A. K. Shirvinskaya permit a different interpretation

832 s.	757 v.s.	ν_{as}GeOGe	570 p.		415 p.	
805 p.	725 m.	$\nu_{as}O^-GeO^-$	557 s.	ν_sGeOGe	403 s	δGeO
782 v.s.		$\nu_s O^-GeO^-$	540 s.			

The existence of three ν_s GeOGe bands differing very little in frequency (with an increase in the number of high-frequency bands as compared with the high-temperature form of $SrGeO_3$) agrees better with the idea of Ge_3O_9 rings which have lost their three-fold axis (symmetry no higher than C_2 or C_s), similar, for example, to the rings in $BaGe(Ge_3O_9)$ [4], than with that of

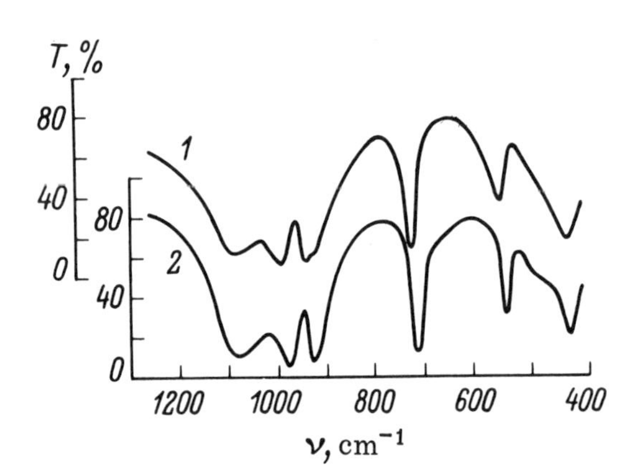

Fig. 61. Infrared spectra of alkaline-earth metasilicates with $[Si_3O_9]^{6-}$ ions. 1) Pseudowollastonite α-$CaSiO_3$; 2) strontium metasilicate $SrSiO_3$. Pressed in KBr, 5 mg.

$[Ge_3O_9]_\infty$ chains. The spectrum of the low-temperature form of $Sr_3Ge_3O_9$ shows a considerable similarity to that of the compound $Ca_2BaSi_3O_9$ (this chapter, §4) containing nonsymmetrical Si_3O_9 rings. The heating of this form of strontium metagermanate to a temperature above 800° causes it to transform into the high-temperature form, which may also be obtained directly from the melt; the spectrum of this was considered earlier.

Let us now move on to pseudowollastonite, the high-temperature form of calcium metasilicate, formed on heating wollastonite above 1160° or on crystallization of the melt. The infrared spectrum of pseudowollastonite [5] (Fig. 61) differs sharply from that of wollastonite in view of the comparatively small number of bands in the Si–O stretching region, which by itself suggests a high symmetry of the complex anion. In the frequency range of the symmetric Si–O–Si stretching vibration there is only a single band at about 725 cm^{-1}, which, together with the high intensity of this band, indicates the existence of annular $[Si_3O_9]^{6-}$ anions. The only alternative for a composition corresponding to the metasilicate might be a structure of the $CuGeO_3$ type with a $[SiO_3]_\infty$ chain containing only one tetrahedron in the identity period. However, this suggestion fails to agree with the cell parameters† of the high-temperature $CaSiO_3$ crystal. Comparison of the lattice parameters of pseudowollastonite, strontium metasilicate, the low-temperature form of barium metasilicate, and the so-called low-temperature form of barium metagermanate with the lattice constants of the high-temperature form of strontium metagermanate, the structure of which has been solved by Hilmer [7], led Liebau [2, 8] to the conclusion that cyclic $[X_3O_9]^{6-}$ anions existed in all these compounds. Table 29 gives the assignments of the frequencies in the infrared spectra of pseudowollastonite and strontium metasilicate [9] on the assumption of a plane configuration of the cyclic $[Si_3O_9]^{6-}$ anions corresponding to the intrinsic symmetry D_{3h}. In addition to the low probability of there being a reduction in the SiOSi angle corresponding to a change to a nonplanar ring shape (although even in a plane ring of this type ∠SiOSi is fairly small, about 130°), we might also present arguments in favor of a planar ring based on an analysis of the spectrum. The nondegenerate (totally symmetric) ν_sSiOSi vibration belongs to an inactive symmetry species for a plane (D_{3h} or C_{3h}) form of the ring; conversely, it is activated in the infrared spectrum for a nonplanar configuration, which permits a symmetry no higher than C_{3v}. In the spectra of pseudowollastonite and strontium metasilicate this vibration is never found; the only band of similar frequency at about 550 to 560 cm^{-1} evidently belongs to a deformation vibration (ρO^-SiO^- of symmetry type A_2''), in good agreement with the corresponding frequency in the spectrum of catapleiite, which certainly contains Si_3O_9 planar rings. The absence of data regarding the Raman spectrum prevents us from establishing the frequency of the ν_s SiOSi vibration of type A_1'. An approximate calculation (considering only the stretching force constants) based on Prima's secular equations leads to the con-

†Crystals of the high-temperature form of $CaSiO_3$ belong to the triclinic system, but are nearly hexagonal. Thus the deviations in the angles are under 1° [6].

TABLE 29. Identification of the Frequencies in the Infrared Spectrum of Pseudowollastonite and Strontium Metasilicate

$CaSiO_3$ (Pseudowollastonite)	$SrSiO_3$	Frequency assignments for the ion $[Si_3O_9]^{6-}$ (D_{3h})
1095—1080 v.s. doub?	1078 v.s.	$\nu_{as}O^-SiO^-(A_2'')$
994 v.s.	980 v.s.	$\nu_{as}SiOSi(E')$
943—928 v.s. doub	930 v.s.	$\nu_s O^-SiO^-(E')$
725 v.s.	714 v.s.	$\nu_s SiOSi(E')$
565 s.	551 s.	$\rho O^-SiO^-(A_2'')$
—	484 p.	Contamination ?
447 v.s.	443 s.	$\delta Si\hat{O}Si(E')$

Note: For comparison also see the spectra of the structurally similar $LnSiO_3$ (Chapter V).

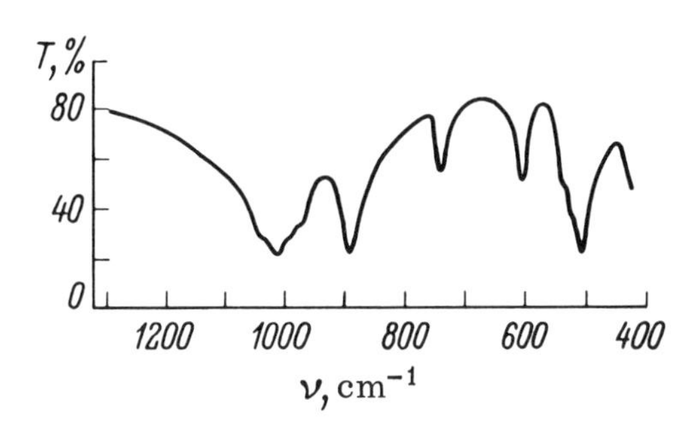

Fig. 62. Infrared spectrum of $BaSiO_3$ (pressed in KBr, 3.5 mg).

clusion that this vibration has a higher frequency than the lowest of the ν_sSiOSi vibrations of the pyroxene chain (by about 10%). This is evidently associated with the rise in ν_sSiOSi on reducing the SiOSi valence angle. Thus there are no grounds for expecting the appearance of this vibration in the frequency range below 650 cm^{-1}. We note further that a change to a nonplanar configuration of the ring should have led to an increase in the number of stretching vibrations of the anion active in the infrared spectrum in the range 1100 to 900 cm^{-1}. A further discussion on the spectra of the Si_3O_9 and Ge_3O_9 rings appears in the last section of this chapter.

A study of the infrared spectrum of the high-temperature form of $BaSiO_3$ [1] carried out before the x-ray solution of the structurally similar crystal of the "high-temperature" form of $BaGeO_3$ clearly established the structural type of the silicon–oxygen anion in this compound.† The infrared spectrum (Fig. 62) differs considerably from the spectra of silicates with ring anions and is reminiscent of the spectra of the pyroxenes or sodium and lithium metasilicates. Together with strong absorption in the range 1050 to 850 cm^{-1} comprising at least five bands, the spectrum shows two medium-strength bands in the range 750 to 550 cm^{-1} characteristic of ν_s SiOSi vibrations. We may suppose that these two bands correspond to the in-phase and antiphase ν_s SiOSi vibrations of two translationally nonidentical Si–O–Si bonds in the anion, i.e., that the anion constitutes a chain $[(SiO_3)_2]_\infty$ with two tetrahedra in the unit cell. The existence of an identity period of about 4.6 Å in this crystal, coinciding with one of the lattice parameters of barium disilicate (sanbornite), enables us to identify it with the identity period of the metasilicate chain. The identification of the frequencies in Table 30 is based on the assumption of a chain symmetry described by a one-dimensional space group, the factor group of which is isomorphic with C_2. The only nontrivial symmetry element is the $\bar{C}_2$ screw axis. This chain

†On the basis of the lattice parameters $[(SiO_3)_2]_\infty$ chains or $[Si_4O_{12}]^{8-}$ rings were suggested [2].

TABLE 30. Interpretation of the Infrared Spectrum of the High-Temperature Form of Barium Metasilicate

Identification of frequencies ($[Si_2O_6]_\infty$ chain)		Frequencies of absorption maxima, cm^{-1}
assignments	symmetry species (factor group C_2)	
$\nu'_{as}SiOSi$	*B*	1038
$\nu_{as}SiOSi$	*A*	1010
$\nu'_{as}O^-SiO^-$	*B*	990
$\nu_{as}O^-SiO^-$	*A*	974
$\nu'_{s}O^-SiO^-$	*B*	891
$\nu_{s}O^-SiO^-$	*A*	
$\nu'_{s}SiOSi$	*B*	740
$\nu_{s}SiOSi$	*A*	604
δ, γ, and ρO^-SiO^-		537 524 508

TABLE 31. Structure of the Complex Anions in the Metasilicates and Metagermanates of Alkaline-Earth Elements

Structure of anion (value of angle XOX)	Compound MXO_3	
	X = Si	X = Ge
Chain $[(XO_3)_2]_\infty$ d ≃ 5.2 Å angle SiOSi ≃ 141°	$MgSiO_3$	$MgGeO_8$
Chain $[(XO_3)_3]_\infty$ d ≃ 7.3 Å angle SiOSi = 149°, 140°, 139°	$CaSiO_3$, low-temperature form	$CaGeO_3$*
Deformed ring $[X_3O_9]^{6-}$	†	$SrGeO_3$, low-temperature form
Plane ring $[X_3O_9]^{6-}$ angle SiOSi ≃ 130°	$CaSiO_3$, high-temperature form $SrSiO_3$ $BaSiO_3$, low-temperature form	$SrGeO_3$, high-temperature form — $BaGeO_3$, low-temperature form
Chain $[(XO_3)_2]_\infty$ d ≃ 4.6 Å angle GeOGe = 110°	$BaSiO_3$, high-temperature form	$BaGeO_3$, high-temperature form

*As compared with the silicate, one of the GeOGe angles is enlarged.

†An anion of this type is realized in $Ca_2BaSi_3O_9$.

configuration is derived from the form of the chains in the layer-like anion of sanbornite, which may be regarded as the result of the condensation of chains of this type (just as the layer in the talc structure is the result of the condensation of pyroxene chains). The solution of the structure of the "high-temperature" form of $BaGeO_3$ [10] confirms these considerations: four $BaGeO_3$ "molecules," i.e., two elementary sections of $[(GeO_3)_2]_\infty$ chains in the cell of a crystal belonging to the space group $P2_12_12_1-D_2^4$, occupy positions with one-dimensional site symmetry† $2_1-\overline{C}_2$.

The sequence of the transitions between the four types of structures in the metasilicate and metagermanate structures here considered may be seen from Table 31. It follows from this that an increase in the ionic radius of the cation is the main factor determining the transformation from one structural type to another. The change in the structure of the anion evidently bears a secondary character and is the result of an adaptation to the dimensions and configuration of the oxygen polyhedra of the cations. Particularly significant in this respect is the displacement in the position of the critical points, i.e., the limits of stability of one kind of structure, on passing from silicates to germanates. Owing to the greater dimensions of the GeO_4 tetrahedra as compared with the SiO_4 tetrahedra, on passing to the germanates there is, relative to the anion, a decrease in the dimensions of the cation; this follows from the fact that OSiO or OGeO angles are fairly rigid, so that oxygen–oxygen distances in the anion, which correspond also to edges of the oxygen polyhedron round the cation, are fairly stable. Thus $CaSiO_3$ crystallizes in two forms, and $CaGeO_3$ only in one, corresponding to the low-temperature form of $CaSiO_3$. On the other hand, strontium germanate is dimorphic, while only one form is known for $SrSiO_3$. It should be noted that, as follows from Table 31, not all types of structure are observed in both the silicates and the germanates, and in addition to this the analogies between the structure of monotypic silicates and germanates are not always complete. This is not only because of the discontinuous character of the variation in the sizes of the cations in the series under consideration, as a result of which only a few of the possible r_{cation}/r_{anion} ratios are represented, but also because of the inadequacy of purely "packing" considerations in explaining all the details of a specific structure. In particular, the tendency of the germanates to form smaller GeOGe angles (less participation of the π bonds?) than the SiOSi helps us to understand why the displacement of the stability boundaries of structures of various types on passing from silicates to germanates is smaller than might be expected from the r_{GeO}/r_{SiO} ratios; it is sufficient to consider that on the average $\angle$ GeOGe = $\angle$ SiOSi − 10°.

Another conclusion directly arising from Table 31 reduces to the fact that a rise in temperature affects the change in structure in the same sense as an increase in the dimensions of the cation, i.e., as the temperature rises the effective radius of the cation does likewise. This was noted earlier in [12], and is clearly a direct consequence of the dynamic characteristics of crystals with complex anions and monatomic cations. The lattice vibrations, i.e., mainly the vibrations of the coordination polyhedra of the cations, have considerably lower frequencies than the internal vibrations of the complex anions. This is valid at least for the stretching vibrations of the anions. The part played by the rather lower-frequency deformation vibrations of the anions in the mechanism of solid-phase transformations will be considered separately.

The energy of the excited levels of the lattice is comparable with the thermal energy even at relatively low temperatures. We remember that at room temperature kT is about 200 cm^{-1} and at 1000°C, 900 cm^{-1}. Hence on raising the temperature the population of the excited lattice levels increases sharply, while the majority of the anion vibrations still continue in the ground state. The considerable anharmonicity of the vibrations of the cation–oxygen bonds, probably much greater than the anharmonicity of the vibrations of the covalent bonds inside the anion, thus leads to an increase in the average interatomic distances $\bar{r}_{M-O}$ in the oxygen polyhedron

†This follows directly from the relation between the number of molecules in the unit cell, equal to 2, and the ratio of the orders of the factor groups of the space group of the crystal and of the site group of the chain.

of the cation, for almost constant dimensions of the anion. If we estimate the increment in $\bar{r}_{M-O}$ from data relating to the linear thermal expansion coefficients, these not exceeding 10^{-5} to 10^{-6} for the majority of silicates, neglecting changes in $\bar{r}_{Si-O}$, then on heating to 1000°C this becomes altogether 0.02 to 0.01 Å. Hence the increase in the effective radius of the cation with increasing temperature can only constitute a reason for the rearrangement of the structure when the ratio of the ionic radii in the original form is in fact close to the critical value for the structure in question.

Up to this point we have considered the laws governing the sequence of the structural types of alkaline-earth metasilicates and metagermanates exclusively in relation to the dimensions of the element–oxygen tetrahedron and the ionic radius of the cation. This enables us to explain the transition from one structural type to another for certain critical relationships of these parameters, but gives no information regarding the factors governing the specific structure of the complex anion in various structural types. Indications of some of the relationships determining the choice of one particular structure for the complex anion (out of a series of virtual structures almost equally satisfying the relationships of the geometrical parameters) may be obtained from the values of the SiOSi and GeOGe angles in Table 31. There is in fact a fairly clear tendency toward a reduction in ∠SiOSi on passing from the structures of the pyroxene or wollastonite type to rings of the pseudowollastonite type and then on to the $BaSiO_3$ chain.† This change in the valence angle of the bridge bonds is hardly a matter of chance and may be ascribed to the weakening of the $d\pi$–$p\pi$ interactions as a result of the increase in the order of the bonds of type $Si-O^-$. The latter, in turn, is a natural consequence of the rise in the basicity of the cations in the series Mg^{2+}, Ca^{2+}, Sr^{2+}, Ba^{2+}, leading to an increase in the effective charge on the side oxygen atoms of the chain or ring.

Thus the series of structures considered confirms the desirability of using relationships of a geometrical character in crystal-chemical investigations, but at the same time indicates the necessity of allowing for differences in basicity, i.e., in the electron-donor capacity of the cations, even when there is a similarity in the structure of their outer electron shells. Later we shall give yet another series of examples of this kind when considering the structure of the silicates of rare-earth elements.

We shall now consider some examples illustrating the possibilities of using even the comparatively sparse information provided by the infrared spectrum of a polycrystalline sample of a silicate of unknown composition when taken together with the results of chemical analysis, information relating to lattice constants, and (if possible) the space group of the crystal. The examples which we shall be giving demonstrate the desirability of using the spectroscopic method as one of the main methods in preliminary study of new synthetic compounds or minerals. A study of this kind by no means replaces a full x-ray diffraction analysis of the structure, but, being less laborious, facilitates the treatment of a wider range of samples and yields some preliminary conclusions as to their structure.

---

† The value of ∠SiOSi in the chain in barium metasilicate is, strictly speaking, unknown, but it probably lies close to 120°, which follows from the similarity between this chain and the analogous chain in $BaGeO_3$, the structure of which has been solved, and from the relatively high ν_s SiOSi frequencies in the infrared spectrum. We may furthermore expect that ∠SiOSi will be slightly greater than ∠GeOGe in the analogous structure, since the identity periods of the two chains are almost the same (4.54 and 4.58 Å). A similar relationship is found in many cases between the SiOSi and GeOGe angles; it is evidently due to the lower efficiency of the π bonds formed by the germanium atoms. It is interesting to note that, corresponding to the nearly tetrahedral value of ∠GeOGe in $BaGeO_3$, there is also a sharp difference in the interatomic distances $r_{Ge-O(Ge)} = 1.87$ Å and $r_{Ge-O^-} = 1.71$ Å [10].

Barium Silicate $2BaO \cdot 3SiO_2$. In the $BaO-SiO_2$ system the existence of three crystalline compounds in the range $BaO:SiO_2$ = 1:1 to 1:2 has been reliably established; these are the metasilicate $BaSiO_3$ and the disilicate $BaSi_2O_5$, the spectra of which have already been discussed, and a silicate of composition 2:3 [15], constituting the sole representative of compounds with such stoichiometry among the alkaline-earth silicates. Since the structure of the silicon–oxygen anion in barium disilicate has been established by direct x-ray diffraction analysis, while the structure of the anion in the metasilicate has been reliably determined by comparing spectroscopic data with the lattice parameters, it would appear possible to draw certain conclusions regarding the structure of the silicate $2BaO \cdot 3SiO_2$ [1].

Existing x-ray information regarding the structure of these crystals has been confined to an attempt at determining the lattice parameters [15]: the crystal was ascribed to the monoclinic system (space group $P2_1-C_2^2$ or $P2_1/m-C_{2h}^2$) with a = 12.51 Å, b = 4.69 Å, c = 6.97 Å, β = 93° 23′. Grebenshchikov's oscillation x-ray diffraction pattern of the crystal gave a value of the identity period along the long axis of the acicular crystals very close to that derived in [15], namely, 4.76 Å [16]. If we assume that in the $2BaO \cdot 3SiO_2$ crystal all the oxygen atoms participate in the formation of the complex anion, the composition of the latter is Si_3O_8. Let us consider possible versions of the structure of the anion satisfying this formula.

Figure 63b gives a schematic representation of a complex $[Si_3O_8]_\infty$ layer proposed by Ito for the structure of epididymite [17] but rejected in subsequent investigations. The infrared spectrum of a polycrystalline $2BaO \cdot 3SiO_2$ sample (Fig. 64, Table 32) contains a large number of bands in the frequency range of the stretching vibrations of the anion, including not less than six bands in the ν_sSiOSi frequency range. This cannot be matched with the $[Si_3O_8]_\infty$

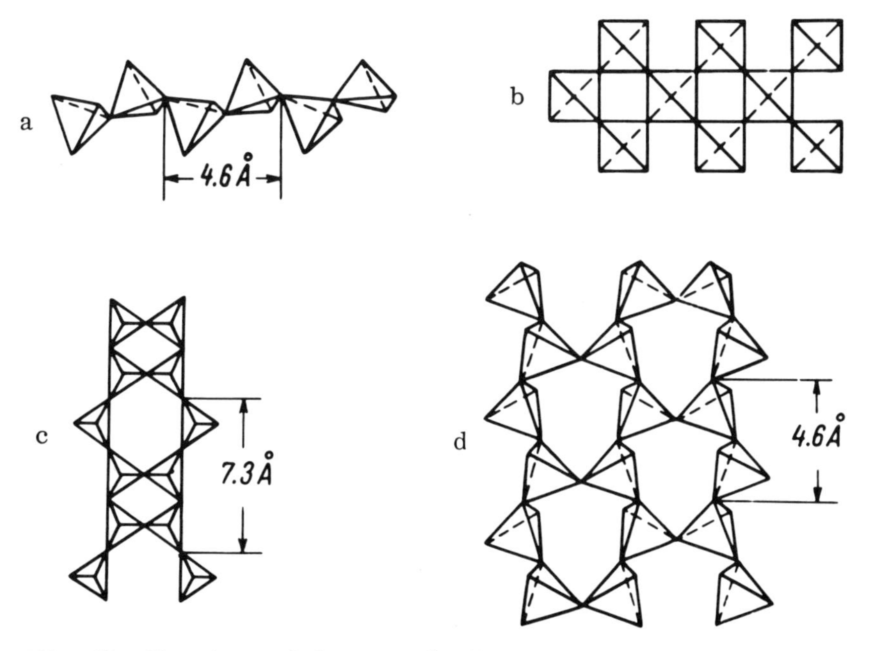

Fig. 63. Structure of chains and ribbons in barium silicates. a) The $(Si_2O_6)_\infty$ chain in crystals of the high-temperature form of barium metasilicate; b) the $(Si_3O_8)_\infty$ chain according to Ito [17]; c) the $(Si_6O_{16})_\infty$ ribbon derived from the wollastonite chain; d) the $(Si_6O_{16})_\infty$ ribbon in barium silicate $Ba_4Si_6O_{16}$ (derived from the chains in barium metasilicate or from the layers of barium disilicate).

TABLE 32. Frequency of the IRS Maxima of $Ba_4(Si_6O_{16})$

Frequency, cm^{-1}	Identification
1112 1055 1020 (doub ?) 986 935 896	ν_{as}SiOSi $\nu_{as}O^-SiO^-$ $\nu_s O^-SiO^-$ νSiO^-
780 ? 755 745 732 710 634 595	ν_sSiOSi
535 512 504 473 442	$\delta(Si_6O_{16})$ and ν_sSiOSi (2)

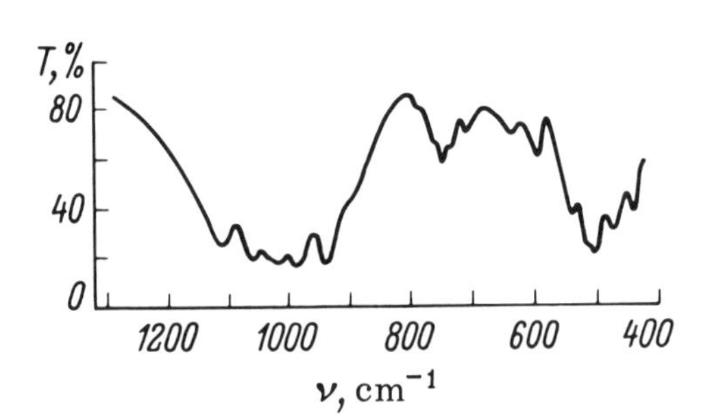

Fig. 64. Infrared spectrum of $Ba_4Si_6O_{16}$. Pressed in KBr, 5 mg.

chain containing only four Si–O–Si bonds nonequivalent with respect to translational symmetry. It is therefore appropriate to consider structures with an elementary link of doubled composition: Si_6O_{16}. Figure 63c and d shows two possible forms of $[Si_6O_{16}]_\infty$ ribbons. The first of these may be considered as the result of the condensation of wollastonite chains by a process differing from the condensation of these chains into xonotlite and epididymite ribbons. The second ribbon, also corresponding to the formula $[Si_6O_{16}]_\infty$, may be obtained by the condensation of three chains of the high-temperature form of barium metasilicate (Fig. 63a) or considered as a ribbon cut from the layer lattice of barium disilicate. Both forms of ribbon contain eight Si–O–Si bonds in the identity period; hence it is impossible to choose between them only on the basis of the number of observed bands of symmetrical Si–O–Si valence vibrations. However, the ribbon formed from wollastonite chains should be characterized by the wollastonite identity period of about 7.3 Å or a multiple of this value.

Thus the $[Si_6O_{16}]_\infty$ ribbon formed of three barium metasilicate chains (Fig. 63d) is evidently the only possible model for an anion structure satisfying the stoichiometry of the crystal, the observed infrared spectrum, and the existence of a repeat distance of 4.6 to 4.7 Å. Ribbons of this type have not been observed in other silicates before. We note that the presence of two closed rings of Si–O–Si bonds in one identity period containing eight such bonds is in exact agreement with the appearance of six S–O–Si bands in the 600 to 750 cm^{-1} range, since two ν_sSiOSi bands may be displaced in the direction of lower frequencies.

The result thus obtained is also interesting because compounds of the $BaO-SiO_2$ system of compositions 1:1, 2:3, and 1:2 are structurally extremely similar, since they are formed by the successive condensation of one and the same metasilicate chain first into ribbons and then into layers. There have been frequent discussions in the literature as to the existence of a series of solid solutions or several discrete compounds in the range of compositions $2BaO \cdot 3SiO_2 - BaO \cdot 2SiO_2$ [15, 16, 18, 19]. In view of the foregoing, the existence of compounds $5BaO \cdot 8SiO_2$ and $3BaO \cdot 5SiO_2$ proposed in [15] receives a natural explanation, since the relationships indicated may presumably be ascribed to analogous ribbons of four $[Si_8O_{21}]_\infty$ and five $[Si_{10}O_{26}]_\infty$ chains of barium metasilicate. At the same time, the structural similarity of all these compounds leads to a certain arbitrariness in the distinction between the concepts of a continuous series of solid solutions in this range of compositions and several compounds, which may even exist, with slight deviations in the composition from stoichiometrical.

The model proposed for the structure of the complex anion on the basis of lattice parameters and infrared spectral data, that is, a ribbon of three chains with an identity period of two tetrahedra, has recently received direct confirmation by the x-ray diffraction analysis of $Ba_4Si_6O_{16}$ [20]. The lattice parameters coincide with those obtained in [15] except for the constant c, which has to be doubled (Z = 2); $[Si_6O_{16}]_\infty$ ribbons extend along the b axis and the configuration of these agrees with that shown in Fig. 63d.

Potassium Europium Pyrosilicate. This constitutes yet another characteristic example of the combined use of the spectroscopic and x-ray diffraction methods for establishing the structural formula and elucidating the structure of a complex anion [21]. This compound was obtained by crystallization of a solution of Eu_2O_3 and SiO_2 in a melt of potassium fluoride in an attempt at growing europium silicate single crystals; it proved to be different from all known compounds in the $Eu_2O_3-SiO_2$ system (see Chap. V). The infrared spectrum of the crystals obtained (Fig. 65) showed groups of bands in the range 1000 to 850 cm^{-1} and 550 to 400 cm^{-1} and also a single band of medium intensity at 685 cm^{-1}. This band clearly corresponds to the symmetric stretching vibration of the Si–O–Si bonds and hence indicates the presence of condensed silicon–oxygen tetrahedra. Moreover, the existence of only one band in the ν_sSiOSi frequency range suggests the presence of diortho groups Si_2O_7. Other possible versions of the anion structure corresponding to the existence of only one ν_sSiOSi band, such as planar rings with symmetry D_{nh}, a chain of the $[SiO_3]_\infty$ (with one tetrahedron in the identity period), or a two-dimensional hexagonal lattice of the talc type, are to be rejected, since they fail to agree with the relatively large number of maxima in the region corresponding to higher-frequency vibrations of the anion. The assumption as to the existence of $[Si_2O_7]^{6-}$ ions agrees with the approximately 1:3.5 ratio of Si and O contents derived from chemical analysis, and leads us to believe that there are no oxygen ions O^{2-} in the structure.

The considerable intensity of the ν_s SiOSi band in the infrared spectrum indicates the noncentrosymmetric nature of the Si_2O_7 groups ($\angle$ SiOSi $\neq$ 180°); correspondingly the symmetry of

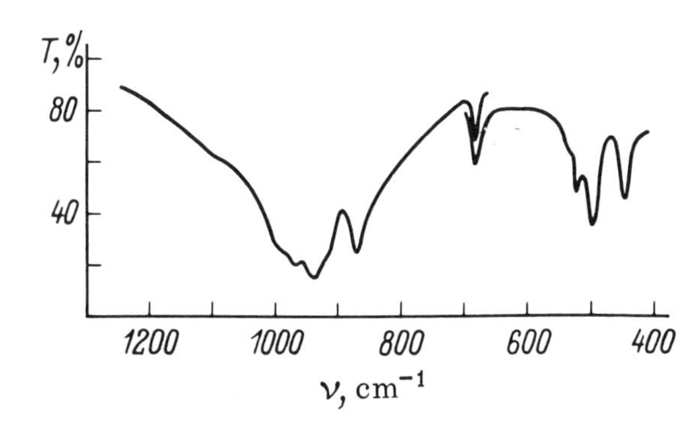

Fig. 65. Infrared spectrum of $K_3EuSi_2O_7$. Pressed in KBr, 5 mg.

TABLE 33. Frequencies of the Absorption Maxima of $K_3EuSi_2O_7$ in the Infrared Spectrum

Freq. (cm^{-1}) and intensity	Assignments (ion $Si_2O_7-C_{2v}$)
(1100 v.w.)	
985 p.	$\nu_{as}SiOSi\ (B_1)$
968 v.s.	$\nu_{as}SiO_3\ (B_2, A_1)$
938 v.s.	$\nu'_{as}SiO_3\ (B_1, A_2)$
920 p.	$\nu_sSiO_3\ (A_1)$
870 v.s.	$\nu'_sSiO_3\ (B_1)$
685 m.	$\nu_sSiOSi\ (A_1)$
531 p.	δSi_2O_7 and $\nu Eu-O$
521 m.	
497 s.	
446 s.	

these groups cannot be higher than C_{2v} (Table 33). According to x-ray diffraction rotation patterns and Weissenberg photographs the crystal belongs to the hexagonal system (Laue class 6/mmm) and the unit-cell parameters are a = 9.98 and c = 14.44 Å. Comparison of these data with the spectroscopically established presence of Si_2O_7 ions suggests a formula for the resultant compound in the form of $3K_2O \cdot Eu_2O_3 \cdot 3SiO_2 = 2K_3Eu(Si_2O_7)$, which agrees satisfactorily with the results of chemical analysis:

$K_3Eu(Si_2O_7)$	K_2O	Eu_2O_3	SiO_2
Calculated, % . . .	32.30	40.24	27.46
Found, %	28.91	43.24	26.61

With the incorporation of six $K_3Eu(Si_2O_7)$ "molecules" in the unit cell, the calculated density is 3.50 g/cm³, in good agreement with the pyknometric value of d^{25} = 3.41 g/cm³.

The presence of systematic extinctions of the $h\bar{h}0l$ type for all $l = 2n+1$ in the x-ray diffraction pattern yields three possible space groups $C_{6v}^3-P6_3cm$, $D_{3h}^2-P\bar{6}c2$ and $D_{6h}^3-P6_3/mcm$ of which the last belongs to the dihexagonal-dipyramidal class D_{6h} established crystallographically. It is thus not hard to convince oneself, by turning to the table of the particular positions (Table 1) in the space groups and considering the fact that a site symmetry of only C_{2v}, C_2, or C_s can apply to the Si_2O_7 ions, that with six such ions in the unit cell of the space group D_{6h}^3 the only arrangement possible is a six-fold set of particular positions with symmetry C_{2v}.

Thus the combination of spectroscopic and x-ray methods, using the results of chemical analysis and crystal-optical investigations, enables us, without any laborious analysis of the structure, to establish the structural formula and discover what kind of structure the complex anion may have as well as the manner in which this is arranged in the crystal. We note, by the way, that there is an interesting analogy between potassium europium pyrosilicate $K_3Eu\,(Si_2O_7)$ and the pyrosilicates and pyrogermanates of the type $X_2Pb_2\,(Y_2O_7)$, where X = K, Rb, Cs and Y = Si, Ge described in the literature [22]. The structure of $K_2Pb_2\,(Si_2O_7)$ studied by Naray-Szabo [23] is characterized by layers of composition $Pb_2Si_2O_7$ with centrosymmetric Si_2O_7 groups united into a layer by Pb atoms lying in trigonal coordination with respect to oxygen. Existing preliminary considerations suggest that the $K_2EuSi_2O_7$ crystals contains layers† of

† Although of course "organized" in a different manner, as there are no grounds for expecting a coordination number of under 6 for the Eu^{3+} ions.

composition $EuSi_2O_7$, save that the Si_2O_7 groups are noncentrosymmetric ($\angle$SiOSi $\neq$ 180°). Starting from the foregoing considerations regarding the dependence of the angle SiOSi on the degree of covalence of bonds between the terminal $Si-O^-$ groups and the cations (p. 45), the reduction in $\angle$SiOSi on passing from $K_2Pb_2(Si_2O_7)$ to $K_3Eu(Si_2O_7)$ may be associated with a reduction in the covalence of the Eu–O bond as compared with that of Pb–O and an increase in the relative content of K^+ cations.

§ 2. Hydroxyls in Silicates

Spectroscopic investigations into hydroxyl groups in silicates, phosphates, and similar compounds have been widely undertaken in view of the fact that the O–H vibrations differ considerably in frequency from other fundamental vibrations of the crystal and are localized chiefly in this bond. This enables us in many cases to limit consideration to the vibrations of the X–O–H groups without proceeding to a complete analysis of all the vibrations of the system. The high sensitivity of the frequencies and intensities of the O–H vibrations to changes in the chemical nature of the X–O(H) bond, to the formation of a hydrogen bond X–O–H . . . O, and to changes in its length may be used in a variety of crystal-chemical investigations. The difficulty of locating the protons in complex silicate lattices by x-ray structural analysis makes it particularly desirable to supplement these methods spectroscopically so as to find in what molecules or ions the hydroxyls lie, in particular, whether the complex silicon–oxygen anion contains "acid" groupings Si–OH, whether there are water molecules present in the structure, how many types of such molecules there are in relation to the strength of their bond with the remaining elements of the structure, and so on. Here we shall briefly consider present existing possibilities of using information regarding the vibrations of the hydroxyl in the crystal-chemical study of silicates, without considering the numerous investigations into the theory of the hydrogen bond, which are set out in a number of reviews and monographs [24-27].

Stretching and Deformation Vibrations of the Hydroxyls in H_2O Molecules and $(OH)^-$ and $(H_3O)^+$ Ions. These exhibit considerable differences in the frequency, strength, and shape of the bands, and offer excellent prospects for the spectroscopic identification of these elements of crystal structure.

In the spectra of crystal hydrates containing water molecules which have not lost their chemical individuality and only form relatively weak hydrogen bonds, one can usually identify the bands of the normal vibrations of H_2O molecules, to some extent modified by intermolecular interactions. Thus the water in beryl that is least coupled to the structure of the crystal shows OH stretching vibrations at about 3598 and 3690 cm^{-1} [28], which agrees closely with the frequencies ν_s and ν_{as} of the OH of an isolated H_2O molecule in a solution of water in CCl_4 [29]. The water molecules in beryl, their maximum concentration being no greater than 1%, occupy positions in the spaces formed by channels of six-membered siloxane rings, and in no way constitute a necessary element of the structure. A similar contour characterizes the bands of the stretching vibrations of the water molecules in dioptase $Cu_6(Si_6O_{18}) \cdot 6H_2O$, as noted by Ryskin [29]; however, the frequencies of the symmetric and antisymmetric O–H vibrations are reduced by about 350 cm^{-1}.

In such compounds as opal ($SiO_2 \cdot nH_2O$) the water hardly interacts at all with the framework of the structure and only fills gaps in it; owing to the large size of the gaps, the water in them exists in a liquid-drop state, and its spectrum differs in no way from that of liquid water (Fig. 66 [30]).

For the majority of crystal hydrates containing relatively weakly coupled water molecules a typical feature is the presence of bands of H_2O stretching in the frequency range below 3600 cm^{-1} but not below 3000 cm^{-1}. The width of the bands is usually no greater than 100 cm^{-1}, which in many cases enables us to resolve the frequencies of the symmetric and anti-

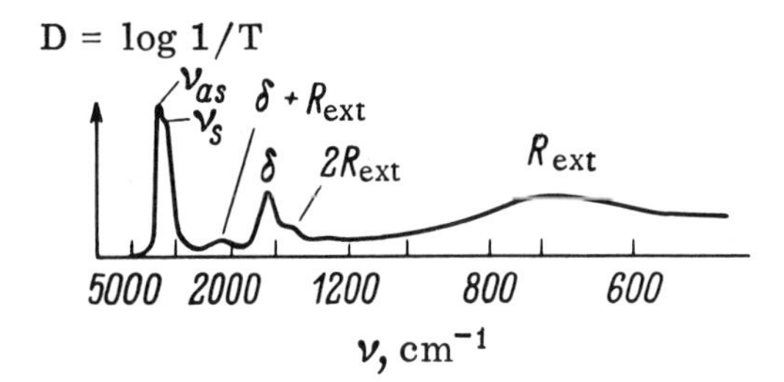

Fig. 66. Absorption spectrum of liquid water at room temperature.

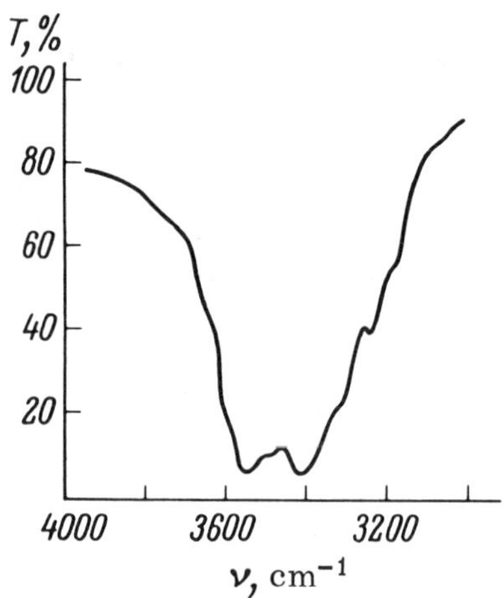

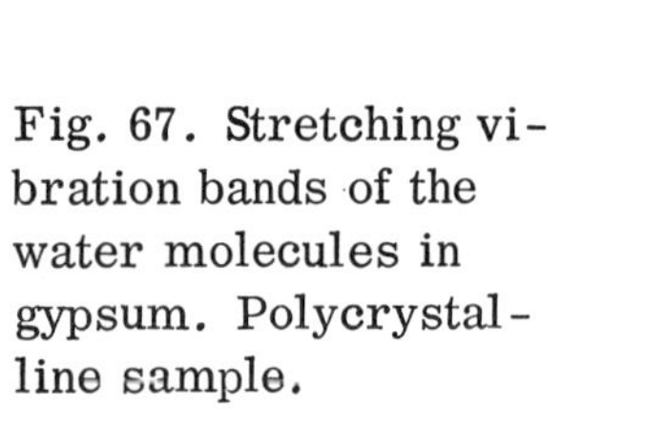

Fig. 67. Stretching vibration bands of the water molecules in gypsum. Polycrystalline sample.

symmetric vibrations. Thus for example in the spectra of single-crystal hydrated nitrates the $\nu_{as}H_2O$ vibration bands are found in the range 3600 to 3400 and the $\nu_s H_2O$ between 3400 and 3200 cm^{-1} [30]. In the spectrum of gypsum $CaSO_4 \cdot 2H_2O$ the band of OH stretching also has the form of an intense doublet with maxima at about 3554 and 3408 cm^{-1} [31]. In addition to this there is also a considerable number of weaker maxima (Fig. 67); an analogous picture occurs in the Raman spectrum [32, 33], and on the basis of a comparison with the low-frequency vibrations of the gypsum lattice it is ascribed to combinations of νH_2O with these vibrations [33]. It is interesting to note that when studying the polarization of the bands in the infrared spectrum of a gypsum single crystal an attempt was made in [34] to identify the Davydov components of the vibrations of the H_2O molecules in the crystal. However, the use of data relating to the dichroism of the ν_{as} and ν_s components of H_2O in order to establish the orientation of the water molecules led to entirely unsatisfactory results, although for the components of the δH_2O vibration good agreement was obtained with the dichroism estimated from structural data.

In crystal hydrates containing several types of water molecules coordinated in different ways and participating in hydrogen bonds of different strengths, the structure of the absorption spectrum in the frequency range of OH stretching may have an extremely complicated character, while the width of the bands may increase considerably. A characteristic example of a silicate containing water molecules participating in fairly strong hydrogen bonds is afwillite $Ca_3(SiO_3OH) \cdot 2H_2O$, in which the spectrum contains broad (about 200 cm^{-1}) bands at around 3340 and 3130 cm^{-1} [35].

In the region adjacent to the frequencies of the water-molecule stretching vibrations, weaker bands at around 2900 cm^{-1} may sometimes be observed in crystal hydrates (tobermorite [29]); the origin of these is not clear. In the range 2100 to 2300 cm^{-1} there may be some weak absorption due to a combination frequency, the over-all combination of δH_2O and the librational vibrations of the water molecule (the corresponding band may be seen in the liquid-water spectrum of Fig. 66).

The deformation vibration δH_2O, situated in the spectrum of liquid water† at 1650 cm^{-1}, in the majority of crystal hydrates occupies a narrow range of the spectrum between 1670 and 1590 cm^{-1} and only in rare cases is it slightly displaced toward lower frequencies [36]. The width of this band is usually not very great, and the isotopic coefficient $\delta OH/\delta OD$ indicates an anharmonicity smaller than for the stretching vibrations. The presence of the δH_2O band in the absorption spectrum is one of the important criteria for the presence of crystal-hydrate water in the specimen under examination. In the lower-frequency range the spectra of crystal hydrates may show the librations of the water molecules in the form of wide bands at 500 to 800 cm^{-1} [30].

The idea of there being $(OH)^-$ ions linked mainly by electrostatic forces to the X atom only makes sense where X is an element of the I or IIA groups of the periodic system, i.e., when the X–OH bond has only a small proportion of covalence. The isolated $(OH)^-$ ion has a single internal degree of freedom, which corresponds to a frequency of about 3700 cm^{-1} [37]. In the spectra of crystals νOH is reduced by 100 to 200 cm^{-1}; the width of the band is not very great and does not exceed 50 cm^{-1}. Thus in the spectra of the calcium hydrosilicates the frequencies of the vibrations of the $(OH)^-$ ions coordinated around the calcium atoms are according to Ryskin [29]:

Xonotlite $Ca_6(Si_6O_{17})(OH)_2$ — 3640 cm^{-1}

α-$Ca_2(SiO_3OH)(OH)$ — 3532 cm^{-1}

Hillebrandite $Ca_{12}(Si_6O_{17})_2(OH)_4 \cdot 12Ca(OH)_2$ — 3617, 3553 cm^{-1}

Here the ratio $\nu OH/\nu OD \geq 1.35$. The hydrogen atoms in the $(OH)^-$ ions are evidently unable to form any significant hydrogen bonds; conversely, the absence of strong hydrogen bonding may serve as an indication that the hydroxyl occurs in the form of an $(OH)^-$ ion. This, however, does not mean that, in complex structures containing hydroxyls belonging to other groups or ions, the oxygen atom in the $(OH)^-$ cannot serve as a proton acceptor for these hydroxyls.

As already mentioned, when studying the spectra of single-crystal hydroxides of the alkali and alkaline-earth metals, near the νOH frequencies a large number of weak maxima were found, and these were ascribed to the interaction of the νOH with librations of the $(OH)^-$ ions or other (including acoustic) lattice vibrations. In the frequency range above 400 cm^{-1} these compounds show no bands which could be ascribed to δXOH vibrations; this agrees with the idea of the purely ionic character of the X–(OH) bonds.

The existence of hydroxonium ions $(H_3O)^+$ is usually regarded as the final stage in the transfer of a proton in a strong hydrogen bond, with complete dissociation of the O–H bond:

$$X{-}O{-}H\ldots O{\Big\langle}{H \atop H} \rightarrow X{-}O^-\ldots\left(H{-}O{\Big\langle}{H \atop H}\right)^+$$

The formation of hydroxonium ions is characteristic of such systems as aqueous solutions or the crystal hydrates of strong acids: $HNO_3 \cdot H_2O \rightarrow (H_3O)^+(NO_3)^-$. The hydroxonium ion has a pyramidal configuration, which for symmetry C_{3v} suggests the appearance of four fundamental vibrations. The bands of the stretching vibrations observed in the range 3400 to 2700 and around 2100 cm^{-1} [38-41] are distinguished by a very great width and insignificant intensity; the use of these in order to identify the $(H_3O)^+$ ions in the structure of the crystals is seldom unambiguous. Of far greater interest is the fairly narrow, strong band of the defor-

†At a slightly lower frequency the first overtone of the librational (external) vibration of the H_2O molecule occurs (see Fig. 66).

mation vibration near 1700 cm^{-1}. The frequency of this fluctuates between 1670 and 1750 cm^{-1}, which exceeds the δH_2O frequencies observed in crystals [36]. Another deformation vibration of the $(H_3O)^+$ ion occurs near 1150 cm^{-1}, but its low intensity in the infrared spectrum and the possibility of the superposition of bands corresponding to the vibrations of the complex anion impede its use. The external degrees of freedom of the H_3O^+ ion in crystals were considered in [42]. The literature only contains very limited information regarding attempts at the spectroscopic identification of $(H_3O)^+$ ions in silicates, particularly in micas [43, 44].

Spectroscopic Manifestations of the Nature of the X–OH Bond. These are of considerable interest for establishing the positions of protons in crystals, for example, in finding to which atoms the hydroxyls are coordinated in the structure. The frequencies of the stretching vibrations of the hydroxyls depend considerably on the strength of the hydrogen bonds in which the hydroxyl groups are participating. This leads to a correlation between the νOH frequency and the O...HO distance obtainable from x-ray data. A number of investigations [45-48] have been devoted to studying the dependence of νOH on $r_{OH...O}$, and the resultant graphs enable us with reasonable accuracy to estimate the lengths of the hydrogen bonds from spectroscopic data. One such relationship mainly based on results obtained for hydroxides [49-51] is shown in Fig. 68. It is interesting to note that the result obtained for ε-Zn $(OH)_2$, deviating from the general law, arose from inaccuracy in the x-ray determination of the OH...O distance. The refinement of the ε-Zn $(OH)_2$ structure [52] led to lengths of the hydrogen bonds agreeing with the relationship of Fig. 68. In addition to the dependence of νOH on $r_{OH...O}$, a relation was also established between the distance O...HO and H–O [53]. The relation between the length of the hydrogen bond and the width and integrated intensity of the νOH band is much more complex [36, 54].

In spectroscopic investigations into the structure of hydroxyl-containing silicates, the relation between the frequencies of the deformation vibrations δXOH and the nature of the atom X is particularly important. The existence of such a relationship enables us to use spectroscopic data in order to decide whether the hydroxyl is connected to the silicon–oxygen anion or the cation, and in the presence of several chemically differing types of cation which of these it is connected to, without having recourse to deuterization, which is often a difficult matter with natural compounds, and also to decide on whether bands corresponding to deformation vibrations of the hydroxyls are likely to appear in the frequency range of the stretching vibrations of the silicon–oxygen anion. It proved impossible to establish any more or less universal dependence of the frequencies of the deformation vibrations of the hydroxyls on the parameters of the hydrogen bridges, although for limited groups of compounds correlations of practical value were obtained.

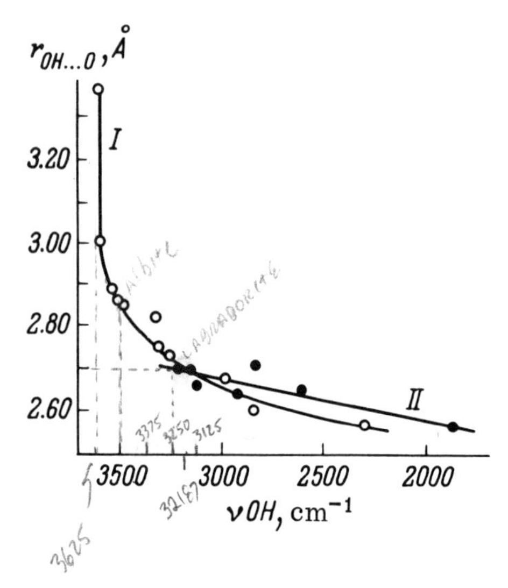

Fig. 68. Relationship between the frequency of the νOH vibration and the OH...O distance [50]. I) Hydroxides with OH...OH bonds; II) hyroxides with OH...O bonds.

The frequency of the δXOH vibration in the X–O–H group is determined not only by the kinematics of this system (mass of the atom X, the angle XOH, etc.) and the superposition of an additional H . . . O bond, but also by the order (degree of covalence) of the X–O bond. In fact, as already mentioned for the hydroxides of elements belonging to group IA of the periodic system with purely ionic X–OH bonds, the deformation vibrations of the hydroxyl reduce to the librations of the OH^- groups as a whole relative to the cation sublattice. An increase in the covalent (directional) character in the X–O bond increases the "rigidity" of the angle XOH and the corresponding δXOH frequency. The formation of a hydrogen bond XOH . . . O, impossible for a purely ionic X–O bond, acts on the frequency of δXOH in the same sense. Thus, whereas in calcium and magnesium hydroxides no absorption bands associated with δOH vibrations were observed in the frequency range above 400 cm^{-1}, in agreement with the high νOH frequencies in these compounds, in the spectra obtained from the hydroxides of IIB- and III-group elements the δXOH frequencies were relatively high. Table 34 shows the δXOH and νOH frequencies in some of these substances, which have been carefully studied by many authors [37, 50, 55-61].

The existence of several νOH and δOH bands was ascribed in a number of cases to the presence of different metal–hydroxyl distances in the crystal and different lengths of the hydrogen bonds.

Thus the frequencies of the stretching and deformation vibrations of the hydroxyl are related to the strength (length) of the hydrogen bonds and (the former inversely and the latter directly) to the proportion of covalence in the X–OH bond. Since in the majority of cases $m_X \gg m_H$, the differences in the mass of the X atoms may be unimportant in comparison with the changes in the degree of covalence of the X–OH bonds, if the latter are large. Hence it would appear possible to establish a numerical relation between the covalence of the X–OH bonds and the frequencies νOH and δOH. Glemser and Hartert [37,56], taking as a characteristic of the nature of the X–O(H) bond the ionic radius of the hydroxyl r^*_{OH}, determined from x-ray data as the difference between the distance $r_{X-O(H)}$ and the ionic radius of the atom X, proposed a simple empirical relation

$$r^*_{OH} = 8.9 \cdot 10^{-4} (4720 - \delta - 0.7\nu)$$

where r^*_{OH} is in Å and δ and ν in cm^{-1}. This formula is reasonably satisfied for many hydroxides, chiefly of elements of groups I to III of the periodic table [59]. Comparison [50] of the values of r^*_{OH} established by x-ray diffraction with those calculated from spectroscopic data shows comparatively slight discrepancies (Table 35).

TABLE 34. Frequencies of the Vibrations of the Hydroxyl in Various Hydroxides

Compound	νOH, cm^{-1}	δOH, cm^{-1}
ε-$Zn(OH)_2$	3226	1092
	—	1033
$B(OH)_3$	3195	1193
γ-$Al(OH)_3$ (gibbsite)	3617	1020
	3520	960
	3428	916
	3380	—
α-AlOOH (diaspore)	2944	1077
	2915	963

TABLE 35. Dimensions of the OH^- Ion from Structural and Spectroscopic Data

Compound	Ionic radius of the hydroxyl, Å	
	x-ray diffraction	spectroscopic
$B(OH)_3$	1.15	1.12
γ-$Al(OH)_3$	1.16	1.22
γ-FeOOH	1.35	1.33
α-AlOOH	1.45	1.47
	1.55	1.57
α-FeOOH	1.51	1.53
	1.59	1.63

Deviations from this relationship and in particular from the relation between νOH and δOH bands ($\Delta\delta$OH = −0.66 $\Delta\nu$OH) used in its derivation, are found in the case of the stronger hydrogen bonds, where the experimental values of the frequencies δOH are much lower than expected [61,62].

As the directional character of the X−O bonds increases we may expect that not only δOH bands but also the bands of the out-of-plane deformation vibration γOH will appear in the frequency range above 400 cm^{-1}. In [57,58] these vibrations were provisionally associated with the bands at 743 cm^{-1} in $Al(OH)_3$ and 690 to 700 cm^{-1} in α-AlOOH and GaOOH.

Among the hydroxyl-containing silicates with OH groups coordinated to cations forming partly covalent bonds with oxygen, most detailed spectroscopic study has been devoted to micas and other silicates with layer structures. Whereas in the talc spectrum the presence of Mg−O−H bonds is only revealed experimentally by the appearance of the νOH band,† for systems containing Al−O−H bonds the δAlOH bands are clearly observed in the region of 900 cm^{-1} [63]; the identification of these is confirmed [64] by the spectra of the deuteroanalogs, as in Fig. 69. The replacement of Al by Ga and Fe leads to the successive reduction of this frequency from 935 to 910 and 827 cm^{-1}. When studying the absorption spectrum of single-crystal muscovite $KAl_2(OH)_2 \cdot (Si_3AlO_{10})$ not only the δAlOH but also the γAlOH band was identified [65]:

	δ	γ
OH	925	405
OD	705	<300

Polarization measurements revealed a different orientation of the moments for these two vibrations. Attention should also be paid to the low value of δOH/δOD = 1.31, which probably indicates an interaction of the δAlOH vibration with the vibrations of the silicon−oxygen layer. The stretching vibrations of the hydroxyls for di- and tervalent cations in silicates with a layer-like structure have been studied by many authors [63-70]. Polarization measurements of the νOH band offer the prospect of deciding the orientation of the moments of the transitions,

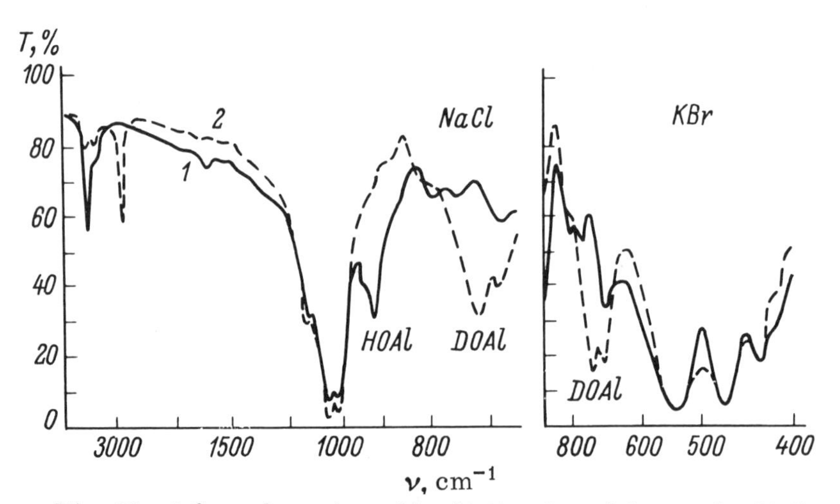

Fig. 69. Infrared spectra of kaolinite (1) and deuterokaolinite (2); bands of the deformation vibrations of the hydroxyl.

† The δMgOH vibration has now been identified; see footnote to Table 26.

and hence (with a little care) the orientation of the OH bonds. In particular, according to [65], the hydroxyls in muscovite are inclined at about 16° to the cleavage plane, and their projection on this plane makes an angle of 32° with the b axis.

"Acid" Anions in Silicates, Sulfates, Phosphates, and Germanates. These anions and the spectroscopic methods of identifying them are particularly important from the point of view of hydrogen-bond theory. The characteristics of the hydrogen bonds in these substances are associated, in particular, with the mechanism underlying the ferroelectric phase transformation observed in many acid salts. Furthermore, a study of the conditions of formation of hydrogen bonds by X–O–H groups provides information regarding the electron structure of the acid anions. Finally, the processes of the hydration and dehydration of silicate anions are essentially connected with the nature of the binding (cementing) properties of a number of silicates, and a study of the role of the hydroxyls attached to silicon in the mechanism of the solid-state reaction may clearly promote an understanding of the nature of many phase transformations. The combined use of the methods of x-ray structural analysis and vibrational spectroscopy, together with nuclear magnetic resonance, substantially eases the problem of establishing the position of the protons in the crystals of acid salts.

For the compounds under consideration containing the X–O–H grouping (X = S, P, Si, Ge), a typical feature is the formation of extremely strong hydrogen bonds, which is reflected by a considerable reduction in the frequencies of the stretching vibrations of the hydroxyl to below 3000 and even right down to 1500 cm^{-1}. The majority of compounds of this type show several νOH bands, usually situated in the ranges 2900–2400 cm^{-1} (A), 2500–1900 cm^{-1} (B), and 1900–1600 cm^{-1} (C) [71, 81]. The width of these bands is usually very great and may reach many hundreds of cm^{-1}. Bands of types A and B are undoubtedly associated with the stretching vibrations of the hydroxyl, although in some cases their interaction with the X–O vibrations of the anion, if the frequencies of the latter are high, may be considerable, as occurs in the spectra of the bicarbonates [72]. The existence of two bands A and B can hardly be ascribed to hydrogen bonds of different lengths, since both bands are observed in the spectra of crystals containing, according to structural data, only identical OH...O distances. The assumption that the two νOH bands belong to the two normal vibrations of the $O_2P(OH)_2$ system in the KH_2PO_4 crystal [73] fails to explain the appearance of both bands in the spectra of crystals of di-basic phosphates with O_3POH ions [74], acid sulfates, and carbonates [75]. The ascription of these two bands to two components formed by the splitting of the energy levels of νOH as a result of the existence of two potential-energy minima with a relatively low barrier (or inflection) has been proposed by a number of authors [31, 76-80]. Attempts have also been made to correlate the νOH splitting with the absorption observed in the long-wave part of the spectrum (about 120 to 150 cm^{-1}), hypothetically ascribed to a transition between two sublevels [82]. However, this interpretation of the A and B bands is hardly unambiguous; in particular, their displacement on deuterization shows no marked reduction in the splitting [83], which we should expect in view of the reduction in the energy of the ground level for approximately the same height of the potential barrier.

All the foregoing attempts at interpretation have practically ignored the collective character of the vibrations of the system of strongly interacting hydrogen bridges and the change in the shape of the potential curve in the course of vibration. A study of the possible role of these factors was first undertaken by Reid [54], and the development of the views which he expressed led to a hypothesis as to the possible origin of the bands of type C. Initially the C band was ascribed to a deformation vibration of the hydroxyl δXOH [71, 77] by the extrapolation of the relationship established for weaker hydrogen bonds between the frequency of δXOH and the strength (length) of the hydrogen bond. However, as the hydrogen bond becomes stronger the rise in the δXOH frequency gradually slows down, and its value at about 1400 cm^{-1} in acid carbonates [72] is evidently close to the maximum value for acid salts. According to Ryskin [75], the C band is associated with the in-phase vibrations of the protons in the systems of infinite chains or lattices of hydrogen bonds uniting the acid anions in KH_2PO_4 and similar crystals. In other words, for the in-phase vibrations of the protons, for example, in the chain

```
O    O                              O    O
 \  /                                \  /
  P                                   P
 /  \                                /  \
O    O—H...O        O—H...O        O    O—H...
             \      /
               P
             /   \
            O     O
```

with the simultaneous transfer of all the H atoms into a symmetrically equivalent position with respect to the proton-accepting oxygen in the neighboring anion, we may expect that the νOH frequency will be lower than for other phase relationships between the displaced protons. This idea agrees with the fact that the frequencies of the C bands are close to the νOH frequencies in systems containing symmetrical hydrogen bonds, i.e., in systems with a symmetrical (according to neutron diffraction) potential well even for uncorrelated motion of the protons. In the spectra of such compounds the A and B bands do not appear.

The frequencies of the in-plane deformation vibrations of the hydroxyls in acid salts, as noted earlier, never exceed 1400 cm^{-1}, while in acid phosphates, silicates, and germanates they occur in the form of fairly narrow bands in the range 1300 to 1200 cm^{-1} [74, 75, 84-87]. The anharmonicity of these vibrations and their interaction with the X–O vibrations are extremely slight $\delta OH/\delta OD = 1.39$ to 1.40 instead of about 1.35 for $\nu OH/\nu OD$.

Thus the Hartert and Glemser relationship between the frequency of the deformation vibration and the strength of the hydrogen bond is only justified for relatively weak hydrogen bonds, and in the case of strong hydrogen bonds the dominant role begins to be played by other factors, leading to a fall in the δXOH frequency as the hydrogen bond strengthens. The identification of the bands of the out-of-plane deformation vibrations γXOH, lying at lower frequencies, close to the X–O vibration frequencies of the complex anion, involves serious difficulties; in the spectrum of the α hydrate of the dicalcium silicate $Ca_2(SiO_3OH)OH$, the vibration γSiOH is associated with the band at 712 cm^{-1}, which is displaced to 526 cm^{-1} on deuteration [84].

The frequencies and intensities of vibrations involving the heavy atoms of complex anions in acid salts show considerable changes when compared with the spectra of the analogous compounds not containing hydroxyl groups in the composition of the complex anion. These changes are also of interest as they may be used in studying the complex anions and in considering the charge distribution and orders of the bonds in the complex anion. In the anions XO_4 with intrinsic symmetry T_d the replacement of one oxygen atom by a hydroxyl leads to a fall in symmetry; it may be considered that the symmetry of the $(HO)XO_3$ ion is approximately described

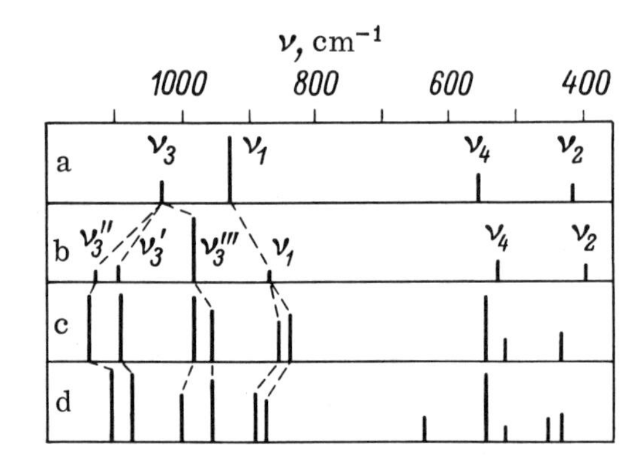

Fig. 70. The $[PO_4]^{3-} \rightarrow [HOPO_3]^{2-}$ transition. Raman spectra: a) the $[PO_4]^{3-}$ ion in solution; b) solution of K_2HPO_4; infrared spectra; c) K_2HPO_4 crystal; d) $K_2HPO_4 \cdot 3H_2O$ crystal.

by the group C_{3v} with a corresponding change in the selection rules and a reduction in the degree of degeneracy:

$$\begin{array}{llllll} XO_4\,(T_d) & \nu_3\,(F_2) & & \nu_1\,(A_1) & \nu_4\,(F_2) & \nu_2\,(E) \\ & \swarrow \quad \downarrow & & \downarrow & \swarrow \quad \searrow & \downarrow \\ HOXO_3\,(C_{3v}) & \overbrace{\nu_3',\ \nu_3''\,(E)} & \nu_3'''\,(A_1) & \nu_1\,(A_1) & \overbrace{\nu_4',\ \nu_4''\,(E)} \quad \nu_4'''\,(A_1) & \nu_2\,(E) \end{array}$$

Figure 70, taken from [74] gives a schematic indication of the change in the frequencies of the P–O stretching vibrations on passing from the $[PO_4]^{3-}$ to the $[HPO_4]^{2-}$ ion in a solution and in a crystal of potassium hydrogen phosphate K_2HPO_4. In addition to the splitting of the vibration of type E observed even in the spectrum of the solution, the spectrum of the crystal also shows splitting of the nondegenerate vibrations. The latter undoubtedly has an intermolecular origin and is due to the effect of the resonance of the acid anion vibrations acting through the directional bonds in the hydrogen bridges, which is stronger than in the crystals of the alkali phosphates with purely electrostatic interionic interactions. Apart from the splitting of the frequencies, one notices the considerable fall in the frequency of the ν_1 vibration, with a simultaneous increase in the average sum of the frequencies of the components of the ν_3 vibration as compared with the frequencies of the PO_4 ion. The intensity of the band of ν_1 vibrations in the absorption spectrum is very great and comparable with the intensity of the bands corresponding to the components of the ν_3 vibration.

Many characteristic features of the spectra of the acid salts, both in the frequency range of the X–O vibrations and in that of the hydroxyl vibrations, may be explained on the basis of existing concepts of the electron structure of the X–O bonds and the donor–acceptor theory of the hydrogen bond [24, 74, 84]. The latter, of course, starts from the polar character of the A–H bond in the A–H . . . B bridge and the existence in this system of an effective positive charge on the hydrogen atom, which also leads to the use of a favorably oriented orbital occupied by an unshared pair of electrons of the B atom for the formation of the H . . . B bond. The idea of the hydrogen bond as an acid–base interaction [24, 26] makes the strength of the hydrogen bond formed by the acid A–H and a specified base B strictly dependent on the degree of protonization of the hydrogen atom. Siebert [88] and Ryskin [84] indicate two possible mechanisms for the relocation of an electron cloud δe released in the "drawing out" of the proton by an acceptor B outside the A–H bond. Both mechanisms lead to the formation of a strong hydrogen bond.

The first mechanism, the formation of yet another bond by the atom A, is represented by

$$R_nX-A-H+B+Y \rightarrow R_nX-\underset{}{\overset{\overset{\textstyle Y \leftarrow \delta e}{|}}{A}}-H\ldots B$$

This process leads to the formation of strong hydrogen bonds in crystal hydrates if the water molecules are not only donors but also acceptors of protons. In the presence of a suitable proton donor, for example an acid, we may even have the formation of an H_3O^+ ion, with complete equality of the O–H bonds. If however the O–H bonds of the oxygen atom in the water molecule are not polarized by interaction with any proton donor, the formation of a strong hydrogen bond is unlikely. Analogous causes (absence of other bonds to the O atom) may explain the ineffectiveness of the $(OH)^-$ ions as proton donors. The formation of X–OH bonds, however, as mentioned earlier, promotes the polarization of the O–H bond and leads to a strengthening of the hydrogen bonds, for example, on passing from the hydroxides of group I elements to those of elements belonging to groups II and III.

The second mechanism is the use of the additional electron cloud to raise the multiplicity of the bonds formed by the A atom:

$$R_nX-A-H+B \rightarrow R_nX \overset{\downarrow\!-\!-\!\delta e}{\cdots} A-H\ldots B$$

This case is typical of carboxylic acids, in which the formation of strong hydrogen bonds is accompanied by an increase in the multiplicity of the C–OH bond with a simultaneous fall in the multiplicity of the C=O bond, which is clearly established spectroscopically. This process thus provides for the existence of a multiple bond at the X atom and the possibility of a redistribution among the orders of the bonds. An analogous version, using the vacant orbital of the X atom and the unshared pair of electrons of the A atom in order to form the additional A–X bond, is realized in silanols (p. 42), thus sharply distinguishing these from carbinols. Similar phenomena characterize many acid salts forming extremely strong hydrogen bonds.

For an unchanged valence of the P atom and a multiplicity of the P–O bonds in the $[PO_4]^{3-}$ ion equal to about 1.4 [88], a transition to the $[O_3POH]^{2-}$ ion leads to a fall in the multiplicity of the P–OH bonds (to approximately unity in the isolated ion) and a rise in the multiplicity of the $P\langle(O^-)_3$ bonds. This is reflected† as a fall in the $\nu_1 PO_4$ frequency and a rise in ν_3 ($\frac{1}{3}[\nu_3' + \nu_3'' + \nu_3''']$ $> \nu_3$). The formation of a hydrogen bond $O_3POH\ldots O$ leads not only to a redistribution of the freed charge δe between the O atoms and PO_3 group, but also to a strengthening of the P–OH bond. Thus as the hydrogen bond becomes stronger there will probably be a certain leveling in the orders of the bonds in the O_3POH ion, and the fact of the formation of strong hydrogen bonds is clearly associated with the favorability of this from the energy point of view. An example of such changes in bond strengths in the O_3POH ion is the $K_2HPO_4 \rightarrow K_2HPO_4 \cdot 3H_2O$ transition [74]. In the latter crystal the slightly lower νOH frequencies indicate stronger hydrogen bonds. At the same time, we see from Fig. 70 that this transition is accompanied by a rise in frequency ν_1 and a fall in frequencies ν_3' and ν_3''.

From considerations based on the displacement of a considerable part of the electron cloud of the hydrogen atom beyond the limits of the O–H group on formation of a strong hydrogen bond, there should be a reduction in the proportion of covalence characterizing the O–H bond. Hence, as noted in [85], in addition to a fall in the frequency of the νOH vibrations, there may be a fall in the δXOH frequency. Under certain conditions the part played by this factor may be more substantial than the rise in the δXOH frequency associated with the superposition of an additional H...O bond.

We noted earlier that the formation of strong hydrogen bonds by the hydroxyl groups of the silanols, accompanied by a strengthening of the Si–OH bond, was a direct indication of the participation of the 3d orbitals of silicon in its bonds with oxygen. Spectroscopic investigations of acid silicates not only enable us to identify the acid silicate anions, until recently still regarded as comparatively rare, but also provide additional confirmation as to the important part played by the $p\pi - d\pi$ bonds in the redistribution of electron density among the bonds of the acid anion. According to a preliminary x-ray investigation of Heller [89], the $Ca_2(SiO_3OH)OH$ hydrate of dicalcium silicate corresponds to the formula O_3SiOH, the distance between the O atoms of the hydroxyl group of the O_3SiOH group and the OH ion being about 2.5 Å, i.e., it suggests the formation of a strong hydrogen bond. Analysis of the infrared spectrum confirms these views [74]: in the OH stretching region there is a narrow band at 3532 cm^{-1} corresponding to the $(OH)^-$ ion and also wide bands at 2450 and 2847 and a weaker one at 2594 cm^{-1}, the first two of which are evidently analogous to bonds of types A and B in other acid salts. It is an interesting fact that the absence of the C-type bands from the spectrum may be explained by the use of the O atom of the $(OH)^-$ ion as proton acceptor, and hence the absence of chains or other-

† By considering the form of the vibrations of the O_3POH ion it is not hard to convince oneself that the greatest effect of the change in the P–OH distance is concentrated in the ν_1 vibration (and partly the ν_3'''), while the changes in the $P\langle(O^-)_3$ distances are chiefly localized in the ν_3' and ν_3'' vibrations.

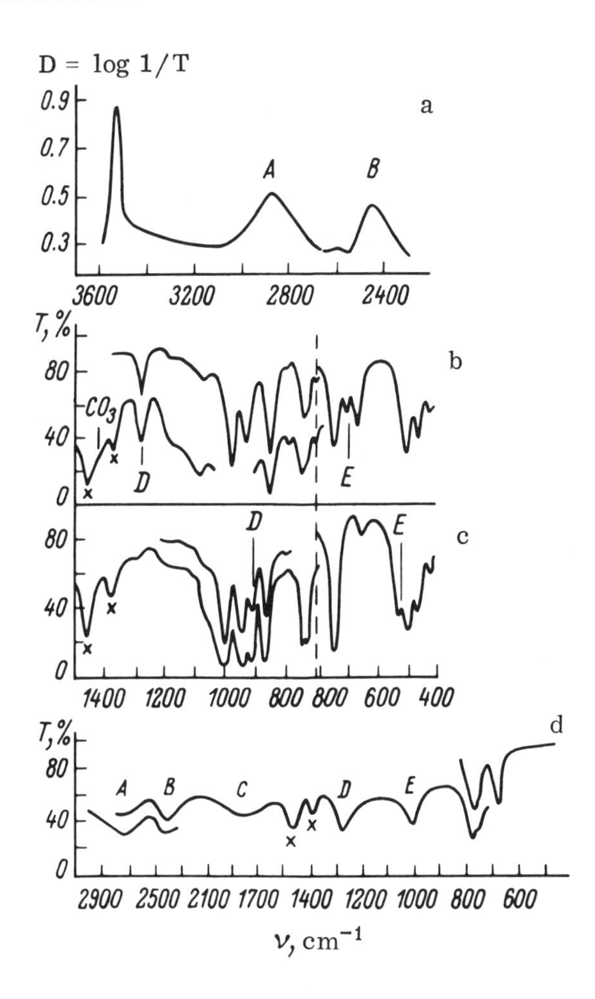

Fig. 71. Infrared spectra of acid silicates and germanates [4, 74]. a) The α-hydrate of tricalcium silicate $Ca_2(HOSiO_3)OH$, range of stretching vibrations of the hydroxyl; b), c) $Ca_2(HOSiO_3)OH$ and $Ca_2(DOSiO_3)OD$, Si–O vibrations and deformation vibrations of the hydroxyl; d) SrH_2GeO_4. Crosses indicate bands of paraffin oil.

wise linked systems of hydrogen bridges. The appearance of δSiOH bands, recognized from the effect of deuterization, at 1282 cm^{-1} clearly indicates the existence of a covalent X–OH bond, which in this crystal can only be Si–OH. The γSiOH band is observed at about 712 cm^{-1}. No less a characteristic form is found for the range of stretching vibrations of the silicate anion; strong bands are present not only in the range of ν_3 frequencies of the SiO_4 ion, but also in the ν_1 range (Fig. 71a, b, c). We remember that in the spectra of the anhydrous ortho- or oxyorthosilicates of calcium the absorption band corresponding to the ν_1 vibration is of low intensity or entirely absent, although formally it is not forbidden by the selection rules. Hence its high intensity in the spectrum of the hydrosilicate indicates a strong asymmetrization of the silicon–oxygen tetrahedron, if not geometrically then at least dynamically.

The spectra of another acid calcium silicate, afwillite, the structure of which, according to Megaw [90], corresponds to the formula $Ca_3(O_3SiOH)_2 \cdot 2H_2O$, was studied in [86, 91]. One interesting feature is the presence, among other νOH frequencies of the H_2O molecules and O_3SiOH ion, of a C-type vibration band in the region of 1920 cm^{-1}, associated in [86] with the existence of chains of hydrogen bonds between silicon–oxygen tetrahedra. A large number of bands in the range of the stretching frequencies of the tetrahedron indicates strong resonance interaction between the vibrations of the tetrahedra. The role of carriers of these interactions may naturally be ascribed to short hydrogen bonds (2.49 to 2.56 Å [90]). These data, indicating the capacity of the O_3SiOH ions to form strong hydrogen bonds, can hardly be explained without assuming the formation of additional dπ–pπ bonds between the Si and O. If only electrons in the sp^3 orbitals take part in the formation of the Si–O bonds, the relocation of some of the negative charge of the hydrogen atom in the SiO_4 groups, necessary in order to form a strong hydrogen bond, is only possible by increasing the multiplicity of the Si–OH bond, with a corresponding reduction in the order of the remaining bonds of the O_3Si group. In other words, the O_3SiOH system would be similar to systems of the R_3COH type, in which, of course, strong hydrogen bonds do not occur. Hence the formation of strong hydrogen bonds by the O_3SiOH ions must be ascribed to the location of the charge δe in bonds involving the 3d orbits of Si. The presence of electronegative oxygen atoms reduces the energy of the 3d orbitals of Si and facilitates the formation of stronger hydrogen bonds than in the silanols with C_3SiOH groups.

Analogous conclusions regarding the possibility of participation by the d orbitals of other Group IV elements and the formation of X–O bonds were drawn by Stavitskaya and Ryskin [4] after studying the spectrum of the dihydrogen germanate SrH_2GeO_4. According to [92, 93], this compound is structurally similar to the well-known potassium dihydrogen phosphate KH_2PO_4. Owing to the low frequencies of the Ge–O vibrations the SrH_2GeO_4 spectrum admirably demonstrates the characteristic bands of the stretching and deformation vibrations of the hydroxyls in acid anions (A, B, C, D, and E in Fig. 71d). Comparison of the frequencies of the hydroxyl vibrations in KH_2PO_4 and SrH_2GeO_4 suggests that the energies of the hydrogen bonds in these crystals are similar. Corresponding to the vibrations of the tetrahedron we have the bands 773 to 752 (ν_3) and 676 cm^{-1} (ν_1), the intensity of the latter being very high.

In conclusion, it should be mentioned that the changes in the spectra of acid silicates and germanates sometimes observed on deuteration in the range of the Si–O and Ge–O bond vibrations probably also indicate a considerable bond lability, allowing a change in the order of the $X{-}O^-$ and X–OH bonds for even a slight change in the strength of the hydrogen bond accompanying the replacement of H by D.

We shall now describe some applications of the foregoing correlations between the spectra of the hydroxyl-containing groups and their structural role in the crystals of silicates, not mentioned earlier.

A study of the infrared spectrum of the so-called hexahydrated sodium metasilicate $Na_2O \cdot SiO_2 \cdot 6H_2O$ carried out by Ryskin et al. [87] confirmed the existence in this compound of an orthosilicate anion containing hydroxyl groups (Si)OH, as proposed in [95], in which the formula was written $Na_2H_2SiO_4 \cdot 5H_2O$. According to [95], the sodium metasilicates with five- and nine-fold hydration correspond to the same formula of the complex anion. The infrared spectra of $Na_2O \cdot SiO_2 \cdot 6H_2O$ and the corresponding deuteroderivative contain five bands of νOH vibrations in the range 3550 to 3044 cm^{-1} (νOD – 2620 to 2300 cm^{-1}), ascribed to molecules of water of crystallization taking part in weak and medium-strength hydrogen bonds. The presence of bands at 2780 and 2270 cm^{-1}, displaced to 2136 and 1680 cm^{-1} on deuteration, indicates the existence of strong hydrogen bonds, which in the compound under consideration can only be formed by hydroxyls linked to silicon atoms. Deuteration† also enables us to identify the bands of other vibrations of the $\geq$Si–O–H groups

	H	D
δSiOH	1187 cm^{-1}	—
	1085 cm^{-1}	783 cm^{-1}
γSiOH	700 cm^{-1}	522 cm^{-1}

The remaining bands associated with the Si–O stretching in the complex anion are located in the range 970 to 775 cm^{-1}, characteristic of the stretching vibrations of the orthosilicate ion, since the condensation of silicon-oxygen tetrahedra leads to the formation of

† A δSiOD band is evidently superposed on the νSiO band at 876 cm^{-1}.

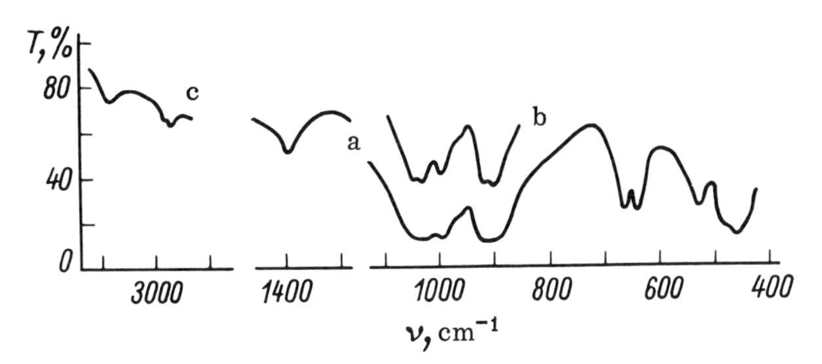

Fig. 72. Infrared spectrum of pectolite $Ca_2Na(Si_3O_8OH)$. Pressed in KBr: a) 5 mg; b) 2 mg; c) 20 mg.

lower ν_s SiOSi frequencies and on the other hand ν_{as} SiOSi frequencies usually exceeding 1000 cm^{-1}. The existence of νSiO bands differing in number and arrangement in compounds containing light and heavy hydrogen received no clear explanation in [87]; the choice between anion formulas $O_2Si(OH)_2$ and $OSi(OH)_3$ was also a difficult matter. The spectrum of the compound $Na_2O \cdot SiO_2 \cdot 5H_2O$ studied in [96, 97], which has also not yet received any explanation, in many essential features resembles the spectrum of "hexahydrated sodium silicate." †

According to Liebau [98], pectolite crystals were assigned a structure containing ribbon anions of the xonotlite type $Ca_4Na_2(Si_6O_{17}) \cdot H_2O$. X-ray analysis of the structure of pectolite carried out by Buerger [99] led to the conclusion that it contained a wollastonite chain, in each identity period of which one of the oxygen atoms was replaced by a hydroxyl: $Ca_2Na(Si_3O_8OH)$. The infrared spectrum of pectolite shown in Fig. 72 [100] demonstrates the desirability of using the correlations derived in this section in order to choose between two versions of the structural formula. In fact, in the range 500 to 700 cm^{-1}, the spectrum contains three bands which may be ascribed to symmetrical Si–O–Si stretching vibrations, agreeing not only with a chain of the $[Si_3O_9]_\infty$ type but also with a ribbon of the xonotlite type, if its symmetry is not effectively lower than C_{2h} (p. 113). However, the absence of bands in the range 1100 to 1200 cm^{-1}, indicating the absence of SiOSi angles close to 180° (characteristic of the xonotlite ribbon), make the wollastonite-type structure more probable. This proposition agrees excellently with the observed hydroxyl frequencies: the absorption in the range 2800 to 2900 cm^{-1} indicates a strong hydrogen bond, which is natural for the Si–OH group and improbable for the H_2O molecule, the more so since the δH_2O frequency is absent from the spectrum.

It should be noted that, according to Mamedov and Belov [101], hillebrandite crystals β-$2CaO \cdot SiO_2 \cdot H_2O$ may contain portland layers $Ca(OH)_2$ and ribbons of the xonotlite type as two structural elements. Studies of the infrared spectrum and proton magnetic resonance [102, 103] confirmed the first of these propositions. In the frequency range of the vibrations of the silicon–oxygen anion there are quite considerable differences, of which the absence of strong bands in the hillebrandite spectrum above 1070 cm^{-1} is particularly noteworthy. Hence the existence of xonotlite ribbons can only be admitted if we assume a considerable deformation of the Si–O–Si bridges "knitting" the two chains into a ribbon: the SiOSi angle in hillebrandite can hardly exceed 140°, instead of 180° in xonotlite.

Among substances containing Si–O–H groupings we should mention silicic acids; equally we note the presence of these groups in crystalline quartz and glasses, clearly indicated by spectroscopy, as in [104–107]. We note that the presence of the SiOH groups in quartz and silicic acid leads to the appearance of absorption near 940 to 950 cm^{-1} [109], not observed in

† The x-ray structural analysis of this crystal gave the formula $Na_2H_2SiO_4 \cdot 4H_2O$ [94].

anhydrous silicon dioxide. The position of the band changes very little on deuteration, and this suggests associating it with νSi–OH vibrations [110]. The similarity in the position of the νSi–OH band and the bands ascribed to the $\nu Si{-}O^{-}(M^{+})$ vibrations in the spectra of alkali silicate glasses may be considered as an indication of the similar orders of these bonds, considerably exceeding the order of the Si–O(Si) bonds. The Si–O–H groupings situated on the surfaces of silicates and the chemical reactions involving them have been widely studied by spectroscopy; however, a discussion of these questions lies outside the scope of this book.

No less an important part in the structure of silicates may also be played by small traces of water acting as catalysts in relation to many solid-state transformations [111]; however the spectroscopic investigation of this field is a very difficult matter.

§3. Spectroscopic Methods of Studying the Coordination of Cations with Respect to Oxygen

Considering the possibility of interpreting the vibrational spectra of silicates, we have so far only turned our attention to lattice vibrations, i.e., vibrations of the coordination polyhedra of the cations, in so far as this has been necessary in order to establish the possibility of separating them from internal vibrations of the complex anion or at any rate its stretching vibrations. We have also considered the rise in the cation–oxygen vibration frequencies on reducing the coordination number of the cation, particularly when the latter is capable of forming partly covalent bonds with oxygen. This enables us to consider using spectroscopic methods not only in studying the structure of complex anions in silicates but also in obtaining certain information regarding the structure of the cation part of the system, primarily the coordination of the cations. At the present time a considerable amount of material has been accumulated, enabling a number of empirical correlations to be established between the position of the absorption bands of the cation–oxygen vibrations and the coordination number of the cation. The fact that this question has, in a number of cases, not been subjected to analysis by the method of normal coordinates does not vitiate the practical value of the correlations established, since the main field of their use is not the calculation of the vibrational spectrum and the force constants (both in the anion and in the cation part of the structure) of a crystal the structure of which has already been established by nonspectroscopic means, but, on the contrary, the study of cation coordination in a crystal of unknown structure. Of special interest is the possibility of using these correlations when studying the mechanisms of isomorphous substitutions or when studying the structure-forming role of certain atoms in glasses.

Although a reduction in the coordination number of the cation leads to a fall in the interatomic distances and a rise in the force constants even in the case of typical ionic crystals, the greater proportion of existing data regarding the dependence of the frequencies in vibrational spectra on the coordination number relates to atoms with amphoteric properties capable of forming at least partly covalent bonds with oxygen. This is chiefly associated with the fact that the frequencies of the vibrations of such bonds are higher than those of purely ionic bonds, and hence they fall in a region accessible for study with ordinary prism instruments. For the same reason comparatively light cations have been most studied.

Further, changes in coordination number for cations with a large ionic radius frequently have less marked effects, so that they cannot be so easily identified spectroscopically. We may nevertheless expect that the rapid development of the technique of long-wave infrared spectroscopy will enable us to estimate the prospects of studying the coordination of heavy atoms with large atomic radii by spectroscopic methods more reliably.

We shall now present some existing data regarding the relation between the frequencies of the vibrational spectrum and the coordination of the atoms most important in silicate structures. The correlations described are very approximate, since the specific features of the kinematics of individual crystals, particularly differences in the kinds of links between the coordination polyhedra, are not taken into account.

Boron. A detailed consideration of the spectra of the borates lies outside the limits of this book, all the more so since the methods of analyzing the spectra of the silicates considered in the preceding chapters are inapplicable to borosilicates in view of the high frequencies of the B–O vibrations. However, the changes observed in the spectra of the borates when the coordination number of the B atom is changed are extremely characteristic. The possible oxygen coordination of boron varies from the plane triangle observed in the orthoborate anion $[BO_3]^{3-}$, ions of type $[B_2O_5]^{4-}$ in the pyroborates, and the $[B_3O_6]^{3-}$ rings or $[BO_2]_\infty$ chains of the metaborates, to the BO_4 tetrahedron in such compounds as $4ZnO \cdot 3B_2O_3$, BPO_4, $BAsO_4$. The interatomic distances in the B–O bonds increase sharply from an average value of 1.36 Å for the trigonal to 1.48 Å for the tetrahedral coordination [112]. The differences in the B–O bond lengths for the same boron atom may also be considerable, particularly between bonds of the B–O(B) and $B\text{–}O^-(\ldots M^+)$ types, reaching approximately 0.15 Å. This is evidently due to the ability of one of the p orbitals (for an sp^2 configuration of the B atom) to form a π bond with the oxygen atom, becoming stronger as the effective negative charge on this atom increases. In many borates the situation is complicated as a result of the coexistence of trigonally and tetrahedrally coordinated boron within a single complex anion. In structures with BO_4 tetrahedra there may also be a considerable asymmetrization of the tetrahedron, culminating in the formation of a markedly elongated pyramid. The lengths of the B=O and B–O bonds in the plane O=B–O–B=O molecule are (according to electron diffraction [113]) 1.20 and 1.36 Å respectively. The latter value exactly agrees with the average B–O distance for trigonally coordinated boron and corresponds, according to an estimate by Krogh-Moe [114] to a bond order of 1.25, based on the proposed single bond length of 1.43 Å.

A large number of investigations have been devoted to the spectroscopic study of borates. Many of these are analyzed by Weir and Schroeder [115], who studied the infrared spectra of about 80 borates and considered the relationship between the spectrum and the structure of the anion. We must also mention work on the spectra of boric acid [116-118], its esters [119-124], and also the B_2O_3 and B_2S_3 molecules [125]. A change in the degree of condensation of the BO_3 groups in inorganic borates may be clearly established spectroscopically owing to the considerable displacement of the stronger absorption bands in the range 1200 to 1300 cm^{-1} for compounds containing isolated $[BO_3]^{3-}$ ions, reaching 1400 to 1450 cm^{-1} in the spectra of structures with condensed BO_3 groups, ring and chain metaborates, pyroborates, etc. In addition to kinematic factors, the increase in the order of the $B\text{–}O^-$ bonds as compared with the orthoborates, which contain three such bonds at the same boron atom, probably makes itself felt here. In compounds containing boron exclusively in the tetrahedral coordination, the range of strong absorption is displaced in the direction of lower frequencies and its upper limit is no higher than 1100 to 1000 cm^{-1} for condensed systems. The position of the B–O vibration frequencies in all the inorganic borates shows a clear dependence on the chemical nature of the cation. The spectra of boron compounds containing complex anions incorporating BO_3 and BO_4 exhibit absorption in the range of frequencies characteristic of both groups.

The spectra of natural borates and borosilicates have also been considered in the literature in connection with the possibility of determining the coordination of boron [126, 127]; however, the identification of the frequencies of the BO_4 groups in silicates in which Si–O vibrations of similar frequency occur and interact with them is difficult. In these cases, spectroscopic data only enable us to decide whether trigonally coordinated boron is absent or not. The existence of sharp differences between the frequencies of the B–O vibrations corresponding to the trigonal and tetrahedral coordination of boron has repeatedly been used for studying the coordination of boron in borates and also borosilicate glasses [128-130]. For these substances infrared spectroscopy, together with nuclear magnetic resonance, is the main direct method of studying the dependence of short-range order on composition.

The approximate values of the force constants (diagonal) are $13.4 \cdot 10^5$ dyn/cm for B–O [125] and 5.3 • and $4.2 \cdot 10^5$ dyn/cm for B–O, depending on the ternary or quaternary coordination of the B atom [131, 132] in the oxygen-containing organic derivatives.† The only inorganic compound in the spectrum of which the frequencies of the "isolated tetrahedral ion" of $[BO_4]^{5-}$ may be identified is $Ta_2O_5 \cdot B_2O_3 = 2TaBO_4$. Judging from the spectrum of Ta_2O_5, Ta–O vibrations do not lie in the frequency range accessible to prism-optical systems, and the four bands observed in the infrared spectrum may be ascribed to vibrations of the BO_4 tetrahedron [115]: 840 cm^{-1} – F_2, 579 cm^{-1} – A_1, 490 cm^{-1} – F_2, 300 cm^{-1} – E. The structure of $TaBO_4$ is similar to that of zircon, and the B atoms lie in positions surrounded by oxygen in roughly the tetrahedral manner. Neither the factor group of the crystal, isomorphic with D_{4h}, nor the site symmetry V_d of the BO_4 tetrahedron give any explanation for the presence of the band of the totally symmetric vibration in the infrared spectrum. The force constant of the B–O bond calculated from the frequency values indicated is much smaller than the value given in the foregoing. This may probably be ascribed to the low order of the B–O bonds (i.e., about $3/4$ according to the ratio of the force constants), which is slightly reduced because we have not taken account of the reduction in the frequencies of the BO_4 group under the influence of the large mass of the Ta atoms. It is interesting to note that in BPO_4, which is structurally similar to low-temperature cristobalite, the length of the B–O bond in the BO_4 tetrahedron is 1.44 Å, i.e., close to the length of the single B–O bond, while in the PO_4 tetrahedra shortening of the interatomic distances occurs on account of the $d\pi$–$p\pi$ bond. In other words, the fact that there is no such mechanism for raising the multiplicity of the bond in the case of B, an element of the second period, is evidently responsible for the fact that the order of the B–O bond in the BO_4 group is in any case no greater than unity. The spectrum of BPO_4 contains bands around 1095, 925, 615, and 550 cm^{-1} [115, 133]. Since the first of these is close in frequency to the ordinary values of $\nu_{as} PO_4$ in orthophosphates, the strong band at 925 cm^{-1} may (with a certain caution) be assigned to the $\nu_{as} BO_4$ vibration, which on estimating the corresponding force constant suggests an order of the B–O bond in the region of unity. In view of the strong interaction of the vibrations, which leads to a substantial displacement of the frequencies in the spectrum of the isostructural $BAsO_4$ [115] (the As–O frequencies are lower than that P–O frequencies and interact more strongly with the B–O frequencies), this estimate cannot be regarded as reliable. We remember, however, that in the spectrum of $AlPO_4$, which in both structure and dynamics‡ is a far closer analog of SiO_2 than BPO_4, there is no similarity to the spectrum of SiO_2: there is a strong band with a frequency slightly higher than $\nu_{as} PO_4$ in the spectrum of BPO_4, and other bands in the region of 700 cm^{-1}, which agrees closely with the $\nu_{as} AlO_4$ frequency.

It follows from the foregoing that compounds containing anions in which boron is trigonally coordinated with respect to oxygen are to some extent similar to those containing the tetrahedral oxy-anions of Si, P, and S: the rise in the order of the $X–O^-$ bond as the effective negative charge on the O atom becomes greater is brought about by the existence of unoccupied orbitals suitable for the formation of π bonds with the central atom. On the other hand, for a

† For the $[BO_3]^{3-}$ ion this value increases by about 30%.

‡ The close structural similarity of $AlPO_4$ and SiO_2 is confirmed by the closeness of the geometrical parameters of the structures, the almost complete analogy between the polymorphic transformations, and so on [134]. On the other hand, the masses of Al and P differ little from that of Si, like the interatomic distances; finally, for Al as an element of the third period we cannot exclude the possibility that the d orbitals participate in the formation of bonds with oxygen.

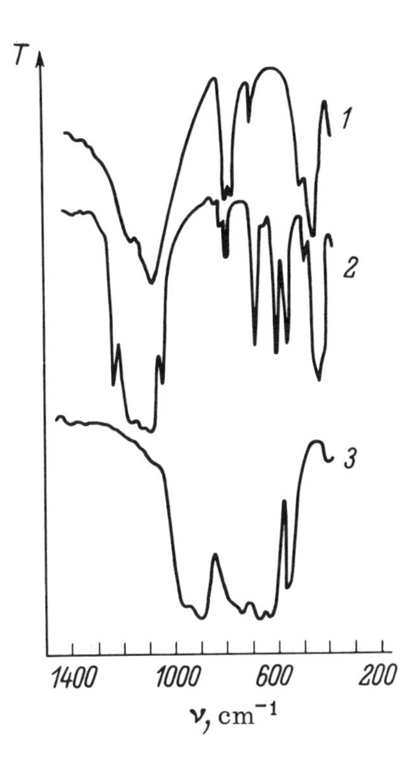

Fig. 73. Infrared spectra of various forms of SiO_2 containing Si in tetrahedral or octahedral coordinations. 1) Quartz; 2) coesite; 3) stishovite.

tetrahedral coordination of boron with respect to oxygen, the accumulation of negative charge on the O atoms does not lead to any substantial rise in the order of the B–O bonds.†

Silicon and Germanium. Although in the overwhelming majority of cases silicon possesses a coordination number of four with respect to oxygen, there are two examples in which Si enters into an oxygen octahedron.‡ The first of these relates to the recently discovered stishovite, a form of SiO_2 with the structure of rutile, in which the Si atom is surrounded by four O atoms at distances of 1.716 Å and two at distances of 1.872 Å [135]. The infrared spectrum of stishovite, studied by Lyon [136], exhibits a clear displacement of the region of strongest absorption to lower frequencies as compared with the spectra of other forms of SiO_2 (Fig. 73). The displacement of the highest-frequency group of bands from 1100 to 900 cm^{-1} is in accordance with the ratio of the force constants 1:2/3, if we neglect the differences in the kinematics of the systems considered. Octahedral coordination of silicon by oxygen has also been established in silicon pyrophosphate SiP_2O_7 [137, 138], where silicon plays the part of the cation with respect to the pyrophosphate anion $[P_2O_7]^{4-}$. The infrared spectrum of SiP_2O_7 [139] is similar to the spectra of the iso-structural cubo-octahedral pyrophosphates§ [139-141] and does not contain bands which could be ascribed to Si–O vibrations in the frequency range above 700 cm^{-1}. With this vibration we associate the band at about 675 cm^{-1}, which moves on replacing Si by Ti, Ge, Zr, or Sn: 625, 550, 548, 537 cm^{-1} [141]. This low frequency for the Si–O vibrations in SiP_2O_7, somewhat lower than the center of the interval 950 to 600 cm^{-1} in stishovite, may be explained not only by kinematic factors but also (probably) by the relatively low force constant: in the presence of pyrophosphate ions with P–O bonds formed by the more

†In structures with very asymmetrical BO_4 tetrahedra, some of the B–O distances indicate, according to x-ray data, an order of the B–O bond exceeding 1.0; however, the other distances are then considerably greater.

‡A third example is thaumasite [232, 233].

§Regarding the interpretation of the ZrP_2O_7 spectrum, see p. 295.

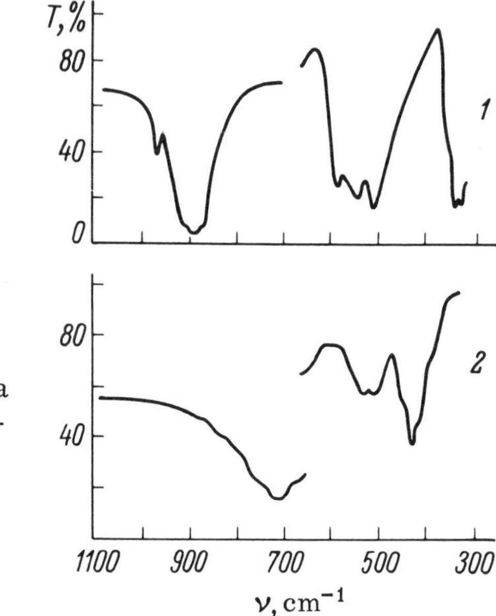

Fig. 74. Infrared spectra of types of GeO_2. 1) Hexagonal; 2) tetragonal.

electronegative P atoms and possessing a higher proportion of covalence, the ionicity of the Si–O bonds may increase. In an analogous way, the frequencies of the Al–O vibrations of the aluminum–oxygen octahedra in aluminum oxide (α-Al_2O_3) or aluminum hydroxides are much higher than the frequencies of similar octahedra in aluminum silicates, i.e., in the presence of the more covalent Si–O bonds.

The larger germanium atom is more frequently surrounded octahedrally by oxygen. Moreover, in certain structures germanium evidently occurs in the octahedral and tetrahedral coordinations at the same time: the structure of the compounds $XO \cdot 4GeO_2$, where X = Sr, Pb, Ba, corresponds to the formula $XGe^{VI}(Ge_3O_9)$ with rings of three germanium–oxygen tetrahedra, and also germanium atoms in the role of cations surrounded octahedrally by oxygen [142]. The coexistence of the tetrahedral and octahedral coordinations of germanium in the compound $2Na_2O \cdot 9GeO_2$ was mentioned in [143].

Figure 74 shows the infrared spectra of two forms of germanium oxide, the hexagonal form, very close in structure to quartz, and the tetragonal form with a rutile-like space lattice of germanium–oxygen octahedra [144]. The displacement of the strong group of bands due to Ge–O stretching vibrations amounts to 165 cm^{-1} on passing from the tetrahedral to the octahedral lattice (885 and 720 cm^{-1}, respectively), i.e., the frequencies of these bands are almost in the same relation as the frequencies in the spectra of the corresponding forms of SiO_2. Comparison of the spectrum of the vitreous germanium dioxide with the spectra of these two crystalline forms leads to the conclusion that the Ge atoms have tetrahedral coordination [144, 145]. No less considerable a shift of the absorption bands was observed in the spectra of the germanates: on passing from magnesium metagermanate $MgGeO_3$, with a structure of the pyroxene type, to the compound of the same composition obtained under high pressure and possessing the structure of ilmenite (octahedral coordination of the Ge), the region of strong absorption in the infrared spectrum moved from between 700 and 900 to between 550 and 600 cm^{-1} [146].

The strong displacement of the bands in the infrared spectra of alkali germanate glasses observed on changing the alkali concentration was also ascribed to a change in the coordination of the germanium [147].

Titanium. This is the only group IVA element exhibiting tetrahedral coordination by oxygen in certain cases. Even for zirconium a coordination number of under six is never found;

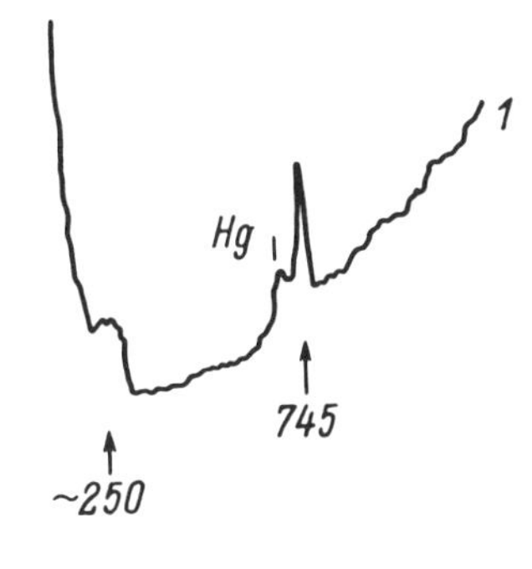

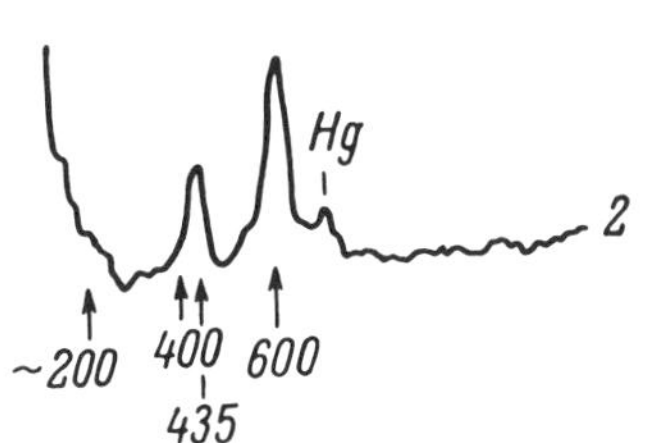

Fig. 75. Raman spectra 1) of barium orthotitanate Ba_2TiO_4; 2) of rutile (TiO_2).

hence reasonably detailed information regarding changes in the vibrational spectrum with coordination are limited, for this group, to titanium compounds, which have recently been frequently studied by means of both their infrared and Raman spectra. Interest in the spectroscopic study of the coordination of titanium has been largely stimulated by a practical problem: that of deciding tht role of titanium dioxide in the mechanism underlying the catalytic crystallization of glass. On the other hand, a number of careful spectroscopic investigations of the metatitanates ($BaTiO_3$ etc.) with the perovskite structure have been carried out in recent years in connection with interest in the mechanism of ferroelectric transformations taking place in these crystals and its dependence on certain lattice vibrations.

It was considered for a long time that the "ion" Ti^{4+}, possessing a comparatively large ionic radius, was always octahedrally coordinated with respect to oxygen, although indications started appearing in the literature to the effect that titanium had a tetrahedral coordination in Ba_2TiO_4 [148] and also that a five-fold coordination might occur [149]. Spectroscopic investigations yielded fairly specific criteria regarding the coordination of titanium. The Raman spectrum of barium orthotitanate showed a strong line at 745 cm^{-1}, i.e., at a considerably higher frequency than in the spectra of rutile TiO_2 [150], barium metatitanate [151, 152], and sphene $CaTi(SiO_4)O$ [153], containing titanium in octahedral coordination (the octahedra have common edges in rutile or common vertices in the metatitanates). The highest-frequency line of rutile lies at 600 cm^{-1} [154] (Fig. 75). The line at 745 cm^{-1} was ascribed to the totally symmetric stretching vibration of the TiO_4 tetrahedron. The infrared spectrum of Ba_2TiO_4 [155] showed strong absorption between 700 and 800 cm^{-1}, ascribed to the stretching vibrations of the TiO_4 group, and differed considerably from the spectra of the strontium and zinc orthotitanates, which absorbed at around 500 to 600 cm^{-1}, in accordance with the octahedral coordination of titanium in these compounds. Thus we perceive characteristic regions of absorption for the vibrations of the Ti–O bonds in the TiO_6 and TiO_4 groups (Fig. 76). These conclusions are in good agreement with existing data relating to the frequencies of Ti–O in the titanium–oxygen octahedra of the metatitanates of Ba, Sr, Pb, and Ca [156].

Another source of information regarding the vibrational frequencies of the titanium–oxygen tetrahedron comes from the infrared spectra of dilute solid solutions of Zn_2TiO_4 and Zn_2SiO_4 [155, 157-158]. The zinc orthosilicate crystal (willemite) does not contain any positions capable of being populated by cations apart from the tetrahedral positions occupied by the Si and Zn atoms. Hence a small number of titanium atoms may occur only in tetrahedral coordina-

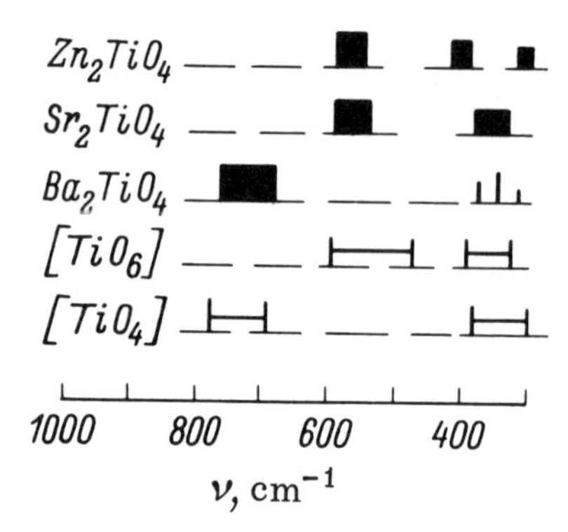

Fig. 76. Distribution of bands in the infrared spectra of the orthotitanates with octahedral (Zn, Sr) and tetrahedral (Ba) coordination of titanium.

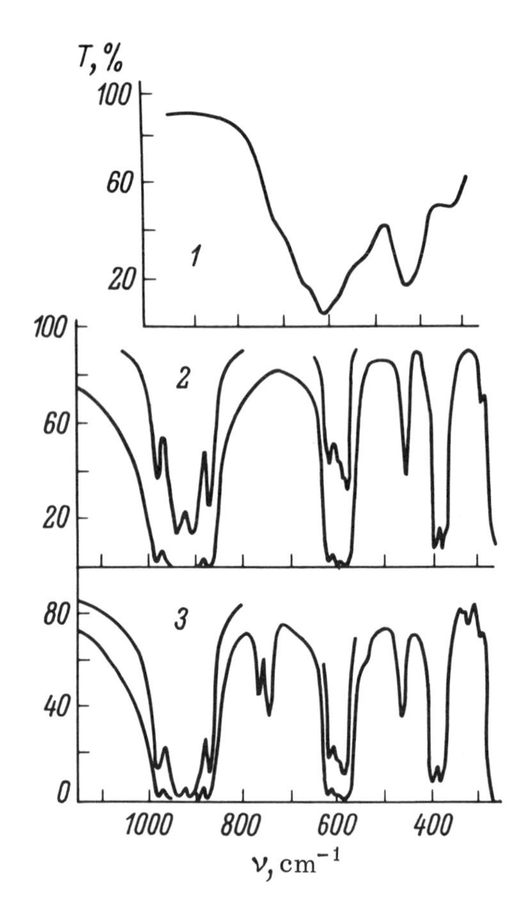

Fig. 77. Identification of the bands of the TiO_4 tetrahedron. Infrared spectra: 1) Zn_2TiO_4; 2) Zn_2SiO_4; 3) dilute solid solution of $Zn_2(Si_{0.95}Ti_{0.05})O_4$ according to Tarte [158].

tion, or only in octahedral coordination if a solid solution is not formed and the Zn_2TiO_4 exists in the form of a second crystalline phase. Comparison between the spectra of the solid solution and those of pure Zn_2TiO_4 and Zn_2SiO_4 (Fig. 77) rejects the latter alternative, and the bands in the range 700 to 800 cm^{-1} (and near 300 cm^{-1}) must be ascribed to the vibrations of TiO_4 tetrahedra, constituting the local vibrations of titanium defects in the willemite lattice. It is interesting to note that a dilute solution of germanium tetrahedra in willemite also absorbs in the same region. Completely analogous results were also obtained for solid solutions of titanates and germanates in Sr_2SiO_4 and Ba_2SiO_4. This indicated the replacement of Si by Ti and Ge atoms (since the Sr and Ba atoms have a coordination differing from tetrahedral) and further confirmed the validity of the correlations indicated in Fig. 76. Existing information regarding the vibrational spectra of the tetra-alkoxy derivatives of titanium $Ti(OR)_4$ (R = CH_3, C_2H_5, etc.) [159-161] in no way contradicts the foregoing assignments of the frequencies of the TiO_4 tetrahedron, since the lower frequencies (about 600 cm^{-1}) of the Ti–O vibrations in these compounds may be ascribed to interaction with C–O vibrations, or the tendency of organic orthotitanates toward intermolecular association with the formation of additional donor–acceptor bonds of titanium with oxygen. This tendency of titanium to increase the oxygen tetrahedron to an octahedron, i.e., to pass into the electron state sp^3d^2, explains, in particular, the difference between the structure of the hydrolytic-condensation products of $Ti(OR)_4$ and that of the corresponding silicon compounds [162, 163].

Slightly different conclusions as to the location of the vibration bands of the titanium–oxygen tetrahedra were drawn when studying the infrared spectra of garnets [164, 165]. The ideal formula of a silicate with the garnet structure may be written as $R_3R'_2Si_3O_{12}$ where usually R = Ca^{2+}, Mg^{2+}, Fe^{2+}, Mn^{2+} and R′ = Al^{3+}, Fe^{3+}, Cr^{3+}. As already mentioned, the high-frequency part of the infrared spectra of garnets has a form typical of orthosilicates, while the vibrational frequencies involving the cation–oxygen bond as well as the deformations of the SiO_4 tetrahedron never exceed 600 cm^{-1}. However, even in this region the changes in the spectrum on replacing some di- and trivalent cations with others are comparatively small. As the amount of titanium increases, the spectra of the garnets develop gradually intensifying absorption at about 650 cm^{-1} (with a certain broadening and spreading of the bands between 800 and 1000 cm^{-1}), which leads to the conclusion that the titanium enters into the garnet in tetrahedral coordination and that the absorption in the region of 650 cm^{-1} is associated with the TiO_4 group. However, it is still not quite clear what causes the vibrational frequencies of the titanium–oxygen tetrahedron to be lower for garnets than for other silicates.

The establishment of spectroscopic criteria for the coordination of titanium in silicates facilitated the study of the state of titanium in silicate and phosphate glasses and products of their crystallization [166-169]. Bobovich, having estimated the force constant of the Ti–O bond in the titanium–oxygen tetrahedron from the $\nu_s TiO_4$ frequency in Ba_2TiO_4, approximately calculated the frequencies of the stretching vibrations of $[Ti_2O_6]_\infty$ chains of the pyroxene type and came to the conclusion that the frequencies observed in the spectra of silicate glasses containing titanium corresponded to the vibrations of condensed titanium–oxygen tetrahedra. At the same time titanium may also exist in the glass in six- and possibly five-fold coordination. The results of the use of infrared spectra for studying the coordination of titanium in glasses were analyzed by Tarte [171, 172]. In this connection it is timely to mention that the sharp distinction frequently made between the structural role of titanium in glasses as a lattice-former when in tetrahedral coordination and, on the other hand, simply a modifying element when in octahedral coordination is hardly well-founded. The Ti–O bonds in the octahedron have a fairly directional character, and the titanium-oxygen octahedra may in many cases be considered as an element of the structure of the complex anion (p. 84 and also [170]).

Aluminum and Gallium. The spectroscopic elucidation of the coordination of aluminum is perhaps of greatest interest for silicates, since aluminum, in view of its great crystal-chemical similarity to silicon, very frequently replaces the latter isomorphously or forms a common anion system with a random or, conversely, ordered distribution of Si and Al in the tetrahedral positions. Just as frequently aluminum also occupies octahedral positions in silicates, playing the part of a cation with respect to the complex silicate or aluminosilicate anion.

Let us first consider existing data regarding the relationship between the Al–O vibration frequencies and the coordination number of aluminum in silicon-free structures, the oxides of aluminum and aluminates, where the aluminum is the only element forming a complex anion together with the oxygen atoms. In corundum α-Al_2O_3 the aluminum–oxygen octahedra form a space lattice, while in the hydroxides, gibbsite $Al(OH)_3$, diaspore α-AlOOH, and boehmite γ-AlOOH, they form layers of octahedra with common vertices linked together by hydrogen bonds. The strongest absorption in the infrared spectrum of α-Al_2O_3 is observed between 600 and 700 [158] or 600 and 750 cm^{-1} [173]. Aluminum hydroxides show absorption associated with the Al–O vibrations between 700 and 780 cm^{-1} and also at lower frequencies [58]. The comparatively high frequencies of the Al–O vibrations in these compounds are probably explained not only by kinematic characteristics, the condensation of aluminum–oxygen tetrahedra, but also by the partly covalent character of at least some of the bonds in the octahedron. This is reflected in the considerable differences between the lengths of the Al–O bonds within a single octahedron: 1.82 to 2.01 Å in α-AlOOH and from 1.87 to 2.06 Å in γ-AlOOH. The high frequencies of the deformation vibrations of the hydroxyls associated with the aluminum atom around 950 to 1080 cm^{-1} indicate the "rigidity" of the AlOH angle and correspondingly the partly covalent character of the Al–O bond [58].

In the spectrum of the so-called β form of $Al_2O_3(Na_2O \cdot 11Al_2O_3)$, containing aluminum–oxygen tetrahedra in addition to octahedra, the region of strong absorption becomes wider than in the spectrum of α-Al_2O_3, extending up to 850 cm^{-1} [158] (Fig. 78). In the spectra of the aluminates of alkali and alkaline-earth metals forming three-dimensional frameworks of AlO_4 tetrahedra, similar in structure to tridymite or cristobalite [176], the highest frequencies of the Al–O vibrations reach 900 to 850 cm^{-1} [177]. The Al–O distances in these structures are fairly small: 1.66 Å in $KAlO_2$ [178], 1.68 to 1.82 Å in $CaAl_2O_4$ [179], 1.79 Å in $BaAl_2O_4$ [180]. Thus the transition to tetrahedral coordination evidently increases the covalence of the Al–O bonds substantially. The similarity between the spectra of these aluminates and the spectra of $LiAlO_2$, $RbAlO_2$, the structures of which had not been solved by x-ray diffraction, led to the conclusion the aluminum occurred in tetrahedral coordination in these compounds. Also of interest is the existence of a low-temperature form of $LiAlO_2$, which was assigned a structure similar to α-$NaFeO_2$, i.e., with the Al in the octahedral coordination [181]. A spectroscopic study of the transformation of the low-temperature form of $LiAlO_2$ into the high-temperature form carried out by Kolesova [173] confirms that this transformation is similar to the α–β transformation in $NaFeO_2$, involving a rearrangement of the complex anion into a framework

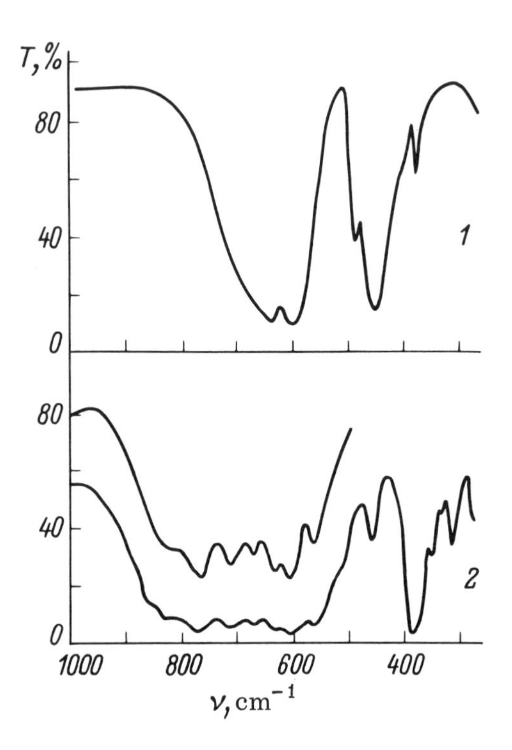

Fig. 78. Infrared spectra of various forms of Al_2O_3. 1) α-Al_2O_3; 2) β-Al_2O_3.

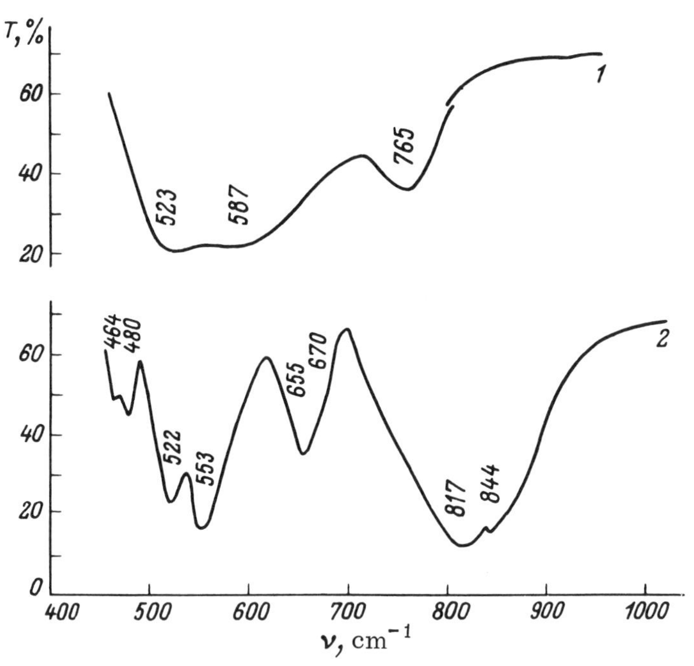

Fig. 79. Infrared spectrum of various forms of $LiAlO_2$. 1) Low-temperature; 2) high-temperature form.

of tetrahedra. There is also an obvious displacement of the absorption bands to higher frequencies (Fig. 79).

Among the simpler structures containing essentially covalent Al–O bonds we should mention the Al_2O molecule, the infrared spectrum of which, according to a study carried out by the gas matrix method, contains bands at about 990 and 710 cm^{-1} [182]. These bands characterize the ν_{as} AlOAl and ν_s AlOAl frequencies for an AlOAl angle of about 140 to 150°, i.e., for an order of the Al–O(Al) bonds probably exceeding unity.

Considerably greater difficulties arise in the identification of vibrations involving the Al–O bonds in the spectra of silicates. In the spectra of garnets containing aluminum atoms in isolated oxygen octahedra, the Al–O vibrations may be associated with the bands in the region of 500 to 400 cm^{-1} [164]. However, in the region of 500 to 650 cm^{-1} the absorption bands in the spectra of garnets are also sensitive to cation substitution. Absorption in the region between 500 and 600 cm^{-1} also occurs in many other aluminum silicates [183]. Hence as already mentioned it is not always possible to obtain a reliable identification of the frequencies of symmetric Si–O–Si stretching vibrations, for example, in pyrosilicates or pyroxenes containing Al^{3+} among the cations. On the other hand, we may assert with fair conviction that the Al–O vibration frequencies of the aluminum–oxygen octahedra in the silicate spectra are in any case no greater than 600 to 650 cm^{-1}.

In the spectra of the aluminosilicates, i.e., compounds containing aluminum atoms isomorphously replacing silicon in the tetrahedral network, the question as to the identification of the vibrations predominantly localized in the Al–O bonds is particularly complex. Kolesova [183] gave a number of arguments in favor of associating the Al–O vibrations in the spectra of the aluminosilicates containing tetrahedrally coordinated aluminum with the absorption in the range 720 to 780 cm^{-1} corresponding to absorption in the range 800 to 1100 cm^{-1} in the spectra of silicates. In particular, the spectrum of natural α-spodumene $LiAl(Si_2O_6)$ containing Al atoms in the octahedral positions and complex silicon–oxygen anions in the form of $[Si_2O_6]_\infty$ chains shows no absorption between 700 and 800 cm^{-1}. The transformation (on heating) of α-spodumene to the β-form with a lattice of the keatite type, i.e., a three-dimensional framework of interlinked silicon– and aluminum–oxygen tetrahedra, is accompanied by marked changes in the spectrum, including the appearance of a band at about 760 cm^{-1}. The presence of aluminum–oxygen tetrahedra in the spectra of aluminosilicate glasses [184] is associated with absorption in the same range of frequencies.

On the other hand, when studying changes in the infrared spectrum of micas and other layer-like silicates, as some of the Si atoms are isomorphously replaced by Al there is a displacement to lower frequencies of the majority of the bands corresponding to the frequency range of the stretching vibrations of the complex anion [67, 69]. On the basis of these results methods of spectroscopically determining the percentage of Si replacement by Al in these substances have even been proposed [185]. Simultaneously, the intensity of many bands is considerably reduced, and all the bands are broadened, giving the spectrum a much more diffuse form. In fact, in view of the random character of the Al and Si distribution, the small differences in the masses of these atoms, and the slightly lower force constants of the Al–O bonds as compared with the corresponding Si–O bonds, the replacement of Si by Al is to a first approximation equivalent to the introduction of a certain lack of order into the distribution of the masses and force constants in the periodic structure, which should clearly lead to a less discrete optical spectrum. The slight reduction in the rigidity of the structure expresses itself as a reduction in the absorption-maxima frequencies. As a result of this, although there are certain differences in the manner in which individual frequencies change on making isomorphous substitutions, these differences being used as a basis for ascribing the frequencies in question mainly to vibrations of bonds of the Si–O, $Si–O–Al^{IV}$, $Si–O–Al^{VI}$, and suchlike type in [64, 67, 69], it is hardly possible to set up any general principles for the identification of the Al–O vibration frequencies in the case of a random distribution of Al atoms in the silicon–oxygen anion.

Further refinement in the location of the bands associated with the Al–O vibrations is, in view of the structural role of the aluminum, only possible in the simplest substances, including several aluminum silicates and aluminosilicates with an ordered distribution of Al and Si over the crystallographic positions. By way of example, we may consider the infrared spectra of three crystalline compounds corresponding to the same chemical formula $Al_2O_3 \cdot SiO_2$ but differing considerably in structure. The spectra of these compounds are shown in Fig. 80 [186].

Structurally, kyanite is a typical oxyorthosilicate $Al_2^{VI}(SiO_4)O$, containing, in addition to SiO_4 tetrahedra not possessing common vertices, additional oxygen anions O^{2-} and aluminum cations octahedrally surrounded by these anions, and O atoms entering into the silicon–oxygen tetrahedra. The spectrum of kyanite is in actual fact similar to the spectra of the orthosilicates: in the frequency range of the stretching vibrations of the SiO_4 tetrahedron there are three bands in the range 1040 to 940 cm^{-1} corresponding to the three components of the $\nu_{as}SiO_4$ vibration, and a weaker ν_sSiO_4 vibration band at 890 cm^{-1}. The multiplet character of the band of the triply degenerate vibration and the activity of the totally symmetric vibration in the infrared spectrum agree with the site symmetry of the tetrahedra C_1; the space group of the crystal is $P\bar{1}-C_i^1$. The large number of strong bands in the range 400 to 700 cm^{-1} cannot be explained simply by deformation vibrations of the tetrahedron. This applies particularly to those lying between the usual regions for νSiO_4 and δSiO_4, i.e., from 600 to 670 and 715 cm^{-1}. These bands are undoubtedly associated with Al–O vibrations; their frequencies are higher than the vibration frequencies of the aluminum–oxygen octahedra in the spectra of aluminum silicates. This rise in the Al–O frequencies may be associated with the presence of oxyions O^{2-} in the kyanite crystals; it suggests the existence of bonds with the aluminum atoms possessing a higher proportion of covalence and higher force constants† than bonds of the $Al^{3+} \ldots O^{-}(-Si)$ type.

†Regarding the structural role of the O^{2-} anions and their bonds with the cations see also Chap. V.

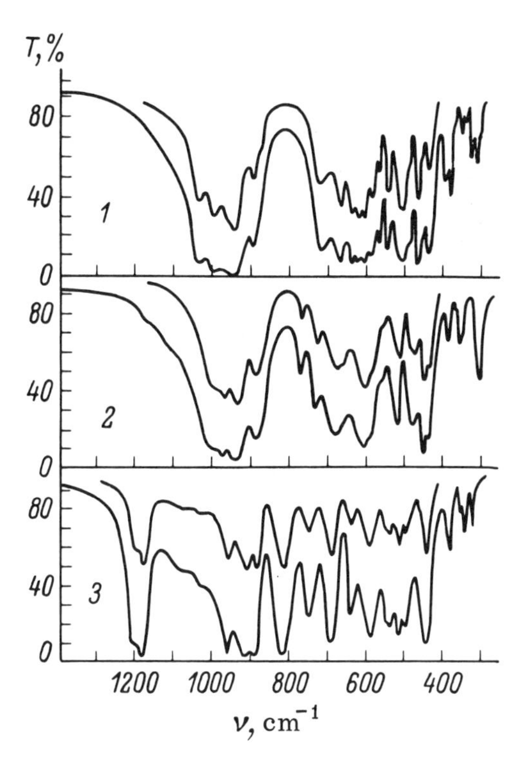

Fig. 80. Infrared spectra of minerals corresponding to the formula $Al_2O_3 \cdot SiO_2$ [191]. 1) Kyanite $Al_2(SiO_4)O$; 2) andalusite $Al_2(SiO_4)O$; 3) sillimanite $Al(SiAlO_5)$.

In the crystal of andalusite $Al^{VI}Al^{V}(SiO_4)O$, also an aluminum oxyorthosilicate, half the aluminum atoms have a coordination number of five relative to oxygen, in contrast to kyanite. The highest-frequency part of the andalusite spectrum has the form characteristic of orthosilicates: the site symmetry of the silicon–oxygen tetrahedra C_s (space group Pnnm–D_{2h}^{12}, Z = 4) is reflected in the three $\nu_{as}SiO_4$ vibration bands at 1020 to 940 cm^{-1} and the $\nu_s SiO_4$ band at about 890 cm^{-1}. In the lower-frequency range, as in the kyanite spectrum, there is a large number of bands; the bands ascribable to the Al–O vibrations embrace not only the range 600 to 700 cm^{-1} but also 730 and 770 cm^{-1}. The latter cannot be ascribed to any fundamental vibrations of the Si–O bonds, since the structure does not contain any condensed silicon–oxygen tetrahedra. The displacement of some of the Al–O vibrations toward higher frequencies on passing from the kyanite spectrum to that of andalusite is evidently associated with the transfer of some of the Al atoms to five-fold coordination and the higher covalence of their bonds with oxygen. The foregoing considerations find confirmation in recent refinements to the structures of andalusite [187] and kyanite [188].

The structure of andalusite may be considered in relation to the coordination of aluminum as intermediate between kyanite and sillimanite. The structural formula of sillimanite in the

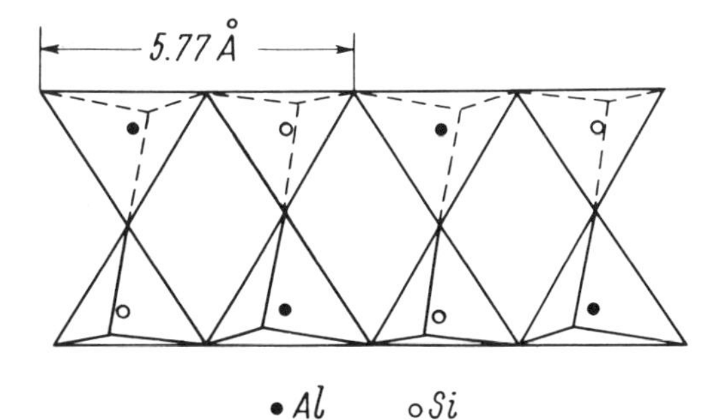

Fig. 81. Structure of the $(Si_2Al_2O_{10})_\infty$ sillimanite ribbon.

form $Al^{VI}(SiAlO_5)$ reflects the fact that half the Al atoms act as cations octahedrally surrounded by oxygen, while the other half enter into the $[Si_2Al_2O_{10}]_\infty$ complex anions in tetrahedral coordination with respect to oxygen. The structure of the sillimanite strip is shown schematically in Fig. 81. The distribution of the Si and Al atoms in the tetrahedra is ordered, and this explains the doubling of the unit link in the strip, which would have only half the identity period if these atoms were indistinguishable. In view of the ordered distribution of the Al^{IV} atoms in the complex anion, the spectrum of the latter is completely discrete and may be approximately interpreted. In addition to the earlier-considered correlations between the spectrum and structure of the silicate anion we must here remember the probability that there is some reduction in the force constants of the Al^{IV}–O bonds as compared with the corresponding Si–O bonds, while the masses of Si and Al are almost identical. Considering that the vibration frequencies of the AlO_6 octahedra forming the spectrum of lattice vibrations in sillimanite are no greater than 600 to 650 cm^{-1} if the crystal contains no O^{2-} ions, we obtain seven or eight strong bands in the range 1200 to 650 cm^{-1}. These bands are to be compared with the fourteen limiting stretching vibrations in the "one-dimensional" $[Si_2Al_2O_{10}]_\infty$ ribbon. The identification of these bands is simplified on considering the precise data of [189] on interatomic distances and valence angles in the sillimanite ribbon (the signs ⊥ and ‖ indicate Si–O–Al bonds in directions across and along the ribbon axis):

$$r_{Si-O(Al^{IV}),\perp} = 1.564 \pm 0.006\ \text{Å}, \quad r_{Al^{IV}-O(Si),\perp} = 1.721 \pm 0.006\ \text{Å}$$
$$r_{Si-O(Al^{IV}),\parallel} = 1.633 \pm 0.004\ \text{Å}, \quad r_{Al^{IV}-O(Si),\parallel} = 1.800 \pm 0.004\ \text{Å}$$
$$r_{Si-O^-} = 1.629 \pm 0.007\ \text{Å}, \quad r_{Al^{IV}-O^-} = 1.758 \pm 0.005\ \text{Å}$$
$$\angle SiOAl_\perp = 171.6^\circ \pm 0.4^\circ, \quad \angle SiOAl_\parallel = 114.4^\circ \pm 0.2^\circ$$

Particularly important here is the nearly 180° value of the SiOAl angle in the bonds "knitting" the two chains into a ribbon and the small interatomic distances in these bonds, much smaller even than the lengths of the $Si-O^-$ and $Al-O^-$ bonds. The Si–O–Al bonds along the direction of the ribbon axis are characterized, on the other hand, by an anomalously small value of ∠SiOAl and corresponding large interatomic distances.

The unit cell of the crystal (space group $Pnma-D_{2h}^{16}$) contains sections of two ribbons possessing intrinsic (and site) symmetry C_{2h}. The ribbons are invariant with respect to reflection in the symmetry planes passing through the $Si-O-Al_\perp$ bonds, to screw translation along the $\bar{C}_2$ axis coinciding with the axis of the ribbon, and to inversion in centers displaced by half the translation defined by the operation $\bar{C}_2$ from the symmetry planes. Table 36 gives a classification of the stretching vibrations of the ribbon with respect to the irreducible representations of the C_{2h} group. The experimental frequency values may be compared with the vibrations ac-

TABLE 36. Identification of the Frequencies of the Stretching Vibrations of a Sillimanite Ribbon $[Si_2Al_2O_{10}]_\infty$

Nature of the vibrations	Symmetry species of the C_{2h} group	Frequency, cm^{-1} [186]
$\nu_{as}SiOAl_\perp$	B_u, A_g	1200, 1180
νSiO^- $\nu_{as}SiOAl_\parallel$	B_u, A_g B_u, A_u, B_g, A_g	960, 913, 888
νAlO^- $\nu_s SiOAl_\parallel$	B_u, A_g B_u, A_u, B_g, A_g	820, 751, 692
$\nu_s SiOAl_\perp$	B_u, A_g	⩽ 650

tive in the infrared spectrum if we consider that ν_{as} SiOAl$_\perp$ probably lies between 1100 and 1200 cm^{-1}, the frequency νSiO$^-$ at about 900 to 1000 cm^{-1}, like the ν_{as} SiOAl$_\parallel$, while the frequencies ν_s SiOAl$_\parallel$ may be higher than the ordinary values of the ν_s SiOSi frequencies owing to the small SiOAl$_\parallel$ angle. The existence of two closed rings of Si–O–Al bonds in the identity period of the ribbon and the closeness of $\angle$SiOAl$_\perp$ to 180° makes us exclude ν_s SiOAl$_\perp$ from the list of vibrations with frequencies above 650 cm^{-1}. Passing to a consideration of the factor group of the space group D_{2h} leads us to expect the splitting of vibrations of type B_u in the site group C_{2h} into two components active in the infrared spectrum. This splitting is only observed experimentally for the ν_{as} SiOAl$_\perp$ vibration. Vibrations of type A_u give only one active component on Davydov splitting, while vibrations of types A_g and B_g remain inactive.

Thus despite the very tentative character of the classification of vibrations here employed it appears possible to explain the main characteristics of the spectra. It must be emphasized, however, that the examples here given are more the exception than the rule, and in the majority of cases the interpretation of the spectra of aluminosilicates involves great difficulties even when the structure of the crystal is known.

According to existing data, oxygen compounds of gallium are largely similar to the corresponding compounds of aluminum, although the Ga–O vibration frequencies should (other things being equal) be displaced into the long-wave region. A study of the spectrum of gallium hydroxide GaOOH [58] established a displacement of the absorption band at 760 cm^{-1} in the spectrum of the structurally similar diaspore α-AlOOH to about 640 cm^{-1}. The same region exhibits absorption by α-Ga_2O_3, the gallium analog of corundum, with a space lattice of gallium-oxygen octahedra [190]. In the spectra of $BaGa_2O_4$ containing GaO_4 tetrahedra and β-Ga_2O_3 containing both tetrahedra and octahedra, like β-Al_2O_3, Kolesova observed absorption bands near 700 cm^{-1} and 680 to 750 cm^{-1}, respectively. These results suggest that $LiGaO_2$ and the two polymorphic forms of $NaGaO_2$, which show strong absorption bands at 770 to 670 cm^{-1}, have a structure formed by GaO_4 tetrahedra.

Here we must also stop to consider existing information regarding the location of the Fe^{3+}–O vibration bands, in view of the great crystal-chemical similarity between the Fe^{3+}

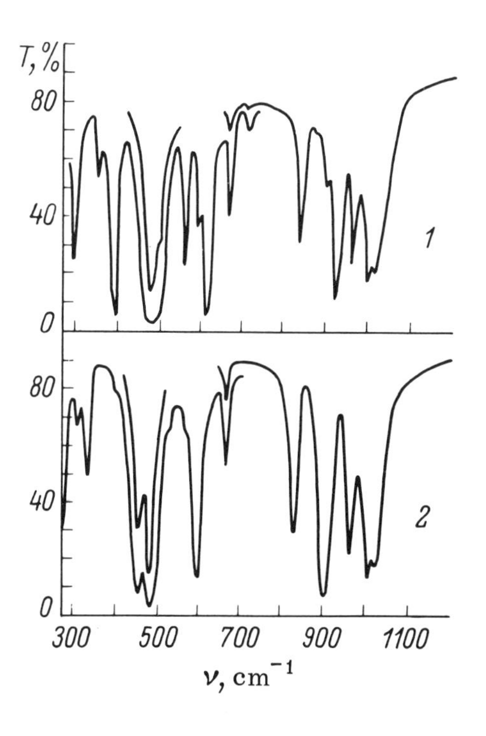

Fig. 82. Infrared spectra of synthetic silicates with the structure of akermanite–hardystonite. 1) $Sr_2Mg(Si_2O_7)$; 2) $Sr_2Zn(Si_2O_7)$ [171].

and Al^{3+} ions. The influence of the coordination of trivalent iron on the vibrational spectrum was studied by Tarte [158, 164, 171]. The strongest absorption in α-Fe_2O_3 was found near 550 cm^{-1}, and approximately in the same region there was absorption from other compounds, $LiFeO_2$ and $CaFe_2O_4$, which form a lattice of iron–oxygen octahedra. In the spectra of garnets (silicates and germanates) containing FeO_6 octahedra, the lower frequencies in the 300 to 400 cm^{-1} range were ascribed to the Fe–O vibrations. In ferrites with the garnet structure and other compounds in which Fe occupies the tetrahedral positions, however, the Fe–O frequencies rise to between 550 and 650 cm^{-1} and to 700 cm^{-1} if there is a system of condensed iron–oxygen tetrahedra.

Magnesium and Zinc. These are the only representatives of cations belonging to groups IIA and B of the periodic table able to exist in both tetrahedral and octahedral coordination with respect to oxygen. In crystals of silicates containing magnesium in six-fold coordination such as $MgSiO_3$, the Mg–O frequencies lie in the range 450 to 300 cm^{-1} [158]. In these compounds, as already mentioned, the Mg–O vibrations interact weakly with the stretching vibrations of the silicon–oxygen anion; the main manifestation of these interactions is the increase in the Davydov splitting of the stretching vibrations of the anion, considerably greater than that noted in the corresponding compounds of calcium, strontium, etc. In the spectrum of akermanite $Ca_2MgSi_2O_7$ and similar compounds of the series $Sr_2XSi_2O_7$ (X = Mg, Mn, Zn, Co) absorption in the range 650 to 1050 cm^{-1} has a pattern characteristic of the pyrosilicates and changes very little on replacing one cation by another.† In the lower-frequency range, however, there are considerable changes enabling us to ascribe the absorption between 600 and 500 cm^{-1} to the Mg–O vibrations of the MgO_4 tetrahedra contained in the akermanite structure. We remember once again that the complete loss of bands in the range 540 to 450 cm^{-1} in the talc spectrum on replacing Si by Ge prevents us from ascribing these bands to vibrations of the Mg–O bonds in the octahedron and indicates a considerable contribution of the degrees of freedom associated with the deformation of the silicate anion to these vibrations [87]. This once again illustrates the arbitrariness of the process of separating the vibrations of the cation polyhedra in the silicate spectra. The frequencies of the Zn–O vibrations for the zinc in tetrahedral coordination lie somewhat lower in these compounds, at 500 to 450 cm^{-1} (Fig. 82), but they may rise considerably in systems containing condensed zinc–oxygen tetrahedra. It is interesting to note that the occurrence of zinc in tetrahedral coordination, despite its comparatively large ionic radius (0.74 Å), is undoubtedly associated with the directional character of its bonds, which is to a certain extent typical of other IIB elements as well.

Lithium. This is the only one of the alkali elements for which a change in the coordination number has been studied spectroscopically. This is due to the relatively high Li–O vibration frequencies arising from the small mass of the Li, the possibility of identifying the bands of these vibrations by means of the isotopic effect, and the existence of crystals containing lithium atoms in tetradral coordination with respect to oxygen. A study of the spectra of compounds containing Li atoms in octahedral coordination revealed no substantial displacements of the bands in the frequency range above 300 cm^{-1} on replacing ^6Li by ^7Li [191]. On the other hand, in the spectra of compounds containing lithium in tetrahedral coordination some of the bands in the range 500 to 400 cm^{-1} show a clear isotopic displacement, reaching 20 to 30 cm^{-1}. Thus in the spectrum of lithium carbonate Li_2CO_3 the replacement of ^6Li by ^7Li, without appreciably displacing the bands of the internal vibrations of the carbonate ion in the range

† The spectrum of $Sr_2MgSi_2O_7$ (Fig. 81) contains, apart from the ν_s SiOSi band at 670 cm^{-1}, a weak band at 710 cm^{-1}; this band cannot be ascribed to the Davydov splitting of ν_s SiOSi, since in the factor group D_{2d} symmetry types A_1 and B_2, of which only one is active in the infrared spectrum, correspond to this vibration. It is possible that this band relates to a combination frequency.

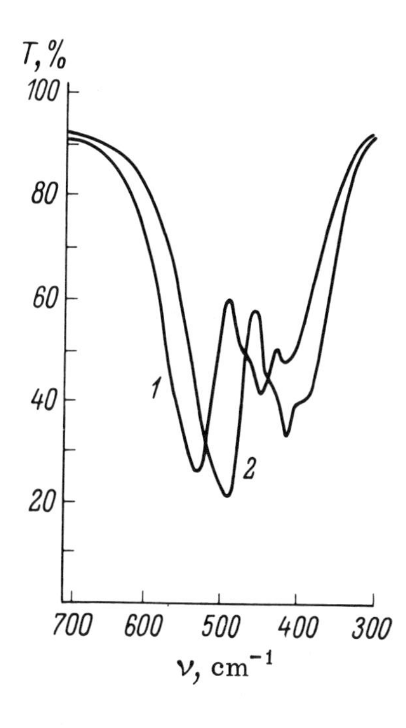

Fig. 83. Isotopic effect in the spectrum of Li_2CO_3 [196]. 1) $^6Li_2CO_3$; 2) $^7Li_2CO_3$.

1500 to 700 cm^{-1}, displaces the bands in the range 540 to 400 cm^{-1} by 34-17 cm^{-1} to lower frequencies (Fig. 83). By analogy, the isotopic shift of 473 to 447 cm^{-1} observed in the spectrum of the compound $LiCrGeO_4$ (with spinel structure [192]) clearly established the presence of Li atoms in the tetrahedral positions and this is supported by the low frequencies of the Ge–O vibrations, indicating octahedral coordination of germanium.

In conclusion, we present a table approximately characterizing the positions of the regions of strongest absorption in the infrared spectra of silicates and other compounds containing coordination polyhedra of types XO_4 and XO_6. In this table (Table 37) we attempt to classify

TABLE 37. Characteristic Regions of Strong Absorption in the Infrared Spectra of Compounds with Coordination Polyhedra XO_n

X	Region of strong absorption, cm^{-1}			
	tetrahedra XO_4		octahedra XO_6	
	isolated	polymerized	isolated	polymerized
Si	1050—800	1200—1000	∼700	∼900
Ge	850—680	900 and less	⩽550	700 and less
Ti	800—690	950—850	⩽500?	600—500
Al	800—650	870—700	500—400	650 and less
Fe^{3+}	650—550	700—550	400—300	550—400
Cr^{3+}	—	—	450—300	650 and less
Ga	700—570	750—600	∼400	600—500
Zn	500—450	600—400	—	—
Mg	600—500	—	—	480 and less
Fe^{2+}	∼450	—	—	320
Mn^{2+}	∼450	—	—	∼320
Li	500—400	600—400	<300?	<300?

Note: Data taken from [158,171] are shown here with insignificant changes.

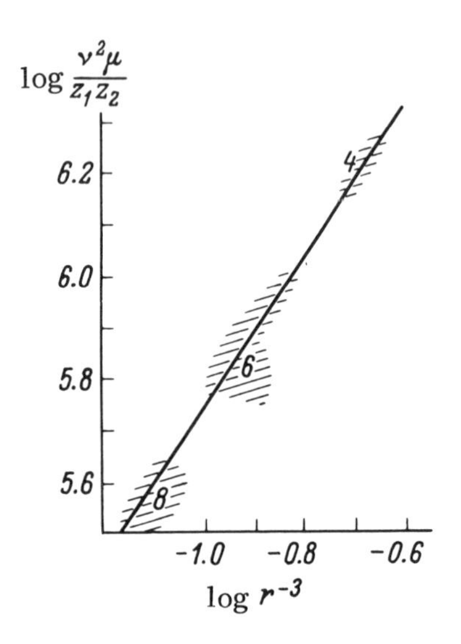

Fig. 84. Relationship between the frequencies of the valence vibrations, the cation–anion distance (r), and the coordination number of the cation. The latter is indicated by the figure on the shaded regions corresponding to the scatter of the experimental points.

the regions of absorption in relation to the degree of polymerization of the XO_n polyhedra; if any $X'O_{n'}$ polyhedra have common O atoms with $X''O_{n''}$ polyhedra, then both types are considered as isolated. We remember, however, that actually the change of X′–O frequencies associated with the degree and mode of polymerization of the $X'O_{n'}$ polyhedra is clearly determined not only by the changes in the kinematics of the system but also by dynamic factors, in particular by the effect of the nature of the neighboring X″–O(X′) bonds on the force constants of the X′–O bond. Hence the classification of the characteristic regions of absorption in Table 37 bears an extremely arbitrary character and should be used with great caution.

The general form of the empirical relationship between the frequencies of the stretching vibrations and the coordination number of the cation may also be represented graphically (Fig. 84), according to the data of White and Roy, who studied a series of oxides and fluorides [193]. These authors showed that in the majority of cases there is an almost inverse proportionality between the value of $\nu^2\mu/z_1z_2$ (ν is the highest of the stretching frequencies, μ is the reduced mass of the cation–anion pair of atoms, z_1 and z_2 are the effective charges on these atoms) and the third power of the distance between the cation and the anion. The coefficient of proportionality depends greatly on the coordination number; the experimental points obtained for compounds with different coordination of the cations form isolated regions, shown in Fig. 84. Hence we may use this relationship in order to study the coordination of the cations, although the deviations may be considerable.

§4. Spectroscopic Manifestations of Disorder in the Structure of Silicates

In considering the possibility of interpreting the vibrational spectra of silicates we have so far based our arguments on an ideally ordered, infinitely extended crystal. However, there is no doubt that in the spectra of real crystals a more careful consideration of all the bands, even weak ones, will require consideration of possible deviations from the ideal structure: isomorphous substitutions, deviations from stoichiometric composition, the presence of isotopes, dislocations, and the limited dimensions of the crystals. From the point of view of the crystal chemistry of silicates, the greatest interest lies in any disruptions of the periodicity with which the lattice points of the geometrically almost undistorted space lattice of the crystal are occupied. This occurs both in mixed crystals, in which (in the majority of cases) one assumes a random distribution of the two components, and also in certain silicates with a complex Al–Si–O framework, in which the siting of the silicon and aluminum atoms may have different degrees of ordering. A specific case of a defect in the occupation of lattice points is the population of one set of particular positions by a smaller number of atoms than that provided for by the multiplicity of the set. Here we shall not consider spectra of such defective structures as glasses and other substances with an almost complete absence of long-range order. In conclusion, we shall consider the possibilities of using the spectra of solid solutions formed between isostructural silicates and germanates in order to increase the reliability of the identification of the frequencies and obtain a more objective estimate of the force constants.

Vibrational Spectra of Mixed Crystals. These have been studied repeatedly; however, a theoretical analysis of the results involves considerable difficulties. It is sufficient to note that, in contrast to an ideal crystal, the departure from translational symmetry of a mixed crystal makes not only the limiting vibrations but also vibrations with a finite wavelength potentially active. A detailed review of theoretical investigations into the dynamics of mixed crystals is given by Maradudin, Montroll, and Weiss [194]. We shall confine ourselves to a short list of the main experimental data regarding the characteristics of the spectra of mixed crystals.

Even in 1928, when studying the infrared spectra of the alkali halide crystals in the series of mixed NaCl – KCl cyrstals by the method of residual rays, it was found that, as in the spectra of the end members of the series, these contained only one band, the maximum of this moving smoothly on changing the composition; decomposition of the solid solutions on prolonged storage led to the appearance of two bands with frequencies corresponding to pure NaCl and KCl in the spectrum [195]. Analogous results were obtained when studying the lattice vibrations of mixed molecular crystals by the Raman method. Thus in the spectra of solid solutions of paradichloro- and paradibromobenzene, as in the spectra of the pure compounds, there were nine lines representing the external vibrations of the molecules (six orientational and three translational); the frequencies of these lines varied with concentration [196-200]. In the frequency range of the internal vibrations of the molecules, the spectra of the mixed molecular crystals are usually a superposition of the spectra of the two components. On the other hand, in the spectra of crystals with complex anions, on partial substitution of some cations by others there is a gradual displacement of the frequencies of the complex anion, depending on the composition [201]; this may be explained by a gradual change in the average field on varying the number of cations of one sort or the other.

Sometimes, a relatively simple form of the spectra of solid solutions is observed for very low concentrations of one of the components. In this case local vibrations of defect centers develop, the frequency of these lying outside the "continuum" of the vibrational frequencies of the crystal, if the masses of the impurity particles are much smaller than those of the particles in the lattice points of the main lattice, or if the force constant is much higher. The amplitude of such vibrations falls exponentially on moving away from the impurity center. Experimentally, local vibrations of defect centers were first observed in the infrared spectrum when studying the absorption of light by U centers (H^- ions in anion vacancies) in alkali halide crystals [202]. The vibrations of impurities falling into the forbidden band of the vibrational spectrum of the matrix crystal behave in a similar way. Saksena [203] considered the local vibrations of impurities in α-quartz.

In recent years the spectra of mixed crystals of comparatively simple structure have been studied by many authors [203a-203d] (see also [203e]), and it has been established that, as a rule, there is either a "one-band" or a "two-band" type of mixed-crystal spectrum. In the first case the band in the spectrum of the mixed crystal is single and occupies an intermediate position between the corresponding bands of the extreme members of the series; in the second case there are two bands in the spectrum of the mixed crystal, similar in frequency to the bands of the pure compounds and little displaced with changing composition. In some cases, both types of band may occur at the same time in the spectra of mixed crystals for different vibrations. Simplified phenomenological theories have been developed in order to predict a "one-band" or "two-band" character for the spectra of various systems, depending on the position of the impurity bands (for a low concentration of one of the components) relative to the optical and acoustic branches of the matrix lattice, or on the mutual disposition of the optical branches for the extreme members of the series. Model theories have also been constructed for the simplest cases. Thus for the model of a one-dimensional crystal comprising atoms of sorts A, B, and C, in which the content of B and C is varied (with a random distribution of these in the chain), a spectrum of the one-band type is obtained for $m_A < m_B$, m_C and a two-band spectrum for $m_A > m_B$, m_C [203f].

Among the mixed crystals in the silicate and similar systems studied by spectroscopic methods we must distinguish the three most important forms of isomorphous substitutions.

1. Solid solutions formed by substitutions in the lattice sites occupied by cations. If the frequencies of the vibrations of the cation–oxygen bonds are fairly low, then, after substitutions of this kind, the changes in the vibrational spectrum in the frequency range of the stretching vibrations of the complex anion are usually insignificant, and their concentration dependence is expressed in the form of a smooth change in frequencies and intensities. In the range of lower frequencies there are considerable changes in the spectrum, enabling us, in many cases, to identify bands associated to various extents with the vibrations of the oxygen polyhedra of the cations.

2. Solid solutions of orthosilicates formed by replacing the Si atoms in the tetrahedra by other atoms (e.g., Ge).† In such solid solutions the frequencies of the internal vibrations of the $X'O_4$ and $X''O_4$ tetrahedra, at any rate in the frequency range of stretching vibrations, retain unaltered values corresponding to the frequencies in the spectra of the extreme members of the series if the differences in the intrinsic frequencies are large. If, however, the frequencies of the vibrations of the $X'O_4$ and $X''O_4$ tetrahedra are comparatively close, then on formation of a solid solution there is a displacement of the vibration frequencies of the two sorts of tetrahedra (the higher falls and the lower rises), varying monotonically with concentration. In systems of this type considerable interest arises in the spectra of samples with small (1 to 10%) contents of one of the XO_4 tetrahedra: effects of resonance interaction between the vibrations of these tetrahedra are almost entirely absent, the bands are narrow, while the splitting of degenerate vibrations (the number of components and the distance between them) and the displacement of the frequencies characterize the symmetry and strength of the internal field of the matrix lattice. In certain cases, however, the number of bands corresponding to the vibrations of the tetrahedra present in only small concentrations exceeds the number expected; possible reasons for such anomalies include interaction with the vibrations of the matrix crystal and local fluctuations in the concentration of the impurity tetrahedra.

3. Solid solutions formed by the substitution of the Si atoms in silicates with complex anions possessing translational symmetry in one, two, or three directions.

Tarte's book [205] describes a large number of spectroscopically studied solid solutions corresponding to the first two forms of isomorphous substitutions. We shall therefore confine ourselves to the consideration of an example [206] relating to the third case.

---

† Substitutions in other complex anions of the island type (Si_2O_7, Si_3O_9, etc.) have been little studied. However, for the $K_2Pb_2(Si_2O_7,Ge_2O_7)$ system [204], for example, the impossibility of the existence of a complex anion of mixed composition $[SiGeO_7]^{6-}$ has been established spectroscopically, while in the $Sc_2(Si_2O_7-Ge_2O_7)$ system we have been able to identify such anions (Chap. V). Some solid solutions with structures of the island type are described later.

X-ray diffraction and refractometric data indicate the existence of a continuous series of solid solutions in the Li_2SiO_3–Li_2GeO_3 system, which is in keeping with the similar crystal structure of the extreme members of the series: both the silicate and the germanate belong to the space group $Cmc2_1$–C_{2v}^{12}, and the differences in lattice constants are small (Li_2SiO_3: a = 9.36, b = 5.39, c = 4.67 Å; Li_2GeO_3: a = 9.60, b = 5.54, c = 4.80 Å). The complex anions in both crystals consist of infinite $(X_2O_6)_\infty$ chains, while the factor group of the one-dimensional space group is isomorphic with C_{2v} (p. 98). Hence mixed $Li_2(Si, Ge)O_3$ crystals are of interest as one of the few comparatively simple substances for which the spectrum in the stretching frequency range may reasonably be considered as the spectrum of a one-dimensional mixed crystal. We note that the presence of Si and Ge atoms in the chain, independently of whether these are distributed in an ordered or random manner, leads to a reduction in the symmetry of the chain from C_{2v} to C_s as a result of the loss of symmetry elements, including the translation along the chain axis. Hence the vibration of the $(X_2O_6)_\infty$ chain of symmetry type A_2 is not forbidden in the infrared spectrum.

The infrared spectra of $Li_2(Si, Ge)O_3$ solid solutions contain only the bands present in the spectra of pure Li_2SiO_3 and Li_2GeO_3 (with slight frequency changes), and also three bands not having any analogs in the spectra of the extreme members of the series (Fig. 85). The frequencies of these bands also depend little on the Si and Ge concentration. The appearance of the three new bands in the spectra of the solid solutions cannot be explained by activation of the A_2 vibrations, since among the stretching vibrations of the chain only one belongs to this symmetry type. Hence it is possible that in the spectrum of the chain containing Si and Ge atoms three possible versions of nearest neighbors occur: Si–O–Si, Ge–O–Ge, and Si–O–Ge; it is to the last of these that the appearance of the new bands in the spectrum of the solid solution corresponds. In other words, we may suppose that for each of the normal vibrations of the metagermanate or metasilicate chain there are three bands in the spectrum of the solid solution, with frequencies close to those of the silicate or germanate chains or a chain with an

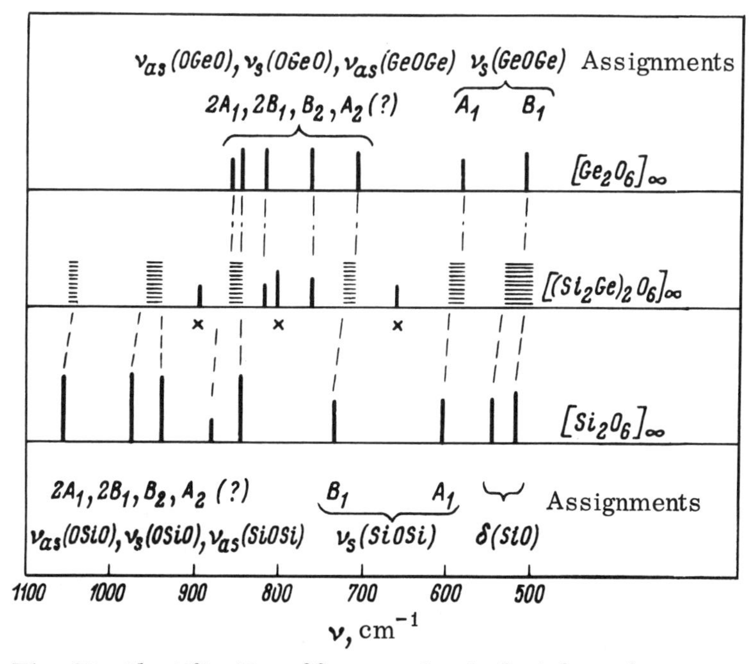

Fig. 85. Identification of frequencies in the infrared spectra of $Li_2(Si, Ge)O_3$ mixed crystals. Crosses indicate bands not having analogs in the Li_2SiO_3 and Li_2GeO_3 spectra.

ordered arrangement of Si and Ge. In fact each of the three frequencies considered lies between the frequencies ν_s'XOX(B_1), ν_{as}XOX(A_1), ν_{as}'XOX(B_1) of the germanate and silicate chains (the frequencies of the ν_sXOX(A_1) vibrations in the two chains are close together and merge into a single band in the spectrum of the solid solution). As regards the vibrations involving predominantly the X–O$^-$ side bonds, in the spectrum of the solid solution a simple superposition of the silicate and germanate bands corresponds to these; this is clearly due to the much weaker interaction of the vibrations of the O$^-$SiO$^-$ and O$^-$GeO$^-$ groups, which are not directly coupled to each other. Since for any Si and Ge concentrations the bands not coinciding with the bands of pure Li_2SiO_3 and Li_2GeO_3 are much weaker than the latter, it is possible that the distribution of silicon and germanium in the solid solutions is not completely uncorrelated.

Degree of Order in the Distribution of the Atoms over the Crystallographic Positions. This is one of the most important aspects of the problem of order and disorder in the structures of silicate minerals. There is considerable interest, in particular, in the distribution of silicon and aluminum over the tetrahedral positions in aluminosilicates. In the majority of cases one is concerned with structures containing complex three-dimensional Si–Al–O frameworks. Complex anions of this type, possessing translational symmetry in three dimensions, were not considered earlier in our discussion of the qualitative classification of the vibrations of silicate anions, because in the present case the separation of the internal vibrations of the anion does not lead to any substantial simplification of the interpretation. Most of the branches of the vibrational spectrum of the crystal (and in SiO_2 all of them) relate to the internal vibrations of the three-dimensional framework, and as a rule the orders of the secular equations remain very high, even on neglecting the vibrations of the cation–oxygen bonds. On the other hand, if we refrain from setting ourselves the problem of determining the frequencies of the internal vibrations and confine ourselves to considering the conclusions which may be drawn simply by counting the number of vibrations of some quasi-characteristic type and considering the symmetry of the system, the three-dimensional Si–Al–O frameworks even possess certain advantages. The absence (or relatively small number) of bonds of the Si–O$^-$ and Al–O$^-$ type in these structures means that we can count on a better separation of the ν_{as}SiOSi and ν_sSiOSi vibrations (as compared with less highly condensed anions) if the frequency ranges characteristic of these are preserved. Then the problem of separating the ν_sSiOSi frequencies, which, as shown earlier, are of greatest interest in establishing correlations between the structure and the spectrum, reduces to the separation of these from the frequencies of the deformation vibrations of the framework.

The feldspars $M^+[AlSi_3O_8]^-$ or $M^{2+}[Al_2Si_2O_8]^{2-}$ form one of the most widespread groups of framework-type silicates. A common feature in the structure of these minerals is the complete incorporation of the aluminum into the composition of the complex anion, which only contains bonds of the (Si, Al)–O–(Si, Al) type and none of the so-called end bonds (Si, Al)–O$^-$. The excess of negative charges associated with the replacement of some of the silicon atoms in the tetrahedra by aluminum atoms is compensated by the incorporation of mono- or divalent cations (Na, K, Ca, Ba) into the spaces in the framework. The structure of the frameworks themselves may be represented as the result of the complete condensation of ribbons consisting of rings formed by four tetrahedra. The three-dimensional lattice of silicon– and aluminum–oxygen tetrahedra formed comprises rings of four or eight tetrahedra each. The crystals of the feldspars have a low symmetry (triclinic or monoclinic) with a large number of tetrahedra in the unit cell. The character of the distribution of Si and Al atoms over the tetrahedral positions may clearly vary from the purely random to the partly or completely ordered. The order–disorder transformations here depend substantially on the heat treatment, and an ordered Si and Al distribution is characteristic of the low-temperature forms, transforming into disordered forms at high temperatures.

Since the unit cells of the monoclinic feldspars contain two tetrahedral atoms, in sets of multiplicity eight, in asymmetrical proportions, while in the triclinic systems each of these

TABLE 38. Possible Arrangements of Si and Al in Feldspars

Space group	Occupation of the sets			
$C2/m - C_{2h}^3$	Si + Al Si		Si + Al Si + Al	
$C\bar{1}$	Si + Al Si Si Si	Si + Al Si + Al Si Si	Si + Al Si + Al Si + Al Si	Si + Al Si + Al Si + Al Al

Note: For convenience of comparison with monoclinic feldspars instead of $P\bar{1}-C_i^1$ we use the centered cell $C\bar{1}$.

TABLE 39. Relation between the Infrared Spectra and the Arrangement of Si and Al in the Frameworks of Feldspars

Mineral (space group)	Distribution of Al according to x-ray structural data				Number of maxima in the IRS between 790 and 720 cm^{-1}
	$Si_1(0)$	$Si_1(m)$	$Si_2(0)$	$Si_2(m)$	
Orthoclase (C2/m)	~0.30		~0.19		4
Sanidine (C2/m)	0.25		0.25		2
Ideal distribution	0.36		0.14		—
Albite ($C\bar{1}$)					
low temperature	0.96	0.00	0.03	0.02	4
high temperature	0.29	0.27	0.20	0.22	2
Microcline					
intermediate	0.65	0.25	0.03	0.01	2
maximum	0.93	0.03	0.01	0.01	—

Note: The ideal distribution is the most stable one in monoclinic feldspars, derived from consideration of the electrostatic charge balance [209].

sets decomposes into two with multiplicity four, Table 38 shows a number of possible versions of the arrangement of the Si and Al with different degrees of order.

For each of these versions the condition that centrosymmetry is to be preserved requires that the Si and Al in the same set should be distributed randomly.

The main method of experimentally studying the distribution of Si and Al in the frameworks of feldspars has been x-ray structural analysis. A large number of investigations† have established a more or less ordered arrangement in the low-temperature forms, transforming into a disordered state in the high-temperature versions. Table 39 presents some quantitative data regarding the distribution of Al and Si in alkali feldspars obtained as a result of these investigations. These data cannot, however, be regarded as exhaustive: the accuracy of determining the Al and Si content in various positions is limited owing to the small difference in the

† The most up-to-date review of the state of this question and references to original papers are contained in an article by W. H. Taylor [207]. A number of articles on the structure of feldspars were translated into Russian in [208].

Si–O and Al–O distances and the almost equal scattering powers of Si and Al. Among other direct methods of studying the distribution of silicon and aluminum in feldspars a certain significance has recently been ascribed to nuclear magnetic resonance [210] and infrared spectroscopy [211, 212]. The use of the latter has been based exclusively on the empirically established relation between the position, intensity, and width of certain bands and the degree of order in the structure. A fall in the degree of order in the Al and Si distribution (for example, after heat treatment of the low-temperature forms) is accompanied by a broadening of the bands and a fall in intensity at the absorption maximum.

Let us consider the possibility of a qualitative interpretation of the spectra of feldspars in the frequency range of the stretching vibrations of the framework and their changes with Si and Al distribution. Both monoclinic and triclinic alkali feldspars contain eight Si and Al atoms, in the ratio 6:2, and sixteen O atoms in the primitive cell. Correspondingly in the crystal there are sixteen translationally nonequivalent (Si, Al)–O–(Si, Al) bonds. Since the O atoms lie in sets of general positions, while the symmetry operations of the space group include that of inversion, of the sixteen vibrations of type ν_s SiOSi half are symmetric and the other half antisymmetric with respect to this operation (this is also valid with respect to the vibrations ν_{as} SiOSi). For both factor groups C_{2h} and C_i, vibrations antisymmetric with respect to the inversion center are active in the infrared spectrum. Figure 86 gives a schematic representation of the spectra of the alkali feldspars according to the data of [211–213]. In accordance with the foregoing, the range 800 to 1000 cm^{-1} contains no absorption bands. Hence the range of ν SiOSi vibrations, lying at somewhat lower frequencies than in the silica spectrum as a result of the partial replacement of Si by Al, i.e., as a result of the lower rigidity of the system (the differences in the reduced masses are negligibly small), may be separated out without difficulty. The number of maxima observed in this region is much smaller than expected, probably as a result of superposition. To vibrations of the ν_sSiOSi type, which embrace a wider region of the spectrum and as a rule overlap very little, we must clearly ascribe the bands between 800 and 500 cm^{-1}, since the replacement of Si by Al also reduces these frequencies.

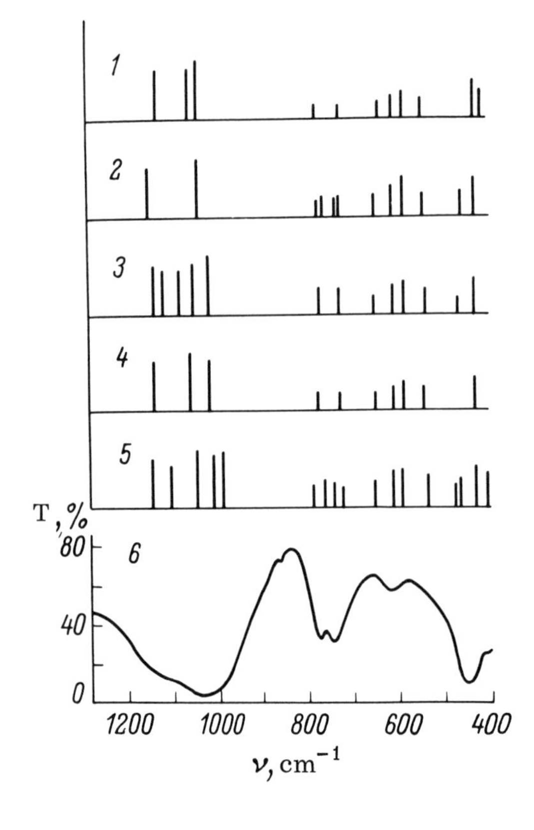

Fig. 86. Infrared spectra of framework-type aluminosilicates. Feldspars: 1) sanadine; 2) orthoclase; 3) microcline; 4) adularia; 5) albite (low-temperature). Zeolite: 6) analcime.

The spectra of low-temperature albite and orthoclase contain eight bands in this region, as we should expect. The spectra of sanadine, adularia, and high-temperature albite show only six maxima, the main differences reducing to a change in the number of bands in the range 790 to 720 cm^{-1}. The existence of four bands in this region is thus characteristic for structures with an ordered arrangement of the Si and Al, while in the spectra of disordered structures there are only two maxima (Table 39). To say that the broadening of the bands is a cause of the reduction in observed multiplicity is not always a satisfactory explanation. We may rather suppose that the fall in the number of bands is a consequence of chance degeneracy associated with the disordered, i.e., uniform distribution of Al and Si in the (Si, Al)—O—(Si, Al) bridges. An exception is the spectrum of microcline, in which there are only two (narrow) bands, despite the ordered distribution of Al and Si, between 720 and 790 cm^{-1}.

A case simpler for interpretation is that of the distribution of Al and Si in those structures in which the transformation from the ordered to the disordered state leads to a true degeneracy of some of the vibrations and a change in the selection rules as a result of the change in the symmetry of the system. Analcime $Na(AlSi_2O_6) \cdot H_2O$ is structurally and chemically largely similar to the feldspars, yet at the same time it is a typical zeolite. X-ray structural analysis gave the space group Ia3d $-O_h^{10}$ with sixteen formula units in the unit cell [214]. The silicon- and aluminum-oxygen tetrahedra form a three-dimensional framework comprising rings of four and six tetrahedra. The Na^+ cations and water molecules are situated in the spaces of the framework. Since the chemical composition, and in particular the ratio of Si and Al contents in analcime, are distinguished by great constancy and only deviate slightly from the ideal formula, we might suppose that the silicon and aluminum atoms were crystallographically different, occupying different sets, and that their distribution in the lattice was ordered. However, in the cubic lattice (Ia3d) 32 Si and 16 Al atoms lie on special positions on two-fold axes, occupying all together one set with a multiplicity 48, and are thus crystallographically indistinguishable. It was therefore suggested that the analcime crystal really had a tetragonal symmetry (with a sharply-expressed pseudocubic tendency): in the space group $I4_1/acd-D_{4h}^{20}$ the 32 Si atoms may be sited in a 32-fold set of general positions, and the Al in a 16-fold set with site symmetry C_2.

We may show that the question as to the type of Si and Al distribution, and hence the real symmetry of analcime, may (at least in principle) be answered by an appeal to spectroscopic data. In the primitive cell of a crystal with symmetry Ia3d (equal to half a centered cell) 48 oxygen atoms lie in a general position. Correspondingly the 48 bonds of type (Si, Al)—O—(Si, Al) give as many ν_s SiOSi vibrations, these being distributed with respect to the irreducible representations of the factor group O_h in the following manner: $3F_{1g}$, $3F_{1u}$, $3F_{2g}$, $3F_{2u}$, $2E_g$, $2E_u$, A_{1g}, A_{1u}, A_{2g}, A_{2u}. Since only vibrations of symmetry type F_{1u} are active in the infrared spectrum, in the ν_s SiOSi frequency range we can only expect the appearance of three bands if we assume a disordered arrangement of the Si and Al. Further, on assuming a space group $I4_1/acd$ with a sixteen-fold (for a primitive cell) set of general positions, the oxygen atoms are sited in three such sets. It is not hard to convince oneself that the ν_s SiOSi and ν_s SiOAl vibrations in this case give six frequencies active in the infrared spectrum and of different magnitudes (as a result of differences in the force constants of the Si—O—Si and Si—O—Al bonds): $3E_u$ and $3A_{2u}$.

The infrared spectrum of analcime (Fig. 86) given in [213] is reminiscent of the spectrum of silica in the region of the stretching vibrations of the framework, although the absorption maxima are considerably displaced to lower frequencies, this evidently being associated with the replacement of some of the Si—O bonds of the framework by Al—O bonds, which are characterized by smaller force constants. The presence of water of a zeolitic character in the structure manifests itself in the almost "unperturbed" values of the frequencies νH_2O 3620 cm^{-1} and δH_2O 1635 cm^{-1}. In the range 800 to 500 cm^{-1} the spectrum of analcime contains just three discrete bands at 775, 746, and 620 cm^{-1}; as indicated in the foregoing, this argues in favor of the indistinguishability of the Si and Al atoms in this lattice. We note that according to the latest data [215] analcime deviates from cubic symmetry at temperatures below 250°C, but very little. No reasonably accurate x-ray determinations of the arrangement of Si and Al in this crystal have been made.

Among the compounds in which transformation from the ordered to the disordered distribution of Si and Al atoms causes sharp changes in the infrared spectrum, we must mention sillimanite and also mullite, which is very similar to the latter according to x-ray diffraction, although different in composition. The disordered arrangement of the Si and Al with respect to the tetrahedral positions in mullite leads to a very strong blurring and changes in the intensity of the infrared bands, while preserving similarity to the sillimanite spectrum in general features [205]. In this case the spectroscopic method is the most convenient for the practically important problem of identifying these two minerals. We note that by comparing the spectra of sillimanite and mullite we may directly deduce that the structure of the latter preserves anomalously large (almost 180°) (Si, Al)O(Si, Al) valence angles; this agrees with existing data relating to the structure of mullite [216].

The ordered or disordered population of oxygen polyhedra larger than tetrahedra also manifests itself clearly in the spectra. Thus, in the compound $MgAl_2O_4$, with the spinel structure, a comparison of the infrared spectra of samples quenched from various temperatures led to the conclusion that there was a transformation from an ordered arrangement with Mg and Al atoms lying in tetrahedral and octahedral positions, respectively, to a disordered distribution at high temperatures [217]. The order and disorder in the population of the tetrahedral and octahedral positions in the spinels $LiAl_5O_8$ and $LiFe_5O_8$ were discussed on the basis of spectroscopic evidence in [205] and [217].

§5. Use of the Spectra of Solid Solutions in Analyzing the Spectra of Silicates

Method of Quasi-Isotopic Substitution in the Analysis of the Spectra of Silicates. In calculating the frequencies in the vibrational spectra of both molecules and crystals it is essential, in order to reproduce the experimentally observed frequencies with reasonable accuracy, to introduce a considerable number of force constants, usually exceeding the number of frequencies of the normal vibrations, not all of which are optically active. Hence any objective determination of the set of force constants can only be carried out after introducing additional assumptions.

In the spectroscopy of molecules of organic compounds, particularly hydrocarbons, one assumption of this kind is usually the reasonably well-justified one that the force field remains unaltered in the molecules of isotope-substituted derivatives. On replacing H by D the value of the isotopic displacement is large (up to 40%), and this, together with the possibility of using a whole series of molecules, differing in the number and position of the hydrogen atoms replaced by deuterium, ensures fair reliability in determining the force constants. Another approach reduces to the assumption that the force constants corresponding to the same structural elements remain unaltered, or that they vary very slightly and monotonically over homologous series of molecules.

Both methods are of limited value when calculating the spectra of inorganic molecules, ions, and crystals. Thus the comparative rarity of homologous series, so typical for organic compounds, and the usually strong mutual influence of the bonds prevents us from taking the force constants obtained by calculating the spectrum of one substance and transferring them to another. A certain analogy to the method of homologous series may be secured by analyzing data obtained as a result of studying and calculating the spectra of the molecules of hetero-organic compounds containing various different "inorganic" structural elements. It is convenient to use this kind of information in order to make a reasonable choice of the force constants when calculating the frequencies of complex inorganic anions (in the present case, silicates).

The method of isotopic substitution in the case of inorganic compounds not containing hydrogen is only applicable to a few atoms ($^{6}Li-^{7}Li$, $^{10}B-^{11}B$, $^{16}O-^{18}O$, $^{40}Ca-^{44}Ca$), the corresponding changes in mass not greatly exceeding 10%, i.e., the displacement in the frequencies may reach 5% of the original value. This is often sufficient when verifying the identification of frequencies in the spectrum, but for refining the values of the force constants the sensitivity is too low.

The difficulties encountered when seeking means of objectively determining the force constants from the spectra of inorganic compounds in some cases suggest using the following fact. If we retain the same spatial configuration of the molecules or complex ions X_nO_m, but replace the X atoms by the atoms of an element of the same subgroup of the periodic table but belonging to a different period, the consequent change in the vibration frequencies is principally a result of the change in the mass of the atom X rather than the force constant of the X–O bond. In fact, on passing, for example, from X = Si to X = Ge, the mass of the atom X changes by a factor of over 2.5, while the change in the force constant of the X–O bond averages about 10%. Similar relationships hold for P and As and for S and Se.

It seems reasonable, for cases in which the salts $M_pX'_nO_m$ and $M_pX''_nO_m$ are isostructural, to use the same values of force constants in the first stages of calculating their spectra (or that of the X'_nO_m and X''_nO_m anions), and only to vary these coefficients slightly and individually in the final stages of the calculation, so as to secure the best agreement between the observed and calculated frequencies. If the crystals of the "isoformula" salts $M_pX'_nO_m$ and $M_pX''_nO_m$ are isostructural, then corresponding solid solutions are frequently formed over a wide range of compositions, and sometimes over the whole range of $M_p(X'_nO_m-X''_nO_m)$. If, moreover, the distribution of the X′ and X″ atoms in the lattice is random or nearly so, the spectra of the solid solutions may reasonably be considered as a superposition of the spectra of the $X'_xX''_{n-x}O_m$ anions with $x = 0, 1, 2, \ldots, n$ (or more precisely of the crystals $M_pX'_xX''_{n-x}O_m$), while the number of anions of different "kinds" is determined by the ratio of the concentrations of the X′ and X″ atoms. Thus, as in the case of isotopically substituted molecules, we may use data relating to the partly substituted derivative in the calculation; this greatly increases the number of experimental frequencies introduced into the analysis and increases the reliability of the results of the calculation.

The main differences distinguishing this process from the ordinary analysis of the spectra of a series of isotopically substituted molecules are: a) the necessity of considering changes in the matrix of kinematical interaction, not only as a result of the change in the mass of the atoms but also as a result of changes in the bond lengths and frequently the valence angles, and b) the impossibility of separating out each of the partly substituted derivatives in pure form. The latter disadvantage is only partly compensated by the possibility of allowing for the concentration dependence of the intensities of the bands, which facilitates the identification of the frequencies of anions of different composition. Hence the use of the proposed method, which we shall subsequently call the method of quasi-isotopic substitution, is not very promising for $n > 3$ or 4, since the rapidly increasing overlapping of the many bands of anions of different compositions (and different isomers of the same composition, differing in the arrangement of the X′ and X″ atoms) makes any complete determination of the set of fundamental frequencies of each of the anions quite impossible.

The principal conditions for the applicability of the proposed method to any particular group of real substances are, as already indicated: a) the isostructural nature of the compounds $M_pX'_nO_m$ and $M_pX''_nO_m$, b) the existence of a continuous (or fairly wide) series of solid solutions between these compounds, and c) the random character of the distribution of the X′ and X″ atoms in the lattice, including positions within the same complex anion. Satisfaction of the first of these conditions, which occurs quite often and in particular for X′ = Si, X″ = Ge, although by no means in every case, usually means that the second will also be satisfied. However, in every particular case it is essential that both conditions should be verified separately, since exceptions may occur; for example, more efficient packing may be possible when polyhedra of different sizes are present, corresponding to the X′ and X″ atoms, than when the polyhedra are all of the same sort. A typical example of this kind is the fact that x-ray structural analysis of the $Li_2Si_2O_5-Li_2Ge_2O_5$ system in the neighborhood of the composition $Li_2(Si_{0.25}Ge_{0.75})_2O_5$ reveals a new type of structure of the $[(Si, Ge)_4O_{10}]_\infty$ anion ribbons, somewhat reminiscent of that in sillimanite, although this structure is not characteristic of either of the end members of the series, which at lower temperatures form continuous solid solutions with complex anions in the form of infinite layers [218].

Regarding the third of the conditions mentioned, there is very little data in the literature enabling us to judge how often and to what degree of accuracy it is satisfied. One of the few papers devoted to the x-ray study of the Si and Ge distribution in solid solutions of silicates and germanates led to the conclusion that there was a correlated distribution of these atoms over three translationally nonequivalent "kinds" of XO_4 tetrahedra in a $[(XO_3)_3]_\infty$ chain of the wollastonite type; thus, in the $Ca(SiO_3-GeO_3)$ system the amount of Ge in one sort of tetrahedra was almost twice that in the other two sorts [219]. An analogous increase in the As content of tetrahedra of the same sort in the wollastonite-like $[(PO_3)_3]_\infty$ chain of the high-temperature form of Madrell's salt was found in $Na(PO_3-AsO_3)$ solid solutions [220, 221]. The deviation from a random distribution of X′ and X″ atoms was attributed in both cases [219, 221] to the fact that, in the original melt and glass, the $(XO_3)_n$ chains were accompanied by monomeric XO_4 groupings which were "built in" to the chain on crystallization.

We may nevertheless consider that the decisive factor in the nonuniformity of the Si and Ge (or P and As) distribution, quite independent of the mechanism underlying its development, is the fact that the crystal structure in question contains a variety of XO_4 tetrahedra, differing considerably as regards their packing characteristics and links with the neighboring polyhedra, and that one particular type of tetrahedron is more suitable for accommodating the large Ge (or As) atoms. We note further that, of the three different X–O–X bonds in the chain, the two forming comparatively small XOX angles (139 to 140° in $CaSiO_3$, the third angle being 149°) are adjacent to this tetrahedron; the tendency toward a reduction in XOX angles on passing from X = Si to X = Ge has already been noted and finds further support in the results of a number of structural analyses (see [222]). It is a characteristic fact that, in the case of the system of solid solutions $K(PO_3-AsO_3)$, crystallizing with the formation of X_3O_9 ring anions, in which all three tetrahedra are approximately equivalent (though not entirely, since in the lattice the ring has no three-fold axis), Grunze et al. [223] came to the conclusion that anions of the $[P_3O_9]^{3-}$, $[AsP_2O_9]^{3-}$, $[As_2PO_9]^{3-}$, $[As_3O_9]^{3-}$ types occurred in proportions corresponding approximately to a random distribution of P and As.

The foregoing considerations by no means suggest that the method of quasi-isotopic substitution cannot be used in the case of a correlated distribution of X′ and X″ atoms in solid solutions; however, its efficiency then becomes more limited, and, in order to use it, it is best to determine the form of the correlation from independent (x-ray diffraction) results. It also clearly follows from the foregoing that, in addition to using the spectra of solid solutions in analyzing and calculating the spectra of the extreme members of the series, it is of considerable interest to use the spectroscopic method for studying the actual character of the Si and Ge distribution in the solid solution. We shall now present some results of the application of the quasi-isotopic-substitution method to a specific system; for the sake of simplicity, analysis of the spectrum is limited to the molecular approximation.

Analysis of the Spectra of Silicates and Germanates Containing Cyclic $(XO_3)_3$ Anions, Using the Quasi-Isotopic-Substitution Method. The empirical identification of the frequencies in the spectra of silicates and germanates with cyclic X_3O_9 anions proposed earlier (Chap. II and this chapter, §1) was mainly based on the as-

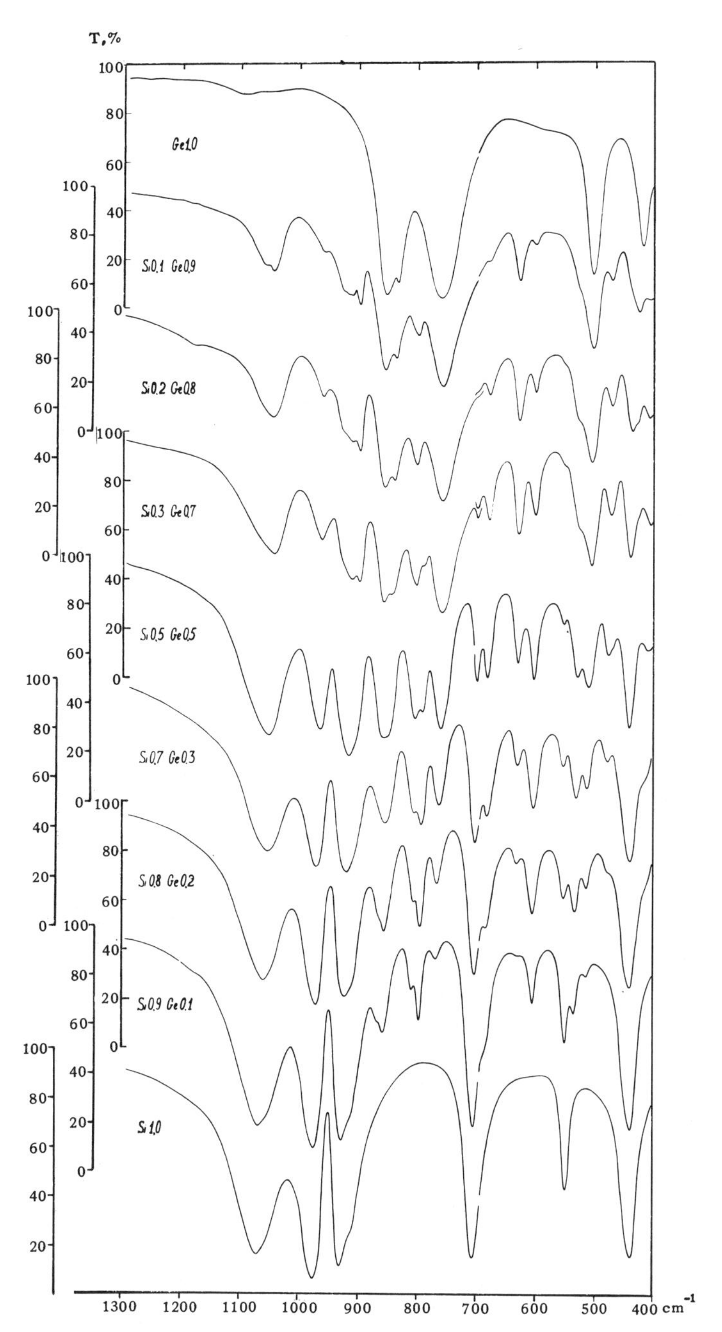

Fig. 87a. Infrared spectra of solid solutions in the $Sr_3Si_3O_9$–$Sr_3Ge_3O_9$ system; the silicate and germanate content is indicated on the curves (pressed in KBr, content of the substance under investigation varies from 3.9 mg for $Sr_3Si_3O_9$ to 5.0 mg for $Sr_3Ge_3O_9$, so as to ensure the constancy of the number of $(Si, Ge)_3O_9$ ions in all the samples).

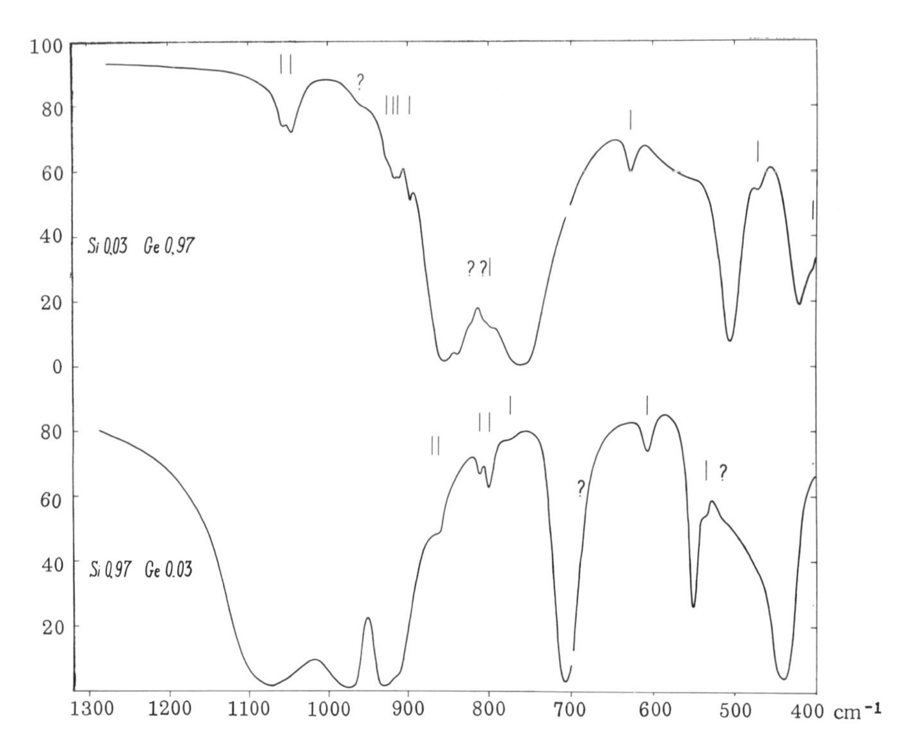

Fig. 87b. Infrared spectra of $Sr_3Si_3O_9$–$Sr_3Ge_3O_9$ solid solutions for a small (3% mol.) concentration of one of the components. The "impurity" bands are marked in the figure. (Pressed in KBr, 8 mg for $Sr_3(Si_{0.97}Ge_{0.03})_3O_9$ and 10 mg for $Sr_3(Si_{0.03}Ge_{0.97})_3O_9$.)

sumption that the stretching frequencies of the $X\genfrac{}{}{0pt}{}{O^-}{O^-}$ and $X{-}O{-}X$ groups were of a relatively characteristic nature and also on data relating to the dichroism of the bands in the reflection spectra of single-crystal samples. Complete agreement was nevertheless not reached with the results of the calculation of the spectra of the Si_3O_9 ion (symmetry D_{3h}) carried out in [224, 225]. The application of the quasi-isotopic-substitution method to the spectra of these anions, containing altogether three (equivalent) tetrahedral positions and with a symmetry no lower than C_3, seemed appropriate and offered the prospect of securing a more reliable interpretation of the spectrum and values of the force constants. The $Sr_3Si_3O_9$–$Sr_3Ge_3O_9$ system was chosen for investigation; according to Grebenshchikov et al. [226], at temperatures of over 900° this contains a continuous series of solid solutions with the structure of the high-temperature form of $Sr_3Ge_3O_9$. The comparatively weak and predominantly ionic bonds between the oxygen atoms and the Sr^{2+} cations to a certain extent justified the use of the quasi-molecular approximation in calculating the majority of the frequencies of the complex anions.

Figure 87a shows the infrared spectra of $Sr_3Si_3O_9$, $Sr_3Ge_3O_9$, and the solid solutions formed by these; in all the samples the total number of X_3O_9 groups remained constant. On the assumption that the extinction coefficients for the bands of each kind of $Si_{3-n}Ge_nO_9$ ions (n = 0, 1, 2, 3) are independent of the concentrations of the other kinds of ions, this enables us to use changes in optical density in order to discuss changes in the proportions of anions of different compositions. The results obtained by using several slightly overlapping bands indicate that the distribution of the Si and Ge atoms is almost random. The spectra of the solid solutions are quite discrete, and the width of the bands does not deviate from values typical of the spectra of poly-

TABLE 40. Identification of the Vibration Frequencies of the Complex Anions in the Spectra of $Sr_3Si_3O_9$–$Sr_3Ge_3O_9$ Solid Solutions*

$Sr_3Si_3O_9$ experimental frequencies		Calculated frequencies of complex anions		Wavenumbers of maxima in the infrared spectra of solid solutions. Relative concentration of Si*									Calculated frequencies of complex anions		$Sr_3Ge_3O_9$ experimental frequencies	
				0.97†	0.90	0.80	0.70	0.50	0.30	0.20	0.10	0.03†				
1	2	3	4	5	6	7	8	9	10	11	12	13	14	15	16	17
Raman spectrum‡ (polycrystalline sample)	Infrared spectrum (polycrystalline sample)	Si_3O_9	Si_2GeO_9	0.969 0.030 0.001 0.000	0.889 0.099 0.011 0.001	0.752 0.188 0.047 0.012	0.591 0.253 0.109 0.047	0.250 0.250 0.250 0.250	0.047 0.109 0.253 0.591	0.012 0.047 0.188 0.752	0.001 0.011 0.099 0.889	0.000 0.001 0.030 0.969	$SiGe_2O_9$	Ge_3O_9	Infrared spectrum (polycrystalline sample)	Raman spectrum (polycrystalline sample)
–	1071	1058 A_2''	1057 B_2	1071	1070 ↖	1061 ←	1055 ←	1052 ↙	1065sh. ←	1065sh. ←	1057	1057 }	1055 B_1 }	–	–	–
1058	–	1053 E''	1058 A_2	–	1060sh? ↙			↖	1044 ←	1044 ←	1045	1045 }		–	–	–
975	–	976 A_1'	–	–	–	–	–	–	–	–	–	–	–	–	–	–
–	975	968 E'	974 A_1 968 B_2 }	975	975 ←	971 ←	969 ←	965 ←	962 ←	961 ←	959	959	972 A_1	–	–	–
(951)	–	–	–	–	–	–	–	–	928 →	926 ←	927	926 }		–	–	–
917 {	930	} 920 E'	919 B_2	930	928 ←	923 ←	919 ←	917 ←	912 ←	912 ←	911-916(†?)	910-917 (2?) }	} 910 A_1	–	–	–
	912 sh.		904 A_1	912	916sh. ←	912sh. ←	907sh? ←	905sh?	–	–	–	–		–	–	–
–	–	–	–	–	–	–	–	–	899 ←	898 ←	899	898	892 B_2	–	–	–
(876)	–	–	867 B_2	870	870sh. →	867sh.	–	–	–	–	–	–	–	–	–	–
865	–	869 A_2'	852 B_1	862	859 →	857 —	857 →	855 (2?) ←	859 →	857 →	856	856	852 A_2	853 A_2'	856	–
(840)	–	–	–	–	–	–	–	– ↖	–	–	–	–	–	852 E''		}
–	–	–	–	–	–	–	–	–	845 →	840 →	838	835	852 B_1	829 E'	835 }	844
–	–	–	–	–	–	–	–	–	–	–	–	821?	–	–	–	–
–	–	–	–	–	–	–	–	–	810sh.	?	?	804?	811 B_2	809 A_1'	–	816
–	–	–	–	811	810 →	807 →	806 →	805 ←	802 ←	801 ←	799	798	805 A_1	–	–	–
–	–	–	803 A_1	799	797 →	795 →	794 ←	792 ←	790	–	–	–	–	–	–	–
–	–	–	–	774	769 →	767 →	764 →	762 →	760 →	758 →	760	759	732 B_2	753 E'	760	750
–	707	719 E'	–	707	706 ←	704 ←	703 ←	701 ←	700 ←	701	–	–	–	731 A_2'	–	–
–	–	–	689 A_1	688?	686sh. →	686 →	684 —	684 ←	680 ←	680 ←	680	–	–	–	–	–
–	–	–	–	–	633 →	633 →	632 →	631 →	630 ←	629 ←	628	–	–	–	–	–
–	–	–	586 B_2	607	606 →	606 —	604 —	604 ←	601 ←	600	–	627	610 B_2	–	–	–
567	–	577 A_1'	–	–	–	–	–	–	–	–	–	–	–	–	–	–
553	551	546 A_2''	538 B_2	551	552 ←	553 ←	553 ←	553 ←	531 ←	550	?	–	–	–	–	–
(537)	–	519 E''	535 A_1	535	536 →	536 —	532 —	530 —	528 ←	526 ←	527	–	–	–	–	–
512	–	507 A_1'	476 B_1	515?	515 →	514 →	514 →	511 →	507 —	507 →	505	505	{514 A_1 {510 A_1	497 E'	505	–
492	–	482 A_2'	484 A_2	–	–	–	–	–	–	–	–	–	–	493 A_1'	–	479
–	–	–	476 A_1	?	?	479sh. →	478 →	477 ↘	–	–	–	–	–	–	–	–
–	–	–	–	–	–	–	–	470sh. —	473 ←	471 ←	471	472	502 B_1	–	–	–
–	–	–	–	–	–	–	–	–	–	–	–	–	471 A_2	–	–	–
–	439	439 E'	442 B_2	439	439 ←	439 —	438 ←	440 ←	439 ←	436	?	–	–	–	–	–
–	–	–	–	–	–	–	–		424sh?	426 →	424	420	447 A_1	442 A_2''	419	–
–	–	–	–	–	–	414sh. →	414 —	409 →	404 ←	405 ←	406	405	416 B_1	420 E''	–	–

*The percentage of Si_3O_9, Si_2GeO_9, $SiGe_2O_9$, and Ge_3O_9 is calculated on the assumption of a purely statistical distribution of Si and Ge in each solid solution.

†The changes of relative intensities of the bands with the composition of the solid solutions are indicated by arrows (the spectra of solid solutions with Si/Ge ratio 97/3 and 3/97 were studied on the more concentrated specimens and, therefore, were not used in the estimations of relative intensities).

‡The wavenumbers of bands which can probably be attributed to admixture of Sr_2SiO_4 are shown in brackets.

crystalline samples of the pure silicates or germanates. (We may suppose that this is to some extent associated with the comparatively weak interaction between the internal vibrations of the complex anions and the lattice vibrations.) On the other hand, the total number of bands observed as the Si–Ge content varies is much smaller than we should expect for the Si_3O_9, Ge_3O_9, Si_2GeO_9, and $SiGe_2O_9$ ions, if we remember that, for the last two, no symmetry higher than C_{2v} is possible, while all degeneracies and the greater proportion of the bans (in the selection rules) are removed. It follows from this that, as a result of superpositions, only some of the vibrations of the "mixed" Si_2GeO_9 and $SiGe_2O_9$ ions can be identified experimentally.

For a more reliable interpretation of the vibration bands of these ions it seemed useful to study the spectra of solid solutions with small (3 mol. %) contents of the silicate or germanate: for a random distribution of Si and Ge in the lattice, the probability of the formation of one of the two mixed rings is negligibly small. Thus all the "impurity" bands in the spectrum of a $Sr_3(Si_{0.97}Ge_{0.03})_3O_9$ sample must be ascribed to the Si_2GeO_9 ion, and on the other hand in the $Sr_3(Si_{0.03}Ge_{0.97})_3O_9$ spectrum the impurity bands may be associated with the $SiGe_2O_9$ ion. The spectra of these samples (obtained by using considerably greater sample weights than the spectra in Fig. 87a) are shown in Fig. 87b; the results of the analysis presented in Table 40 confirm the foregoing considerations, with one exception. At the same time it should be noted that, even for small concentrations, many of the impurity bands of the Si_2GeO_9 and $SiGe_2O_9$ ions capable of being ascribed to only o n e internal vibration of the anion possess a complex (at least doublet) structure in these spectra. This can hardly be made to agree with the foregoing simple scheme of the spectrum of the solid solution, since "Davydov" (i.e., resonance) splittings of the internal vibrations of the anions only present in low concentrations are improbable. One possible explanation is that considerable fluctuations may develop in the distribution of the anions of mixed type (possibly with the formation of a local superstructure), or (and this is more likely) the frequencies of the $X_2'X''O_9$ impurity anion may depend appreciably on the composition of a fairly large part of the crystal surrounding a specific anion.†

The resultant data may be used, if we consider the spectra of the solid solutions (for the sake of simplicity) as the result of the superposition of the spectra of "isolated" Si_3O_9, Si_2GeO_9, $SiGe_2O_9$, and Ge_3O_9 ions, in order to calculate the frequencies of these ions with the aid of a larger number of experimental frequencies than could be obtained from the spectra of the pure silicate or germanate alone. The frequencies and the forms of the vibrations were calculated using internal vibrational coordinates. Figure 88 shows the scheme of the spatial configuration of these complex anions and indicates the notation of the vibrational coordinates. The following are the geometrical parameters used in the calculation:

$$\text{Ion } [Si_3O_9]^{6-}\ (X_1 = X_2 = X_3 = Si),\ r_{SiO(Si)} = 1.64 Å,\ r_{SiO^-} = 1.60 Å,$$

$$\angle O^-SiO^- = \angle O^-SiO = \angle OSiO = 109°28',\ \angle SiOSi = 130°32',$$

planar ring, symmetry D_{3h}.

$$\text{Ion } [Ge_3O_9]^{6-}\ (X_1 = X_2 = X_3 = Ge),\ r_{GeO(Ge)} = 1.77 Å,\ r_{GeO^-} = 1.73 Å,$$

$$\angle O^-GeO^- = \angle O^-GeO = \angle OGeO = 109°28',\ \angle GeOGe = 130°32',$$

planar ring, symmetry D_{3h}.

Ion $[Si_2GeO_9]^{6-}$ (X_1 = Ge, $X_2 = X_3$ = Si); the bond lengths and the angles at the germanium (silicon) atoms are as in the Ge_3O_9 (Si_3O_9) ion, planar ring, angles GeOSi = 127°53′ and angles SiOSi = 135°50′ (from the condition that the ring is closed), symmetry C_{2v}.

† Another possible explanation for the band structure is based on ascribing the "excess" maxima to vibrations with $\vec{k} \neq 0$, which are potentially optically active in the mixed crystal.

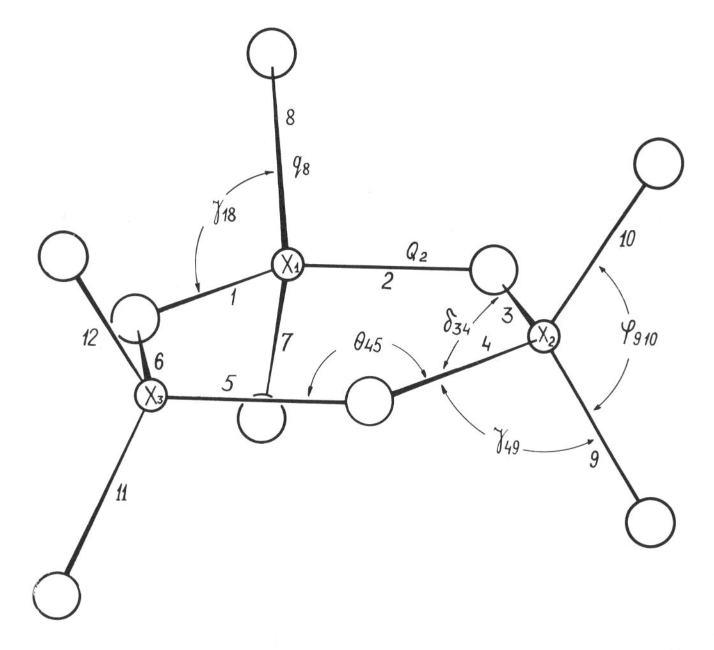

Fig. 88. Spatial configuration and internal vibrational coordinates of the X_3O_9 rings.

Ion $[SiGe_2O_9]^{6-}$ (X_1 = Si, $X_2 = X_3$ = Ge); the bond lengths and angles at Si(Ge) are taken from the Si_3O_9(Ge_3O_9) ion, planar ring, angles GeOSi = 132°56′ and angles GeOGe = 125°44′ (from the condition that the ring is closed), symmetry C_{2v}.

As internal vibrational coordinates we used the changes in the bond lengths X–O (Q_i, i = 1, 2, . . . , 6) and $X-O^-$ (q_i, i = 7, 8, . . . , 12), the changes in the angles OXO (δ_{ik}), the angles O^-XO^- (φ_{ik}), the angles OXO^- (γ_{ik}), and the angles (θ_{ik}), and also the three coordinates of the nonplanar vibrations of the ring χ_{ik} describing the deviation of the O atoms from the plane of the X atoms.† Of these 36 coordinates 30 are independent. Three dependent coordinates are eliminated by means of the relations connecting the angular coordinates in the three tetrahedra; three additional conditions relate the coordinates Q_i, δ_{ik}, θ_{ik} in the ring. The first of these is the same for all four rings ($\delta_{12} + \delta_{34} + \delta_{56} + \theta_{23} + \theta_{45} + \theta_{16} = 0$) and the other two are presented below.

A general form of quadratic potential function has been used for the description of the force field. To make the estimations of the force constants less arbitrary, most of the off-diagonal terms, primarily those likely to be of low absolute value, have been set at zero in the process of adjustment of the force constants, and the force constants describing each SiO_4 or GeO_4 group were supposed to be the same regardless of the composition of the complex anion they enter. Only a limited number of parameters has been used to describe the interaction between these groups in the complex anion. Consequently, the number of experimentally determined frequencies exceeds the number of force constants used to describe the force fields in the four types of complex anions.

Additional experimental frequencies and more reliable assignments were obtained by the use of the Raman spectra of $Sr_3Si_3O_9$ and $Sr_3Ge_3O_9$. Both compounds were studied in the form of polycrystalline powders using He–Ne laser excitation. The Raman spectrum of $Sr_3Si_3O_9$ was

† The coordinates χ_{ik} are defined by the angle between the plane of the ring and the plane formed by bonds i and k with a common O atom.

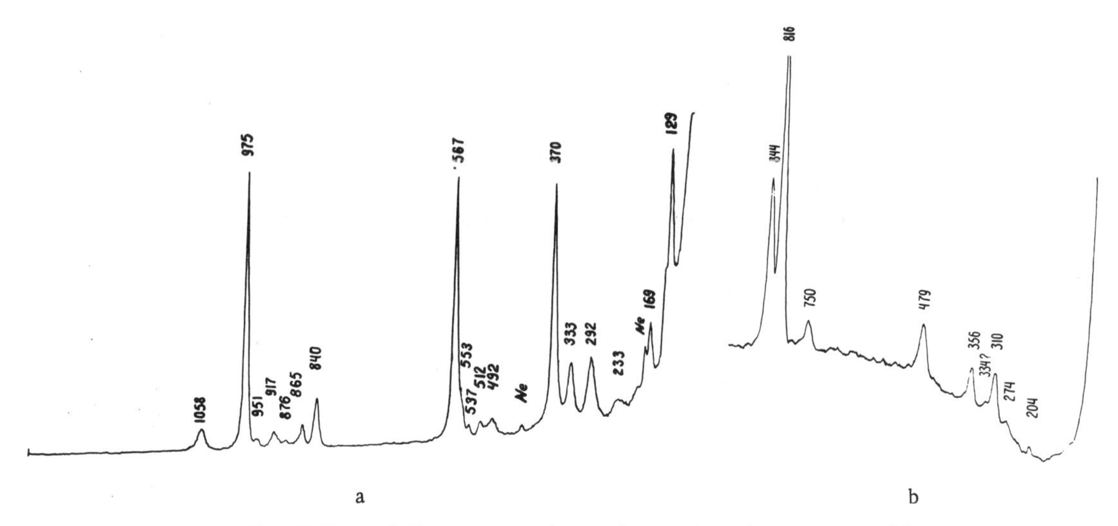

Fig. 89. Infrared Spectrum of $Sr_3Si_3O_9$ (a) and $Sr_3Ge_3O_9$ (b).

kindly supplied by Professor G. Michel (Liège University, Belgium). The Raman spectrum of $Sr_3Ge_3O_9$ was investigated by L. A. Leites (the Institute of Organo-Element Compounds, Acad. Sci. USSR, Moscow). Both spectra are shown in Fig. 89. In the spectrum of $Sr_3Si_3O_9$ the two most intense bands in the region of Si–O stretching vibrations at 975 and 567 cm^{-1} are readily identified as the two vibrations of the A_1' species, ν_s SiO_2 and ν_s SiOSi, respectively. The total number of bands in the Raman spectrum of $Sr_3Si_3O_9$ between 1100 and 800 cm^{-1} (seven) considerably exceeds the expected value: only four fundamentals (A_1', E″, and two E′) are Raman-active for D_{3h} symmetry in this frequency region. If a possible lowering of the symmetry of the ring anion to C_{3h} (as in benitoite and probably also in catapleiite – see the considerations following from the analysis of the infrared spectrum in Chap. 11) is taken into account, one more vibration becomes active in the Raman spectrum. The vibration ν_{as} SiOSi belonging to the A_2' species for D_{3h} symmetry, can be assigned to the band at 865 cm^{-1}, if the threefold symmetry of the ring is retained,† the largest possible number of fundamental bands between 1100 and 800 cm^{-1} is six, but the sixth – ν_{as} SiO_2, A_2'' for D_{3h} – corresponds to the strong infrared band at 1071 cm^{-1} and this has no counterpart in the Raman spectrum.

The increased number of bands in the Raman spectrum of $Sr_3Si_3O_9$ might be explained by the Davydov splitting of some internal vibrations of the complex anion in the spectrum of the crystal, which is also a probable cause of the doublet structure of the infrared band at 930 to 912 cm^{-1}. (The space group of the hexagonal or pseudohexagonal crystal $Sr_3Si_3O_9$ is as yet unknown, but similar splitting of the corresponding band in the spectrum of catapleiite has been discussed above.) It is doubtful, however, that Davydov splitting of the nondipolar A_2' vibration could be large enough to give rise to the two widely separated bands at 840 and 865 cm^{-1} in the Raman spectrum. It is likely that the sample of $Sr_3Si_3O_9$ studied in the Raman spectrum was contaminated by Sr_2SiO_4. If admixture of Sr_2SiO_4 is supposed, the band at 840 cm^{-1} may be assigned to ν_s SiO_4 in accordance with the infrared spectrum of this compound. Some of the weak

† The threefold symmetry of the ring anion in $Sr_3Si_3O_9$ is proved by the absence of splitting of the majority of bands corresponding to degenerate vibrations. Especially significant is the simple shape of the band at 707 cm^{-1} in the infrared spectrum because its assignment to E′ (ν_s SiOSi) vibration is certain, and its sensitivity to ring distortion is verified by calculations of the spectra of nonsymmetrical rings.

bands – 876 cm^{-1} and, probably, 537, and 951 cm^{-1} – may be assigned to Sr_2SiO_4 as well. On the other hand, the band at 951 cm^{-1} may arise from an upper-stage transition of $\nu_s\,SiO_2(A_1')$ of the ion Si_3O_9 since the ground-state transition corresponding to this vibration produces the very strong Raman band at 975 cm^{-1}. The band at 537 cm^{-1} may be equally successfully assigned to the E″ vibration of the ion Si_3O_9 as can be seen in Table 40.

In the Raman spectrum of $Sr_3Ge_3O_9$ only four bands were found in the interval 850 to 400 cm^{-1}, where the molecular approximation seems to be meaningful for band assignments. An extremely intensive band at 816 cm^{-1} certainly corresponds to the vibration $\nu_s\,GeO_2$ (A_1' for D_{3h}). The band at 479 cm^{-1} may be assigned to ν_s GeOGe of species A_1', the frequency of which nearly coincides with the ν_s GeOGe of species E′ seen in the infrared spectrum at 505 cm^{-1}. The absence of ν_s GeOGe of species A_1' in the infrared spectrum may be regarded as an indication of the flat configuration of the ring (with a horizontal plane of symmetry), though this is not required by the symmetry of the crystal.†

Let us now return to the results of normal coordinate calculations. All available experimental data are compared with calculated frequencies for the complex anions Si_3O_9, Si_2GeO_9, $SiGe_2O_9$, and Ge_3O_9 in Table 40. Only frequencies higher than 400 cm^{-1} are considered, not because of incompleteness of the experimental data in the low-frequency region, but primarily because of the inadmissibility of the molecular approximation for frequencies near lattice vibrations (i.e., Sr–O vibrations) due to their possible mixing with some of the internal vibrations of the complex anion. Thus, little attention was paid to the coincidence or noncoincidence of calculated and experimental frequencies in the long-wavelength part of the spectrum. (For example, in the Raman spectrum of $Sr_3Si_3O_9$ a very intense band lies at 370 cm^{-1}, but the nearest calculated A_1' frequency of Si_3O_9 is at 312 cm^{-1}.)

The assignment is also shown schematically in Fig. 90. The assignments of the vibrations of "mixed" anions are indicated by arrows only if the corresponding vibration can be identified in the spectra of solid solutions as a separate band not overlapped by bands of anions of other compositions. The regions where the bands of a given mixed anion are obscured by the absorptions of other anions (primarily of pure Si_3O_9 and Ge_3O_9 ions) are shaded. The correspondence between calculated and observed frequencies in the short wavelength part of the spectrum is quite satisfactory. The results of calculations remove some difficulties arising when a purely empirical assignment (based only on symmetry considerations and characteristic group frequencies) is attempted. For instance, only one of the three vibrations of type ν_s SiOGe and ν_s GeOGe of the anion $SiGe_2O_9$ had been identified experimentally (species B_2 for C_{2v} symmetry), but the positions of the other two remained uncertain. The calculated frequencies of these two vibrations (both species A_1 for C_{2v}) lie between 500 and 530 cm^{-1}, and neither can be seen on the strong background arising from the bands at 505 cm^{-1} of Ge_3O_9, 551 cm^{-1} of Si_3O_9, and 535 to 525 cm^{-1} of Si_2GeO_9. It is noteworthy that one of the ν_s SiOGe frequencies (species A_1) of Si_2GeO_9 is of the same magnitude or lower than the rocking frequency $\rho\,O^-SiO^-$ of species B_2.

The forms of vibrations (unnormalized, in the symmetry coordinates) are given in Tables 41 and 42 for Si_3O_9 and Ge_3O_9 ions. This gives an opportunity to judge the reliability of the

† Strictly speaking, in the crystal $Sr_3Ge_3O_9$ with space group $R\bar{3}-C_{3i}^2$ and two molecules in the unit cell [7], the Ge_3O_9 ions are in sites with symmetry C_3. This makes potentially active in the Raman spectrum six vibrations lying in the narrow interval 850 to 700 cm^{-1}: the degeneracy is not removed, but each of the internal vibrations of a Ge_3O_9 ion is split into a doublet, one component of which is active in the Raman spectrum and the other in the infrared. In both spectra up to 10 internal vibrations of Ge_3O_9 ion may be, in principle, registered between 850 and 400 cm^{-1}.

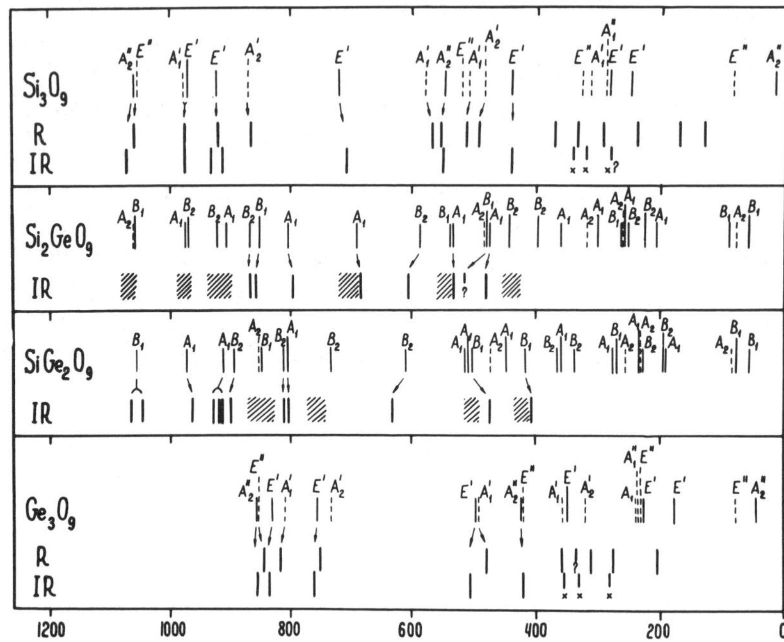

Fig. 90. Identification of frequencies of Si_3O_9, Si_2GeO_9, $SiGe_2O_9$, and Ge_3O_9 anions; top line: calculated frequencies; bottom line: experimental data (spectra of $SrSiO_3 - SrGeO_3$ solid solutions). Diagonal shading indicates ranges of frequencies for those bands of Si_2GeO_9 and $SiGeO_9$ whose exact location is prevented by superposition of bands belonging to ions of different composition. Crosses denote bands in logwave spectrum of $Sr_3Si_3O_9$ and $Sr_3Ge_3O_9$ according to [205] and [227]; note that large number of maxima in short-wave part of spectrum of $Sr_3Ge_3O_9$ in [227] probalby indicates presence of impurities, to which some of the bands in the long-wave region may also belong.

TABLE 41. The Calculated Frequencies and Forms* of Vibrations of Si_3O_9 Ion.†

Description of vibration‡	E′	q	Q_1	Q_2	φ	δ	γ_1
ν_s O⁻SiO⁻	968	1	−0.249	0.006	−0.394	0.359	−0.007
ν_{as} SiOSi	920	−0.013	1	0.786	0.238	−0.613	0.331
ν_s SiOSi	719	0.020	−0.718	1	−0.316	−0.004	1.511
δ O⁻SiO⁻	439	0.074	−0.084	−0.083	1	−0.217	0.035
δ SiOSi	282	−0.074	0.354	−0.117	−0.052	1	0.274
γ O⁻SiO⁻	247	0.019	−0.106	−0.552	−0.308	−0.627	1

Description of vibration	A_2''	q	γ	χ
ν_{as} O⁻SiO⁻	1058	1	0.678	0.636
ρ O⁻SiO⁻	546	−0.106	1	1.021
$\delta_\perp$ SiOSi	16	−0.005	−0.113	1

Description of vibration	A_2'	Q	γ
ν_{as} SiOSi	869	1	−0.313
γ O⁻SiO⁻	482	0.086	1

Description of vibration	E″	q	γ_1	γ_2	χ
ν_{as} O⁻SiO⁻	1053	1	0.332	−0.477	−0.320
ρ O⁻SiO⁻	519	−0.055	1	−0.073	−0.912
τ O⁻SiO⁻	326	0.060	0.121	1	0.120
$\delta_\perp$ SiOSi	83	−0.001	0.095	0.022	1

Description of vibration	A_1'	q	Q	δ	φ
ν_s O⁻SiO⁻	976	1	−0.262	0.454	−0.413
ν_s SiOSi	577	−0.016	1	0.600	0.921
δ SiOSi	507	−0.119	0.057	1	−0.564
δ O⁻SiO⁻	312	0.033	−0.542	0.347	1

*Unnormalized, in symmetry coordinates.

†τ O⁻SiO⁻ (A_1'') = 287 cm^{-1} (γ).

‡The following designations are used to give a qualitative description of vibrational modes: ν – bond stretch, δ – bending (scissoring) SiOSi or O⁻SiO⁻, γ – wagging (fan-like motions of Si<O⁻O⁻ groups), ρ – rocking (pendulum-like motions of Si<O⁻O⁻ groups), τ – twisting (torsional vibrations of Si<O⁻O⁻ groups), $\delta_\perp$ – puckering vibrations of the ring.

TABLE 42. The Calculated Frequencies and Forms* of Vibrations of Ge_3O_9 Ion†

Description of vibration‡	E′	q	Q_1	Q_2	φ	γ_1	δ
$\nu_s\,O^-GeO^-$	829	1	−0.826	−0.404	−0.226	−0.029	0.444
$\nu_{as}\,GeOGe$	753	0.657	1	0.708	−0.018	0.120	−0.402
$\nu_s\,GeOGe$	497	−0.069	−0.569	1	−0.191	1.341	−0.441
$\delta\,O^-GeO^-$	348	0.021	−0.002	0.007	1	−0.038	−0.168
$\delta\,O^-GeO^-$	227	0.014	0.311	−0.295	−0.190	1	0.638
$\delta\,GeOGe$	176	−0.015	−0.011	0.343	0.021	−0.495	1

Description of vibration	A_2''	q	γ	χ
$\nu_{as}\,O^-GeO^-$	853	1	−0.295	−0.309
$\rho\,O^-GeO^-$	422	0.034	1	1.160
$\delta_\perp\,GeOGe$	44	0.003	−0.133	1

Description of vibration	A_2'	Q	γ
$\nu_{as}\,GeOGe$	731	1	0.017
$\gamma\,O^-GeO^-$	319	−0.071	1

Description of vibration	E″	q	γ_1	γ_2	χ
$\nu_{as}\,O^-GeO^-$	852	1	−0.147	0.229	−0.154
$\rho\,O^-GeO^-$	420	0.017	1	0.007	1.002
$\tau\,O^-GeO^-$	228	−0.018	0.035	1	−0.320
$\delta_\perp\,GeOGe$	75	0.001	−0.116	0.041	1

Description of vibration	A_1'	q	Q	φ	δ
$\nu_s\,O^-GeO^-$	809	1	−0.174	−0.182	0.114
$\nu_s\,GeOGe$	493	0.109	1	0.353	1.438
$\delta\,O^-GeO^-$	355	0.039	0.056	1	−0.605
$\delta\,GeOGe$	235	−0.075	−0.682	−0.567	1

* See footnote to Table 41.
† $\tau\,O^-GeO^-$ (A_1'') = 234 cm^{-1} (γ).
‡ See the footnote to Table 41.

qualitative description of vibrational modes, frequently used in this book, based on the assumption of fully characteristic group vibrations (i.e., vibrations fully localized in a given $Si\langle{}^{O^-}_{O^-}$ or $Si{\diagup}^{O}{\diagdown}Si$ unit). Such a description is also given in these tables. The forms of vibrations in mixed rings (Tables 43 and 44) indicate that considerable shifts of vibrational frequencies, as compared with the spectra of the unsubstituted rings Si_3O_9 or Ge_3O_9 are displayed mainly by vibrations localized in Si–O–Si or Ge–O–Ge bridges, and the vibrations of the exocyclic groups $Si\langle{}^{O^-}_{O^-}$ or $Ge\langle{}^{O^-}_{O^-}$ do not change markedly with change in composition of the complex anion. As the latter vary little in the spectra of mixed anions the method of quasi-isotopic substitution seems to be, in general, less fruitful for structures with a low degree of condensation of element–oxygen tetrahedra.

The set of force constants for the complex anions Si_3O_9 and Ge_3O_9, with some additional constants used to describe the force fields of mixed anions, are listed in Table 45. The near equality of the ratios $K_{SiO^-}/K_{SiO(Si)}$ and $K_{GeO^-}/K_{GeO(Ge)}$ obtained from these values may be probably treated as an indication of the limited role of $(d \leftarrow p)\pi$ bonding in the strengthening of lateral $(Si-O^-)$ bonds, otherwise a considerable difference between the relative strengths of $Si-O^-$ and $Ge-O^-$ bonds would be expected because of the lesser tendency of 4d-orbitals to form π-bonds.

An independent confirmation of the assignment of the vibrations of the flat Si_3O_9 anion used above may be obtained by comparison with the spectrum of a similar but nonplanar ring. It is worth mentioning that the existence of such rings was first suggested by examination of the infrared spectrum of walstromite $Ca_2BaSi_3O_9$; the symmetry and lattice parameters of this crystal and of the similar crystal margarosanite, $(Ca, Mn)_2PbSi_3O_9$, were also in accordance with the presence of Si_3O_9 ions [228, 229] but no conclusions on their configuration could be derived from the data. The structures of both crystals have now been determined by x-ray diffraction [230, 231], and the existence of essentially asymmetric Si_3O_9 ions (without a C_3 axis, site symmetry C_1) is definitely established. The infrared spectrum of $Ca_2BaSi_3O_9$ (Fig. 91) contains 12 bands in the region of Si–O stretching vibrations. This is in accordance with the

TABLE 43. The Calculated Frequencies and Forms of Vibrations of Si_2GeO_9 Ion

Assignments

A_1	q^{II}	Q^{II}	q^I	Q^{III}	Q^I	δ^{II}	φ^I	φ^{II}	γ^{III}	δ^I
974	1	-0.253	-0.008	-0.178	0.058	0.452	0.004	-0.417	-0.032	0.008
904	0.082	1	0.250	-0.199	-0.741	-0.360	-0.091	0.124	0.337	0.408
803	-0.013	-0.317	1	0.119	-0.015	0.095	-0.166	-0.049	-0.189	0.034
689	0.013	-0.169	-0.156	1	-0.559	-0.076	-0.039	0.320	-1.241	-0.027
534	-0.057	0.766	0.119	0.656	1	1.108	0.231	1.119	-0.526	0.841
476	-0.133	0.139	-0.029	0.006	-0.106	1	-0.062	-0.885	-0.186	0.458
358	-0.003	0.077	0.036	0.020	0.131	-0.169	1	-0.266	0.049	-0.170
299	0.015	-0.301	-0.035	-0.432	-0.609	0.708	0.862	1	-0.611	-0.373
257	-0.046	-0.137	-0.009	0.744	-0.375	0.383	0.250	-0.370	1	-1.048
205	0.017	-0.323	-0.026	0.159	-0.162	-0.256	0.101	0.050	0.354	1

A_2	q^{II}	γ^{II}	γ^{III}	γ^I	χ^{II}
1058	1	0.531	-0.408	-0.025	0.310
484	-0.059	1	-1.000	-0.630	0.836
316	0.051	-0.031	1	-0.402	0.004
259	-0.011	0.637	0.263	1	-0.200
76	-0.002	0.084	0.018	0.082	1

B_1	q^{II}	q^I	γ^{III}	γ^{II}	γ^I	χ^{II}	χ^I
1057	1	0.003	-0.456	-0.426	-0.023	-0.318	-0.532
852	-0.006	1	0	-0.017	-0.280	0	-0.209
538	0.097	0.007	1	0.274	0.139	0.367	1.206
476	0.039	0.026	-0.397	1	0.715	1.222	-0.324
260	-0.023	0.023	0.068	-0.590	1	0.570	-0.212
87	0.003	0.002	-0.049	-0.063	-0.098	1	0.402
57	0.002	0	-0.100	0	0.061	-0.405	1

B_2	q^{II}	Q^{III}	Q^{II}	Q^I	φ^{II}	γ^{III}	δ^{II}	γ^I
968	1	-0.162	-0.213	-0.031	-0.397	0.011	0.370	0.015
919	-0.005	1	0.210	0.273	0.204	-0.017	-0.468	0.004
867	0.035	-0.545	1	0.665	0.085	0.328	-0.140	0.080
586	0.031	-0.154	-0.484	1	-0.331	-0.813	-0.232	-0.576
442	0.067	-0.098	0.031	-0.118	1	-0.614	-0.174	0.106
396	0.106	-0.099	-0.523	0.541	0.753	1	-0.735	-1.885
251	-0.065	0.186	0.225	0.016	0	-0.482	1	-0.872
224	-0.016	0.160	-0.361	0.593	0.296	0.001	0.706	1

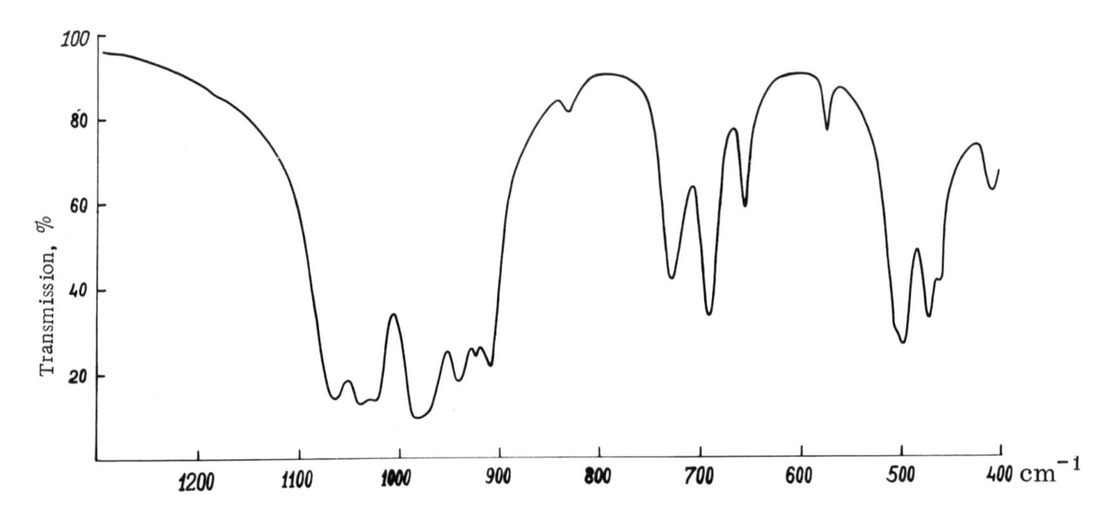

Fig. 91. The infrared spectrum of $Ca_2BaSi_3O_9$ (KBr pellet, 4 mg.).

TABLE 44. The Calculated Frequencies and Forms of Vibrations of $SiGe_2O_9$ Ion

A_1	q^{I}	Q^{I}	q^{II}	Q^{III}	Q^{II}	φ^{I}	φ^{II}	δ^{I}	γ^{III}	δ^{II}
972	1	-0.344	-0.007	-0.002	0.082	-0.420	0.003	0.441	0	0.012
910	0.115	1	0.207	0.004	-0.938	0.220	-0.078	-0.523	-0.055	0.287
805	-0.011	-0.196	1	-0.137	0.052	-0.067	-0.172	0.095	0.045	0.078
514	-0.118	0.225	0.051	1	-0.376	-0.411	0.133	0.026	-1.035	0.589
510	-0.048	0.798	0.119	0.505	1	1.066	0.310	0.934	-0.334	1.237
447	0.122	-0.192	0.030	0.145	0.015	1	0.094	-0.796	-0.178	0.466
358	-0.018	0.059	0.038	0.032	0.110	-0.232	1	-0.165	-0.092	-0.380
277	-0.011	-0.342	-0.065	-0.376	-0.782	0.742	0.906	1	-0.628	0.029
232	0.052	-0.434	-0.007	0.329	-0.289	0.283	-0.008	-0.044	1	0.649
191	0.025	-0.103	-0.028	-0.398	0.162	-0.019	0.122	-0.536	-0.673	1

A_2	q^{II}	γ^{I}	γ^{II}	γ^{III}	χ^{II}
852	1	0.013	-0.200	-0.189	0.164
471	-0.020	1	-0.770	0.004	1.286
254	0.014	0.663	1	-0.131	-0.532
230	0.016	0.068	0.049	1	-0.322
82	-0.001	-0.050	0.110	0.036	1

B_1	q^{I}	q^{II}	γ^{I}	γ^{III}	γ^{II}	χ^{II}	χ^{I}
1055	1	0.004	-0.602	-0.001	-0.033	-0.493	-0.002
852	-0.007	1	-0.011	-0.207	-0.200	-0.164	-0.199
502	0.100	0.020	1	0.009	0.632	1.360	0.053
416	0	0.022	-0.013	1	-0.018	0.057	0.965
270	-0.032	0.016	-0.497	-0.064	1	0.283	0.229
76	0.003	0	-0.086	0.082	-0.064	1	-0.826
55	0.003	0.003	-0.057	-0.118	-0.092	0.806	1

B_2	Q^{I}	q^{II}	Q^{III}	Q^{II}	φ^{II}	γ^{III}	γ^{I}	δ^{II}
892	1	0.198	0.173	0.625	0.014	0.061	-0.411	-0.268
811	-0.369	1	0.514	-0.063	-0.131	0.049	0.239	0.090
732	-0.068	-0.296	1	-0.101	0.113	0.113	0.042	-0.389
610	-0.407	-0.110	-0.092	1	0.167	0.158	1.580	0.084
365	-0.214	-0.006	0	0.322	1	0.374	-0.760	0.162
336	-0.073	-0.021	-0.073	0.112	-1.029	1	-0.198	-0.176
226	0.577	0.062	-0.120	-0.690	0.665	0.969	1	-0.766
193	0.188	-0.013	0.143	-0.130	-0.230	-0.239	0.177	1

expected value for a Si_3O_9 ion, if all degeneracies are removed and all vibrations are infrared-active. It is significant that a single weak band appears at 830 cm^{-1}, slightly lower than a group of 7 or 8 closely spaced bands between 1100 and 900 cm^{-1} corresponding to six ν_{as} and ν_s SiO_2 vibrations and to two ν_{as} SiOSi vibrations generated by the splitting of the E′ vibration (for D_{3h}). The band at 830 cm^{-1} may be attributed to the third ν_{as} SiOSi vibration, species A_2' for D_{3h}, which is in good agreement with the assignment already given for the planar Si_3O_9 ring.† As could be expected, three bands are situated in the region of ν_s SiOSi vibrations because of degeneracy removal and activation of a vibration belonging to the A_1' species in the planar ring. At lower frequencies a single band corresponding to an A_2'' vibration (ρ O^-SiO^-) of the planar ring is readily identified in practically the same position. The doublet near 500 cm^{-1} is

† Remember, that in the infrared spectrum of the low-temperature form of $Sr_3Ge_3O_9$, where the existence of nonplanar and asymmetrical Ge_3O_9 anions is supposed (see §1 of this chapter), a weak band at 725 cm^{-1} may be assigned to the same vibration of the germanate anion.

TABLE 45. The Force Constants of the Complex Anions Si_3O_9, Ge_3O_9, Si_3GeO_9, and $SiGeEO_9$ (units × 10 ^{6} cm^{-2})

X_3O_9*	X = Si	X = Ge
K_{XO^-}	10.7	9.0
$K_{XO(X)}$	5.9	5.1
$K_{(X)OXO(X)}$	1.8	1.4
$K_{(X)OXO^-}$	2.25	1.75
$K_{O^-XO^-}$	2.25	2.0
K_{XOX}	0.4	0.3
$K_{\perp XOX}$	0.2	0.2
$H_{(X)OX, XO(X)}$	0.6	0.4
$H_{XO(X), (X)OX}$	0.3	0.3
h_{O^-X, XO^-}	0.7	0.0
$h_{O^-X, XO(X)}$	1.1	0.0
$A_{XO(X), (X)OXO(X)}$	−0.2	0.0
$A_{XO(X), XOX}$	0.2	0.2
$a_{XO(X), (X)OXO^-}$	0.2	0.2
$a_{XO^-, (X)OXO^-}$	0.4	0.2
a_{XO^-, O^-XO^-}	0.4	0.2

*For the description of the force fields of Si_2GeO_9 and $SiGe_2O_9$ the following force constants were added to the set of corresponding constants of Si_3O_9 and Ge_3O_9 anions: K_{SiOGe} = 0.35, $H_{GeO,OSi}$ = −0.3, $A_{SiO,SiOGe}$ = $A_{GeO,SiOGe}$ = 0.2. All force constants were calculated in units × 10^{-6} cm^{-2} using a scale with unit of masses equal to 1/16th of the oxygen mass and 1 Å as a unit of length.

explained by the splitting and activation in the infrared spectrum of the ρO^-SiO^- vibration of species E″. Three maxima between 400 and 500 cm^{-1} correspond to the two SiOSi and SiO_2 bending vibrations of species E′ and A_1' for the planar ring. Thus, the assignments in the spectrum of the nonplanar Si_3O_9 ring with no three-fold symmetry axis do not contradict the conclusions drawn from the analysis of the spectrum of the more symmetrical anion given above.

Possibility of Using the Method of Quasi-Isotopic Substitutions for the Spectra of Systems Possessing Translational Symmetry. Up to this point we have considered the application of the method of quasi-isotopic substitution for analyzing the spectra of crystals in which the interpretation of the spectra may be confined (with a certain degree of certainty) to the interpretation (on the quasi-molecular approximation) of the spectrum of a complex anion comprising a finite number of atoms. Let us now consider the possibility of employing this method for analyzing the spectra of systems possessing translational symmetry. Such systems include both the "infinite" chain anions of the metasilicates, the layer anions of the disilicates, and the three-dimensional lattice of quartz and the feldspars, and also crystals with anions of the "island" type, but with such high vibrational frequencies of the cation–oxygen bonds that the quasi-molecular approximation is inapplicable to the interpretation of their spectra. As already mentioned, the calculation of the vibrational spectra of such systems, carried out on the assumption that only the limiting frequencies of the branches of the vibrational spectrum are optically active is effected by introducing translational symmetry coordinates transforming the secular equation of infinitely high order into an equation of order 3n−3 (3n−4 for the one-dimensional case). In order to increase the objectivity of the calculations it seems desirable here also to introduce data relating to the spectra of the quasi-isotopically substituted derivatives; however, the necessity of preserving translational symmetry (as a condition of the applicability of the limiting-frequency method) demands that a model representing an "ordered mixed" crystal should be used in the calculation. This expression means a mixed crystal, in each cell of which some of the positions formerly occupied by X′ atoms are filled (with a whole-number occupation factor) by X″ atoms in such a fashion that all the cells remain translationally equivalent.

It should be noted that the dynamics of isotopically disordered crystals have not been theoretically studied in adequate measure, particularly for cases in which the impurity concentration is not small; most investigations have been restricted to a consideration of simple one-dimensional models. A natural experimentally studied example of the spectroscopic effects of quasi-isotopic substitution in a simple silicate system is given by the spectra of Li_2SiO_3–Li_2GeO_3 solid solutions described earlier.

It follows from the results of this investigation that, despite the fact that the spectra of two- and three-dimensional tetrahedral lattices with partial replacement of Si by Ge, or chains with more than two tetrahedra in the identity period, have not been studied experimentally, we may yet expect that the spectra of such systems will have a reasonably discrete character if two fundamental conditions are satisfied: 1) the number of Si(Ge) atoms in the unit cell of the one-, two-, or three-dimensional lattice considered is not large, and 2) all these atoms lie in the same crystallographic set of positions. If these conditions are satisfied, the method of quasi-isotopic substitution may probably be used, if not for refining the force constants, then at least for checking the interpretation of the spectrum. Substances which might well be studied spectroscopically by this method include, for example, the plane hexagonal $(Si_2O_5)_{\infty,\infty}$ lattice of the talc type and possibly the three-dimensional $(Si_3O_6)_{\infty,\infty,\infty}$ lattice of quartz. As regards the error in frequency calculations associated with the use of the limiting vibration frequencies of a regularly mixed crystal instead of the experimentally observed bands in the spectrum of a crystal having a disordered distribution of the two types of atoms, some idea of the magnitude of this error may probably be obtained from the range of variation in the frequencies of the "impurity" bands observed on varying the concentrations of the X′ and X″ atoms over a wide field; the error committed in replacing the disordered model by the ordered–mixed model is always smaller than this range of variation.

References

1. A. N. Lazarev, T. F. Tenisheva, and R. G. Grebenshchikov, Dokl. Akad. Nauk SSSR, 140:811 (1961).
2. F. Liebau, Neues Jb. Miner., 94:1209 (1960).
3. A. N. Lazarev and T. F. Tenisheva, Opt. i Spektr., 11:584 (1961).
4. G. P. Stavitskaya and Ya. I. Ryskin, Opt. i Spektr., 10:343 (1961).
5. A. N. Lazarev, Opt. i Spektr., 12:60 (1962).
6. J. W. Jeffery and L. Heller, Acta Crystallogr., 6:807 (1953).
7. W. Hilmer, Naturwiss., 45:238 (1958).
8. F. Liebau, Acta Crystallogr., 10:790 (1957).
9. A. N. Lazarev, T. F. Tenisheva, Mao Chzhi-Tsyun, and N. A. Toropov, Zh. Strukt. Khim., 2:741 (1961).
10. W. Hilmer, Acta Crystallogr., 15:1101 (1962).
11. G. Eulenberger, A. Wittmann, and H. Nowotny, Mh. Chem., 93:1046 (1962).
12. F. Liebau, Z. Phys. Chem., 206:73 (1956).
13. A. E. Ringwood and M. Seabrook, J. Geophys. Res., 68:4601 (1963).
14. Yu. I. Smolin, Dokl. Akad. Nauk SSSR, 181:595 (1968).
15. R. S. Roth and E. M. Levin, J. Res. Nat. Bur. Standards, 62:193 (1959).
16. R. G. Grebenshchikov and N. A. Toropov, Dokl. Akad. Nauk SSSR, 142:392 (1962).
17. T. Ito, Z. Kristallogr., 88:142 (1934).
18. R. A. Thomas, J. Amer. Ceram. Soc., 33:35 (1950).
19. R. G. Grebenshchikov and N. A. Toropov, Izv. Akad. Nauk SSSR, Otd. Khim. Nauk, 545 (1962).
20. H. Katscher and F. Liebau, Naturwiss., 52:512 (1965).
21. I. A. Bondar', T. F. Tenisheva, Yu. F. Shepelev, and M. N. A. Toropov, Dokl. Akad. Nauk SSSR, 160:1069 (1965).
22. A. Durif-Varambon and J. F. Lajzerowicz, Bull. Soc. Franc. Miner. Crist., 86:88 (1963).
23. I. Naray-Szabo, Phys. Chem. of Glasses, 4:38A (1963).
24. N. D. Sokolov, Usp. Fiz. Nauk, 57:205 (1955).
25. G. C. Pimentel and A. L. McClellan, The Hydrogen Bond, Freeman & Co., San Francisco (1960).
26. A. R. Ubbelohde and K. J. Gallagher, Acta Crystallogr., 8:71 (1955).
27. C. G. Cannon, Spectrochim. Acta, 10:341 (1958).
28. W. Lyon and E. L. Kinsey, Phys. Rev., 61:482 (1942).

29. Ya. I. Ryskin, G. P. Stavitskaya, and N. A. Toropov, Zh. Neorg. Khim., 5:2727 (1960).
30. H. G. Hafele, Z. Phys., 148:262 (1957).
31. A. N. Lazarev, Izv. Akad. Nauk SSSR, Ser. Fiz., 21:322 (1957).
32. J. Cabannes, L. Couture, and J.-P. Matheiu, J. Chem. Phys., 50:C89 (1953).
33. A. I. Stekhanov, Dokl. Akad. Nauk SSSR, 92:281 (1953).
34. M. Haas and G. B. B. M. Sutherland, Proc. Roy. Soc., A236:427 (1956).
35. Ya. I. Ryskin and G. P. Stavitskaya, Izv. Akad. Nauk SSSR, Otd. Khim. Nauk, 793 (1963).
36. G. V. Yukhnevich, Usp. Khim., 32:1397 (1963).
37. E. Hartert and O. Glemser, Z. Elektrochem., 60:746 (1956).
38. C. C. Ferriso and D. F. Hornig, J. Chem. Phys., 23:1464 (1955).
39. M. Falk and P. A. Giguere, Canad. J. Chem., 35:1195 (1957).
40. R. C. Taylor and G. L. Vidale, J. Amer. Chem. Soc., 78:5999 (1956).
41. D. J. Millen and E. G. Vaal, J. Chem. Soc. (L.), 2913 (1956).
42. R. Savoie and P. A. Giguere, J. Chem. Phys., 41:2698 (1964).
43. A. F. Burns and J. L. White, Science, 141:800 (1963).
44. G. B. Bokii and D. K. Arkhipenko, Zh. Strukt. Khim., 3:697 (1962).
45. R. C. Lord and R. E. Merrifield, J. Chem. Phys., 21:166 (1953).
46. K. Nakamoto, M. Margoshes, and R. E. Rundle, J. Amer. Chem. Soc., 77:6480 (1955).
47. R. E. Rundle and M. Parasol, J. Chem. Phys., 20: 1487 (1952).
48. G. C. Pimentel and C. H. Sederholm, J. Chem. Phys., 24:639 (1956).
49. E. Hartert and O. Glemser, Naturwiss., 40: 199 (1953).
50. O. Glemser, Angew. Chem., 73:785 (1961).
51. E. Schwarzmann, Z. Anorgan. und Allgem. Chem., 317:176 (1962).
52. H. G. Schnering, Z. Anorgan. und Allgem. Chem., 330:170 (1964).
53. H. K. Welsh, J. Chem. Phys., 26:710 (1957).
54. C. Reid, J. Chem. Phys., 30:182 (1959).
55. O. Glemser, Nature, 183:1476 (1959).
56. O. Glemser and E. Hartert, Z. Anorgan. und Allgemein. Chem., 283:111 (1956).
57. V. A. Kolesova and Ya. I. Ryskin, Opt. i Spektr., 7:261 (1959).
58. V. A. Kolesova and Ya. I. Ryskin, Zh. Strukt. Khim., 3:680 (1962).
59. R. F. Klevtsova and P. V. Klevtsov, Zh. Strukt. Khim., 5:860 (1964).
60. Ch. Cabannes-Ott, Compt. Rend., 242:2825 (1956).
61. P. Tarte, Spectrochim. Acta, 13:107 (1958).
62. D. Hadzi and M. Pintar, Spectrochim. Acta, 12:162 (1958).
63. H. Beutelspacher, Sixth Internat. Conf. on Soil Science, Report B (1956), p. 329.
64. V. Stubican and R. Roy, Z. Kristallogr., 115:200 (1961).
65. W. Vedder and R. S. McDonald, J. Chem. Phys., 38:1583 (1963).
66. V. C. Farmer, Miner. Mag., 31:829 (1958).
67. V. C. Farmer and J. D. Russell, Spectrochim. Acta, 20:1149 (1964).
68. J. M. Serratosa and W. F. Bradley, J. Phys. Chem., 62:1164 (1958).
69. V. Stubican and R. Roy, J. Amer. Ceram. Soc., 44:625 (1961).
70. W. Vedder, Amer. Mineralogist, 49:736 (1964).
71. J. T. Braunholtz, G. E. Hall, F. G. Mann, and N. Sheppard, J. Chem. Soc. (L.), 868 (1959).
72. P. Tarte, Hydrogen Bonding, Pergamon Press (1959), p. 119.
73. P. S. Narayanan, Proc. Indian Acad. Sci., A33:240 (1951).
74. Ya. I. Ryskin and G. P. Stavitskaya, Opt. i Spektr., 8:606 (1960).
75. Ya. I. Ryskin, Opt. i Spektr., 12:518 (1962).
76. M. A. Kovner and V. N. Kapshtal', Izv. Akad. Nauk SSSR, Ser. Fiz., 17:561 (1953).
77. R. Blinc and D. Hadzi, Molec. Phys., 1:391 (1958).
78. A. I. Stekhanov, Dokl. Akad. Nauk SSSR, 106:433 (1956).
79. R. G. Snyder and J. A. Ibers, J. Chem. Phys., 36:1356 (1962).
80. E. A. Pshenichnov and N. D. Sokolov, Opt. i Spektr., 11:16 (1961).

81. H. Siebert, Z. Anorgan. und Allgemein Chem., 303:162 (1960).
82. D. Hadzi, J. Chem. Phys., 34:1445 (1961).
83. A. N. Lazarev and A. S. Zaitseva, Fiz. Tverd. Tela, 2:3026 (1960).
84. Ya. I. Ryskin, Opt. i Spektr., 7:278 (1959).
85. Ya. I. Ryskin and G. P. Stavitskaya, Opt. i Spektr., 7:834 (1959).
86. Ya. I. Ryskin and G. P. Stavitskaya, Izv. Akad. Nauk SSSR, Otd. Khim. Nauk, 793 (1963).
87. Ya. I. Ryskin, G. P. Stavitskaya, and N. A. Mitropol'skii, Izv. Akad. Nauk SSSR, Ser. Khim., 416 (1964).
88. H. Siebert, Z. Anorgan. und Allgemein. Chem., 275:225 (1954).
89. L. Heller, Acta Crystallogr., 5:724 (1952).
90. H. D. Megaw, Acta Crystallogr., 5:477 (1952).
91. H. E. Petch, N. Sheppard, and H. D. Megaw, Acta Crystallogr., 9:29 (1956).
92. H. Nowotny, Usp. Khim., 27:996 (1958).
93. H. Nowotny and G. Szekely, Monatshefte Chem., 83:568 (1952).
94. K. H. Jost and W. Hilmer, Acta Crystallogr., 21:583 (1966).
95. E. Thilo and W. Miedreich, Z. Anorgan. und Allgemein. Chem., 267:76 (1951).
96. F. Miller and C. Wilkins, Anal. Chem., 24:1253 (1952).
97. F. A. Miller, G. L. Carlson, F. F. Bentley, and W. H. Jones, Spectrochim. Acta, 16:135 (1960).
98. F. Liebau, Z. Phys. Chem., 206:73 (1957).
99. M. Buerger, Z. Kristallogr., 108:248 (1956).
100. A. N. Lazarev and T. F. Tenisheva, Opt. i Spektr., 12:215 (1962).
101. Kh. S. Mamedov and N. V. Belov, Dokl. Akad. Nauk SSSR, 123:741 (1958).
102. Ya. I. Ryskin, Trans. Fifth Conf. on Experimental and Technical Mineralogy and Petrography, Izd. AN SSSR, Moscow (1958), p. 56.
103. N. M. Bazhenov, A. I. Kol'tsov, N. P. Kirpichnikova, Ya. I. Ryskin, G. P. Stavitskaya, A. I. Boikova, and N. A. Toropov, Izv. Akad. Nauk SSSR, Ser. Khim., 409 (1964).
104. H. Scholze, Glastechn. Ber., 32:81, 142 (1959).
105. H. Scholze, Glastechn. Ber., 34:278 (1961).
106. H. Scholze and H. Franz, Glastechn. Ber., 35:278 (1962).
107. A. F. Fray and S. Nielsen, Infrared Phys., 1:175 (1962).
108. R. V. Adams and R. W. Douglas, J. Soc. Glass Techn., 43:147 (1959).
109. V. A. Florinskaya and R. S. Pechenkina, in: Structure of Glass, Izd. AN SSSR, Moscow-Leningrad (1955), p. 70.
110. A. N. Lazarev, in: The Vitreous State, Izd. AN SSSR, Moscow-Leningrad (1960), p. 239.
111. G. Donnay, J. Wyart, and G. Sabatier, Z. Kristallogr., 112:161 (1959).
112. Internat. Tables for X-Ray Crystallography, Vol. III, Kynoch Press, Birmingham (1962).
113. P. A. Akishin and V. P. Spiridonov, Dokl. Akad. Nauk SSSR, 131:557 (1960).
114. J. Krogh-Moe, Acta Chem. Scand., 17:843 (1963).
115. C. E. Weir and R. A. Schroeder, J. Res. Nat. Bur. Standards, 68A:465 (1964).
116. D. Bethell and N. Sheppard, Trans. Faraday Soc., 51:9 (1955).
117. R. R. Servoss and H. M. Clark, J. Chem. Phys., 26:1175 (1957).
118. D. F. Hornig and R. C. Plumb, J. Chem. Phys., 26:637 (1957).
119. H. Becher, Z. Phys. Chem., 2:276 (1954).
120. R. Werner and K. O'Brien, Austral. J. Chem., 8:355 (1955).
121. R. R. Servoss and H. M. Clark, J. Chem. Phys., 26:1179 (1957).
122. L. J. Bellamy, W. Gerrard, M. F. Lappert, and R. L. Williams, J. Chem. Soc., (L.), 2412 (1958).
123. J. Goubeau and J. W. Ewers, Z. Phys. Chem. (BRD), 25:276 (1960).
124. H. J. Becher, W. Sawodny, H. Noth, and W. Meister, Z. Anorgan. und Allgemein. Chem., 314:226 (1962).
125. V. M. Tatevskii, G. S. Koptev, and A. A. Mal'tsev, Opt. i Spektr., 11:724 (1961).

126. M. V. Akhmanova, Zh. Strukt. Khim., 3:28 (1962).
127. H. Moenke, Silikattechn., 13:287 (1962).
128. P. Jellyman and J. Procter, J. Soc. Glass Technol., 39:173 (1955).
129. J. Krogh-Moe, Arkiv for Kemi, 14:567 (1959).
130. J. Krogh-Moe, Phys. Chem. Glasses, 3:1 (1962).
131. J. Goubeau and D. Hummel, Z. Phys. Chem. (BRD), 20:15 (1959).
132. J. Goubeau and H. Kalfass, Z. Anorgan. und Allgemein. Chem., 299:160 (1959).
133. F. Dachille and R. Roy, Z. Kristallogr., 111:462 (1959).
134. J. R. Van Wazer, Phosphorus and Its Compounds, Vol. I., Chemistry, Interscience, New York (1958).
135. S. M. Stishov and N. V. Belov, Dokl. Akad. Nauk SSSR, 143:951 (1962).
136. R. J. P. Lyon, Nature, 196:266 (1962).
137. G. R. Levi and G. Peyronel, Z. Kristallogr., 92:190 (1935).
138. G. Peyronel, Z. Kristallogr., 94:311 (1936).
139. E. Steger and G. Leukroth, Z. Anorgan. und Allgemein. Chem., 303:169 (1960).
140. J. Lecomte, A. Boulle, C. Doremieux-Morin, and B. Lelong, Compt. Rend., 258:1447 (1964).
141. J. Lecomte, A. Boulle, C. Doremieux-Morin, and B. Lelong, Compt. Rend., 258:131 (1964).
142. C. R. Robbins and E. M. Levin, J. Res. Nat. Bur. Standards, 65A:127 (1961).
143. N. Ingri and G. Lundgren, Acta Chem. Scand., 17:607 (1963).
144. E. R. Lippincott, A. Van Valkenburg, C. E. Weir, and E. N. Bunting, J. Res. Nat. Bur. Standards, 61:61 (1958).
145. V. V. Obukhov-Denisov, N. N. Sobolev, and V. P. Cheremisinov, Opt. i Spektr., 8:505 (1960).
146. D. Tarte and A. E. Ringwood, Nature, 201:819 (1964).
147. M. K. Murthy and E. M. Kirby, Phys. Chem. Glasses, 5:144 (1964).
148. H. P. Rooksby, Nature, 159:609 (1947).
149. S. Andersson and A. D. Wadsley, Nature, 187:499 (1960).
150. P. S. Narayanan, Proc. Indian Acad. Sci., 37:411 (1953).
151. Ya. S. Bobovich and E. V. Bursian, Opt. i Spektr., 11:131 (1961).
152. S. Ikegami, J. Phys. Soc. Japan, 12:46 (1964).
153. Ya. S. Bobovich, Opt. i Spektr., 10:418 (1961).
154. Ya. S. Bobovich, Opt. i Spektr., 13:459 (1962).
155. P. Tarte, Nature, 191:1002 (1961).
156. J. T. Last, Phys. Rev., 105:1740 (1957).
157. P. Tarte, Bull. Acad. Royale Sci. Belge, 169 (1960).
158. P. Tarte, Bull. Soc. Frac. de Ceram., No. 58, 13 (1963).
159. H. Kriegsmann and K. Licht, Z. Elektrochem., 62:1163 (1958).
160. T. Takatani, T. Yoshimoto, and Y. Mashiko, J. Chem. Soc. Japan; Indust. Chem. Sect., 60:1382 (1958).
161. V. A. Zeitler and C. A. Brown, J. Phys. Chem., 61:1174 (1957).
162. D. C. Bradley, R. Gaze, and W. Wardlaw, J. Chem. Soc. (L.), 3977 (1955).
163. A. N. Nesmeyanov, O. V. Nogina, and R. Kh. Freidlina, Izv. Akad. Nauk SSSR, Otd. Khim. Nauk, 373 (1956).
164. P. Tarte, Nature, 186:234 (1960).
165. P. Tarte, Silicates Industriels, 25:171 (1960).
166. Ya. S. Bobovich, Dokl. Akad. Nauk SSSR, 145:1028 (1962).
167. Ya. S. Bobovich, Opt. i Spektr., 13:492 (1962).
168. Ya. S. Bobovich, Opt. i Spektr., 14:647 (1963).
169. Ya. S. Bobovich and G. T. Petrovskii, Zh. Strukt. Khim., 4:765 (1963).
170. V. V. Bakakin and N. V. Belov, Geokhimiya, 91 (1964).
171. P. Tarte, Silicates Industriels, 28:345 (1963).

172. P. Tarte, Revue Universelle des Mines, 18:384 (1962).
173. V. A. Kolesova, Izv. Akad. Nauk SSSR, Otd. Khim. Nauk, 2082 (1962).
174. W. R. Busing and H. A. Levy, Acta Crystallogr., 11:798 (1958).
175. P. R. Reichertz and M. J. Yost, J. Chem. Phys., 14:495 (1946).
176. M. J. Buerger, Amer. Mineralogist, 39:600 (1954).
177. V. A. Kolesova, Opt. i Spektr., 10:414 (1960).
178. T. F. Barth, J. Chem. Phys., 3:323 (1935).
179. M. W. Dougill, Nature, 180:292 (1957).
180. S. Wallmark and A. Westgrem, Arkiv. Kemi, Mineralogi, och Geologi, 12B:1 (1937).
181. H. A. Lehmann and H. Hesselbarth, Z. Anorgan. und Allgemein. Chem., 313:117 (1961).
182. M. Linevsky, D. White, and D. Mann, J. Chem. Phys., 41:542 (1964) (see also: A. A. Mal'tsev and V. F. Shevel'kov, Teplofiz. Vys. Temp., 2:650 (1964).
183. V. A. Kolesova, Opt. i Spektr., 6:38 (1959).
184. V. A. Kolesova, in: The Vitreous State, Izd. AN SSSR, Moscow-Leningrad (1960), p. 203.
185. R. J. P. Lyon and W. M. Tuddenham, Nature, 185:374 (1960).
186. P. Tarte, Silicates Industriels, 24:7 (1959).
187. C. W. Burnham and M. J. Buerger, Z. Kristallogr., 115:269 (1961).
188. C. W. Burnham, Z. Kristallogr., 118:337 (1963).
189. C. W. Burnham, Z. Kristallogr., 118:127 (1963).
190. V. A. Kolesova, Izv. Akad. Nauk SSSR, Ser. Khim., 669 (1966).
191. P. Tarte, Spectrochim. Acta, 20:238 (1964).
192. P. Tarte, Spectrochim. Acta, 16:228 (1963); 21:313 (1965).
193. W. B. White and R. Roy, Amer. Mineralogist, 49:1670 (1964).
194. A. A. Maradudin, E. W. Montroll, and G. H. Weiss, Theory of Lattice Dynamics in the Harmonic Approximation, Academic Press, New York (1963).
195. F. Krueger, O. Reinkober, and E. Koch-Holm, Ann. Physik, 85:111 (1928).
196. M. F. Vuks, Acta Phys.-Chim. URSS, 6(11):327 (1937).
197. A. V. Korshunov and V. A. Sel'kin, Zh. Eksp. Teor. Fiz., 23:576 (1952).
198. S. Bhagavantam, Proc. Ind. Acad. Sci., A13:543 (1941).
199. B. D. Saksena, J. Chem. Phys., 18:1653 (1950).
200. N. N. Profir'eva, Zh. Eksp. Teor. Fiz., 27:439 (1954).
201. M. K. Raju, Proc. Ind. Acad. Sci., A22:150 (1945).
202, G. Schaefer, J. Phys. Chem. Solids, 12:233 (1960).
203. B. D. Saksena, J. Phys. Chem. Solids, 26:1929 (1965).
203a. H. W. Verleur and A. S. Barker, Phys. Rev., (2), 164:1169 (1967).
203b. I. F. Chang and S. S. Mitra, Phys. Rev., (2), 172:924 (1968).
203c. A. S. Barker, J. A. Ditzenberger, and H. J. Guggenheim, Phys. Rev., 175:1180 (1968).
203d. H. B. Rosenstock and R. E. McGill, Phys. Rev., 176:1004 (1968).
203e. R. F. Wallis (ed.), Proc. of Internat. Conf. on Localized Excitation in Solids, Plenum Press, New York (1968).
203f. W. Hass, H. B. Rosenstock, and R. F. McGill, Solid State Commun., 7:1 (1969).
204. J. Lajzerowicz, Bull. Soc. Franc. Miner. Crist., 87:520 (1964).
205. P. Tarte, Étude Experimentale et Interpretation du Spectre Infra-rouge des Silicates et des Germanates. Application à des Problèmes Relatifs a l'État Solide, Acad. Royale de Belgique, Mémoires, Vol. XXXV, Nos. 4a and 4b, Brussels (1965).
206. A. N. Lazarev and T. F. Tenisheva, Opt. i Spektr., 13:708 (1962).
207. W. H. Taylor, "Framework Silicates: the Feldspars" in: Crystal Structures of Minerals (L. Bragg and G. F. Claringbull, eds.), G. Bell and Sons, London (1965), p. 293.
208. Physics of Minerals (Collection of articles), Mir, Moscow (1964).
209. R. B. Ferguson, R. J. Traill, and W. H. Taylor, Acta Crystallogr., 11:331 (1958).
210. E. Brun, St. Hafner, P. Hartmann, F. Laves, and H. Staub, Z. Kristallogr., 113:65 (1959).

211. F. Laves and S. Hafner, Z. Kristallogr., 108:52 (1956).
212. S. Hafner and F. Laves, Z. Kristallogr., 109:204 (1957).
213. H. Moenke, Mineralspektren, Acad. Verlag, Berlin (1962).
214. W. H. Taylor, Z. Kristallogr., 74:1 (1930).
215. D. S. Coombs, Miner. Mag., 30: 705 (1955).
216. R. Sadanaga, N. Tokonami, and V. Takeuchi, Acta Crystallogr., 15:65 (1962).
217. D. Hafner and F. Laves, Z. Kristallogr., 115:321 (1961).
218. H. Völlenkle, A. Wittman, and H. Nowotny, Z. Kristallogr., 126:37 (1968).
219. K. H. Jost, H. Wolf, and E. Thilo, Z. Anorg. und Allgemein. Chem., 353:42 (1967).
220. K. H. Jost, H. Worzala, and E. Thilo, Z. Anorg. und Allgemein. Chem., 325:99 (1963).
221. H. Worzala, Z. Anorg. und Allgemein. Chem., 353:48 (1967).
222. D. R. Peacor, Z. Kristallogr., 126:299 (1968).
223. I. Grunze, K. Dostal, and E. Thilo, Z. Anorg. und Allgemein. Chem., 302:222 (1959).
224. B. D. Saksena, K. S. Agarwal, and G. S. Jauchri, Trans. Faraday Soc., 59:276 (1963).
225. A. M. Prima, Opt. i Spektr., 9:452 (1960).
226. R. G. Grebenshchikov, A. K. Shirvinskaya, and V. I. Shitova, Neorgan. Materialy, 4:1376 (1968).
227. J.-P. Labbe, Thesis, Univ. of Paris (1965).
228. F. P. Glasser and L. S. Dent Glasser, Amer. Mineralogist, 49:781 (1964).
229. F. P. Glasser and L. S. Dent Glasser, Z. Kristallogr., 116:263 (1961).
230. L. S. Dent Glasser and F. P. Glasser, Amer. Mineralogist, 53:9 (1968).
231. R. L. Freed and D. R. Peacor, Z. Kristallogr., 128:213 (1969).
232. H. Moenke, Naturwiss., 51:239 (1964).
233. R. A. Edge and H. F. W. Taylor, Nature (London), 224:363 (1969).

CHAPTER V

SPECTRA AND CRYSTAL CHEMISTRY OF THE SILICATES OF THE RARE-EARTH ELEMENTS

Let us continue our consideration of the prospects of using spectroscopic methods in crystal-chemical investigations, taking as an example one of the newest classes of compounds, the silicates of the rare-earth elements. Systematic investigations into these compounds (as yet only a few have been published) have been carried out for a number of years in the Institute of Silicate Chemistry of the Academy of Sciences of the USSR. The infrared spectra have been used as one of the principal methods of identifying new compounds, establishing the structure of complex anions, classifying materials as regards crystal structure, studying isomorphous relationships and polymeric transformations, and qualitatively analyzing mixtures of several crystal phases. Interest in the compounds of rare-earth elements has risen particularly rapidly in recent years. For this reason, on the one hand, many new types of rare-earth compound have been synthesized and studied, and on the other hand more importance has become attached to the polymorphism of these compounds and their isomorphism with the salts of other cations, this being vital in connection with the introduction of rare-earth cations into the lattices of the latter. It is thus essential not only to extend current x-ray research into the salts of the rare-earth elements but also to develop less troublesome methods suitable for at least a preliminary study and classification of a large number of new compounds. In this respect infrared spectroscopy is not only in no way inferior to x-ray phase analysis but in some ways better, as it enables the structure of the complex anion to be established.

In addition to their potential practical value, the silicates of the rare-earth elements are of considerable interest in elucidating certain general laws of silicate crystal chemistry. Compounds of the rare-earth elements have for a long time been traditional objects of crystal-chemical investigations. The similarity between the chemical properties of the triply-charged Ln^{3+} cations (Ln = one of the lanthanide elements), due to the similar structure of their outer electron shells, together with the successive diminution in their sizes on passing along the La^{3+} to Lu^{3+} series associated with the so-called lanthanide contraction, arouses particular interest in the salts of these elements in connection with elucidating criteria for the stability of various structures and types of coordination as functions of the ratio of the ionic radii. Owing to the effect of the lanthanide contraction, we can compare the structure of rare-earth compounds with those of the corresponding elements in the preceding periods, yttrium and scandium, which may also be of substantial interest. The variation in the ionic radii of the Ln^{3+} cations, which is almost exactly linearly dependent on the atomic number of the element, may be illustrated by the sequence of values in the Zachariasen system of ionic radii (Table 46). As the ionic radii diminish, the binding energy of the valence electrons increases, and this is reflected by a rise in the ionization potentials, heats of formation, and heats of hydration of the Ln^{3+} ions, and hence by a fall in the basicity of the lanthanide series. The basicity of yttrium and scandium agrees with the values of their atomic radii. A change in basicity, i.e., a change in the proportion of covalence in the Ln–O bonds, is one of the factors (together

TABLE 46. Ionic Radii of the Rare-Earth Elements, Yttrium, and Scandium, According to Zachariasen

Ln^{3+} ion	Ionic radius, Å	Ln^{3+} ion	Ionic radius, Å
La	1.04	Dy	0.91
Ce	1.02	Ho	0.89
Pr	1.00	Er	0.87
Nd	0.99	Tm	0.86
Sm	0.97	Yb	0.85
Eu	0.96	Lu	0.84
Gd	0.94	Y	0.88
Tb	0.92	Sc	0.68

with the change in ionic radii) determining the structural characteristics of various types of salts of the rare-earth elements.

The existence of an internal periodicity in the properties of the rare-earth elements, depending on the sequence governing the occupation of the 4f states and their competition with the 5d and (partly) the 6s levels, has comparatively little influence on the structure of the rare-earth salts, owing to the strong screening of the 4f electrons, although this periodicity is clearly manifested, for example, in their paramagnetism or in the distribution of anomalous (di- and quadrivalent) states of the rare-earth atoms. The effect of this factor on the structure of the rare-earth compounds probably increases as the bonds become more and more covalent.

It follows from the discussion of the previous chapters that the use of spectroscopic methods for studying the crystals of silicates and similar compounds is appropriate at two stages of investigation: 1) in order to obtain an initial characterization of new materials so as to classify these and estimate the structure of the complex anion (particularly important if for any reason subsequent x-ray structural analysis proves impossible), and also in order to gain a rough estimate of the force constants, and 2) after establishing the exact spatial configuration of the crystal by diffraction methods, to analyze the spectrum by the method of normal coordinates in order to obtain more reliable values of the force constants, to establish the relation between the characteristics of the lattice dynamics and other physical properties of the crystal, and so on. The majority of the spectroscopic investigations of rare-earth silicates described in this chapter are confined to problems of the first type.

At the present time x-ray structural investigations conducted by the staff of the Institute of Silicate Chemistry [1-6a] and a number of other authors have established the structures of the majority of silicates and some of the germanates of the rare-earth elements. Hence wherever possible conclusions obtained from a preliminary study of the infrared spectra are compared with the results of x-ray structural analyses, so as to be able to judge the advantages of the spectroscopic method more objectively and also appreciate the limitations of its use.

The Use of Infrared Spectra in Studying the Oxides and Other Oxide Compounds of Rare-Earth Elements (R.E.) is usually concerned with establishing the nature of these compounds, in particular identifying different types of structure formed in the La-to-Lu series. In only a few cases are attempts made at interpreting the spectra; these are usually restricted to identifying the frequencies of the internal vibrations of the complex anion and to a qualitative analysis of their dependence on the nature of the Ln^{3+} cation.

The infrared spectra of polycrystalline samples of all three forms (A, B, and C) of Ln_2O_3 oxides stable at relative low temperatures were studied in [7-10], including the long-wave part of the spectrum (up to 50 cm^{-1}). The strong absorption bands in these materials lie at frequencies below 600 cm^{-1}; among compounds of the same structural type some of them show a systematic displacement toward higher frequencies as the ionic radius of the cations diminishes, almost inversely proportional to the unit-cell dimensions. No attempts have been made at interpreting the spectra of the R.E. oxides. The spectra of the R.E. hydroxides have mainly been aimed at determining the nature (degree of coupling, length of the hydrogen bond, etc.) of the hydroxyl groups.

Among the salts of oxygen-containing acids with R.E. cations, those studied in most detail have been hydrated and anhydrous nitrates, the infrared spectra of which have also been obtained in the long-wave region, some of the measurement being made on single crystals [11-16]. The most important result of these investigations is concerned with the influence of the Ln^{3+} cation on the structure of the complex anion: passing from the alkali and alkaline-earth nitrates to the nitrates of the R.E. leads to a distortion in the configuration of the NO_3^- anion, and a reduction in symmetry from D_{3h} to C_{2v}, associated with the rise in the polarization of the anion by the Ln^{3+} cations forming partially covalent bonds with the oxygen. As a measure of this effect one usually employs the splitting of the degenerate N–O stretching vibration and the frequency displacement of the totally symmetric vibration. The marked increase in the distortion of the configuration of the NO_3^- ion with increasing atomic number of the R.E. cation, i.e., with decreasing ionic radius, cannot be completely explained as being simply due to an increase in the electrostatic field, but also involves the fact that the Ln–O bonds become more and more covalent on passing along the series from La to Lu.

Recently infrared spectra have been extensively employed in studying the borates of the R.E. [17-21]. The compounds $LnBO_3$ (Ln = La–Lu, Y, Sc) form three types of crystal structures at relatively low temperatures; in the range La to Nd these borates are structurally similar to aragonite; in the range Sm to Lu there is another structure, known as pseudovaterite from x-ray data, while $ScBO_3$ (and $LuBO_3$ below 1310°C) crystallize with the calcite structure. This has led to the conclusion that the principal part in structure determination is played by the ratio $r_{Ln^{3+}}/r_{O^{2-}}$, which gives the aragonite structure for a value of over 0.71, pseudovaterite between 0.71 and 0.61, and calcite for 0.60_7 or under [22]. Raising the temperature is equivalent to increasing the ionic radius of the cation as regards its effect on the structure, as indicated by the dimorphism of $LuBO_3$. (In addition to this, at high temperatures $LaBO_3$ and $NdBO_3$ form differing high-temperature types of crystal [18], while borates of the pseudovaterite type pass into a different form with a structure similar to vaterite.) The absence of bands with frequency of over 1100 cm^{-1} from the spectra of pseudovaterite borates indicated a transformation to tetrahedral coordination of boron with respect to oxygen, in contrast to borates of other types, while further analysis of the spectra suggested the existence of B–O–B bonds between the BO_4 tetrahedra. On the basis of a combined consideration of x-ray and spectroscopic data, a structure containing rings of three interconnected tetrahedra was proposed for these compounds, although no detailed identification was presented for the bands in the spectra.† In the spectra of the borates of the pseudovaterite type, embracing a wide range of Ln^{3+} cations, there is a substantial displacement (up to several tens of cm^{-1}) of the bands to higher frequencies on passing from Sm to Yb, even in the range of the B–O stretching frequencies. There is also a practically linear dependence of the frequencies on the radius of the cation; the increase in the mass of the cation in this series has no visible effect on the spectrum.

Anhydrous $LnPO_4$ orthophosphates form two structural types with structures similar to monazite $CePO_4$ and xenotime YPO_4. Their infrared spectra [23] in the frequency range of the $[PO_4]^{3-}$ stretching vibrations enable the two types of compounds to be clearly identified, while the number of bands corresponds to that expected from symmetry considerations: in the monazite crystals with $P2_1/c–C_{2h}^5$ symmetry the PO_4 ions lie in a general position (C_1), which leads to the complete removal of degeneracy for the $\nu_{as} PO_4(F_2)$ vibration and the activation of the $\nu_s PO_4(A_1)$ vibration; in the xenotime crystal, with symmetry $I4_1/amd–D_{4h}^{19}$, the $[PO_4]^{3-}$ ions have local symmetry V_d, which leads to a partial removal of degeneracy for $\nu_{as} PO_4$ ($F_2 \rightarrow E + B_2$), both components being active in the infrared spectrum, while the $\nu_s PO_4$ vibration remains

† These views were criticized in [22a].

inactive. Correspondingly, in the spectra of phosphates with the structure of monazite there are four, but for phosphates of the xenotime type only two, νP–O bands. The influence of the Ln^{3+} cations appears as an increase in the vibration frequencies of the spectra of phosphates of both types on passing from La to Eu and Gd to Lu, and also as an increase in the asymmetry of the PO_4 tetrahedra, which changes in the same sequence and manifests itself as an increase in the $\nu_{as} PO_4$ splitting; in the case of the monazite-type crystals the intensity of the $\nu_s PO_4$ band also becomes greater.

The infrared spectra of anhydrous orthosulfates $Ln_2(SO_4)_3$, which have not been subjected to x-ray study, also enable us to separate them into two structural types in the ranges La to Gd and Tb to Lu (including Y) [23]. The displacement of the stability boundaries of the two structural types of phosphates and sulfates (insignificant compared with that expected from geometrical considerations, i.e., the ratios of $r_{Ln^{3+}}$ and the lengths of the P–O and S–O bonds) was also considered as a possible indication of the dependence of the effective ionic radius of the cation Ln^{3+} on the strength of the acid, leading to a relative fall of several percent on passing from the salts of phosphoric to those of the stronger sulfuric acid.

The infrared spectra of the oxysulfates of the R.E., $Ln_2(SO_4)O_2$, which have been studied for the range La to Yb [23-25], show no indications of the existence of several structural types. The stability of the same structure in $Ln_2(SO_4)O_2$ compounds over a wide range of ionic radii of the Ln^{3+} cations may be considered as an indication of the relative "looseness" of the packing; as a result of this there are no critical conditions requiring a transformation to a structure of another type for any member of the series. These ideas agree with the fact that, for a wide range of Ln^{3+} cation radii, there is a relatively loose hexagonal form of the rare-earth orthosulphates, and also with the absence (characteristic for the spectra of oxysulfates) of any resonance ("Davydov") splittings of the vibration bands of the $[SO_4]^{2-}$ ion, despite the large splitting of the degenerate vibrations (and the activation of vibrations forbidden for a regular tetrahedron) associated with the fall in local symmetry; the absence of "Davydov" splitting, as indicated by the relatively narrow and symmetrical contour of the bands of internal vibrations, may be ascribed to the weakening of the coupling between the vibrations of the complex anions as a result of the increase in the $SO_4 \ldots Ln$ distance. The strengthening of the field of the cations in the order La to Yb is reflected as a rise in the vibration frequencies of the SO_4 ion with an increase in the splitting of the degenerate vibrations and a rise in the intensity of the vibrations inactive for symmetry T_d. It is interesting to note that a count of the number of bands, three $\nu_{as} SO_4$ bands in the region 1250 to 1050 cm^{-1}, one $\nu_s SO_4$ band at about 1000 cm^{-1}, three $\delta_{as} SO_4(F_2)$ bands at 700 to 600 cm^{-1}, and two $\delta_s SO_4(E)$ bands at 550 and 400 cm^{-1}, enables us to identify one further band observed above 700 cm^{-1} with vibrations of the Ln–O bonds involving the "additional" oxygen anions O^{2-}, while the remaining lattice vibrations evidently lie below 400 cm^{-1}. The comparatively high $\nu Ln{-}O^{2-}$ frequency indicates the partly covalent character of the bonds formed by the O^{2-} "ions" with the lanthanides, but it is clearly insufficiently high to suggest the formation of complex "lanthanyl cations" with a multiple Ln–O bond. (An analogous situation in the orthosilicates and germanates of the R.E. is discussed in the following section.)

Among other R.E. salts studied by the method of infrared and (partly) Raman spectroscopy, we may mention certain hypophosphites [26] and condensed phosphates [27] and also carbonates [28], chromites [29], perchlorates [30], and perrhenates [31]. In these last, three types of structures have been established by x rays and spectroscopically in the ranges La to Pr, Pr to Dy, and Ho to Lu (including Y).

It follows from this brief review that the infrared spectra of the R.E. serve as an effective means of studying the polymorphism of these compounds, enabling us in some cases to draw specific conclusions regarding the structure of the complex anion, which constitutes an undoubted advantage of the method over x-ray phase analysis. On the other hand, investigations into the spectra of R.E. salts, particularly the relation between the frequencies of the complex anion and the nature of the Ln^{3+} cation, indicate the necessity of allowing for the partial (although probably slight) covalence of the Ln–O bond, which changes substantially over the series La to Lu, despite the fact that the polymorphic transformations of these compounds are primarily associated with the change in the coordination of the Ln^{3+} ions resulting from the "packing" relationships.

§1. Orthosilicates and Germanates of the Rare-Earth Elements

Here we shall consider spectroscopic data regarding the structure and structural classification of silicates with "isolated" SiO_4 tetrahedra and the corresponding germanates, many of which are direct structural analogs of the silicates.

The oxyorthosilicates of the rare-earth elements were studied spectroscopically in [32] and divided into three groups, within each of which the number of bands in the spectrum remained constant, while the frequencies moved regularly on passing from some cations to others. Figure 92a shows the spectra of individual representatives of each of the groups, while Table 47 shows the frequencies of the absorption maxima. The spectra of yttrium and scandium orthosilicates are also shown in Fig. 92b and identify these compounds with the third group. Except for scandium orthosilicate, the compounds of all three groups show no absorption between 600 and 800 cm^{-1}, which indicates the absence of condensed silicon–oxygen tetrahedra (Si–O–Si bonds). Of the stretching vibrations of the SiO_4 ion, the spectra of compounds of the second and third types show only the band of the triply degenerate vibration, the structure of which differs in the spectrum of each of the groups of orthosilicates, indicating considerable differences between the structures of these compounds.

It was assumed in the initial investigation that all the oxyorthosilicates corresponded to the simple formula with an equimolar content of oxides $Ln_2O_3 \cdot SiO_2 = Ln_2(SiO_4)O$, while the

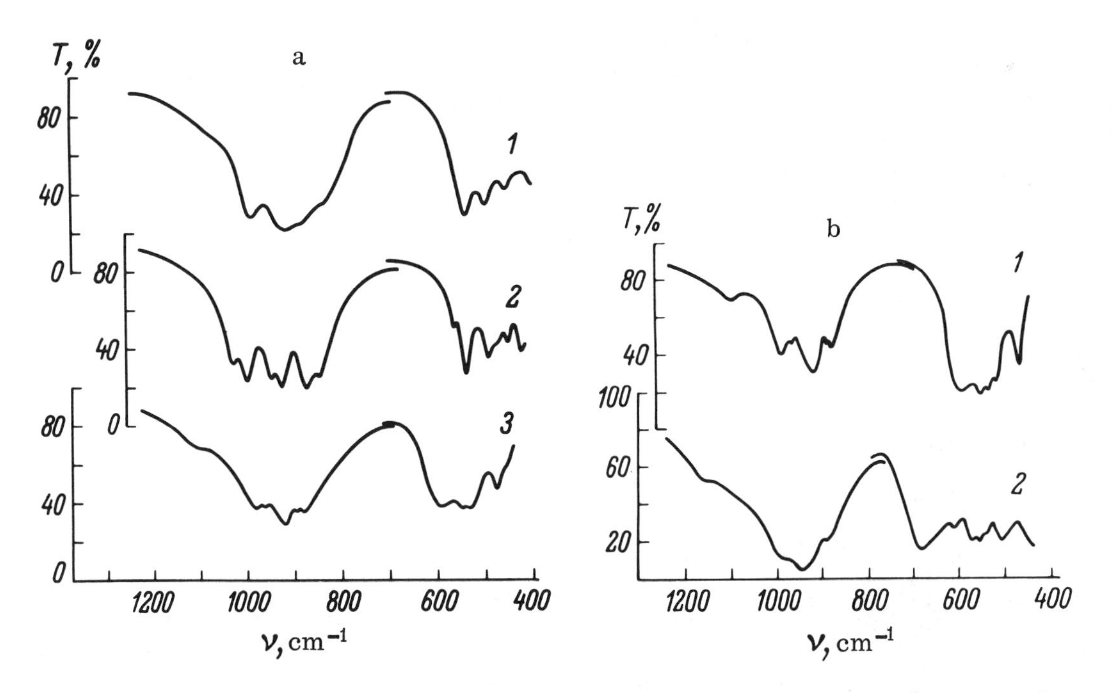

Fig. 92. Infrared spectra of the rare-earth oxyorthosilicates. a) Three types of oxyorthosilicates (pressed in KBr, 5 mg); 1) $La_{4\frac{2}{3}}(SiO_4)_3O$, 2) $Gd_2(SiO_4)O$, 3) $Yb_2(SiO_4)O$; b) oxyorthosilicates of 1) yttrium, 2) scandium (pressed in KBr, 5 mg).

TABLE 47. Frequencies of the Absorption Maxima in the Spectra of Rare-Earth Oxyorthosilicates

Type of structure				Assignments
I	II	III		
$La_{4\frac{2}{3}}(SiO_4)_3O$	$Gd_2(SiO_4)O$	$Yb_2(SiO_4)O$	$Sc_2(SiO_4)O$	
–	1031 v.s.	982 s.	980 v.s.	
990 s.	1000 v.s.	964 s.		
915 v.s.	950 v.s.	920 v.s.	938 v.s.	ν_{as} SiO_4
889 sh.?	928 v.s.			
–	875 v.s.	889 s.	888 s.	
–	849 v.s.	875 s.		
842 sh.	–	–	–	ν_s SiO_4
–	–	593 v.s.	681 v.s.	νLn^-O^{2-}
–	566 m.	–	610 m.	
542 v.s.	540 v.s.	550 s.	571 s.	δSiO_4 (and νLnO?)
499 v.s.	495 v.s.	553 s.	556 s.	
457 s.	477 sh.	475 s.	539 sh.	
402 s.	452 s.	454 sh.	507 v.s.	
–	427 s.	–	440 v.s.	

Notation: v.w. = very weak, w. = weak, m. = medium, s. = strong, v.s. = very strong, sh. = shoulder on the side of a stronger band.

transformation from one type to another constituted a polymorphic transformation mainly due to "dimensional" relationships, and was unambiguously determined by the radius of the cation and the temperature. From this point of view the dimorphism of dysprosium oxyorthosilicate received a natural explanation: the spectrum of this compound indicated that it had a structure of the third type, typical of silicates with cations of small dimensions, if the synthesis temperature were no greater than 1500°C, but transformed to a structure of the second type at higher temperatures. On the other hand, no such clear boundary could be established between the regions of stability of structures of the first and second types; in particular, on growing single crystals of oxyorthosilicates from solution in molten KF, i.e., at temperatures below 1300°C, structures of the first type were obtained over a wide range of cation radii, from La^{3+} to Er^{3+},

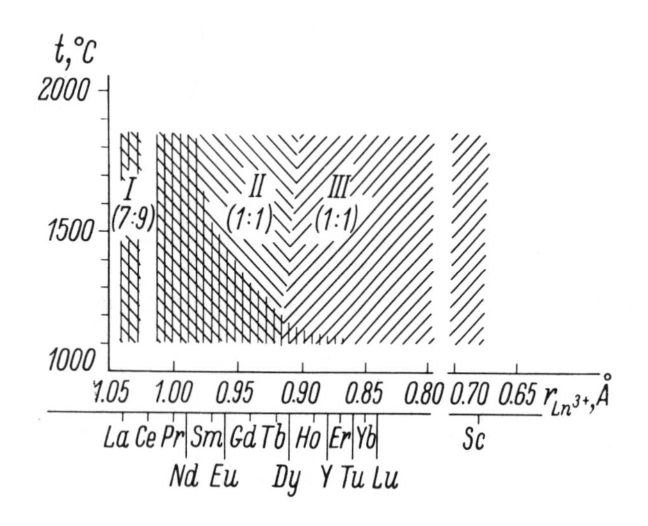

Fig. 93. Range of stability of rare-earth oxyorthosilicates of various compositions and structure.

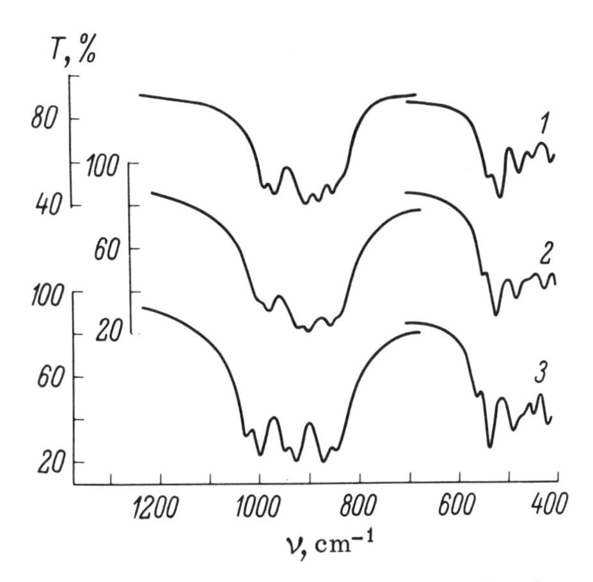

Fig. 94. Infrared spectra of oxyorthosilicates of type II (pressed in KBr). 1) $La_2(SiO_4)O$ (4 mg); 2) $Nd_2(SiO_4)O$ (5 mg); 3) $Gd_2(SiO_4)O$ (5 mg).

overlapping the initially established ranges of ionic radii characteristic of structures of the second (and partly the third) types.

A more detailed investigation [33] into the ranges of temperature and cation radius corresponding to stable oxyorthosilicates of the three types led to the results shown schematically in Fig. 93. The reason for the "superposition" of the range of stability of oxyorthosilicates of the first type on those of the two other types may be understood from the result of an x-ray structural analysis carried out for crystals of the first type by Kuz'min and Belov [34] (see also [3]). This showed that oxyorthosilicates of the first type were isostructural with apatite $Ca_5(PO_4)_3F$, leading to the formula $Ln_{4\frac{2}{3}}(SiO_4)_3O$. For two "molecules" of this composition in the unit cell of a crystal with symmetry $P6_3/m-{}^{2}_{6h}$, six Ln atoms lie in a set with site symmetry C_s, and another $3\frac{1}{3}$ Ln atoms statistically occupy a four-fold set with symmetry C_3, which ensures electrical neutrality of the structure as a whole. The formula indicated for the oxyorthosilicate of type I corresponds to a molar oxide ratio of $LnO_3 : SiO_2 = 7 : 9$. On the other hand, x-ray diffraction studies of erbium and yttrium oxyorthosilicates, belonging to the third type [35, 36], and recent solutions of the structure of the gadolinium oxyorthosilicate (type II [1]) and ytterbium oxyorthosilicate (type III [2]), confirmed that the structures of these compounds corresponded to the formula $Ln_2O_3 \cdot SiO_2$. Thus the transformation from the oxyorthosilicates of type I to compounds of types II or III is not a polymorphic transformation; rather the diagram in Fig. 93 represents the superposition of two diagrams corresponding to compositions $7Ln_2O_3 \cdot 9SiO_2$ and $Ln_2O_3 \cdot SiO_2$.

Let us give some more detailed consideration to each of the three types of oxyorthosilicates, starting with compounds of composition $Ln_2O_3 \cdot SiO_2$. In accordance with spectroscopic data, compounds isostructural, or at least isotypic with $Gd_2(SiO_4)O$ may be obtained over the range Dy^{3+}–La^{3+}. Actually the spectra given in Fig. 94 for the spectra of the corresponding compounds of neodymium and lanthanum as regards the number of bands differ in no way from the spectrum of $Gd_2(SiO_4)O$, while the frequencies of the internal vibrations of the SiO_4 ion fall regularly with increasing ionic radius of the cation (i.e., as the field of the cations weakens), within the limits of experimental error satisfying a linear dependence on $1/r^2_{Ln^{3+}}$:

	$La_2(SiO_4)O$	$Nd_2(SiO_4)O$	$Gd_2(SiO_4)O$
$\nu_{as}SiO_4$ (F_2)	992	1006	1031
	971	983	1000
	911	925	950
	885	906	928
	858	863	875
	838	846	849
$\delta_{as}SiO_4$ (F_2)	547	558	566
	523	530	540
	485	490	495
	459	468	477
	421	433	452
	—	403 ?	427

The identification of the frequencies of the internal vibrations of the SiO_4 anion in the spectrum of $Gd_2(SiO_4)O$ and similar compounds, and also the explanation for the multiplet structure of the bands, follow directly from the symmetry of the crystal $P2_1/c-C^5_{2h}$, there being four "molecules" per unit cell. The SiO_4 tetrahedra are situated in a set of general positions, the degeneracy being completely removed, and each (nondegenerate) vibration is split into four components, two being active in the infrared spectrum:

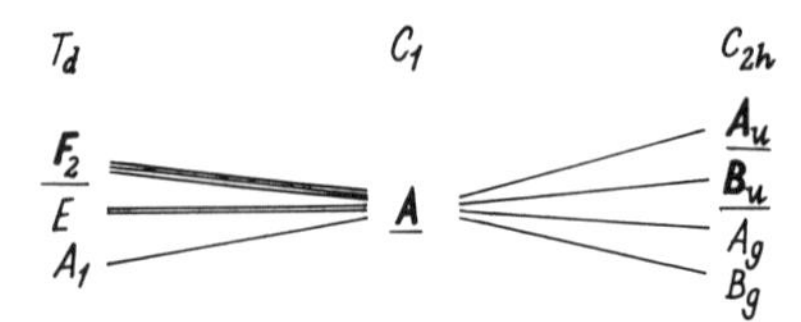

(The types of symmetry active in the infrared spectrum are underlined.)

In the frequency range studied only the $\nu_{as}SiO_4$ and $\delta_{as}SiO_4$ vibrations are observed experimentally; the multiplicity of these is exactly as expected. The $\nu_s SiO_4$ vibration is apparently not strong enough for its bands to be observed in the spectrum of a powder, despite the fact that according to x-ray structural data the tetrahedron is essentially asymmetric. (The reduction in the distance between the three $\nu_{as}SiO_4$ components, formed as a result of the removal of degeneracy, on passing from $Gd_2(SiO_4)O$ to silicates with the larger cations Nd^{3+} and La^{3+} probably indicates a reduction in this asymmetry.)

According to the results of [35, 36, 36a], the crystals of oxyorthosilicates of the third type belong to the $C2/c-C^6_{2h}$ or $Cc-C^4_s$ space group (Z = 8). Analysis of the correlations between the representations of the factor group, site group, and point group of the tetrahedron lead to the conclusion, if one confines attention to the C^6_{2h} group, that there are six components of each of the triply degenerate vibrations of the tetrahedron active in the infrared spectrum, independently of the "means" of siting the SiO_4 ions in the lattice (formally these may have a site symmetry C_1 or C_2). Experimentally only five maxima can be resolved in the $\nu_{as}SiO_4$ frequency range, of which the central one is clearly asymmetric and probably constitutes an unresolved doublet. As in the case of the oxyorthosilicates of type II, there are no signs of the $\nu_s SiO_4$ vibration band in the spectrum; two components of this are formally allowed by the selection rules in the infrared spectrum. The number of bands observed in the range of deformation vibrations in the spectrum of $Sc_2(SiO_4)O$ exceeds the expected number of $\delta_{as}SiO_4$ components (Fig. 92b); there is also a characteristic appearance of the relatively high frequency 681 cm^{-1}. Passing to the spectrum of $Y_2(SiO_4)O$ reduces this frequency by 15%, which indicates a considerable contribution of the degrees of freedom of the cation (i.e., the Ln–O bonds) to the vibration considered, and hence a considerable value for the force constant of the Ln–O bonds in oxyorthosilicates of this type.

The high force constants of the Ln–O bonds, indicating a considerable proportion of covalence in the bonds, should in all probability be primarily ascribed to the presence of Ln–O bonds involving "oxyions" O^{2-} not constituting any part of the composition of the silicon–oxygen tetrahedra. We note that in the spectrum of thortveitite $Sc_2Si_2O_7$ the frequencies of the deformation vibrations of the anion interacting most strongly with the vibrations of the Sc–O bonds are no greater than ~ 550 cm^{-1} (more details follow). Thus the view that the O^{2-} ions in oxyorthosilicates are connected to the rest of the structure by purely electrostatic forces is hardly acceptable. In actual fact the effective charge on these atoms is probably much smaller, while the bonds formed by them are partly covalent. From this point of view the relative rarity of structures with "oxyions" among the silicates of the alkali metals is no chance matter, although such structures have often been observed for silicates with cations belonging to elements of Groups II and III in the periodic system. We may suppose that these structures form an "intermediate step" between crystals with complex anions and monovalent cations, on the one hand, and crystals containing complex cations (such as, for example, the uranyl ion) or crystals with a single essentially covalent lattice (such as aluminosilicates) on the other.

Some indications as to the substantial interaction between the vibrations of the Ln–O bonds and the internal vibrations of the anion are also observed in the spectra of oxyorthosilicates with the structure of apatite (type I). The six SiO_4 tetrahedra in the cell of these crystals occupy a set with site symmetry C_s. For a crystal factor group isomorphic with C_{6h}, the $\nu_{as} SiO_4$ vibration corresponds to three crystal vibrations active in the infrared spectrum, two of type E_{1u} and one A_u, while the $\nu_s SiO_4$ vibration corresponds to one vibration of type E_{1u}:

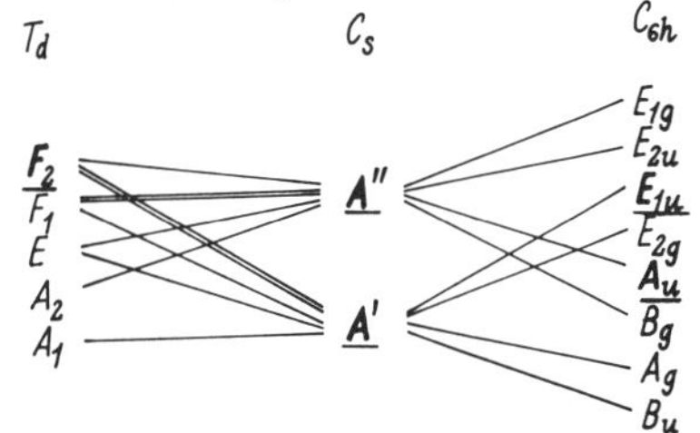

Of the four maxima observed in the νSiO frequency range, the three of highest frequency probably correspond to $\nu_{as} SiO_4$, while the $\nu_s SiO_4$ vibration may probably be associated with the weakest and lowest-frequency of the bands of stretching vibrations of the tetrahedron around 840 to 850 cm^{-1}.†

$La_{4\,2/3}(SiO_4)_3O$: 990,915,889 sh?; 842 sh; 542,499,457,402 cm^{-1}

$Nd_{4\,2/3}(SiO_4)_3O$: 997,921,880 sh; 853 sh; 551,504,465,418 cm^{-1}

In the range 550 to 400 cm^{-1}, instead of the expected three components of $\delta_{as} SiO_4$ ($2E_{1u}$, A_u) there are four bands (Fig. 95). The appearance of the "excess" band may be associated either with the direct appearance of lattice vibrations, or more probably with the strong interaction between these vibrations and the $\delta_s SiO_4$ vibration of the tetrahedron, the frequency of which, for weak bonds in the lattice, fails to reach 400 cm^{-1}.

It should be noted that, as indicated by the scheme in Fig. 93, among the rare-earth silicates with the apatite structure only compounds of elements between La and Nd were stable over the whole range of temperatures studied. Silicates with anions of smaller radius appar-

† L. D. Kislovskii's investigations into the reflection spectra of single crystals of this type agree with the identification presented: in the frequency range of SiO stretching, of the four bands present three have the electric vector in a plane perpendicular to the optic axis and one band has the electric vector along the optic axis.

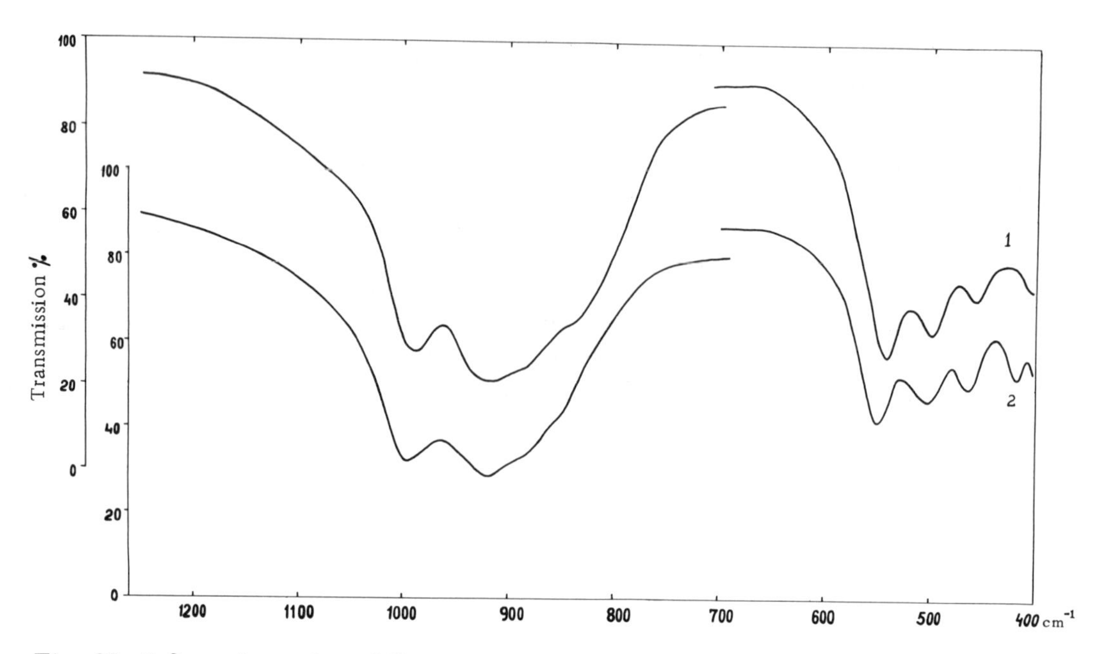

Fig. 95. Infrared spectra of the rare-earth oxyorthosilicates with the apatite structure (pressed in KBr, 4 mg); 1) $La_{4\frac{2}{3}}(SiO_4)_3O$, 2) $Nd_{4\frac{2}{3}}(SiO_4)_3O$.

ently decompose when held at high temperatures, with the formation of the compounds $Ln_2(SiO_4)O$ and $Ln_2Si_2O_7$, the presence of which may be identified spectroscopically.

The infrared spectra of the rare-earth germanates containing "isolated" GeO_4 tetrahedra have not hitherto been systematically studied, although the spectra of certain compounds of composition $Ln_2O_3 \cdot GeO_2 = Ln_2(GeO_4)O$ were obtained in [37], and it was suggested that they were structurally similar to the corresponding silicates. Some x-ray diffraction data regarding compounds of this structure were presented in [38]. However, the literature also contains certain indications regarding the existence of germanium analogs of rare-earth silicates with the apatite structure [39]; the composition of these, like that of the silicates, corresponds to the formula $7Ln_2O_3 \cdot 9GeO_2 = 3Ln_{4\frac{2}{3}}(GeO_4)_3O$, and, for Nd and Y compounds of composition $2Ln_2O_3 \cdot GeO_2$ [40], apparently have no analogs in the corresponding silicate systems. The production of single crystals of the compound $2Y_2O_3 \cdot GeO_2$ was recently described in [41].

We shall now describe the results of a successful study of germanates of composition $7Ln_2O_3 \cdot 9GeO_2$, $Ln_2O_3 \cdot GeO_2$, and $2Ln_2O_3 \cdot GeO_2$. The greater proportion of the compounds studied were synthesized at temperatures between 1100 or 1200 and 1400°C. The infrared spectra of the rare-earth germanates are presented in Figs. 96 to 100. Let us consider these in succession, starting from the composition richest in germanium oxide, $7Ln_2O_3 \cdot 9GeO_2 = 3Ln_{4\frac{2}{3}}(GeO_4)_3O$. The general form of the spectra of these compounds (Fig. 96) and the number of bands in the range of Ge–O stretching frequencies are in good agreement with the assumption that these are isostructural with the corresponding silicates: for symmetry $P6_3/m-C_{6h}^2$ the $\nu_{as}GeO_4$ vibrations correspond to three vibrations of the crystal active in the infrared spectrum, two of type E_{1u} and one of type A_u, while the $\nu_s GeO_4$ vibrations correspond to one E_{1u} type only. In all we see that there are four νGeO bands (Table 48). In contrast to the spectra of the orthosilicates $\nu_{as}GeO_4 < \nu_s GeO_4$, which is not surprising, since for comparatively slight distortions of the tetrahedral configuration of the XO_4 ions in the crystal the change in the frequency ν_s (A_1 for the regular tetrahedron) is almost entirely determined by the change in the force field, while the $\nu_{as}(F_2)$ frequency also depends substantially on the mass of the central atom.

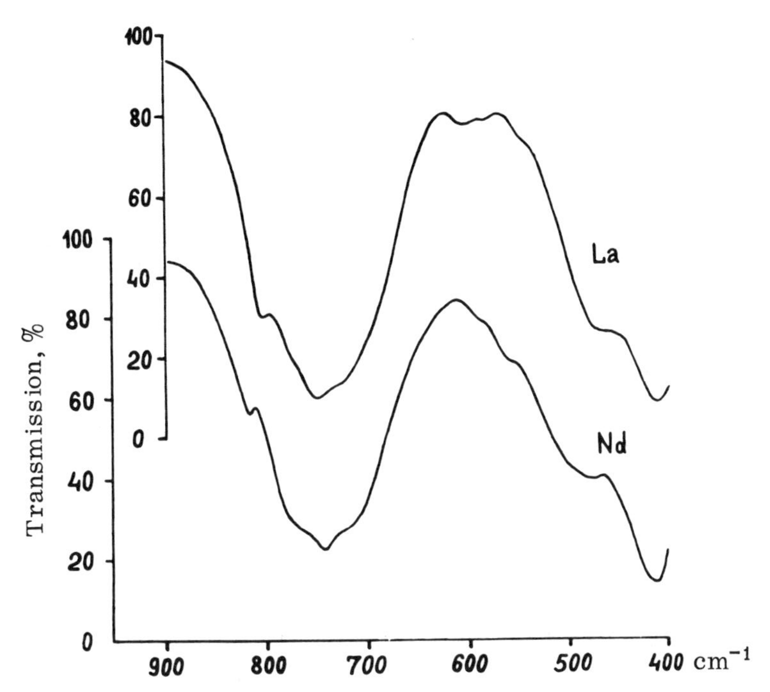

Fig. 96. Infrared spectra of rare-earth germanates with the apatite structure $Ln_{4\frac{2}{3}}(GeO_4)_3O$ (pressed in KBr, 5 mg).

TABLE 48. Frequencies of the Absorption Maxima in the Spectra of $Ln_{4\frac{2}{3}}(GeO_4)_3O$

Description of vibrations	La	Nd
$\nu_s GeO_4$	805 m.	818 m.
	776 sh.	775 sh.
$\nu_{as} GeO_4$	750 v.s.	742 v.s.
	725 sh.	720 sh.
	(604 v.w.)	–
	(586 v.w.)	(590 sh.)
	545 sh., w.	560 sh.
$\delta_{as} GeO_4$	466 s.	470-500 s.
	413 v.s.	416 v.s.

Note: Brackets show the frequencies of weak bands not belonging to the fundamental vibrations.

The spectra of orthogermanates of composition $Ln_2O_3 \cdot GeO_2 = Ln_2(GeO_4)O$ also show a clear similarity to those of the corresponding silicates, the existence of two structural types of these germanates being reliably established spectroscopically (Figs. 97 and 98). The structural similarity of the germanates of the first type to the silicates $Ln_2(SiO_4)O$ with Ln = La to Dy follows from counting the number of maxima in the frequency range of the stretching vibrations of the tetrahedron (although all these maxima can only be completely resolved in the spectra of $La_2(GeO_4)O$ and $Nd_2(GeO_4)O$, while in the spectra of the other germanates the presence of some of them can only be deduced from the asymmetry of the absorption bands around 750 and 700 cm^{-1}). For the symmetry $P2_1/c-C_{2h}^5$ established for $Gd_2(SiO_4)O$, the infrared spectrum should exhibit altogether six components of the $\nu_{as} GeO_4$ vibration (by virtue of the complete removal of the degeneracy and the "Davydov" splitting of each of the components into four,

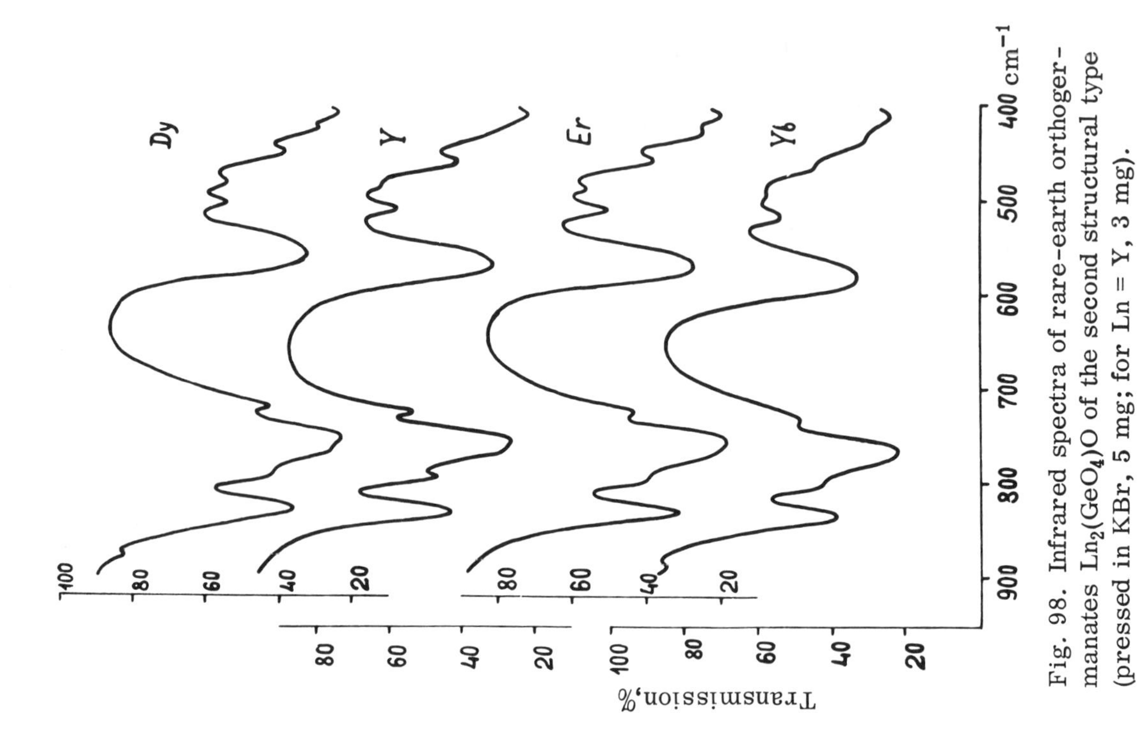

Fig. 98. Infrared spectra of rare-earth orthogermanates $Ln_2(GeO_4)O$ of the second structural type (pressed in KBr, 5 mg; for Ln = Y, 3 mg).

Fig. 97. Infrared spectra of rare-earth orthogermanates $Ln_2(GeO_4)O$ of the first structural type (pressed in KBr, 5 mg).

TABLE 49. Frequencies of the Absorption Maxima in the Spectra of $Ln_2(GeO_4)O$ Compounds

Description of vibrations	Type 1					Type 2			
	La	Nd	Eu	Gd	Tb	Dy	Y	Er	Yb
$\nu_s GeO_4$	818 v.s.	828 v.s.	831 v.s.	835 v.s.	837 v.s.	827 v.s.	830 s.	831 s.	835 s.
	801 s.	803 s.	805 sh.	811 s.	816 s.				
$\nu_{as} GeO_4$	774 v.s.	780 v.s.	783 v.s.	789 v.s.	795 v.s.	793 sh.	795 s.	795 sh.	802 sh.
	749 v.s.	747 v.s.	748 v.s.	750 v.s.	753 v.s.				
	740 sh.	740 sh.	?	?	–	763 sh.	768 sh.	?	770 –
	743 s.	727 sh.	—	–	–	749 v.s.	753 v.s.	756 v.s.	–764 v.s.
	706 v.s.	699 v.s.	698 v.s.	702 v.s.	704 v.s.	718 s.	725 s.	727 s.	734 s.
	688 sh.	~683 sh.	~685 sh. ?	?	?				
$\nu Ln^{-}O^{2-}$	(540 v.w.)	(555 v.w.)	(550 v.w.)	–	(570 v.w.)	–	–	–	–
	–	–	–	–	(529 w., sh.)	558 v.s.	568 v.s.	570 v.s.	582 v.s.
$\delta_{as} GeO_4$ and νLnO?	469 s.	480–490 sh.	485 sh.	489 sh.	495 sh.	503 m.	509 m.	510 m.	516 m.
	424 v.s.	443 sh.	460 v.s.	466 v.s.	469 v.s.	481 m.	485 sh.	488 m.	493 w.
	406 v.s.	426 v.s.	450 v.s.	457 v.s.	463 v.s.	448 s.	458 s.	458 s.	465 sh.
	?	408 v.s.	418 v.s.	430 v.s.	434 v.s.	423 v.s.	~430 sh.	427 v.s.	436 sh.
	?	?	?	407 sh.	424 sh.	405 v.s.	406 v.s.	408 v.s.	411 v.s.

of which two are active in the infrared spectrum) and two $\nu_s GeO_4$ components, as observed experimentally (Table 49). It should be noted that the $\nu_s GeO_4$ vibration bands (situated, as in other cases, at higher frequencies than the $\nu_{as} GeO_4$ bands) are quite strong, whereas in the spectrum of the silicate analog no $\nu_s SiO_4$ bands were observed, despite the absence of any formal ban due to symmetry. It remains uncertain, however, in advance of an x-ray analysis, whether this is associated with the greater configurational (geometrical) asymmetrization of the germanium tetrahedron or with a change in the electrooptical parameters on passing from the silicate to the germanate.

In a completely analogous manner the tetrahedral pulsation appears in the spectra of orthogermanates $Ln_2(GeO_4)O$ of the second type (Dy to Yb) (this is not forbidden, but it is never observed experimentally in the spectra of silicates); these spectra are evidently similar to the corresponding silicates with Ln = Dy to Sc. The symmetry of these silicates allows six bands of $\nu_{as} SiO_4$ vibrations to appear in the infrared spectrum (of these, as already mentioned, five may be resolved experimentally), together with two $\nu_s SiO_4$ bands, which have never been observed. In the spectra of the germanates four $\nu_{as} GeO_4$ components have been resolved, although the intensity distribution over the band as a whole is very close to that observed in the silicates. The $\nu_s GeO_4$ band is strong in the infrared spectrum, although the expected doublet structure has never been resolved. In the spectra of germanates of this type the most interesting feature is the broad, strong band at 560 to 580 cm^{-1}, the frequency of which is hardly any lower than that of the corresponding band in the spectra of the isostructural silicates. This band was attributed to a $\nu Ln–O$ vibration involving the participation of O atoms not forming part of the complex anion. The slightness of the change in frequency on passing from the silicate to the germanate supports this view. In the spectra of the $Ln_2(GeO_4)O$ germanates of the first type just described and those of germanates with the apatite structure, the vibrations of the cation–oxygen bonds are not observed in the form of individual bands; their influence on the spectrum in the frequency range above 400 cm^{-1} is apparently limited simply to raising the anion vibration frequencies (particularly those of the deformation type), this effect becoming stronger as the radius of the cation diminishes. Hence a partial "mixing" of the vibrations appears probable.

It is interesting to compare the regions of stability (on a diagram plotted in coordinates of the ionic radius of the cation and the temperature) of orthogermanates of various compositions and structures with the regions of stability of the corresponding silicates. A similar comparison for silicates and phosphates with X_2O_7 anions described in a subsequent section

TABLE 50. Frequencies of the Absorption Maxima in the Spectra of $Ln_4(GeO_4)O_4$ Compounds

Description of vibrations	Type 1				Type 2				
	La	Nd	Eu	Gd	Tb	Dy	Y	Er	Yb
$\nu_s GeO_4$	744 m.	756 m.	766 w.	761 w.	804 m.	820 m.	836 m.	824 m.	827 m.
	–	–	–	–	784 m.	–	–	–	–
$\nu_{as} GeO_4$	706 v.s.	715 sh.	725 sh. ?	725 sh.	748 sh.	773 sh.	775 m.	780 sh.	787 m.
	689 v.s.	701 v.s.	711 v.s.	715 v.s.	718 s.	723 s.	725 s.	726 s.	736 s.
	673 sh.	~680 sh.	698 sh.	694 v.s.	–	–	–	–	–
	–	–	682 sh. ?	–	–	–	–	–	–
$\delta_{as} GeO_4$ and νLnO	505 s.	513-524 s., doub.	543 m.	558 s.	555 sh.	550 sh.	555 sh.	561 sh.	~573 sh.
			523 m.	524 s.	518 sh.	–	—	–	–
	425 v.s.	429 v.s.	438 v.s.	449 v.s.	464 v.s.	474 v.s.	487 v.s.	482 v.s.	486 v.s.
	410 v.s.	414 v.s.	427 v.s.	438 v.s.	426 v.s.	428 v.s.	428 v.s.	426 v.s.	428 v.s.
	–	–	–	406 v.s.	412 v.s.	413 v.s.	413 v.s.	414 v.s.	414 v.s.
	(impurity bands of La_2GeO_5 at 775, 800, 818)	(impurity bands of Nd_2GeO_5 at 780, 802, 828)	(impurity bands of Eu_2GeO_5 at 783, 805, 830)						

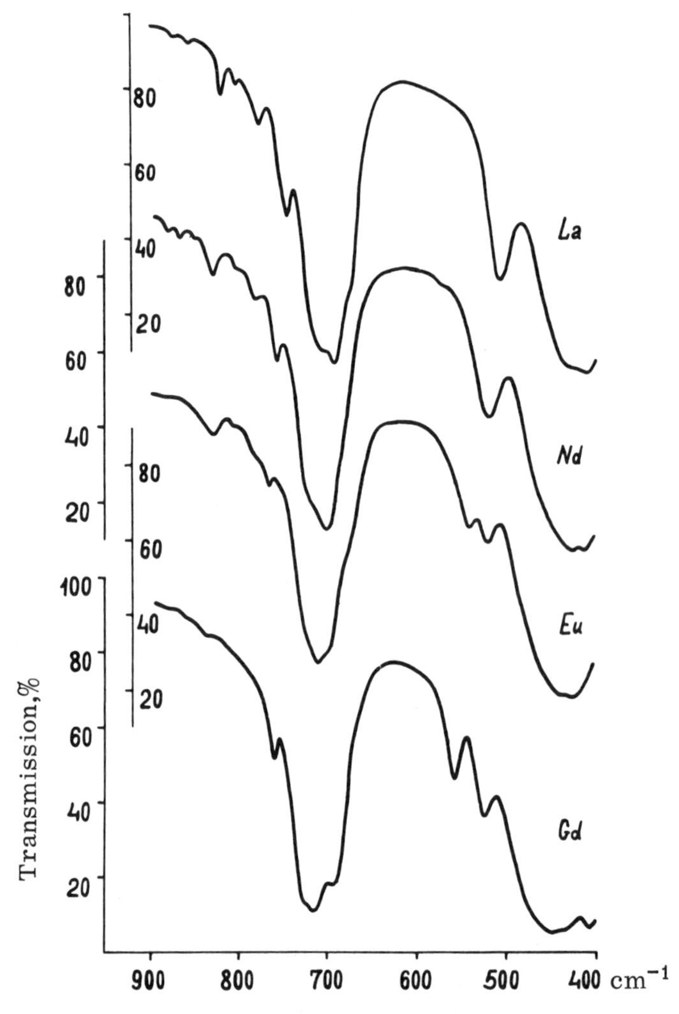

Fig. 99. Infrared spectra of rare-earth germanates $Ln_4(GeO_4)O_4$, first structural type (pressed in KBr, 5 mg).

shows that an approximate correspondence between the diagrams of the alkaline-earth pyrophosphates and rare-earth pyrosilicates may be secured if we introduce a correcting factor approximately equal to the ratio r_{SiO}/r_{PO}, increasing the relative radius of the M^{2+} cation in the phosphates, thus allowing for the fact that the dimensions of the complex anion are smaller than in the silicates. If we compare the stability diagrams of the orthosilicates in Fig. 93 with the results obtained for germanates, we find that for compounds of the apatite type $Ln_{4\frac{2}{3}}(XO_4)_3O$ the range of stability is shortened, so that the stability boundary (on the side of the small-radius cations) at 1100 to 1200°C falls from between Gd and Tb for silicates to between Nd and Sm for germanates. For europium (and all subsequent) germanates, attempts at synthesizing samples of this composition only led to the formation of a mixture of $Ln_2(GeO_4)O$ and a germanate containing more germanium oxide than $7Ln_2O_3 \cdot 9GeO_2$. This displacement of the stability boundary of the apatite structure indicates the necessity of introducing a "dimensional" correction to

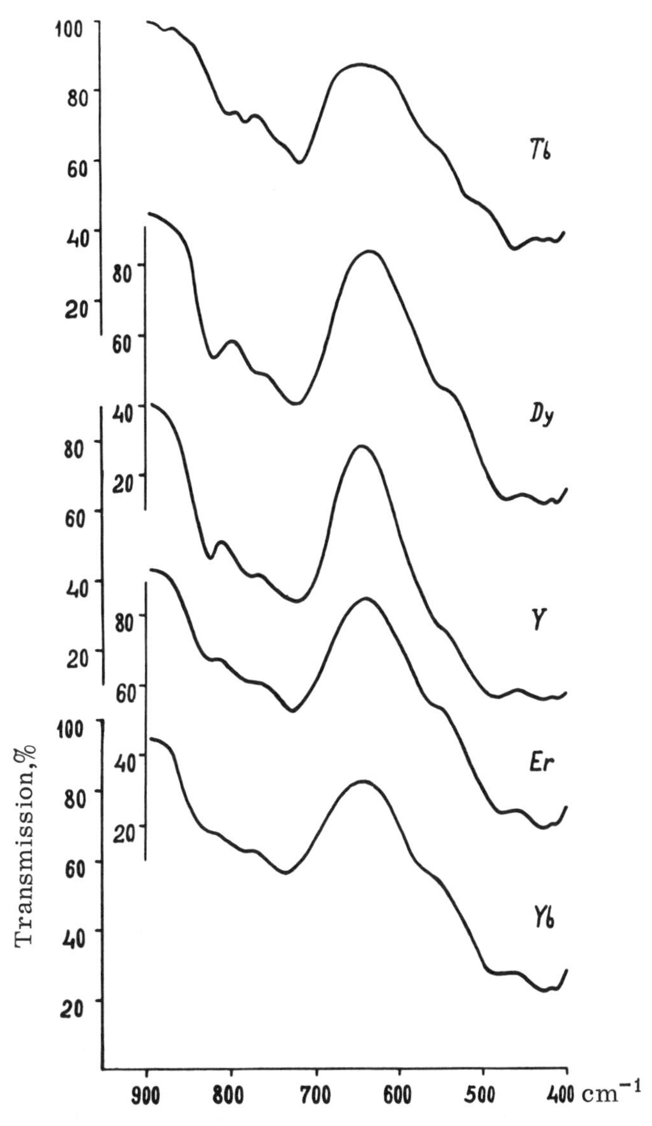

Fig. 100. Infrared spectra of rare-earth germanates $Ln_4(GeO_4)O_4$, second structural type (pressed in KBr, 5 mg; for Ln = Y, 4 mg).

allow for the reduction in the relative dimensions of the Ln^{3+} ion in the germanates; however, the magnitude of this correction is approximately 3% instead of the 7% expected from the r_{GeO}/r_{SiO} ratio. For compounds $Ln_2(XO_4)O$ the displacement of the stability boundary for the two types of structure becomes negligibly small: dysprosium silicate is dimorphic, exhibiting the structure characteristic of silicates with small cations at a low temperature and that of silicates with large cations at high temperature, while for the germanates this boundary (for temperatures of 1200 to 1400°C) is situated between Tb and Dy, i.e., the "dimensional" correction is no greater than 1%. The decline in the part played by this correction as the Si or Ge content diminishes and the number of O atoms not forming part of the complex anion increases is quite natural and indicates a transition to another dominating factor, also "dimensional" in nature, namely, the ratio of the sizes of the rare-earth ion and oxygen ion.

In conclusion, we may consider the spectra of compounds of composition $2Ln_2O_3 \cdot GeO_2$ (Figs. 99 and 100 and Table 50). Spectroscopic data confirm the chemical individuality of these and clearly indicate the existence of two structural types respectively found in the ranges La to Gd and Tb to Yb. The splitting of the single band at around 500 to 520 cm^{-1} in the spectra of the lanthanum and neodymium germanates into a sharp doublet in europium and gadolinium germanates indicates that there may be a change in the symmetry of the crystals, even within the first structural type. Both structural types are characterized by the presence of "isolated" GeO_4 tetrahedra; the frequencies of Ge–O stretching vibrations are high enough for the transition of the Ge into the octahedral coordination to be regarded as improbable (i.e., the transformation of the salt into a complex oxide). At the same time, the considerable reduction in these frequencies (particularly in compounds of La to Gd), as compared with the spectra of the orthogermanates with a lower content of rare-earth oxides considered earlier, indicates a fall in the rigidity of the germanium–oxygen tetrahedra (and probably an increase in the Ge–O distance) in these structures. It should be noted that consideration of the spectroscopic data enables us to eliminate the possibility that discrete complex cations might be formed in accordance with the scheme $2Ln_2O_3 \cdot GeO_2 \rightarrow (LnO)_4GeO_4$, which might have been supposed on the basis of the empirical formula of these compounds: the frequencies of the absorption maxima in the region of the deformation vibrations of the GeO_4 ions differ little from the corresponding frequencies in the spectra of the rare-earth orthogermanates of different composition, and there are no signs of an increase in the Ln–O vibration frequencies such as might be expected when complex LnO cations are formed. Thus the structural formula of these compounds should be written in the form $Ln_4(GeO_4)O_4$ as orthogermanates with additional oxygen anions.

§2. Pyrosilicates of Rare-Earth Elements and Their Crystal-Chemical Analogs

Among the R.E. silicates, compounds of composition $Ln_2O_3 \cdot 2SiO_2$, obtained over the whole range Ln = La to Lu (and also for Y and Sc), are distinguished by a very great variety of types of crystal structure. The use of spectroscopic methods enables us not only to set up a preliminary structural model of these compounds by establishing the presence of complex Si_2O_7 anions with regularly varying configuration but also to trace the dependence of their polymorphic transformations on temperature and the ionic radius of the cation and to discover their probable crystal-chemical analogs among certain phosphates and germanates.

Infrared Spectra and Types of Structure of Rare-Earth Pyrosilicates. An initial spectroscopic study of the compounds $Ln_2O_3 \cdot 2SiO_2 = Ln_2Si_2O_7$ established the formation of three types of these compounds, differing considerably in structure, called I, II, and III [42-44]. The first two were characterized by "bent" Si_2O_7 groups with ∠SiOSi sharply increasing on passing from I to II, while the third type was distinguished by the presence of "straightened" (centrosymmetric) Si_2O_7 groups. This type, as later established by analyzing the structure of the vibration bands of the Si_2O_7 ion, may be divided into two of fairly similar

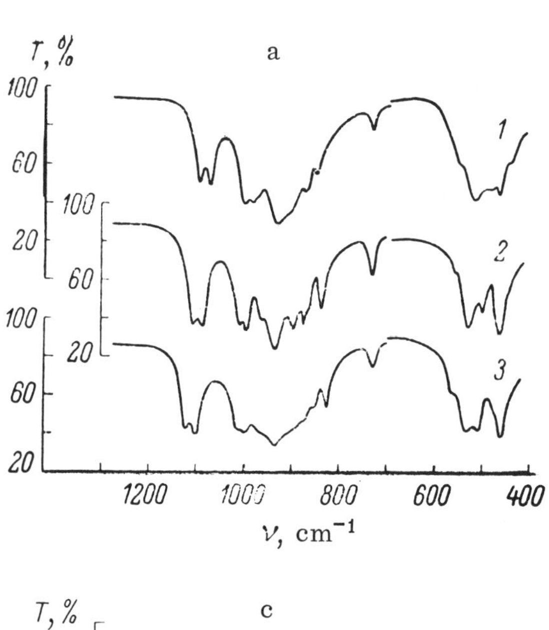

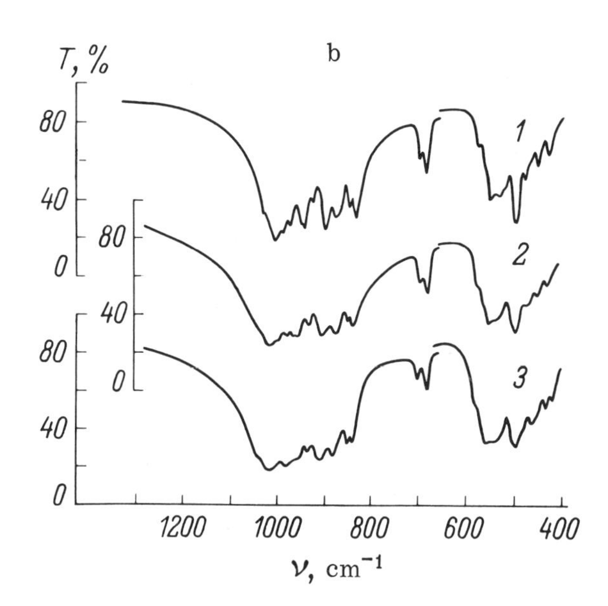

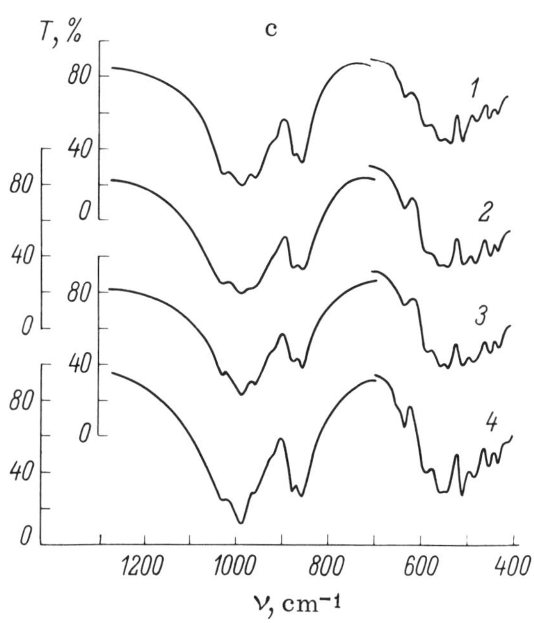

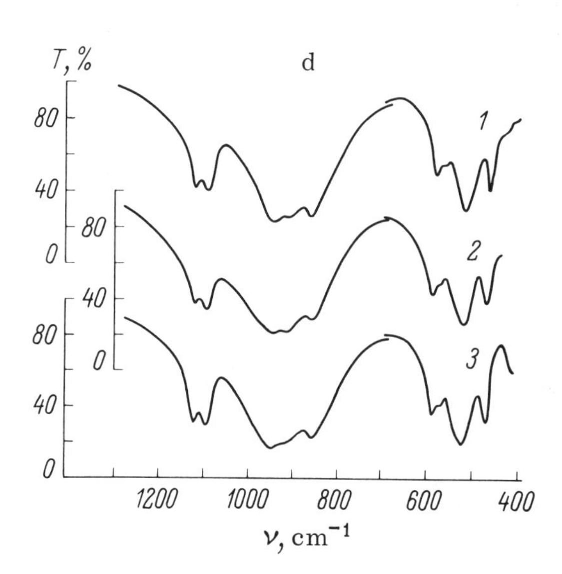

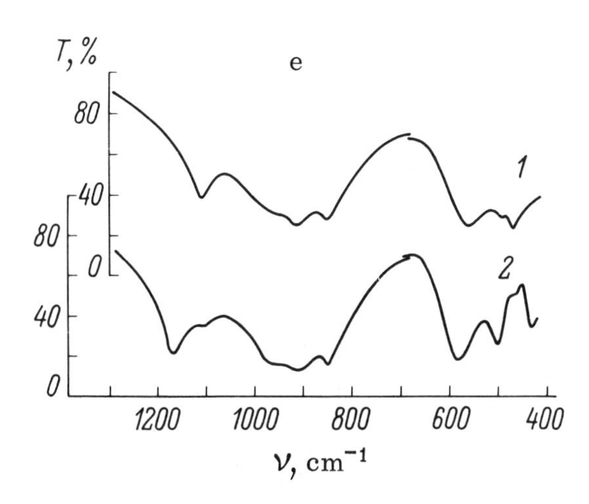

Fig. 101. Infrared spectra of the rare-earth pyrosilicates. a) Type I: 1) $La_2Si_2O_7$, 2) $Nd_2Si_2O_7$, 3) $Eu_2Si_2O_7$; b) Type I′: 1) $Nd_2Si_2O_7$, 2) $Sm_2Si_2O_7$, 3) $Eu_2Si_2O_7$; c) Type II: 1) $Gd_2Si_2O_7$, 2) $Dy_2Si_2O_7$, 3) $Ho_2Si_2O_7$, 4) $Y_2Si_2O_7$; d) Type IIIb: 1) $Ho_2Si_2O_7$, 2) $Y_2Si_2O_7$, 3) $Er_2Si_2O_7$; e) Type IIIa: 1) $Yb_2Si_2O_7$, 2) $Sc_2Si_2O_7$.

structure, IIIa and IIIb. Roy et al. studied these for relatively low temperatures of synthesis [46] and indicated the existence of yet another type of structure of pyrosilicates with large Ln^{3+} cations; compounds of this type, subsequently denoted by I′, are characterized spectroscopically in [47].

Figure 101a gives the infrared spectra of pyrosilicates of structural type I, which corresponds to compounds of La, Ce, Pr, Nd, Sm, Eu, synthesized at a temperature of about 1500°C. The spectra of these compounds contain a number of strong bands in the frequency range 1100 to 800 cm^{-1} and a single band near 730 cm^{-1}, which may be ascribed to the ν_s SiOSi vibration of the Si_2O_7 group. Another region of strong absorption lies in the range 570 to 400 cm^{-1} and corresponds to deformation vibrations of the Si_2O_7 anion; the identification of these is impeded by the possibility of strong interaction with vibrations of the cation–oxygen bond. The number of bands of stretching vibrations of the Si_2O_7 ion in the frequency range above 800 cm^{-1} is no greater than seven, excluding doubling of "interionic" (i.e., resonance) origin, characterized by a distance of 15 to 20 cm^{-1} between the maxima. It follows from this that the symmetry of the Si_2O_7 is lower than C_{2v}, since a vibration of type A_2 is active in the infrared spectrum (Table 10); the number of Si_2O_7 ions in the primitive cell is furthermore not less than two. (The comparatively high strength and frequency of the ν_s SiOSi band clearly indicates that the Si_2O_7 ions are noncentrosymmetric.) If we suppose quite arbitrarily that all the components of the "Davydov" multiplets allowed by the selection rules could be resolved in the spectrum, then, on considering the singlet structure of ν_s SiOSi and the presence of three of four doublets in the higher-frequency range (including the case of ν_{as} SiOSi), we have a choice between three

TABLE 51. Frequencies of the Absorption Maxima in the Infrared Spectra of Rare-Earth Pyrosilicates; Structures with Noncentrosymmetric ("Bent") Si_2O_7 Ions

Identification of frequencies; ion Si_2O_7 (C_{2v})	Type I			Type I′			Type II			
	$La_2Si_2O_7$	$Nd_2Si_2O_7$	$Eu_2Si_2O_7$	$Nd_2Si_2O_7$	$Sm_2Si_2O_7$	$Eu_2Si_2O_7$	$Gd_2Si_2O_7$	$Dy_2Si_2O_7$	$Ho_2Si_2O_7$	$Y_2Si_2O_7$
ν_{as} SiOSi (B_1), ν_{as} SiO_3 (B_2), ν_{as} SiO_3 (A_1), ν'_{as} SiO_3 (A_2), ν'_{as} SiO_3 (B_1), ν'_s SiO_3 (B_1), ν_s SiO_3 (A_1)	1093 s.	1106 v.s.	1121 v.s.	1026 s.	1038 sh.	1042 sh.	—	—	—	—
	1071 s.	1086 v.s.	1098 v.s.	1002 v.s.	1015 v.s.	1017 v.s.	—	—	—	—
	999 v.s.	1010 v.s.	1012 v.s.	985 v.s.	997 v.s.	1000 sh.	—	—	—	—
	981 v.s.	994 v.s.	997 v.s.	970 v.s.	979 v.s.	982 v.s.	1022 v.s.	1023 v.s.	1027 v.s.	1028 v.s.
	970 s.	964 s.	966 sh.	948 v.s.	963 v.s.	959 sh.d.?	982 v.s.	983 v.s.	983 v.s.	987.v.s.
	—	—	—	940 v.s.	954 v.s.		953 v.s.	953 sh.	955 v.s.	957 v.s.
	932 v.s.	934 v.s.	933 v.s.	923 s.	934 s.	937 v.	—	—	—	—
	908 sh.	896 v.s.	908 sh.	—	—	915 sh.	911 sh.	912 sh.	915 sh.	916 sh.
			893 sh.	898 v.s.	906 v.s.	908 v.s.	868 s.	871 s.	872 v.s.	873 v.s.
	872 s.	875 v.s.	874 sh.	877 s.	877 v.s.	883 v.s.	849 s.	850 s.	851 s.	855 v.s.
	866 sh.	865 sh.	854 s.	868 sh.	—	—	—	—	—	—
	850 s.	838 s.	828 s.	846 s.	852 s.	852 s.	—	—	—	—
	—	—	—	833 s.	837 s.	840 s.	—	—	—	—
ν_s SiOSi (A_1)	732 m.	732 m.	730 m.	698 m.	698 m.	702 m.	639 sh.	—	?	645 sh.
	—	—	—	687 m.	683 m.	682 m.	627 w.	630 w.	632 w.	632 w.
Deformation vibrations of Si_2O_7 (and partly νLn–O?)	—	—	—	578 m.	580 sh.	582 sh.	580 s.	582 s.	584 s.	587 s.
	—	—	—	565 sh.	568 sh.	567 sh.	550 v.s.	553 v.s.	553 v.s.	556 v.s.
	552 sh.	556 m.	562 sh.	556 v.s.	556 v.s.	558 v.s.	528 v.s.	535 v.s.	537 v.s.	543 v.s.
	520 v.s.	529 v.s.	531 v.s.	535 v.s.	540 v.s.	543 v.s.	502 v.s.	504 v.s.	506 v.s.	508 v.s.
	492 s.	502 s.	507 v.s.	524 sh.	530 sh.?	532 sh.	469 m.	478 s.	480 s.	482 s.
	483 s.	494 sh.		502 v.s.	500 v.s.	499 v.s.	444 m.	447 m.	448 m.	452 m.
	468 v.s.	470 sh.	474 sh.	481 s.	478 s.	484 sh.	427 m.	430 m.	430 m.	432 m.
		463 v.s.	462 v.s.	473 sh.		465 s.	—	—	—	411 m.
	447 sh.	446 sh.	—	455 m.	457 m.	457 sh.	—	—	—	—
	—	—	—	432 m.	436 m.	435 m.	—	—	—	—
	—	—	—	—	—	423 m.	—	—	—	—

possible site symmetry groups of the Si_2O_7 ion, C_1, C_2, and C_s, and of these the last two are more likely.† This conclusion is tentative, since only by studying the spectra of single crystals under high resolution can we be certain that all the components of the resonance multiplets active in the spectra are resolved experimentally.

Compounds of elements between lanthanum and europium synthesized at lower temperatures form a different structural type, I′ (Fig. 101b). The spectra of these compounds show a distinct doublet at 680 to 700 cm^{-1}, which evidently corresponds to ν_s SiOSi, and up to 11 or 12 maxima in the range above 800 cm^{-1}. The increase in the number of bands of Si–O stretching, including the splitting of ν_s SiOSi, may be considered as an indication of a fall in the site symmetry of the Si_2O_7 group to C_1 (if we suppose C_2 or C_s for type I), or else an increase in the number of these groups in the cell as compared with the pyrosilicates of type I. On the other hand, the considerable frequency difference between the two components of ν_s SiOSi and the large number of bands in the spectrum suggests that there might be more than one type of pyro-ions in the lattice, i.e., that these ions might be situated on at least two crystallographic sets. In this case, independently of resonance interactions, the complexity of the spectrum may be due to a difference in the configuration of the pyrogroups filling different sets.

The next structural type is formed by the pyrosilicates of Gd, Dy, Ho, and Y, synthesized at 1400 to 1500°C. In the spectra of these compounds the number of bands in the Si–O stretching region above 800 cm^{-1} is small (Fig. 101c): the presence of altogether six maxima suggests a site symmetry of C_{2v} for the Si_2O_7 ions (the vibration of type A_2 remains inactive). The number of Si_2O_7 ions in the cell is not less than two, since the ν_s SiOSi band in the spectra of $Gd_2Si_2O_7$ and $Y_2Si_2O_7$ is split into two components, while the number of bands between 580 and 400 cm^{-1} reaches seven or eight. It follows from this, however, that the simple form of the spectrum in the high-frequency range is due to the superposition of bands, and the assumption as to a symmetry of C_{2v} for the Si_2O_7 group is insufficiently well-founded. The considerably lower frequency and low intensity of the ν_s SiOSi band indicates an increase in the SiOSi angles in the Si_2O_7 ions as compared with pyrosilicates of types I and I′.

The use of the earlier-mentioned empirical correlation between the quantity $\Delta = (\nu_{as} - \nu_s \text{SiOSi})/(\nu_{as} + \nu_s \text{SiOSi})$ and the SiOSi angle in the Si_2O_7 group (Chap. II) for discussing the shape of these groups in rare-earth pyrosilicates gives $\angle$SiOSi = 120 to 130° for types I and I′ and 150 to 160° for type II, if we suppose, as earlier, that ν_{as}SiOSi corresponds to the highest-frequency maximum in the range 800 to 1100 cm^{-1}.‡

Let us now compare the conclusions drawn from a consideration of the infrared spectra of these silicates with the results of the x-ray structural analyses carried out by Yu. I. Smolin et al. The conclusions regarding the presence of Si_2O_7 groups and the considerable increase in the SiOSi angle in these groups on passing from silicates of types I and I′ to type II are supported by the results of x-ray structural analysis, although the spectroscopic estimate of $\angle$SiOSi for structures I and I′ gives values rather too low. This is primarily associated with the lack of experimental data available for deriving the dependence of $\angle$SiOSi on Δ, particularly in the case of small values of $\angle$SiOSi, for which only results obtained by an x-ray structural

† This conclusion is a consequence of the fact that only for site symmetry C_2 or C_s (with the SiOSi plane of the Si_2O_7 group oriented perpendicular to the crystal symmetry plane) does the representation A_1 of the intrinsic symmetry group (C_{2v}) decompose into one set of representations of the factor group of the crystal and B_1 into another, so that the number of active components of the ν_s SiOSi(A_1) and ν_{as} SiOSi(B_1) vibrations may correspondingly differ.

‡ The use of this criterion for estimating the SiOSi angles in the structure of type I′ on the assumption that the two ν_s SiOSi maxima correspond to two geometrically different pyrogroups leads to the conclusion that these two groups differ by some 10° in their SiOSi angles.

analysis of ilvaite (of rather low accuracy) were available. In Fig. 102 an attempt is made to reconsider the $\angle$SiOSi/Δ relationship in the light of all available data. We must nevertheless consider that, even if a great deal of information is assembled, a relationship of this kind can hardly be expected to give the SiOSi angles in pyrosilicates spectroscopically to an accuracy of better than 10°, since an approach of this kind does not allow for differences in the coordination of the O atom in the Si–O–Si bridge, and these have a considerable effect on the force constants and vibration frequencies even if the value of $\angle$SiOSi remains constant.

We note further that, in pyrosilicate crystals of type I, according to x-ray structural data, the four Si_2O_7 groups occupy a set of general positions, and hence all the bands of vibrations of these groups should have the same multiplicity (two active components each in the infrared spectrum) as a result of resonance interactions. Hence the simple form of the ν_s SiOSi band (and certain other vibrations observed in the infrared spectrum of a polycrystalline sample) is a consequence of inadequate resolution, and not of differences in the selection rules for the "Davydov" components of ν_{as} and ν_s SiOSi. The supposition as to the relation between the substantial ν_s SiOSi splitting in the spectra of pyrosilicates of type I′ (and also the large number of bands in the region above 800 cm^{-1}) and the existence of two "kinds" of configurationally differing Si_2O_7 groups is supported by the results of x-ray structural analysis. However, here also both the number of ν_s SiOSi bands and the total number of bands of the stretching vibrations are nearly twice the number of internal vibrations of the Si_2O_7 group, although on allowing for resonance splittings this would have to be quadrupled (more details regarding the spectra of structurally similar phosphates follow later). Thus the increase in the number of bands of stretching vibrations of the Si_2O_7 groups observed in the spectra of pyrosilicates of type I′ is mainly associated with the existence of two symmetrically (and configurationally) nonequivalent sets of Si_2O_7 groups, while the resonance splittings are small. In pyrosilicates of type II, for a symmetry $Pna2_1-C_{2v}^9$, the four Si_2O_7 groups in the cell also possess a minimum symmetry C_1 [4]. It follows that for each internal vibration of the Si_2O_7 ion there are four vibrations of the crystal (A_1, B_1, A_2, B_2), three of which are active in the infrared spectrum. However, as noted earlier, the number of bands of Si–O stretching vibrations in the spectrum of polycrystalline samples is small. This may partly be explained by the clearly expressed pseudosymmetry of the crystal, which enables us to describe it approximately on the basis of a group of higher symmetry (not differing as regards extinctions), $Pnam-D_{2h}^{16}$, where the Si_2O_7 groups have site symmetry C_s. On considering the correlations with the factor group of the crystal, this gives two "Davydov" components active in the infrared spectrum for ν_s SiOSi, as observed experimentally:

Intrinsic symmetry — Site symmetry — Factor group†

C_{2v} — C_s — D_{2h}

A_1, B_1 — A' — A_g, B_{1g}, B_{2g}, B_{3g}, B_{3u}, B_{2u}, B_{1u}, A_u

A_2, B_2 — A''

† Correlation between the representations of the groups $C_{2v}-C_s-D_{2h}$ demands information regarding the orientation of the element common to all three groups, the symmetry plane; in order to effect this correlation in the present case, we note that the symmetry plane defining the site group of the Si_2O_7 ion in a crystal with symmetry D_{2h}^{16} coincides with the SiOSi plane.

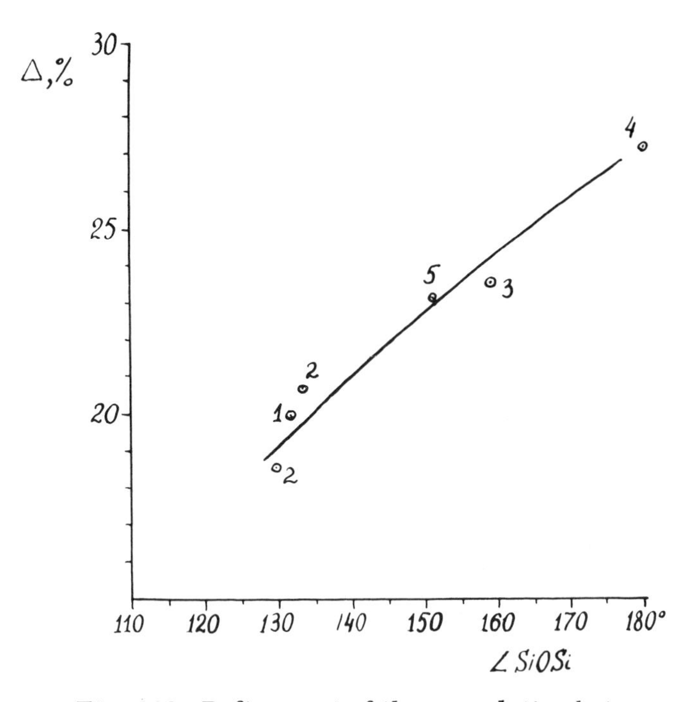

Fig. 102. Refinement of the correlation between $\Delta = (\nu_{as} - \nu_s \text{SiOSi})/(\nu_{as} + \nu_s \text{SiOSi})$ and the angle SiOSi in the Si_2O_7 groups. Structural and spectroscopic data are used for: 1) $Nd_2Si_2O_7$ (I), 2) $Sm_2Si_2O_7$ (I′), 3) $Gd_2Si_2O_7$ (II), 4) $Sc_2Si_2O_7$ (IIIa), 5) hemimorphite. (The value of ν_s SiOSi for $Sc_2Si_2O_7$ is approximate, taken from data relating to solid solutions of $Sc_2Si_2O_7$–$Sc_2Ge_2O_7$.)

However, the number of bands observed experimentally above 800 cm^{-1} still remains only half that permitted by this approximate symmetry (allowing for resonance splittings). This is probably associated with the fact established by x-ray diffraction, that the configuration of the Si_2O_7 group deviates little from the higher symmetry (C_{2v}) and that the resonance splittings of the stretching frequencies are slight.

The spectra of the pyrosilicates of Er, Tm, and Yb, and also $Ho_2Si_2O_7$ and $Y_2Si_2O_7$ synthesized below 1300°C, differ considerably from the spectra of compounds of types I, I′, and II in respect of the small number of bands above 800 cm^{-1} and the absence of any ν_s SiOSi vibration band. The bands below 600 cm^{-1} in these spectra are too strong to be ascribed to ν_s SiOSi. The reduction in this frequency taking place on increasing ∠SiOSi should be accompanied by a fall in intensity. We see from Figs. 101d and 101e that the spectra of these compounds are similar to that of thortveitite ($Sc_2Si_2O_7$), the structure of which is well known [48]; the presence of centrosymmetric Si_2O_7 groups in this structure (i.e., those with ∠SiOSi = 180°) is indicated both in the loss of the ν_s SiOSi band, forbidden by the selection rules, and in the increased frequency of ν_{as} SiOSi. The spectrum of $Yb_2Si_2O_7$, which is equivalent as regards the number of bands to that of $Sc_2Si_2O_7$, may be interpreted by analogy with the latter (Table 52). The appearance of four Si–O stretching bands instead of the three allowed by the selection rules for symmetry D_{3d} is associated, as in the case of thortveitite, with the fall in the site symmetry (to C_{2h}) with the removal of degeneracy: $E_u \rightarrow B_u + A_u$. A certain difference in the spectra of $Yb_2Si_2O_7$ and $Sc_2Si_2O_7$ below 600 cm^{-1} is associated with the change in the mass of the cation, which as a result of the strong interaction of the vibrations is also reflected in the internal vibrations of the

TABLE 52. Frequencies of the Absorption Maxima in the Infrared Spectra of the Rare-Earth Pyrosilicates; Structures with Centrosymmetric ("Straightened") Si_2O_7 Ions

Type IIIb			Identification of frequencies,* ion Si_2O_7				Type IIIa	
$Ho_2Si_2O_7$	$Y_2Si_2O_7$	$Er_2Si_2O_7$	C_{2h} ←	C_i ←	D_{3d} →	C_{2h}	$Yb_2Si_2O_7$	$Sc_2Si_2O_7$
1117 s.	1119 s.	1119 s.	A_u	A_u	ν_{as} SiOSi, A_{2u}	B_u	1110 s.	1169 s.
1087 s.	1091 s.	1095 s.	B_u				—	1104 sh.
—	970 sh.	—					—	—
943 v.s.	947 v.s.	948 v.s.	$2A_u$	$2A_u$	ν'_{as} SiO_3, E_u	A_u	955 v.s.	975 v.s.
910 v.s.	914 v.s.	919 sh.	$2B_u$			B_u	915 v.s.	907 v.s.
860 v.s.	857 v.s.	857 v.s.	$B_u + A_u$	A_u	ν'_s SiO_3, A_{2u}	B_u	852 v.s.	852 v.s.
582 s.	589 s.	588 s.	A_u	A_u	δ'_s SiO_3, A_{2u}	B_u	566 v.s.	587 v.s.
561 sh.	572 sh.	570 sh.	B_u				—	—
—	—	542 sh. }					—	—
518 v.s.	520 v.s.	523 v.s. }	$2A_u$	$2A_u$	ρ' SiO_3, E_u	A_u	498 s.	501 v.s.
467 v.s.	470 v.s.	470 v.s.	$2B_u$			B_u	476 v.s.	468 sh.
436 sh.	—	—	$2A_u$	$2A_u$	δ'_{as} SiO_3, E_u	A_u	—	434 s.
~410 sh.	—	~415 m.	$2B_u$			B_u	—	—

*Only the vibrations of the Si_2O_7 ion active in the infrared spectrum are shown.

anion. It should be noted that the frequency ν_{as} SiOSi also changes considerably from 1169 cm^{-1} in the $Sc_2Si_2O_7$ spectrum to 1110 cm^{-1} for $Yb_2Si_2O_7$. The identification of the frequencies of the deformation vibrations in Table 52 is simply based on symmetry considerations and cannot be reliable if we remember that the lattice frequencies are likely to be quite high in these crystals. A more rigorous analysis of the vibrations of thortveitite and compounds isostructural with it is presented in the next section.

The spectra of $Er_2Si_2O_7$ and the low-temperature forms of $Y_2Si_2O_7$ and $Ho_2Si_2O_7$ differ from the spectrum of $Yb_2Si_2O_7$ in the doublet structure of the ν_{as} SiOSi band and the greater number of bands of the deformation vibrations. In the spectrum of thortveitite the ν_{as} SiOSi band is single, as in the spectrum of $Yb_2Si_2O_7$, although it has a weak satellite at 1100 cm^{-1}. Crystals of $Sc_2Si_2O_7$ contain one Si_2O_7 ion in the unit cell, and at least at low temperatures (i.e., for an unexcited lattice) this eliminates the possibility of a resonance structure of the bands. The satellite of the ν_{as} SiOSi band in the spectrum of $Sc_2Si_2O_7$ is greatly weakened on passing to $-150°C$, which suggests ascribing it to transitions from excited states (see Chap. VI).

In the spectrum of erbium pyrosilicate the components of the ν_{as} SiOSi band are similar in intensity, and their ratio remains constant on lowering the temperature, indicating that both bands belong to transitions from the ground state. In the region of the deformation vibrations of the anion in the spectra of these silicates the highest-frequency band, evidently associated with the δ'_s SiO_3 vibration, is split; this vibration, like the ν_{as} SiOSi, belongs to the nondegenerate representation A_{2u} of the group D_{3d}. Since the ionic radii of the cations increase in the sequence Sc^{3+}, Yb^{3+}, Tm^{3+}, Er^{3+}, Y^{3+}, Ho^{3+}, we may suppose that, while preserving the similar structure (centrosymmetry) of the pyrogroups, passing from $Yb_2Si_2O_7$ and $Tm_2Si_2O_7$ to $Er_2Si_2O_7$ leads to an increase in the volume of the primitive cell by at least a factor of two, and the doublet structure of the bands has a resonance origin. Thus we may assume the existence of two types of crystal structure of the rare-earth pyrosilicates with centrosymmetric anions. Subsequently the thortveitite type ($Sc_2Si_2O_7$, $Yb_2Si_2O_7$, $Tm_2Si_2O_7$) is denoted IIIa and the $Er_2Si_2O_7$ type IIIb.

These considerations are supported by crystallographic data ([5] and also [49]) regarding the pyrosilicates of type IIIb ($Y_2Si_2O_7$, $Er_2Si_2O_7$): for a symmetry of $P2_1/c-C^5_{2h}$ these crystals contain two Si_2O_7 "molecules" in the primitive cell, which leads to the doubling of the bands of vibrations active in the infrared spectrum for a site symmetry C_i, as shown in the left-hand part of Table 52, without removing the mutual exclusion rule. The x-ray analysis of $Yb_2Si_2O_7$ carried out by Yu. I. Smolin confirms its isostructural character with respect to thortveitite; the observed fall in the ν_{as} SiOSi frequency on making the transition $Sc_2Si_2O_7 \rightarrow Yb_2Si_2O_7$ is explained by a certain lengthening of the Si–O(Si) bonds.

Changes in the structure of the rare-earth pyrosilicates with temperature and composition [44,45,47] are of considerable interest, primarily in order to explain the reasons for the sequence of the structural types in the $Ln_2Si_2O_7$ compounds and the accompanying changes in the structure of the Si_2O_7 ion. Thus a study of solid solutions in the $Ln'_2Si_2O_7$–$Ln''_2Si_2O_7$ systems, in which the end members of the series are of different structural types, should enable us to judge the nature of the transition from one structure to the other, even independently of whether any x-ray diffraction data regarding the structure of the crystals under consideration are available. The use of infrared spectroscopy, as one of the most accessible and trouble-free methods of phase analysis, would in these cases appear most justified.

Figures 103a and b show the infrared spectra of a number of samples of different compositions relating to the systems $(La_{1-x}, Yb_x)_2Si_2O_7$ and $(Y_{1-x}, Er_x)_2Si_2O_7$, prepared at temperatures of about 1500°C. The values of x are indicated on the curves. In the first of the systems, for x values between 0.5 and 0.9 the spectrum exhibits bands characteristic of both end members of the series. On the other hand, in the range $0.0 < x < 0.5$ and $0.9 < x < 1.0$, the spectrum contains unmixed, though somewhat diffuse, bands of one of the components only. These results suggest the existence of a region of decomposition of the solid solutions. In the second system, over the range $0.4 < x < 0.8$, the spectra also indicate the presence of two phases of different structures. The frequencies of each of the phases remain almost constant as the concentration of the second phase varies. In particular, the fact that the ν_s SiOSi band is displaced

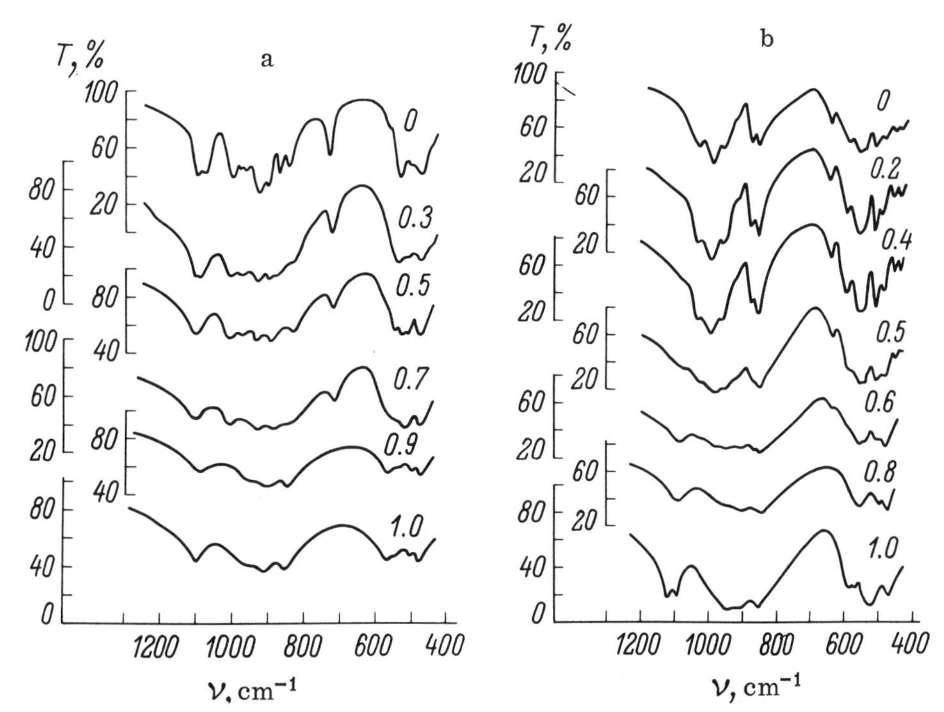

Fig. 103. Infrared spectra of solid solutions of rare-earth pyrosilicates. a) $(La_{1-x}, Yb_x)_2Si_2O_7$; b) $(Y_{1-x}, Er_x)_2Si_2O_7$. Value of x for each sample shown on the curves.

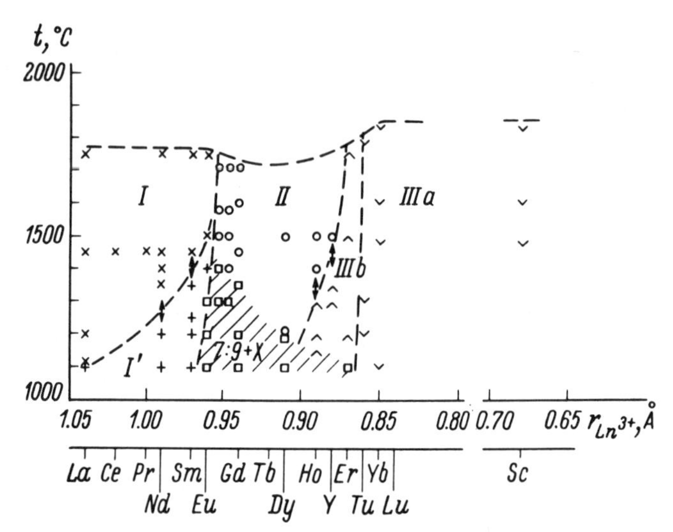

Fig. 104. Isomorphism of rare-earth pyrosilicates. Ranges of stability of five structural types of compounds $Ln_2Si_2O_7$ in relation to the temperature and ionic radius of the cation Ln^{3+} (two-phase region shaded).

very little (between 732 and 721 cm^{-1} for the La–Yb and between 631 and 634 cm^{-1} for the Y–Er system) indicates that the configuration of the Si_2O_7 groups remains unchanged. Thus the transition from one structural type to another takes place suddenly without the formation of any "intermediate" structures.

An analogous kind of transition from one structural type to another occurs for certain pyrosilicates on introducing a small quantity of a pyrosilicate crystallizing with a structure of a different type. Thus in the case of $Eu_2Si_2O_7$ for a synthesis temperature of 1500°C it was found that the compound could be obtained not only with a structure of type I but also one of type II, depending on the purity of the original Eu_2O_3. On adding to the purest $Eu_2Si_2O_7$ crystallizing with a structure of type I some 5 mol. % of $Gd_2Si_2O_7$ there were no changes in the structure, whereas for the same amount of $Dy_2Si_2O_7$ or $Y_2Si_2O_7$ (i.e., pyrosilicates with much smaller cations) the $Eu_2Si_2O_7$ was almost entirely converted into a pyrosilicate of type II. No intermediate structures were found in this case.

The classification of rare-earth pyrosilicates with respect to structural types presented in the previous section indicates that many of these compounds are dimorphic. Hence in addition to the ionic radius of the cation the factors determining the type of crystal structure must also include the temperature. For this reason the spectra of rare-earth silicates synthesized at various temperatures were studied systematically. The results may conveniently be presented in the form of a diagram characterizing the ranges of stability of various structural types in coordinates of temperature and ionic radius of the cation Ln^{3+} (Fig. 104).

Such a diagram, representing essentially a set of cross sections of particular phase diagrams of the Ln_2O_3–SiO_2 systems for the composition $Ln_2O_3 \cdot 2SiO_2$, reveals certain general laws. The lines separating the regions of stability in the coordinates chosen are not vertical, but in all cases are inclined to a varying extent, in the direction of cations of smaller radius for higher temperatures; that is, for the same cation, the structure characteristic of pyrosilicates with larger cations may be obtained at higher temperatures. This is evidently a consequence of the previously identified increase in the effective radius of the cation with increasing temperature. It should be noted that all the transitions of this kind are evidently reversible;

the transitions for which reversibility has been verified experimentally are indicated in Fig. 104 by double arrows. In every case the transition takes place quite slowly. Thus for $Y_2Si_2O_7$ the transformation of the low-temperature form (type IIIb) into the high-temperature form (type II) takes place in about an hour at 1600 to 1500°, while the reverse transformation is achieved by holding for about a day at 1300 to 1350° or a week or more at 1200°C.† We may thus suppose that the transformations under consideration are associated not with deformation but with a radical rearrangement of the crystal structure.

The fact that the structure of scandium and yttrium pyrosilicates, like the polymorphic transformation of the latter, fits within this simple general scheme based on temperature and the radius of the cation supports the view that the transformations from one structural type to another are associated with a change of coordination (coordination number or shape of the coordination polyhedron) of the cation. Such a transformation should, firstly, be accompanied by a considerable change in the whole structure ("packing") of the crystal, and, secondly, exclude the possible formation of "intermediate" structures or smooth transitions from one type to another.

The boundaries of stability between types I and I′ or I′ and II at high temperatures approach one another and both lie in the range of ionic radii Eu^{3+}–Gd^{3+}. A study of samples with compositions $2Eu_2Si_2O_7 \cdot Gd_2Si_2O_7$ and $Eu_2Si_2O_7 \cdot 2Gd_2Si_2O_7$, also represented in the scheme of Fig. 104, brings these two lines still closer together if one considers that these substances resemble pyrosilicates with ionic radii intermediate between Eu^{3+} and Gd^{3+}. However, there are no grounds for asserting that the lines coincide completely.

It should be noted that the region shown in Fig. 104 by the shading indicates the existence of two crystalline phases instead of the pyrosilicate. One of these is an orthosilicate of composition $7Ln_2O_3 \cdot 9SiO_2$, while the other, still not separated in pure form, is evidently a silicate with a higher degree of condensation of the tetrahedra than the pyrosilicates.‡ The spectra of samples of composition $Gd_2O_3 \cdot 2SiO_2$ synthesized at 1500° and 1200° are shown in Fig. 105. The first of these is the spectrum of gadolinium pyrosilicate, while the second, judging from the presence of two fairly strong bands, 31 cm^{-1} apart, in the ν_s SiOSi frequency range, may be considered, for example, as the spectrum of a metasilicate of chain structure with two tetrahedra in the identity period of the chain, or a trisilicate with Si_3O_{10} groups containing two Si–O–Si bridges in one complex anion. Although this compound has not been obtained in the form of single crystals, the second proposition seems the more probable, since it agrees better with the considerable number of maxima in the range 850 to 1050 cm^{-1}, while the synthesis of samples with an oxide ratio of $4Gd_2O_3 \cdot 9SiO_2 = Gd_8(Si_3O_{10})_3$ yielded a product purer in phase composition (and with a more discrete infrared spectrum) than the synthesis of samples of composition $Gd_2O_3 \cdot 3SiO_2$ corresponding to the metasilicate. Similar spectra were obtained for samples of composition $Ln_2O_3 \cdot 2SiO_2$ in the range Ln = Eu to Er on synthesis (by coprecipitation) at low temperatures. On holding these samples at higher temperatures the corresponding pyrosilicates were identified spectroscopically (type I′, II, or IIIb), although the reverse transformation could not be effected. However, before examining single crystals of phase X we cannot establish reliably whether the composition of this phase corresponds to the formula $Ln_2O_3 \cdot 2SiO_2$, i.e., associate the existence of anions more highly condensed than Si_2O_7 with the

† The dimorphic nature of synthetic $Y_2Si_2O_7$ suggests that the polymorphism of natural yttrium pyrosilicates is associated not with composition only but also with the temperature of formation, and reduces to a transformation from structures of type IIIb to type II.

‡ It is difficult to distinguish the frequencies of the compound $7Gd_2O_3 \cdot 9SiO_2$ on the background of the spectrum of this compound, so far provisionally known as the X phase, as they fall into a region of strong absorption and only appear by virtue of the spread in the bands; the presence of the compound may easily be identified, however, by x-ray diffraction.

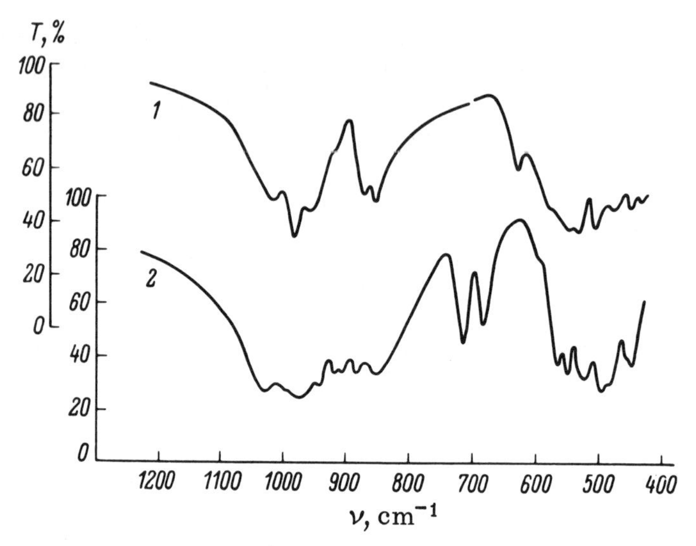

Fig. 105. Isomorphism of rare-earth pyrosilicates. Spectrum of a sample of composition $Gd_2O_3 \cdot 2SiO_2$ (pressed in KBr, 5 mg): 1) compound $Gd_2Si_2O_7$ (type II), 2) spectrum of the same sample synthesized at low temperatures (in the two-phase region).

simultaneous presence of "additional" O^{2-} anions or SiO_4 ions, as observed in certain R.E. pyrogermanates containing Ge_3O_{10} and GeO_4 ions, or whether the presence of condensed systems of tetrahedra (Si_3O_{10} or $[Si_2O_6]_\infty$) is directly reflected in its composition. In the first case the appearance of compounds $7Ln_2O_3 \cdot 9SiO_2$ should be considered as a chance occurrence, while in the second it is a necessary consequence of the difference of the initial oxide ratio ($Ln_2O_3 \cdot 2SiO_2$) from the composition of phase X. It is possible that the similarity in the position of the high-temperature boundary of stability of phase X in Fig. 104 to the corresponding boundary for the compound $7Ln_2O_3 \cdot 9SiO_2$ (Fig. 93) is no chance matter.

In a recent paper [50] Ito systematically studied the polymorphic transformations of R.E. silicates of composition $Ln_2O_3 \cdot 2SiO_2$ in the range Ln = Gd to Lu (Sc). The results were similar in general features to the corresponding part of the diagram in Fig. 104. The phases denoted by β, γ, and δ in [50] evidently correspond to those known here as $Ln_2Si_2O_7$ of types IIIa, IIIb, and II. Certain discrepancies in the positions of the boundary lines for comparatively low temperatures (the existence of a form with the structure of thortveitite for $Er_2Si_2O_7$–$Ho_2Si_2O_7$, not observed in our own work) may be associated with differences in the method of synthesis, which in the case of [50] enabled equilibrium to be reached more rapidly for the specified temperature. Regarding the α phase, the range of stability of which roughly coincides with our own range of existence of the phase X and a silicate of composition $7Ln_2O_3 \cdot 9SiO_2$, there were no direct indications in [50] regarding its possible multiphase nature or any difference in the composition from $Ln_2Si_2O_7$. The characteristics of this phase are limited to the interplane distances for α-$Y_2Si_2O_7$; proofs of the reversibility of the transformations $\alpha \rightarrow \beta$ or $\alpha \rightarrow \delta$ (X to IIIa or X to II in our own notation) are not presented (reversibility is only demonstrated for the transformation $\gamma \rightarrow \delta$, i.e., IIIb to II).

Let us finally discuss the form of the lines separating the ranges of stability of various structural types on the diagrams plotted in coordinates of cation radius and temperature. If we suppose that a factor determining the transformation from one structural type to another (polymorphic transformation) is the change in the coordination of the cation on reaching a certain "critical" ratio of cation and anion sizes, then the boundary line corresponds to the ratio

r_K/r_A = const. Both fundamental characteristics of the form of these lines may be qualitatively explained on the basis of very general considerations of the dynamics of crystals with monatomic cations and polyatomic anions, the bonds within which are mainly covalent. The slope of the boundary line in the direction of small r_K at high temperatures, i.e., the rise in $r_{K\,eff}$ with increasing temperature (in other words the fact that $\partial(r_K/r_A)/\partial T > 0$), is associated with the relatively lower frequencies of the vibrations of the coordination polyhedra of the cations; the "population" of the excited states for these vibrations is more substantial than for the stretching vibrations of the complex anion, so that temperature has a greater influence on the effective dimensions of the cation. The second characteristic, the reduction in the slope in question with rising temperature (i.e., $\partial^2(r_K/r_A)/\partial T^2 < 0$), may be associated with the rise in the population of the excited vibrational states of the anion, which for fairly high temperatures largely compensate further increases in the effective dimensions of the cation.

The isomorphism of $M_2^{\cdot\cdot\cdot}Si_2O_7 - M_2^{\cdot\cdot}P_2O_7$ for the particular case of relatively small radii of the cations is well known: $Sc_2Si_2O_7$–β-$Mg_2P_2O_7$ and also $Mn_2P_2O_7$, β-$Cu_2P_2O_7$, and β-$Zn_2P_2O_7$.† We have seen that the sequence of structural types of the compounds $Ln_2Si_2O_7$ is mainly determined by "packing" factors, i.e., by the temperature-corrected ionic radius of the cation, the increase in which between Sc^{3+} and La^{3+} leads to a complete rearrangement of the structure at certain critical values as a result of a change in coordination. This suggests the possibility of a similar sequence of structural types for the pyrophosphates $M_2P_2O_7$ as the radii of the cations M^{2+} and the temperature vary. Since existing structural data were inadequate for verifying this directly, it seemed useful to compare the diagram of stability regions for the polymorphic forms of the rare-earth pyrosilicates with the corresponding diagram for the pyrophosphates of the alkaline-earth elements obtained in [51] on the basis of data relating to calcium, strontium, and barium pyrophosphates and their solid solutions. For this group of

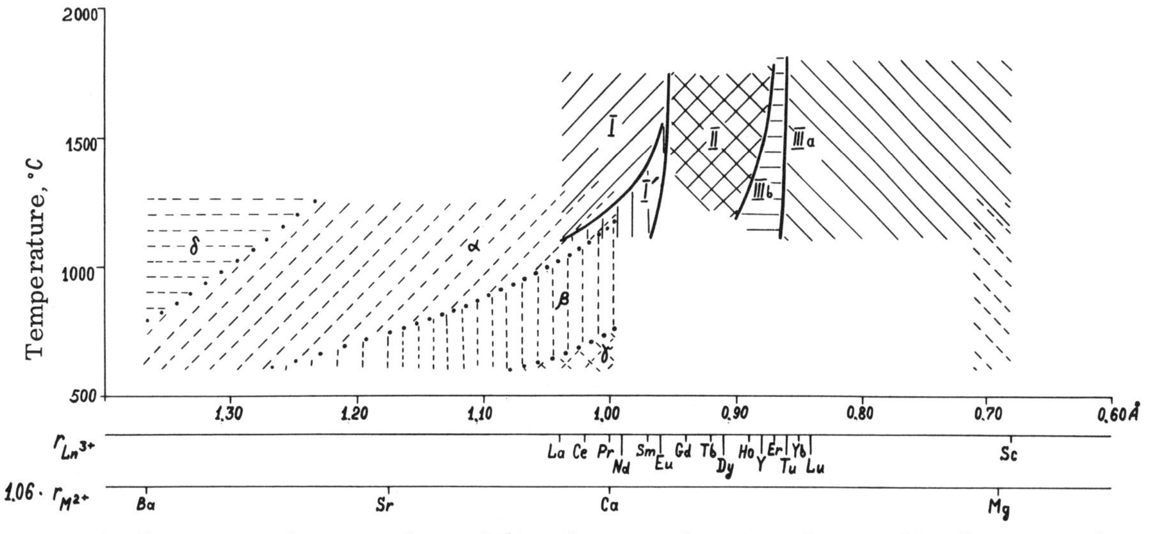

Fig. 106. Comparison between the stability diagram of various forms of $Ln_2Si_2O_7$ (see also Fig. 104) and that of polymorphic forms of the pyrophosphates of alkaline-earth elements $M_2P_2O_7$.

† Here, however, the similarity of the structures is achieved not by straightening ∠POP to 180° like the SiOSi angle in $Sc_2Si_2O_7$, but statistically, by averaging over the orientations of the bent P_2O_7 groups (see Chap. VI and also [54, 55]).

compounds altogether four polymorphic forms, designated δ, α, β, γ, have been described, of which the first has only been obtained for $Ba_2P_2O_7$, the second for the pyrophosphates of Ba, Sr, and Ca, the third for those of Sr and Ca, and the fourth only for $Ca_2P_2O_7$ (at low temperatures).

Figure 106 [52] shows both diagrams, for $Ln_2Si_2O_7$ and $M_2P_2O_7$, the latter also having been reduced to the Zachariasen 1954 system of ionic radii. In order to allow for differences in the dimensions of the silicate and phosphate anion, since the type of structure is determined not so much by the absolute as by the relative dimensions of the cation and complex anion, the scale of the ionic radii of the cations M^{2+} in the phosphates is expanded, i.e., it is considered that $r_{eff.\,M^{2+}} = 1.06\,r_{M^{2+}}$. The coefficient 1.06 is obtained from the ratio of the average distances in the SiO_4 and PO_4 groups as indicated by Cruickshank [53]: $r_{SiO}/r_{PO} = 1.63\,Å/1.54\,Å$. On allowing for this correction the boundaries of the regions of stability of pyrosilicates of types I and I′ and pyrophosphates of types α and β almost coincide, and the same holds for pyrosilicates of types I′ and II and pyrophosphates of types β and γ. Exact agreement between the boundaries is achieved on increasing the ("scale") correction factor to 1.08.

It was concluded from the foregoing that $Ln_2Si_2O_7$ compounds of type I were probably isomorphous or at least isotypic with α-pyrophosphates, those of type I′ with β-pyrophosphates, and those of type II with γ-$Ca_2P_2O_7$. The x-ray data now available relating to the rare-earth pyrosilicates and also β-$Ca_2P_2O_7$ [56], α-$Ca_2P_2O_7$ [57], and α-$Sr_2P_2O_7$ [58, 59] support at least the first two of these analogies. (The fourth δ form of the alkaline-earth pyrophosphates, only found for $Ba_2P_2O_7$, lies outside the limits of the dimensional relationships established in the rare-earth pyrosilicates.)

Comparison of the infrared spectra of the pyrosilicates and pyrophosphates by itself cannot serve as a proof of similarity in their structure, since not only the frequencies but also the forms of the vibrations of the Si_2O_7 and P_2O_7 groups are very different;† however, if in identifying the frequencies we consider the calculation given in [60] for the spectra of the alkali pyrophosphates and also count the number of bands in various quasicharacteristic ranges of frequency, we find that the spectroscopic data in no way contradict the proposed structural analogy. The starting point for this lies in the structural analysis of β-$Ca_2P_2O_7$, which explains the more complicated spectrum of this form of pyrophosphates.

Figure 107 shows the infrared spectra of the pyrophosphates, while the frequencies of the absorption maxima are given in Table 53 (similar results are obtained for the spectra of these compounds in [61]).

The β-$Ca_2P_2O_7$ crystal with symmetry $P4_1-C_4^2$ contains eight P_2O_7 groups in the cell; these occupy two four-fold sets of general positions [56]. Each internal vibration of the P_2O_7 ions occupying one of the sets contributes to crystal vibrations of symmetry types E, A, and B, of which the first two are active in the infrared spectrum.

† The slight changes in masses and bond lengths on replacing Si by P cause no great changes in the kinetic energy matrices (if $\angle SiOSi \simeq \angle POP$), but the force constants are substantially different, with $K_{PO^-}/K_{PO(P)} > K_{SiO^-}/K_{SiO(S)}$, which leads to a change in the form of the vibrations, particularly for the strongly-interacting vibrations $\nu_{as}XOX$ and $\nu_s'XO_3$. As a result of this the $\nu_{as}SiOSi$ vibration primarily corresponds to the highest-frequency band in the spectrum of the Si_2O_7 ion, while in the case of the P_2O_7 ion (at least for $\angle POP < 180°$) the $\nu_{as}POP$ vibration is more associated with the band in the range 900 to 1000 cm^{-1}, i.e., lower than the ν_{as}, $\nu_{as}'PO_3$, and ν_s, $\nu_s'PO_3$ frequencies. In considering the "Davydov" structure of the vibration bands of the P_2O_7 ion, greatest interest lies in the $\nu_s POP$, and partly the $\nu_{as}POP$ range of frequencies, as being freer from overlapping with the vibration bands of other types.

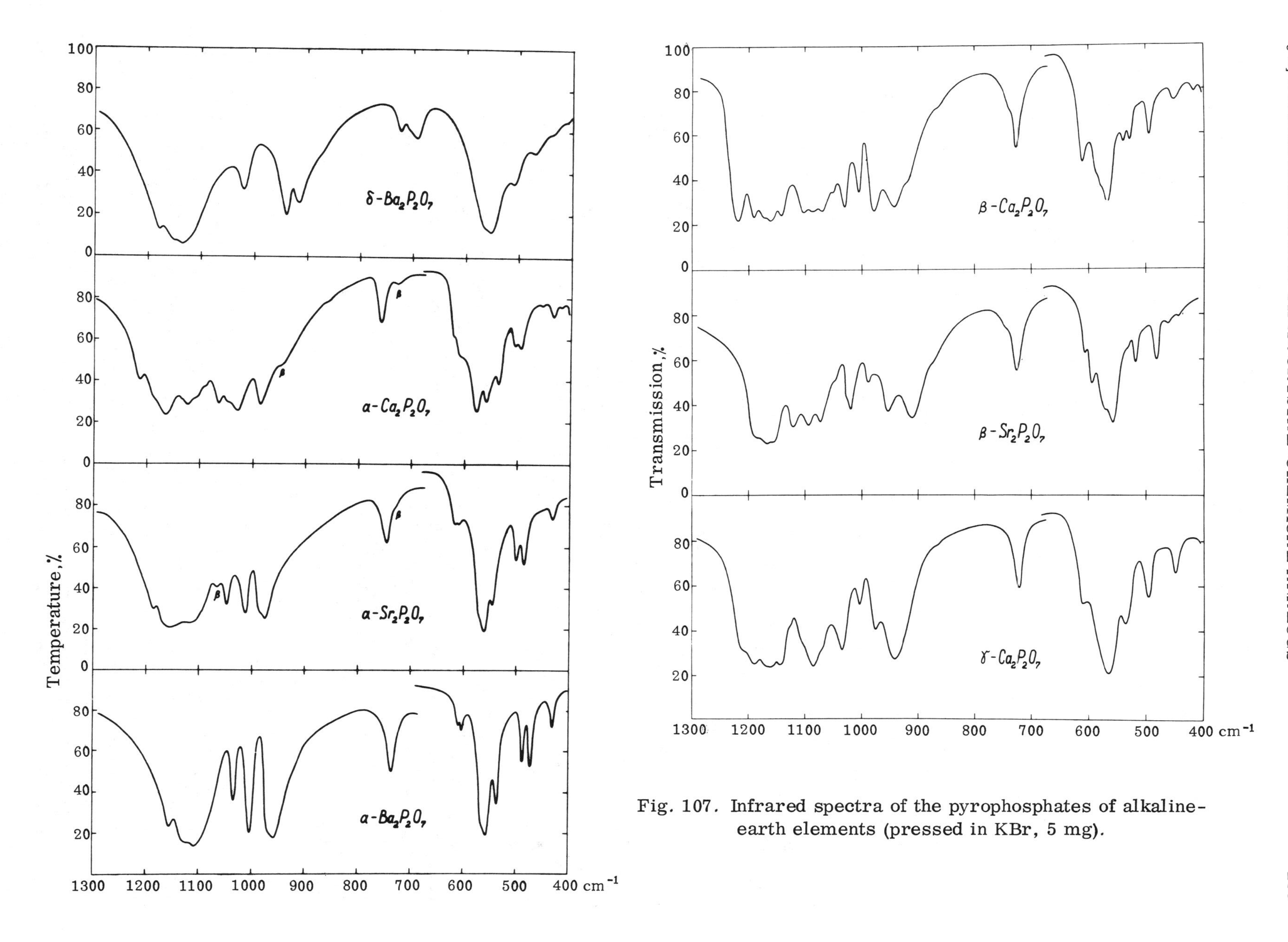

Fig. 107. Infrared spectra of the pyrophosphates of alkaline-earth elements (pressed in KBr, 5 mg).

TABLE 53. Frequencies of the Absorption Maxima in the Infrared Spectra of the Pyrophosphates of the Alkaline-Earth Elements

Assignments (and symmetry types for C_{2v})	δ-form	α-form			β-form		γ-form
	$Ba_2P_2O_7$	$Ba_2P_2O_7$	$Sr_2P_2O_7$	$Ca_2P_2O_7$	$Sr_2P_2O_7$	$Ca_2P_2O_7$	$Ca_2P_2O_7$
				1213 s.		1218 v.s.	1211 sh.
	1179 v.s.		1187 s.	1188 sh.	1187 v.s.	1191 v.s.	1191 v.s.
	1154 sh.	1156 s.	1168 sh. ?	1166 v.s.	1168 v.s.	1173 sh.	1163 v.s., wide
ν_{as}, ν'_{as} PO_3	1136 v.s., wide	1126 v.s.	1155 v.s., wide	1135 sh. ?	1156 v.s.	1161 v.s.	1143 v.s.
(A_1, A_2, B_1, B_2)		1109 v.s.	1115 v.s., wide	1123 v.s.	1122 s.	1143 v.s.	1126 sh.
				1106 sh.	1099 sh. ?	1103 v.s.	1104 sh.
				1092 sh.	1094 s.	1087 v.s.	1085 v.s.
			1066 m. [β?]		1073 s.	1070 v.s.	1071 sh.
		1034 s.	1048 s.	1065 s.	1049 sh.	1051 m.	
ν_s, ν'_s PO_3	1021 s.			1047 sh.	1026 s.	1031 v.s.	1035 s.
(A_1, B_1)		1003 v.s.	1013 s.	1029 v.s.	1020 s.	1006 s.	1004 m.
	940 v.s.	969 sh.	985 sh.	986 v.s.	989 m.	979 v.s.	976 s.
ν_{as} POP	920 sh. ?	958 v.s.	976 v.s.	~950 sh. [β?]	954 s.	943 v.s.	942 v.s.
(B_1)	916 s.				910 v.s.	923 sh.	
	870 sh.			865 sh.	876 sh.	866 sh.	866 sh.
	723 w.	736 m.	746 m.	758 m.	745 sh.	738 sh.	
ν_s POP	709 sh. ?		724 sh. [β?]	727 w. [β?]	726 m.	727 m.	721 m.
(A_1)	693 m.						
		609 w.	617 w.	620 sh.			
δ_{as}, δ'_{as},		603 w.	610 w.	608 sh.	607 m.	612 s.	610 s.
δ_s, δ'_s PO_3				577 v.s.	596 m.	587 sh.	
($2A_1$, A_2, $2B_1$, B_2)	565 sh.	564 sh.	571 sh.	559 v.s.	570 sh.	579 sh.	565 v.s.
and $\rho PO_3(B_2)$	554 v.s.	556 v.s.	562 v.s.	536 s.	558 v.s.	568 v.s.	
					532 sh.	540 m.	535 s.
	510 s.	487 m.	500 m.	505 m.	518 m.	528 m.	
	470 m.	472 m.	486 m.	493 m.	482 m.	495 s.	496 s.
ρPO_3				453 v.w.	460 v.w.		
(A_1, A_2, B_1)	440 w.	431 w.	432 w.	431 w.	443 v.w.	452 w.	449 m.
	410 v.w.			418 v.w.		417 v.w.	

Note: The sign "β?" indicates bands ascribable to the presence within the α-forms of small quantities of the low-temperature β-forms.

Since the P_2O_7 groups in different sets are geometrically different, the frequencies of the vibrations of types E and A for the groups of one set may differ considerably from the corresponding frequencies of the P_2O_7 groups of the other set, independently of whether the resonance exchange energy between the vibrations of the pyro-groups of different sets belonging to the same representation of the factor group of the crystal (E or A) is large or small [63]. Hence each vibration of the P_2O_7 ion should appear in the IR spectrum as a doublet corresponding to the two "kinds" of P_2O_7 ions, and each of its components may itself have a doublet structure (E and A vibrations). Altogether in the νPO range of frequencies there may be from 16 (2× 8) to 32 (4× 8) maxima. The number of maxima observed experimentally in the spectra of the β-pyrophosphates is closer to the first of these figures. It is an important point, however, that in the ν_{as} POP range there are no less than three maxima, while in the ν_s POP range only two are resolved. In agreement with the similarity of the structures, the rare-earth pyrosilicates of type I′ are also distinguished by a relatively complex spectrum; the doublet observed in the ν_s SiOSi range of frequencies, having a fairly large distance between its components, suggests that the Si_2O_7 groups are arranged in two crystallographic sets, as mentioned earlier.

As in the spectra of the pyrosilicates of type I, in the spectrum of α-$Ca_2P_2O_7$ there are 11 or 12 maxima in the frequency range of the stretching vibrations of the anion; a fall in this number to eight or nine in the spectra of α-$Sr_2P_2O_7$ and α-$Ba_2P_2O_7$ was initially ascribed to a reduction in the resonance splitting of the frequencies on increasing the interionic distances in the phosphates of Sr and Ba. The solution of the structures of α-$Sr_2P_2O_7$ [58, 59] and α-$Ca_2P_2O_7$ [57] also associates this difference with the change in the multiplicity of the bands on passing from the orthorhombic α-$Sr_2P_2O_7$ (for which only some of the vibration bands of the P_2O_7 ion can have two active components) to the monoclinic $P2_1/n$ (although similar to α-$Sr_2P_2O_7$) calcium pyrophosphate, in the spectrum of which the selection rules lead to a doublet structure of all the anion bands. In the spectra of all these phosphates the ν_s POP band, like the ν_s SiOSi in the spectra of type-I silicates, shows no doublet structure, in contrast to the case of ν_{as} POP, although this would follow for both bands from the selection rules. This again demonstrates the inadequate reliability of any conclusions regarding the absence of Davydov splitting if the structure of the band is not resolved in the spectrum of a polycrystalline sample.

Pyrosilicates of type II contain four Si_2O_7 groups (in a general position) in the cell of a crystal with symmetry $Pna2_1$-C_{2v}^9, which gives three components active in the infrared spectrum for each internal vibration. However, the experimentally-observed spectrum is simple (possible reasons were discussed earlier); thus, the presence of fine structure in the ν_s SiOSi band could only be detected in the spectra of a few silicates of this type. Hence the absence of fine structure from the ν_s POP band in the spectrum of γ-$Ca_2P_2O_7$ does not exclude the possibility of its structural similarity to type-II pyrosilicates, all the more so because there are two maxima in the ν_{as} POP region, one of them being broad.

Owing to the absence of M^{2+} cations of appropriate size, in the diagram for the phosphates in Fig. 106 there is a gap between Mg^{2+} and Ca^{2+}. An attempt at filling this by means of solid solutions formed between $Ca_2P_2O_7$ and $Mn_2P_2O_7$ was unsuccessful owing to the heterophase nature of the resultant samples. A study of the spectrum of $Sn_2P_2O_7$ revealed no similarity with the spectra of pyrosilicates of types IIIa, IIIb, or II; this may be attributed to the substantially directional character of the bonds formed by the tin atoms and hence a deviation from the predominantly "packing" relationships determining the structures here considered. It should be noted that the proposed isomorphism of the rare-earth pyrosilicates with the alkaline-earth pyrophosphates is also of interest from the point of view of the well-known properties of the latter as phosphors.

Infrared spectra of the pyrogermanates of the rare-earth elements partly described in [37, 63-65] show no similarity to the spectra of the corresponding silicates (except for $Sc_2Ge_2O_7$ considered in the previous section). The main difficulty in identifying the frequencies in the spectra of these compounds is the strong interaction of the lattice vibrations with the internal vibrations of the complex anions (of lower frequency than silicates), which not only impedes the identification of the deformation vibrations but also complicates the problem of separating them from the stretching vibrations of the anion. Hence when studying condensed R.E. germanates the identification of the ν_s GeOGe vibration bands, the most important indicator of the degree of condensation of the germanium–oxygen tetrahedra and the configuration of the complex anion, is unreliable.† Additional difficulties are associated with the absence of experimental data regarding the spectra of alkali or alkaline-earth pyrogermanates, which would have enabled us to estimate the range of ν_s GeOGe frequency values.

In an initial investigation it was established that the spectra of the R.E. pyrogermanates differed sharply for compounds of La to Gd and Dy to Yb; for the first of these groups the spectrum of lanthanum germanate differed somewhat from the rest, suggesting the possibility of distinguishing this compound as a separate structural type. The infrared spectra of the R.E. pyrogermanates given in Fig. 108a, b are similar to those obtained earlier [63, 65], but in a number

† Difficulties of this kind are typical of the spectra of condensed germanates with divalent cations if the mass of the latter is small.

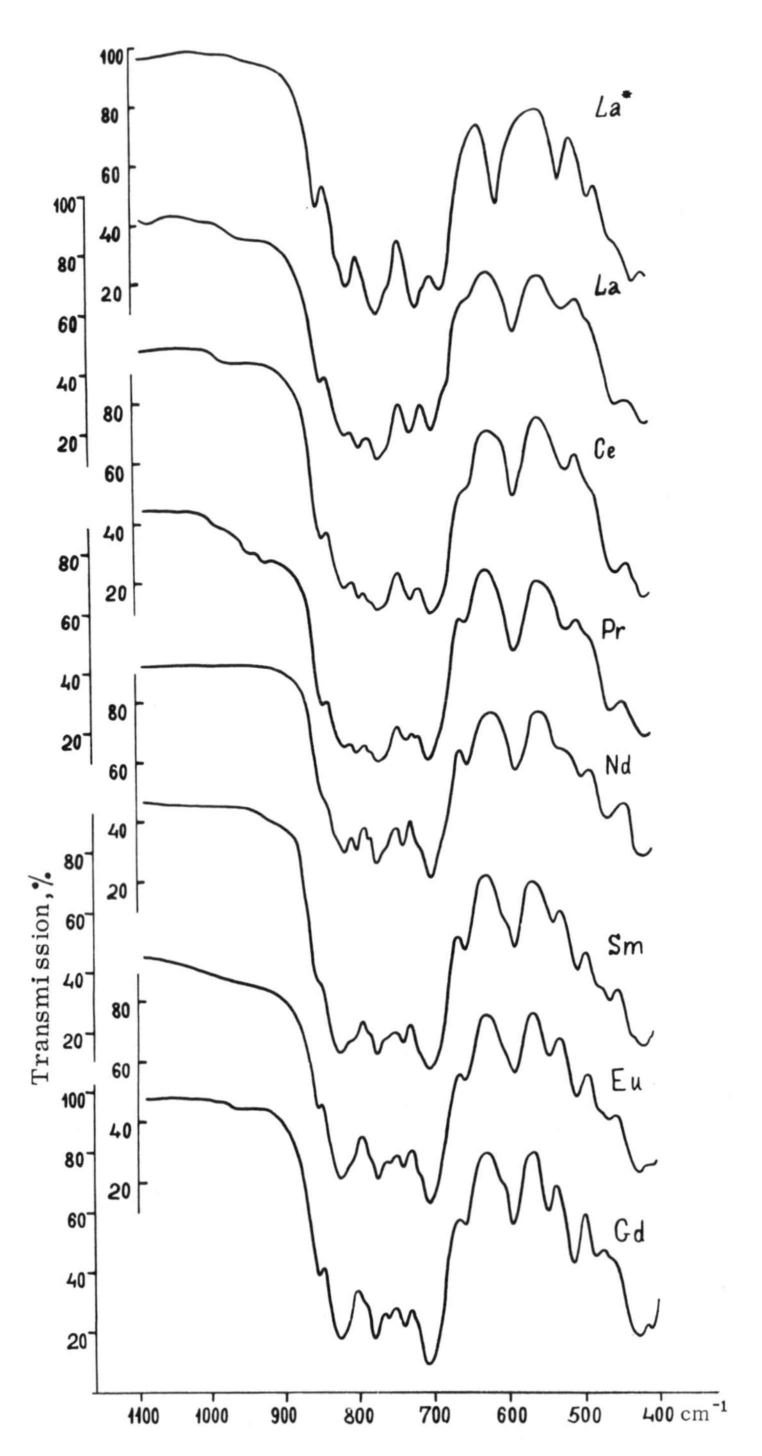

Fig. 108a. Infrared spectra of the pyrogermanates of the rare-earth elements referred to structural types 1 to 3 (pressed in KBr, 5 mg): first structural type, La* synthesis at 1430°; second structural type, La synthesis at 1000°, Ce synthesis at 1250°, Pr synthesis at 1200°, Nd synthesis at 1200°; third structural type, Sm synthesis at 1200°, Eu synthesis at 1200°, Gd synthesis at 1200°.

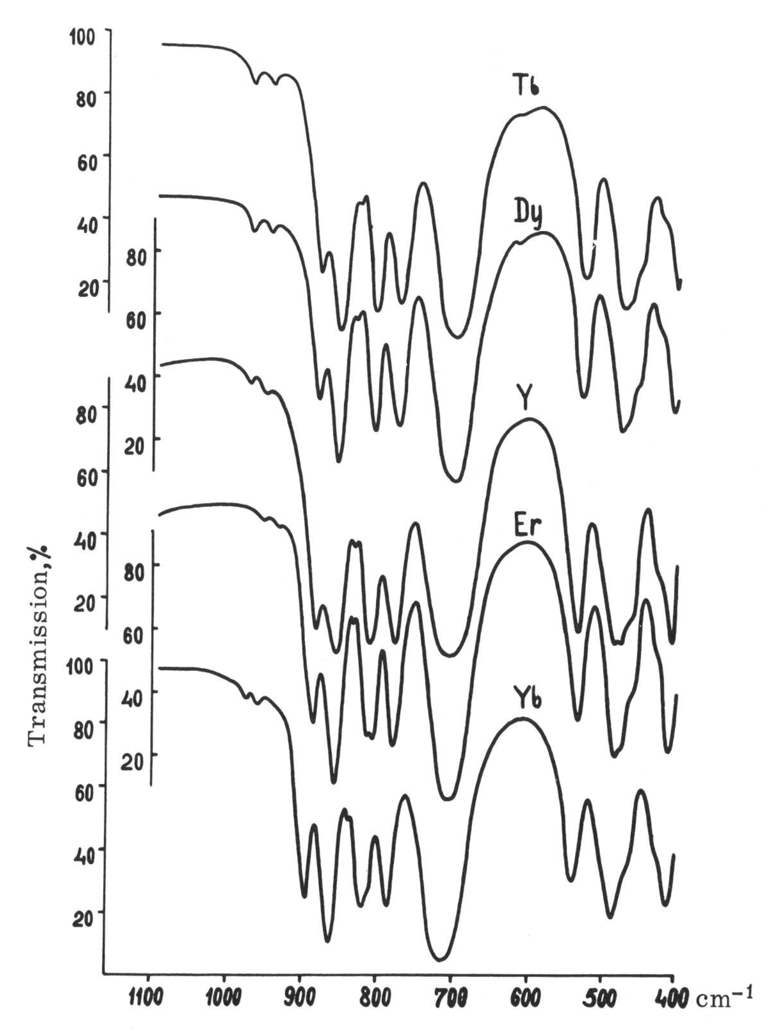

Fig. 108b. Infrared spectra of the pyrogermanates of the fourth structural type (pressed in KBr, 5 mg; cations indicated in the figure, synthesis at 1200°).

of cases are distinguished by better differentiation of the bands. This fact, in conjunction with data obtained for samples synthesized at various temperatures, firstly established the dimorphism of lanthanum germanate and secondly suggested the subdivision of germanates earlier ascribed to structural type 2 into two different types, one for germanates between La (low-temperature form) and Nd, and the other between Sm and Gd. The latter conclusion, despite the similarity between the spectra of the two types of pyrogermanates and between their Debye photographs, is mainly based on differences in the structure of the band lying in the neighborhood of 600 cm^{-1}, the splitting of one of the deformation vibrations at 450 to 480 cm^{-1} in the spectra of the compounds of elements between Sm and Gd, and slight anomalies in the relationship between the frequencies and the radius of the Ln^{3+} cations (Fig. 109); it is nevertheless tentative and requires confirmation by structural investigations.

The identification of the frequencies in the spectra of the first group of germanates (types 1 to 3), which was based on the assignment of the isolated (sometimes doublet-structured) band in the 580-to-600-cm^{-1} range to the ν_s GeOGe vibrations and that of the many relatively close bands below 520 to 540 cm^{-1} to the deformation vibrations of the anion, suggested

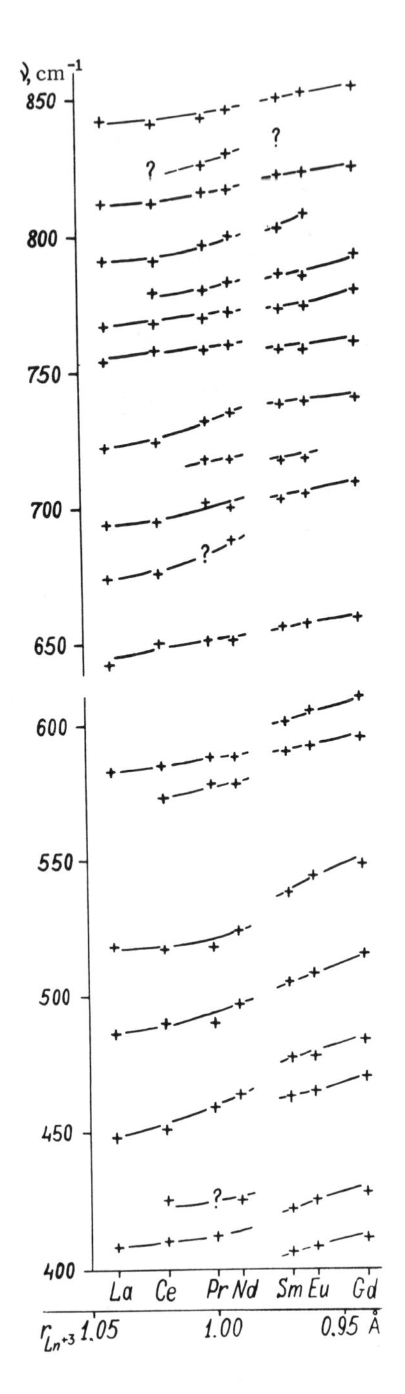

Fig. 109. Frequencies of the absorption maxima in relation to the ionic radius of the cation as a basis for separating the second and third structural types of pyrogermanates (the frequencies in the spectrum of lanthanum germanate of the first structural type are not shown).

the existence of Ge_2O_7 groups. But the x-ray structural analysis of lanthanum germanate crystals (the low-temperature form† according to the classification used here) led to the observation of GeO_4 and Ge_3O_{10} groups, with the retention of the empirical pyrogermanate formula [6]: $2La_2Ge_2O_7 = La_4[Ge_3O_{10}][GeO_4]$. This compels us to reconsider the identification of the bands, and in particular to ascribe the band at 520 to 540 cm^{-1} to the ν_s GeOGe vibrations, since for any configuration of the trigermanate groups Ge_3O_{10} both ν_s GeOGe frequencies are active in the infrared spectrum (for C_{2v} symmetry of these groups, the ν_s GeOGe and ν_s' GeOGe vibra-

† The high-temperature form is evidently similar in structure, but judging by the simplification of the spectrum is probably more symmetrical.

TABLE 54. Frequencies of the Absorption Maxima in the Infrared Spectra of the Pyrogermanates of the Rare-Earth Elements

Assignments	Type 1	Type 2				Type 3			Type 4				
	La	La	Ce	Pr	Nd	Sm	Eu	Gd	Tb	Dy	Y	Er	Yb
	845 s.	842 s.	841 s.	843 s.	846 sh.	850 sh.	852 s.	854 s.	877 s.	880 s.	883 s.	886 s.	895 s.
	819 sh.	?	?	826 sh. ?	830 sh.	835 sh. ?	?	–	852 v.s.	854 v.s.	856 v.s.	858 v.s.	864 v.s.
ν_{as} GeOGe,	805 v.s.	812 v.s.	812 v.s.	816 v.s.	817 v.s.	822 v.s.	823 v.s.	825 v.s.	827 m.	830 m.	831 m.	834 m.	837 m.
ν_{as} GeO_3,	–	791 v.s.	791 v.s.	797 v.s.	800 v.s.	803 v.s.	808 sh.	–	805 s.	807 s.	812 s.	816 s.	821 s.
ν_s GeO_3,	775 sh. ?	–	779 sh.	780 sh.	783 s.	786 sh.	785 sh. ?	793 sh.	–	–	–	810 s.	814 sh.
ν_{as} GeO_2,	765 v.s.	767 v.s.	768 sh.	770 v.s.	772 v.s.	773 v.s.	774 v.s.	780 v.s.	774 s.	775 s.	780 s.	782 s.	787 sh.
ν_s GeO_2,	750 sh.	754 sh.	758 sh.	758 sh.	760 sh.	758 sh.	758 s.	761 s.	714 sh.	712 sh.	720 sh. ?	–	–
ν_{as} GeO_4,	–	722 v.s.	724 v.s.	732 v.s.	735 v.s.	738 v.s.	739 v.s.	740 v.s.	700 v.s.	699 v.s.	703 v.s.	706 v.s.	714 v.s.
ν_s GeO_4	711 v.s.	–	–	718 s.	718 sh.	717 sh.	718 sh.	–	–	–	–	–	–
	701 sh. ?	694 v.s.	695 v.s.	702 v.s.	700 v.s.	703 v.s.	705 v.s.	709 v.s.	–	–	–	–	–
	681 v.s.	674 sh.	676 sh. ?	?	688 sh. ?	–	–	–	–	–	–	–	–
	–	642 w.	650 w.	651 m.	651 m.	656 m.	657 m.	659 m.	–	–	–	–	–
	–	–	–	–	–	601 sh.	605 sh.	610 sh.	–	–	–	–	–
ν'_s GeOGe	605 s.	583 m.	585 m.	588 m.	588 m.	590 m.	592 m.	595 m.	–	–	–	–	–
	–	–	573 sh.	578 sh. ?	578 sh.	–	–	–	–	–	–	–	–
ν_s GeOGe	521 m.	518 w.	517 w.	518 w.	524 w.	538 w.	544 m.	548 m.	528 s.	532 s.	537 s.	537 s.	542 s.
	480 m.	486 sh.	490 sh.	490 sh.	497 m.	505 m.	508 m.	515 s.	476 sh.	478 v.s.	486 v.s.	485 v.s.	489 v.s.
	–	–	–	–	–	477 sh.	478 sh.	484 m.	467 sh.	468 sh.	478 v.s.	480 sh.	–
δGeO	455 –	488 s.	451 s.	459 s.	464 s., d.?	463 s.	465 s.	470 sh.	450 sh.	455 sh.	461 sh.	463 sh.	467 sh.
and ν LnO	– 440 sh.	–	–	–	–	434 sh.	?	–	421 sh.	424 sh.	426 sh.	428 sh.	428 sh.
	422 v.s.	–	425 sh. ?	?	425 –	422 v.s.	425 v.s.	428 v.s.	403 v.s.	405 v.s.	409 v.s.	412 v.s.	414 v.s.
	< 400	408 v.s., d.?	410 v.s.	412 v.s.	– 405 v.s.	406 sh.	408 sh.	411 v.s.	–	–	–	–	–

tions belong to symmetry types A_1 and B_1). The large number of bands in the range 650 to 850 cm^{-1} is thus associated not with the "Davydov" splitting of vibrations of the Ge_2O_7 groups, but with the many stretching vibrations of the GeO_4 tetrahedra and $Ge\langle{}^{O^-}_{O^-}$, $Ge\langle{}^{O^-}_{O^-}O^-$ groups (and also ν_{as} GeOGe) in the Ge_3O_{10} anion. It should be noted that, in the absence of structural data, it would be difficult to advance arguments in favor of the assignment of the bands at 520 to 540 cm^{-1} to the ν_s GeOGe vibrations, since these are closely approached by the bands of deformation vibrations (probably "mixed" with lattice vibrations), and it is not very useful to employ the ν_s GeOGe frequency range in order to draw conclusions regarding the structure of the complex anion because of the relative high frequencies of the lattice vibrations.

An analogous situation occurs on attempting to identify the frequencies in the essentially differing spectra of the other group of pyrogermanates (known as the fourth structural type). The low frequency of the ν_s GeOGe band, close to the frequencies of the deformation vibrations, and its comparatively high intensity (probably associated with interactions with the lattice vibrations) made its identification unreliable; the possibility of centrosymmetric Ge_2O_7 groups was considered as an alternative. Thus the ascription of the band at 530 to 540 cm^{-1} (Table 54) to the ν_s GeOGe vibration only becomes possible after establishing the existence of noncentrosymmetric Ge_2O_7 groups by x-ray diffraction [6a].

On comparing the stability diagrams (in coordinates of ionic radius of the cation and temperature) for the R.E. pyrosilicates and the alkaline-earth pyrophosphates, we observe an analogy in the arrangement of the lines representing "packing catastrophes" leading to the rearrangement of the cation polyhedra if we insert corrections for the different dimensions of the complex anions by using the "effective" cation radii $r_{eff.M^{2+}} = r_{M^{2+}} (\bar{r}_{SiO}/\bar{r}_{PO})$ (i.e., those referred to the interatomic distances in the anion). We may also make an analogous comparison of the pyrosilicates with the pyrogermanates of the R.E. if we equalize the differences in the Si–O and Ge–O bond lengths by introducing effective radii $r_{Ln^{3+}(Ge)}$ from the relations $r_{Ln^{3+}}/r_{Ln^{3+}(Ge)} = \bar{r}_{GeO}/\bar{r}_{SiO}$. However, a similarity in the structure of the complex anion only appears for the single case $Ln_2Si_2O_7$ (IIIa)–$Sc_2Ge_2O_7$ (type 5) and for a limited range of $r_{Ln^{3+}}$. Moreover, x-ray data show that, over a considerable range, a peculiar disproportionation of the complex anions $Ge_2O_7 \rightarrow Ge_3O_{10} + GeO_4$ is energetically favorable.† Finally, the actual sequence of the types of coordination of the cations evidently differs for silicates and germanates [1-6a]. Thus the use of analogies which ignore the chemical individuality of the atoms forming the complex anion, and are not entirely correct even in the geometrical respect (in the $M_2X_2O_7$ systems the dimensions of M and X vary while those of O remain almost constant), is clearly only useful when there are no serious changes in the dimensional relationships, as in salts of elements belonging to the same period (Si and P), whereas it loses its value on passing from one period to another (Si to Ge), even if the cations are the same.

§3. Possibilities of a More Detailed Analysis of the Spectra: Structures of the Thortveitite $Sc_2Si_2O_7$ Type

In the preceding sections, when interpreting the spectra of the rare-earth silicates, we have restricted ourselves to considerations based on the quasicharacteristic nature of certain types of vibrations and to considerations of symmetry properties; this is natural when using spectroscopic methods in order to obtain an initial characterization and classification of new types of compound. However, the existence of x-ray structural analyses for these materials now raises the possibility of a more detailed analysis of the spectra, in particular, the calcu-

† It still remains uncertain whether the silicates of type X mentioned earlier may be similar in structure to these germanates.

lation of the normal-vibration frequencies in order to estimate the force constants. The results of an experimental study of the spectra of the rare-earth silicates indicates comparatively high frequencies for the lattice vibrations (i.e., the Ln–O bonds), and hence the possibility of there being a strong interaction between these vibrations and those of the deformation type, and to some extent the stretching vibrations of the complex anions. Any attempts at calculating the vibrational frequencies of these anions without allowing for lattice vibrations accordingly offers little promise. We shall now describe the preliminary results of a calculation of the spectra of one of the structural types of rare-earth pyrosilicates, thortveitite and its analogs, containing a comparatively small number of atoms in the primitive cell, allowing for all the vibrational degrees of freedom of the crystal. As already mentioned, one of the main difficulties in calculating the vibrational spectra of silicates, in addition to the necessity of allowing for the lattice vibrations, is the inadequacy of the number of observed frequencies for obtaining an objective determination of the set of force constants of the system. The existence of not only pyrosilicates (for example, ytterbium pyrosilicate) isostructural with $Sc_2Si_2O_7$ but also an isostructural scandium pyrogermanate somewhat reduces the arbitrariness encountered when choosing values for the force constants. A calculation of the vibration frequencies of the Si_2O_7 anions in thortveitite carried out on the basis of the quasimolecular approximation enables us to estimate the order of magnitude of the errors in this approximation for compounds of the class considered.

The Infrared Spectra of $Sc_2Si_2O_7-Sc_2Ge_2O_7$ [66]. The probable isostructural character of $Sc_2Ge_2O_7$ with respect to its silicon analog, thortveitite $Sc_2Si_2O_7$, was indicated by Goldschmidt [67]. The infrared spectrum of scandium pyrogermanate was first given in [64], although without any data regarding its structural relationship to $Sc_2Si_2O_7$. A study of $Sc_2Si_2O_7-Sc_2Ge_2O_7$ solid solutions is of particular interest both in connection with elucidating the possible isostructural relationship between scandium silicate and germanate and also in connection with identifying certain frequencies of the Si_2O_7 ion, in particular the ν_s SiOSi vibration, which is inactive in the $Sc_2Si_2O_7$ IR spectrum owing to the centrosymmetry of the Si_2O_7 group and the crystal as a whole. The physicochemical constants and crystal-optical and x-ray structural data indicated the existence of a continuous series of solid solutions, samples of which were obtained by the repeated sintering of the corresponding oxides at 1350°C.

Figure 110 shows the IR spectra of $Sc_2Si_2O_7$ and $Sc_2Ge_2O_7$ and those of the solid solutions. In the Si–O (Ge–O) stretching frequency range the number of bands in the $Sc_2Si_2O_7$ and $Sc_2Ge_2O_7$ spectra is the same, the bands being considerably displaced on passing from the silicate to the germanate. The frequency displacement of the deformation vibrations of the anion is insignificant, which indicates interaction with the lattice vibrations. A characteristic feature of the spectrum of $Sc_2Ge_2O_7$ is a band at 1030 cm^{-1}, the frequency of which is anomalously high as compared with the spectra of the majority of the germanates.† However, a calculation of the vibration frequencies of the Si_2O_7 and Ge_2O_7 ions with $\angle$SiOSi (GeOGe) = 180° shows that this frequency corresponds to the expected value of ν_{as} GeOGe if in the spectrum of the isostructural silicate ν_{as} SiOSi lies at 1170 cm^{-1}.

The IR spectra of the $Sc_2(Si, Ge)_2O_7$ solid solutions firstly prove the existence of a continuous series of these, thus confirming the isostructural relationship of the silicate and germanate, and secondly suggest a high degree of "microhomogeneity" of these solid solutions, which expresses itself in the form of a large number of "mixed" $SiGeO_7$ as well as Si_2O_7 and Ge_2O_7 ions. In the νSiO and νGeO frequency ranges in the spectra of the solid solutions there are bands corresponding to the bands in the spectra of pure $Sc_2Si_2O_7$ and $Sc_2Ge_2O_7$; these move slightly with changing Si and Ge concentration. In addition to this, whereas in the spectra of

† Probably it was because of the unusually high frequency of this band that doubt was cast on its belonging to the $Sc_2Ge_2O_7$ spectrum in [64].

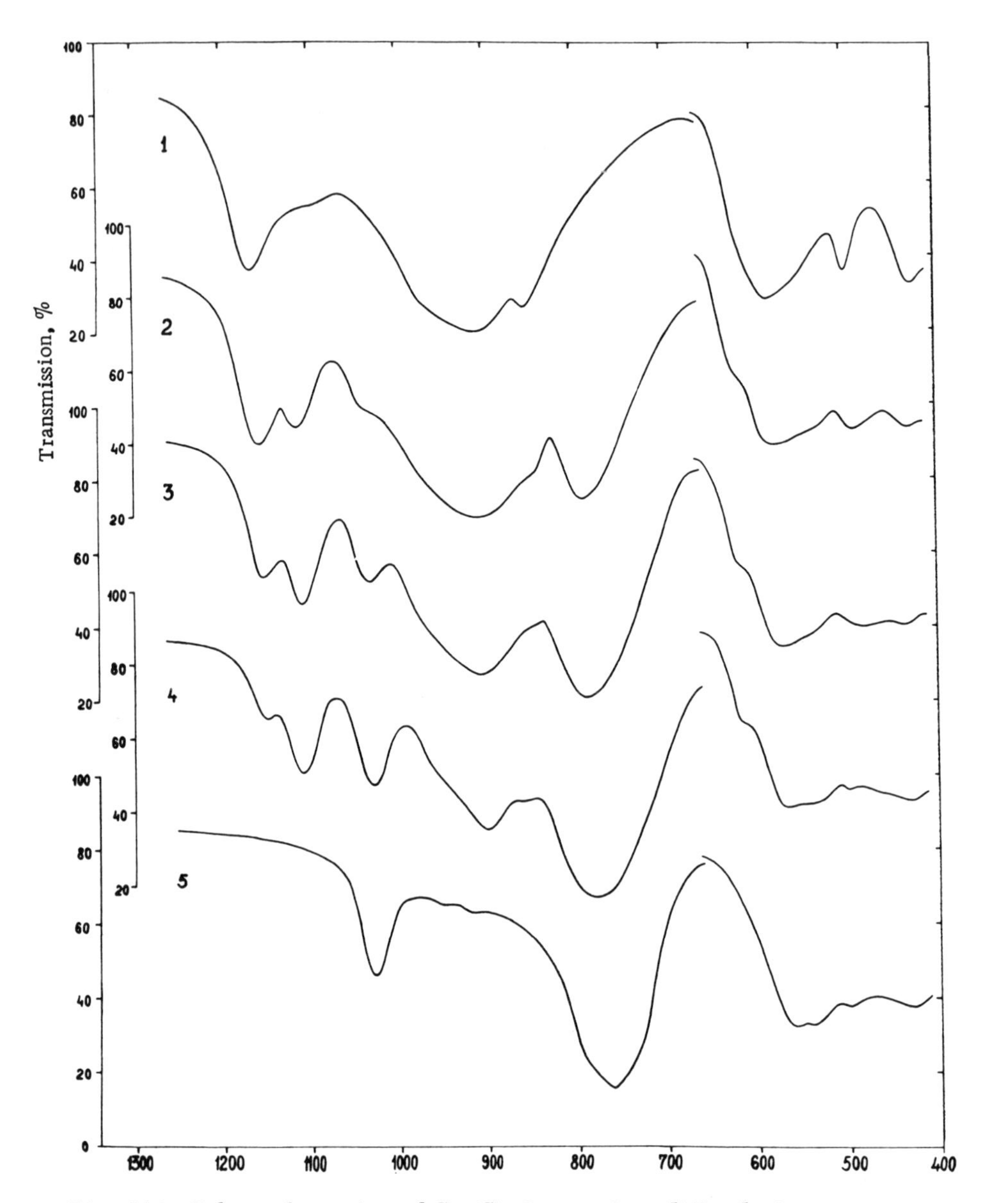

Fig. 110. Infrared spectra of $Sc_2(Si_xGe_{1-x})_2O_7$ solid solutions: 1) x = 1.00, 2) x = 0.75, 3) x = 0.50, 4) x = 0.25, 5) x = 0.00.

pure $Sc_2Si_2O_7$ and $Sc_2Ge_2O_7$ there is a frequency range free from absorption between ν_{as} SiOSi and ν_{as} GeOGe, in the spectra of the solid solutions there is a strong band at 1109 to 1116 cm^{-1} (depending on concentration), and this cannot be ascribed to any vibrations of the Si_2O_7 or Ge_2O_7 ions. It is natural to identify this with the ν_{as} SiOGe vibrations of the $SiGeO_7$ ion, which also agrees with a calculation of the frequencies (Table 55). Still more indicative is the appearance of a band at about 615 to 622 cm^{-1}, i.e., higher than the frequencies of the deformation vibrations of the anion and the lattice vibrations in both $Sc_2Ge_2O_7$ and $Sc_2Si_2O_7$. The intensity of this band is quite high, and its appearance must be associated not with any reduction in the strictness of the selection rules on disrupting the translational symmetry in the mixed crystal (which, generally speaking, leads to the removal of the ban on ν_s SiOSi and ν_s GeOGe in the IR spectrum), but with the disruption of the local centrosymmetry of the structure as a result of the formation of $SiGeO_7$ ions: the mechanical and electro-optical nonequivalence of the $O_3Si(O)$ and $(O)GeO_3$ groups in such an anion yields a fairly high dipole moment for the ν_s SiOGe vibration.

TABLE 55. Vibration Frequencies of the Complex Anions in the Spectra of $Sc_2Si_2O_7$–$Sc_2Ge_2O_7$ Solid Solutions [66]

Frequencies of the absorption maxima in the IR spectrum					Calculated frequencies of the complex anions†			
$Sc_2Si_2O_7$	$Sc_2(Si_{0.75}Ge_{0.25})_2O_7$	$Sc_2(Si_{0.5}Ge_{0.5})_2O_7$	$Sc_2(Si_{0.25}Ge_{0.75})_2O_7$	$Sc_2Ge_2O_7$	Si_2O_7 (D_{3d})	$SiGeO_7$ (C_{3v})	Ge_2O_7 (D_{3d})	
1166	1159	1155	1150	–	1159 A_{2u}	–	–	ν_{as} SiOSi
1104 sh.	–	–	–	–	–	–	–	ν_{as} SiOSi (from excited level)
–	1116	1111	1109	–	–	1108 A_I	–	ν_{as} SiOGe
–	1032	1033	1030	1030	–	–	1035 A_{2u}	ν_{as} GeOGe
978	?	980	965	– }	964 E_u	964 E	–	ν'_{as} SiO_3 (ν_{as} SiO_3 in $SiGeO_7$)
912	911	907	900	– }			–	
–	–	–	–	–	964 E_g	–	–	ν_{as} SiO_3
–	–	–	–	–	883 A_{1g}	–	–	ν_s SiO_3
855	848	844?	858?	–	820 A_{2u}	842 A_1	–	ν'_s SiO_3 (ν_s SiO_3 in $SiGeO_7$)
–	–	–	–	793 sh.	– }	783 E	783 E_u	ν'_{as} GeO_3 (ν_{as} GeO_3 in $SiGeO_7$)
–	791	789	780	762	– }			
–	–	–	–	–	–	–	783 E_g	ν_{as} GeO_3
–	–	–	–	–	–	–	767 A_{1g}	ν_s GeO_3
–	–	–	–	727 sh.	–	766 A_I	765 A_{2u}	ν'_s GeO_3 (ν_s GeO_3 in $SiGeO_7$)
–	–	–	–	–	611 A_{1g}	–	–	ν_s SiOSi
–	622 sh.	618 sh.	615 sh.	–	–	583 A_I	–	ν_s SiOGe
–	–	–	–	–	–	–	416 A_{1g}	ν_s GeOGe
586	581	570	567	558	437 A_{2u}	351 A_I	331 A_{2u}	δ'_s SiO_3, δ_s SiO_3, δ'_s GeO_3
545 sh. ?	560–535 sh.	543 sh.	539	539	414 E_u	450 E	327 E_u	ρSiO_3, ρSiO_3, ρGeO_3
498	490	490–480	498	499	414 E_g	256 E	327 E_g	ρSiO_3, ρGeO_3, ρGeO_3
–	–	–	461	–	380 E_u	395 E	332 E_u	δ'_{as} SiO_3, δ_{as} SiO_3, δ'_{as} GeO_3
430	429	434	430 sh. ?	430	380 E_g	329 E	332 E_g	δ_{as} SiO_3, δ_{as} GeO_3, δ_{as} GeO_3
–	–	–	–	–	257 A_{1g}	237 A_1	228 A_{1g}	δ_s SiO_3, δ_s GeO_3, δ_s GeO_3

† The low-frequency vibrations δSiOSi(E_u) and τSiO_3(A_{1u}) are not taken into account in the calculation. Comparison of the experimental frequencies of the deformation vibrations with those calculated for the isolated anions is impossible owing to interaction with lattice vibrations.

The reduction in local symmetry from D_{3d} to C_{2v} (or more precisely from C_{2h} to C_s; in a crystal with symmetry $C2/m-C_{2h}^3$ the formation of a solid solution with the occupation of the tetrahedral positions by atoms of two sorts, even for a regular arrangement of the Si and Ge, reduces the symmetry to $Cm-C_s^3$) should also lead to activation of the $\nu_s XO_3(A_{1g})$ vibration for the $SiGeO_7$ ion, so that both, $\nu_s XO_3$ and $\nu_s' XO_3$, should be active in the IR spectrum and approximately correspond to $\nu_s GeO_3$ and $\nu_s SiO_3$; equally the $\nu_{as} XO_3(E_g)$ vibration, or the two components into which this splits in the crystal, should be active in the IR spectrum and correspond primarily to $\nu_{as} GeO_3$. These bands, in contrast to the vibrations of the bridge, probably differ little from the corresponding bands of the Si_2O_7 and Ge_2O_7 ions owing to the weak interaction between the vibrations of the SiO_3 and GeO_3 groups, as confirmed by calculation, and it is practically impossible to observe them on the background of the strong bands of the Si_2O_7 and Ge_2O_7 between 700 and 1000 cm^{-1}.

The changes in relative intensity of the band at 1109 to 1116 cm^{-1} with changing Si and Ge concentration agree with its ascription to ν_{as} SiOGe, the number of Si_2O_7, Ge_2O_7, and $SiGeO_7$ ions at each concentration evidently being determined by the random (uncorrelated) distribution of Si and Ge in the lattice. On assuming a statistical occupation of the tetrahedral positions in the $X'X''O_7$ groups by Si and Ge atoms and the constancy of the absorption coefficient in the vibration bands of the Si_2O_7, Ge_2O_7, and $SiGeO_7$ ions, the relation $D/n =$ const should hold for each of the bands, where D is the optical density and n is the number of $X'X''O_7$ ions of a specified sort, proportional to c_1c_2N (c_1 and c_2 are the concentrations of the atoms occupying the X′ and X″ positions in the anion of specified sort, while N is the total number of diortho ions in the sample). Of the bands of the Si_2O_7, Ge_2O_7, and $SiGeO_7$ ions, only the ν_{as}XOX vibration bands can be resolved for all compositions of the solid solutions. The ratio D_{max}/n remains constant for each of the bands to within 20%, which may be regarded as satisfactory if we remember that these bands are partly superimposed on one another and on the stronger absorption edge at 975 to 750 cm^{-1}. Thus the distribution of Si and Ge in the lattice and in the diortho ions themselves is almost random.

The calculation of the vibration frequencies and forms of the Si_2O_7, Ge_2O_7, and $SiGeO_7$ ions was carried out for an idealized configuration with a three-fold axis, group D_{3d} for Si_2O_7 and Ge_2O_7 and C_{3v} for $SiGeO_7$; a transformation to the site symmetry of these ions in the crystal (C_{2h} for the first two and C_s for the last) leads to the splitting of the degenerate vibrations without changing the selection rules. The existence of only one Si_2O_7 group in the primitive cell of thortveitite has the effect that a transformation from the local symmetry group of the anion in the lattice to the factor group of the crystal space group does not change the number of internal vibrations of the anion. The following geometrical parameters were employed: $r_{SiO^-} = 1.61$ Å, $r_{GeO} = 1.75$ Å, $\angle OSiO = \angle OGeO = 109°28'$; $\angle SiOSi = \angle GeOGe = \angle SiOGe = 180°$. The values of the force constants were chosen with due allowance for existing data relating to organosilicon molecules with Si–O bonds, while for the Ge–O bonds values 5 to 10% smaller than for Si–O were used. Variation was carried out by means of the partial derivatives of the frequencies with respect to the force constants, it being considered, in accordance with the relation between the Si–O⁻ and Si–O(Si) bond lengths in thortveitite, that $K_{SiO^-} < K_{SiO(Si)}$.† The frequencies given in Table 55 were obtained by using the following set (in 10^6 cm^{-2}) $K_{SiO(Si,Ge)} = 9.0$, $K_{SiO^-} = 8.5$, $H_{SiO,OSi} = 0.7$, $h_{SiO,SiO} = 0.4$, $K_{O^-SiO^-} = K_{O^-SiO(Si,Ge)} = 2.0$, $K_{GeO(Ge,Si)} = 8.5$, $K_{GeO^-} = 8.0$, $H_{GeO,OGe} = 0.7$, $h_{GeO,GeO} = 0.4$; $K_{O^-GeO^-} = K_{O^-GeO(GeSi)} = 1.8$; all interactions between bonds and angles A–B = 0.2, all other interactions taken as zero. The quantities $K_{SiO(SiGe)}$, $K_{GeO(Ge,Si)}$ and $H_{SiO,OSi}$ and $H_{GeO,OGe}$‡ were chosen so as to obtain agreement be-

† For pyrosilicates with bent Si–O–Si bridges the opposite relation $K_{SiO^-} < K_{SiO(Si)}$ holds, in agreement with the bond lengths (Chap. II).

‡ The strong dynamic interaction of the bonds in the Si–O–Si bridges with ∠SiOSi close to 180° has already been noted when analyzing the spectra of the siloxanes (Chap. II).

tween the calculated and experimental values of ν_{as}XOX for as high as possible a value of ν_sXOX. However, the calculated frequencies ν_sXOX remain extremely low: on allowing for the fact that in the $SiGeO_7$ ion ν_sSiOGe is rather more associated with Ge–O than with Si–O, and the observed value of ν_sSiOGe is close to 620 cm^{-1}, the probable values of ν_sSiOSi and ν_sGeOGe lie at around 670 and 580 cm^{-1}. Still lower are the frequencies of the deformation vibrations; the ascription of which in Table 55 is of an extremely arbitrary character. Moreover, for a vibration of the A_{2u} type, for example, the observed frequency falls by 5% on passing from Si_2O_7 to Ge_2O_7 instead of the calculated 25% (even for equal force coefficients of OSiO and OGeO). This indicates the inapplicability of the quasimolecular approximation in the present case for considering the low-frequency vibrations of the anion.

The application of molecular-vibration-theory methods to the vibrational spectra of crystals, i.e., systems possessing translational symmetry, was described most consistently in [68]. The El'yashevich–Wilson method used in the theory of molecular vibrations solves the problem of finding the frequencies and forms of the normal vibrations of the system by calculating the eigenvalues and eigenvectors of the matrix **GF**, where G is the reciprocal matrix of the kinetic energy and F is the potential-energy matrix. The secular equation from which the frequencies of the normal vibrations are determined is written in the form

$$|\mathbf{GF} - \mathbf{E}\lambda| = 0, \tag{1}$$

where **E** is the unitary matrix of the same order as G and **F** while $\lambda = 4\pi^2c^2\nu^2$, c is the velocity of light, and ν is the wave number.

In calculating the normal vibrations of the crystal, i.e., a system of a very large (we may suppose infinitely so) number of atoms, the order of the matrices **G** and **F** also becomes very large. However, since the crystal lattice possesses translational symmetry, the problem of determining its normal vibrations may be reduced to the solution of secular equations of finite order by using the symmetry properties of the system.

Consideration of symmetry properties also plays an essential part in the solution of the normal vibrations of the molecule. We remember that the transformation from the internal or Cartesian coordinates (changing the bond lengths and interbond angles and displacing the atoms, respectively) to symmetry coordinates partly solves the problem of orthogonalization, by bringing the matrices **G** and **F** to a block form, and thus reduces the order of the secular equations to be solved. The rule for constructing the symmetry coordinates from the internal ones (S_0) or the Cartesian coordinates (X_0) may be written in the following form

$$S^{(\gamma)} = M \sum_R \chi_R^{(\gamma)} R S_0, \tag{2}$$

$$X^{(\gamma)} = K \sum_R \chi_R^{(\gamma)} R X_0, \tag{2'}$$

where $S^{(\gamma)}$ and $X^{(\gamma)}$ are the internal and Cartesian symmetry coordinates in the irreducible representation γ, R are the operations of symmetry, $\chi_R^{(\gamma)}$ are the characters of the operation R in the irreducible representation γ, while M and K are normalizing factors.

Turning to the question of the crystal-lattice vibrations, we note that the lattice symmetry is described by a space group the symmetry operations of which constitute combinations of translations with a finite number H of other symmetry operations. The space group Γ of a crystal of $N_1N_2N_3$ translationally identical cells comprises $N_1N_2N_3H$ symmetry operations. The group of pure translations

$$T_{001}, T_{002}, \ldots, T_{N_1N_2N_3}$$

is an invariant subgroup of the group Γ. Expanding the group Γ into conjugate sets with respect to the group T

$$ET,\ R_2T,\ R_3T,\ \ldots,\ R_HT, \tag{3}$$

we obtain a factor group of the space group. Since T plays the part of a unitary element in the factor group, we may rewrite (3) in the form

$$E,\ R_2,\ R_3,\ \ldots,\ R_H.$$

The order of the factor group does not depend on the order of the group of translations, and the factor group of the infinite or finite space group is the same. The factor group is isomorphic with one of the point groups, and hence its characters coincide with those of the corresponding point group.

Thus if by way of characterizing the motion of the atoms in the crystal lattice we use coordinates having an intrinsic symmetry described by the group T (such coordinates are subsequently called "active") then the set of these coordinates forms the basis of the representation of the factor group H. In other words, after obtaining the matrices **G** and **F** for the "active" coordinates, subsequent solution of the problem as to the vibrations of the crystal is similar to that arising in the case of a molecule, the next stage being the transformation to the coordinates of (point) symmetry of the point group which is isomorphic with respect to the factor group H.

The rule for constructing the "active" natural (internal) or Cartesian coordinates, and hence the matrices $\mathbf{G}^{act}$ and $\mathbf{F}^{act}$, may be written by analogy with (2, 2′) in the form of formulas defining these coordinates as linear combinations of the natural or Cartesian coordinates:

$$S^{act} = \lim_{N_1, N_2, N_3 \to \infty} [N_1N_2N_3]^{-1/2} \sum_{i=0}^{N_1} \sum_{j=0}^{N_2} \sum_{k=0}^{N_3} \chi_T T_{ijk} S_0 \tag{4}$$

$$X^{act} = \lim_{N_1, N_2, N_3 \to \infty} [N_1N_2N_3]^{-1/2} \sum_{i=0}^{N_1} \sum_{j=0}^{N_2} \sum_{k=0}^{N_3} \chi_T T_{ijk} X_0 \tag{4'}$$

where χ_T is the character of the operation T_{ijk} defined by the expression

$$\chi T_{ijk} = e^{-i(\mathbf{k}_i t_i + \mathbf{k}_j t_j + \mathbf{k}_k t_k)}$$

the wave vector $\mathbf{k}_m$ taking the values $L/N_m a_m$ for $L = 0, 1, 2, \ldots, N_{m-1}$ (a_m is the lattice identity period along the specified direction).

In this way we may obtain the $PN_1N_2N_3$ of the "active" coordinates (P is the number of vibrational coordinates per cell) describing the $3m \cdot N_1N_2N_3$ degrees of freedom of the crystal (m is the number of atoms per cell) and construct the 3m vibrational branches.

Supposing that the optically active fundamental vibrations (i.e., those leading to macroscopic changes of dipole moment or polarizability) of a fairly large crystal only comprise the limiting vibrations with $\mathbf{k} = 0$, we may rewrite (4) and (4′), remembering that in this case always $\chi T_{ijk} = 1$, in the form of the following rule for obtaining each of the P natural or 3m Cartesian "optically active" coordinates:

$$S^{opt\ act} = \lim_{N_1, N_2, N_3 \to \infty} [N_1N_2N_3]^{-1/2} (S_{000} + S_{001} + \ldots + S_{N_1N_2N_3}) \tag{5}$$

$$X^{opt\ act} = \lim_{N_1, N_2, N_3 \to \infty} [N_1N_2N_3]^{-1/2} (X_{000} + X_{001} + \ldots + X_{N_1N_2N_3}) \tag{5'}$$

In other words for $\mathbf{k} = 0$ each "optically active" coordinate is an infinite sum of translationally equivalent coordinates. Then the derivation of the matrices $\mathbf{G}^{\text{opt act}}$ and $\mathbf{F}^{\text{opt act}}$ reduces to a determination of the coefficients of kinematic and dynamic interaction between the "optically active" coordinates. We note that the rule for obtaining these coefficients is analogous to the rule for constructing the elements of the matrices **G** and **F** of the totally symmetric vibrations of a molecule; in fact the problem of constructing the matrices $\mathbf{G}^{\text{opt act}}$ and $\mathbf{F}^{\text{opt act}}$ is equivalent to the problem of determining the coefficients of interaction between the coordinates totally symmetric with respect to the group of translations of the crystal. The elements of the matrices $\mathbf{G}^{\text{opt act}}$ and $\mathbf{F}^{\text{opt act}}$, being finite quantities, describe the interaction of the coordinates of a specified cell not only with the coordinates of the same cell but also with those of the surrounding cells (for zero phase shift). We remember that of the 3m limiting vibrations of the crystal 3m − 3 correspond to optical vibrations and three to translations of the crystal as a whole, T_X, T_Y, T_Z.

In solving the problem of the vibrations of a crystal, the derivation of the matrix **G** in internal coordinates is often extremely cumbersome, the existence of a large number of "branches" (in the crystals of present interest not only those of the tetrahedral but also, for example, those of the octahedral type) and equally of closed rings leads to the occurrence of many redundant coordinates,† while the description of the form of the normal vibrations in internal coordinates is not adequately reliable in the case of crystals. Hence quite often the use of Cartesian coordinates is more appropriate. In Cartesian coordinates the matrix $\mathbf{G}^{\text{opt act}}$ has the form of a diagonal matrix of reciprocal masses, i.e., the matrix elements corresponding to the three Cartesian displacements of the a-th atom are equal to its reciprocal mass; the dimensions of the matrix $\mathbf{G}_c^{\text{opt act}}$ equal 3m×3m.‡ The matrix $\mathbf{F}_c^{\text{opt act}}$ may be obtained from the corresponding matrix given in internal vibrational coordinates $\mathbf{F}_i^{\text{opt act}}$ by means of a matrix **B** transforming from the internal to the Cartesian coordinates, defined by the relation

$$\mathbf{S}^{\text{opt act}} = \mathbf{B}^{\text{opt act}}\mathbf{X}^{\text{opt act}} \tag{6}$$

The potential energy of the crystal may be written, on allowing for (6), in the form

$$2V = \tilde{\mathbf{S}}\mathbf{F}_i\mathbf{S} = \tilde{\mathbf{X}}\tilde{\mathbf{B}}\mathbf{F}_i\mathbf{B}\mathbf{X} = \tilde{\mathbf{X}}\mathbf{F}_c\mathbf{X}$$

from which we have

$$\mathbf{F}_c^{\text{opt act}} = \tilde{\mathbf{B}}^{\text{opt act}}\mathbf{F}_i^{\text{opt act}}\mathbf{B}^{\text{opt act}} \tag{7}$$

The dimensions of the matrices in (7) are for $\mathbf{B}^{\text{opt act}}$ 3m×P, for $\mathbf{F}_i^{\text{opt act}}$ P×P, and for $\mathbf{F}_c^{\text{opt act}}$ 3m×3m.§

The next operation after obtaining the matrices $\mathbf{G}_c^{\text{opt act}}$, $\mathbf{B}^{\text{opt act}}$ and $\mathbf{F}_i^{\text{opt act}}$ is that of bringing these to block form by transforming to symmetry coordinates with the help of the characters of the factor group (the point group isomorphic with it). As in the problem of the

† In considering the limiting vibrations of systems possessing translational symmetry, additional relations also arise between the coordinates, associated with the requirement that the shape and size of the cell should be kept constant during the vibrations.

‡ The indices "c" and "i" denote "Cartesian" and "internal"; the latter corresponds to "natural" coordinates in Soviet terminology.

§ This is only true if all the atoms are situated in sets of general positions. In other cases symmetry considerations require the introduction of a large number of coordinates in order to be able to use the (point) symmetry for partial diagonalization of the secular equation. The latter correspondingly gives a certain number of zero roots.

vibrations of the molecule, this is effected by means of corresponding transformation matrices $\mathbf{X}^{\text{opt act sym}} = \tilde{\mathbf{U}}_X \mathbf{X}^{\text{opt act}}$ and $\mathbf{S}^{\text{opt act sym}} = \tilde{\mathbf{U}}_S \mathbf{S}^{\text{opt act}}$:

$$\mathbf{G}_c^{\text{opt act sym}} = \tilde{\mathbf{U}}_X \mathbf{G}_c^{\text{opt act}} \mathbf{U}_X \quad (8)$$

$$\mathbf{F}_i^{\text{opt act sym}} = \tilde{\mathbf{U}}_S \mathbf{F}_i^{\text{opt act}} \mathbf{U}_S \quad (9)$$

$$\mathbf{B}^{\text{opt act sym}} = \tilde{\mathbf{U}}_S \mathbf{B}^{\text{opt act}} \mathbf{U}_X \quad (10)$$

$$\mathbf{F}_c^{\text{opt act sym}} = \tilde{\mathbf{B}}^{\text{opt act sym}} \mathbf{F}_i^{\text{opt act sym}} \mathbf{B}^{\text{opt act sym}} \quad (11)$$

After transformations (8-11) the problem reduces to the solution of secular equations in the form

$$| \mathbf{G}_c^{\text{opt act sym}} \mathbf{F}_c^{\text{opt act sym}} - \mathbf{E}\lambda | = 0 \quad (12)$$

These equations in all have 3m roots, of which the three zero roots (corresponding to the translations of the crystal) may be eliminated in advance in the corresponding irreducible representations by striking out the three columns in the matrix $\mathbf{B}^{\text{opt act sym}}$ and subtracting the corresponding three columns from the rest in the matrix $\mathbf{G}_c^{\text{opt act sym}}$ (which then loses diagonal form).

The main disadvantage of the method described for calculating the vibration spectra of crystals lies in considering the frequencies of the maxima observed in the optical spectrum as the frequencies (for $\mathbf{k} \approx 0$) of the free vibrations of a mechanical system of material points linked by elastic interaction forces. Actually the lattice vibrations active in the infrared spectrum are associated with changes of dipole moment, and the mechanical vibrations are accompanied by changes in the electrical field of the crystal. Hence for the same mechanical vibration the resonance frequency depends on interaction with a field of electrical polarization due to the deformation.

In the simplest cases, for isotropic (cubic) crystals and for certain of the most symmetric directions in uniaxial crystals, there is a complete separation of the dipole vibrations into purely longitudinal and transverse. The frequency of the longitudinal vibration is higher than that of the transverse vibration corresponding to the same mechanical mode; this is due to the effect of the Coulomb field, since for the longitudinal vibration Coulomb forces are added to the elastic forces of interaction, increasing the rigidity of the system.†

The identification of longitudinal and transverse vibrations may be verified by calculation: the frequencies of the longitudinal vibrations may be calculated from the frequencies of the corresponding transverse vibrations if the latter are determined, together with the oscillator strengths and the high-frequency dielectric constant, from the dispersion analysis of data relating to the reflection spectrum of the single crystal. This is done, for example, in the case of quartz by Scott and Porto [70], who showed that the splitting of a number of bands of doubly degenerate (E) vibrations in the Raman spectrum was due to the considerable frequency differences of the transverse and longitudinal vibrations, and not to combination frequencies as proposed by many previous authors. The frequencies of the longitudinal and transverse vibrations may be obtained by a nonphenomenological calculation if we introduce the field of Coulomb

† This effect is frequently called the Lyddane–Sachs–Teller splitting after the authors who first considered it for cubic crystals. For a more rigorous approach it is also essential to allow for the interaction of photon and phonon fields. We note that on studying the interaction of phonons with photons, i.e., the resonance of purely mechanical and electromagnetic vibrations in the crystal in the range of the intersection of their dispersion curves, the concept of quasiparticles of mixed origin, polaritons, has proved fruitful [69].

TABLE 56. Analysis of the Limiting Lattice Vibrations of Thortveitite by Means of the Crystal Factor Group

Types of symmetry \ Characters	E	C_2	σ_h	i	N	T	T′	R′	n_i	n
A_g	1	1	1	1	8	0	1	1	6	8
A_u	1	1	−1	−1	7	1	1	0	5	6
B_g	1	−1	−1	1	7	0	2	2	3	7
B_u	1	−1	1	−1	11	2	2	0	7	9

N = total number of degrees of freedom (branches)
T = number of translations (acoustic branches)
T' = number of optical lattice vibrations of a translational type
R' = number of optical lattice vibrations of an orientational type
n_i = number of internal vibrations of the complex anion
n = total number of optical vibrations.

forces into consideration. Elcombe attempted a calculation of this kind for quartz [71], using a model of rigid Si and O ions with effective charges of about + 2 and − 1, in addition to several parameters describing the interaction field.

For the majority of silicates, which constitute considerably more complicated systems, such calculations involve substantial difficulties. The amount of reasonably complete experimental data obtained for single crystals of the most symmetrical (cubic or uniaxial) silicates is as yet quite inadequate to allow systematic calculations. The fact that the majority of the chemically most important structures of the silicates possess a low symmetry hardly justifies our considering such an analysis of the spectra as immediately promising. In what follows we shall therefore confine ourselves to a calculation involving only short-range forces in order to show that allowing for the mechanical interaction of the internal vibrations of a complex anion with the lattice vibrations yields a satisfactory interpretation for the principal absorption max-

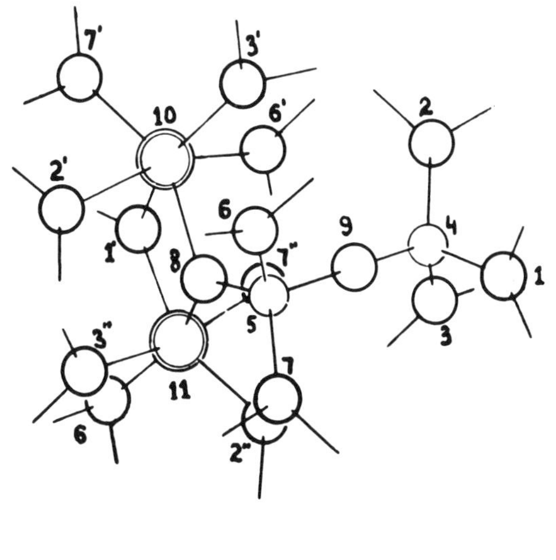

Fig. 111. Fragment of the lattice of thortveitite embracing one translational unit: two Sc coordination polyhedra and an Si_2O_7 group (the number of the atoms indicated by one or two primes correspond to atoms of one of the neighboring primitive cells repeated by translation).

ima and enables us to obtain chemically more significant force constants of the silicate anion than in the quasi-molecular model.

A calculation of the vibrational spectra of $Sc_2Si_2O_7$ and its structural analogs was carried out by the method described, using Cartesian vibrational coordinates.† For all three substances considered, $Sc_2Si_2O_7$, $Yb_2Si_2O_7$ and $Sc_2Ge_2O_7$, the same spatial configuration was used; this meant that on passing from one substance to another only the corresponding elements of the matrix $G_c^{opt\ act\ sym}$ and some of those of the matrix $F_i^{opt\ act\ sym}$ had to be changed. The spatial configuration taken for the calculation was based on the lattice constants and cell atomic coordinates of the thortveitite crystal $Sc_2Si_2O_7$ given in [48]. The crystal belongs to the $C2/m-C_{2h}^3$ space group, the primitive (noncentered) cell containing only one formula unit. The results of an analysis of the limiting vibrations with the aid of the factor group in the form taken when considering molecular (or nearly molecular) crystals are presented in Table 56. The most important interatomic distances and valence angles calculated from the atomic coordinates given in [48] are indicated below.

The enumeration of the atoms corresponds to that taken in Fig. 111, which shows a fragment of the $Sc_2Si_2O_7$ structure embracing the atoms entering into two translationally nonequivalent Sc octahedra and an Si_2O_7 group.

Bond lengths (Å)	Valence angles (deg)	
SiO (4,9) = 1.606	OSiO (9,4,2) = 107°34'	OScO (1,10,8) = 78°36'
SiO (4,1) = 1.620	OSiO (9,4,1) = 103°54'	OScO (7,10,1) = 93°53'
SiO (4,2) = 1.630	OSiO (2,4,1) = 113°35'	OScO (7,10,2) = 74°21'
ScO (10,3) = 2.083	OSiO (2,4,3) = 110°03'	OScO (7,10,8) = 152°31'
ScO (10,2) = 2.203	SiOSi (5,9,4) = 180°	OScO (7,10,6) = 116°02'
ScO (10,8) = 2.117	OScO (7,10,3) = 103°46'	OScO (8,10,6) = 88°43'
	OScO (2,10,6) = 172°35'	OScO (8,10,2) = 79°03'

The most complex question was that of choosing a reasonable form for describing the force field. In order to reduce the number of parameters and secure an objective estimate of these, the part of the potential-energy matrix corresponding to the interaction of the anion with the cations was considered as being diagonal, i.e., only the coefficients of the Sc–O bonds were introduced, and their values were taken as identical for all bonds of this type. (Only at the final stage of the calculation were slight differences in the force constants of the nonequivalent Sc–O or Yb–O bonds, corresponding to differences in their lengths as indicated by x-ray data, introduced.) However, in calculating the partial derivatives of the frequencies with respect to the force constants, in addition to the constants of the Sc–O bonds the interaction constants of the Sc–O bonds with each other (those with a common Sc atom and those with a common O atom) and with Si–O bonds having a common O atom were introduced. The admissibility of the zero values of the cross terms of the potential function on varying the force constants was taken as a criterion of the suitability of such a field for approximately describing the lattice dynamics, and at the same time as a criterion of the admissibility of the assumption as to the predominantly kinematic interaction of the lattice vibrations with the internal vibrations of the anion. The description of the internal force field of the complex anion was based on a quadratic potential function of general form; however, at any rate prior to obtaining more complete data relating to the vibrational frequencies (for example, by studying the Raman spectra), on varying the force constants a considerable proportion of these (presumably smaller in absolute value) were taken as zero so as to reduce arbitrariness in choosing the remaining coefficients. In order to estimate the errors thus introduced, the partial derivatives of the frequencies were calculated with respect to all the force constants, including those equated to zero.

† Calculation by A. P. Mirgorodskii.

Some additional limitations introduced when choosing the values of the force constants were based on the assumption that K_{ScO} remained constant on passing from the silicate to the germanate and that $K_{YbO} < K_{ScO}$. The force constants of the Si–O bonds were taken as equal not only for bonds of the Si–O(Si) and $Si-O^-$ types but also for the two symmetrically nonequivalent kinds of the latter, forming one set with multiplicity 2 in the factor group (intrinsic symmetry C_s, bonds of the 4–1 type in Fig. 111) and one with multiplicity 4 (intrinsic symmetry C_1, bonds of the 4–2 type in Fig. 111). It should be noted that the separate determination of the force constants $K_{SiO(Si)}$ and $H_{SiO(Si), SiO(Si)}$ or $K_{GeO(Ge)}$ and $K_{GeO(Ge), GeO(Ge)}$ is impossible, if only the frequency ν_{as} SiOSi or ν_{as} GeOGe is known. The approximate separation of these constants for $Sc_2Si_2O_7$ and $Sc_2Ge_2O_7$ was based on the earlier-mentioned estimate of the values of ν_s SiOSi and ν_s GeOGe from the frequency of the ν_s SiOGe vibration of the $SiGeO_7$ ion observed experimentally in the infrared spectra of solid solutions. More reliable values could only be obtained by studying the Raman spectra of these crystals. Since for $Yb_2Si_2O_7$ even approximate estimates of the frequency ν_s SiOSi were lacking, while the observed change in the ν_{as} SiOSi frequency on passing from $Sc_2Si_2O_7$ to $Yb_2Si_2O_7$ could not be ascribed (as calculations confirm) to interaction with lattice vibrations, only the value of $K_{SiO(Si)}$ was subjected to variation, while the interaction of the two Si–O bonds sharing a common O atom ($H_{SiO(Si), SiO(Si)}$) was taken as being the same as in the $Sc_2Si_2O_7$ crystal.

Table 57 presents the sets of force constants obtained as a result of varying these for $Sc_2Si_2O_7$, $Yb_2Si_2O_7$, and $Sc_2Ge_2O_7$. Comparison of the calculated and observed frequencies may be regarded as satisfactory if we remember the simplified description of the force field, the limited applicability (particularly at low frequencies) of the harmonic approximation and the concept of limiting vibrations (with **k** = 0), a certain arbitrariness in the assumption as to the

TABLE 57. Force Constants for Crystals of the $M_2X_2O_7$ Type with the Structure of Thortveitite (in units of 10^6 cm^{-2})

	$Yb_2Si_2O_7$	$Sc_2Si_2O_7$	$Sc_2Ge_2O_7$
K_{M-O}	1.6; 1.6; 1.45	2.0; 1.9: 1.75	1.9
$K_{X-O(X)}$	8.7	9.5	8.2
K_{X-O^-}	8.6	8.4	7.0
K'_{X-O^-}	7.9	7.6	6.6
$H_{XO(X), XO(X)}$	0.8	0.8	0.5
$h_{XO^-, XO(X)}$	0.8	0.8	0.0
h_{XO^-, XO^-}	0.8	0.8	0.0
$K_{O^-XO^-}$	1.9	2.0	2.0
$K_{O^-XO(X)}$	2.1	2.2	2.0
K_{XOX}	0.05	0.05	0.00
$a_{O^-XO^-, XO^-}$	0.3	0.3	0.2
$a_{O^-XO(X), XO^-}$	0.3	0.3	0.2
$a_{O^-XO(X), XO(X)}$	0.3	0.3	0.2

Notes: 1) K_{XO^-} relates to the $Si-O^-$ ($Ge-O^-$) bonds lying in the plane of symmetry and K'_{XO^-} to the same bonds occupying a set of general positions.
2) The three values of K_{MO} correspond to the three translationally nonequivalent kinds of these bonds.

constancy of K_{ScO} in the pyrosilicate and pyrogermanate, and the errors involved in using the spatial configuration of $Sc_2Si_2O_7$ to describe the kinematics of $Sc_2Ge_2O_7$ and $Yb_2Si_2O_7$.

Let us finally discuss the principal results of the calculation in question. Firstly, the vibration frequencies of the crystals under consideration are reproduced satisfactorily over a fairly wide range (1220 to 200 cm^{-1}) using very simple assumptions as to the form of the force field. Moreover one may predict the existence of extremely low-frequency vibrations, with $h\nu < kT$, essentially related to deformations of the Si–O–Si (Ge–O–Ge) bridges and internal rotations in these bridges, which may be of interest in any study of the characteristics of thermal motions in these and similar structures, solid-phase transformations, and so on (see Chap. VI). It is thus useful to study the spectra of these crystals in the far infrared.

Secondly, the reliability of the force constants of the complex anion characterizing the chemical bonds formed by the latter is somewhat increased, in particular constituting an improvement on the quasimolecular approximation. The resultant values of the force constants of the Si–O bonds correlate qualitatively with the relations between the lengths of the three types of these bonds. A particularly important conclusion is that of the weakening of the Si–O–Si bridge and the strengthening of the $Si–O^-$ bonds as a result of the $Sc_2Si_2O_7 \rightarrow Yb_2Si_2O_7$ transition, which agrees closely with Smolin and Shepelev's data regarding the structure of ytterbium pyrosilicate. It is to be expected that similar calculations of the spectra of a whole series of silicates will enable us to study the correlation between the force constant and the length of the Si–O bond in more detail. Thirdly, the supposition as to the strong interaction

TABLE 58. Comparison between the Calculated Vibration Frequencies of $Sc_2Si_2O_7$, $Yb_2Si_2O_7$, and $Sc_2Ge_2O_7$ Crystals and Those Observed in the Infrared Spectra

$Yb_2Si_2O_7$				$Sc_2Si_2O_7$				$Sc_2Ge_2O_7$			
Observed frequencies		Calculated frequencies	Symmetry types	Observed frequencies		Calculated frequency	Symmetry types	Observed frequencies		Calculated frequencies	Symmetry types
pressed in KBr	suspension in oil			pressed in KBr	suspension in oil			pressed in KBr	suspension in oil		
1110	–	1110	B_u	1169	–	1167	B_u	1030	1030	1033	B_u
955	–	961	B_u	975 sh.	–	956	B_u	793 sh.	820 sh.	801	B_u
922	–	917	A_u	907	–	907	A_u	762	760	769	A_u
853	–	853	B_u	852	–	849	B_u	727 sh.	720 sh.	716	B_u
–	–	–	A_g		≈ 670†	670	A_g	≈ 580†		580	A_g
556–540	{ –	546	A_u	587–560	{ –	576	B_u	558–539 sh.	558–523 sh.	{ 554	B_u
	–	540	B_u		–	564	A_u			530	A_u
502	–	498	B_u	–	–	540	A_u	–	–	516	A_u
474	–	475	A_u	501	–	512	B_u	499?	–	–	–
428?	432 w.	–	–	468 sh. ?	–	–	–	430	430	446	B_u
—	386	409	B_u	434	429	426	B_u	–	396	393	B_u
–	370 v.w. ?	–	–	–	–	–	–	–	–	–	–
–	336	331	A_u	–	390	410	A_u	–	336	363	A_u
–	280 sh. ?	–	–	–	–	–	–	–	–	–	–
–	272	268	B_u	–	365	341	B_u	–	308	298	B_u
–	245?	–	–	–	–	–	–	–	–	–	–
–	215	232	A_u	–	288	305	B_u	–	280 sh.	272	B_u
–	–	203	B_u	–	258	267	A_u	–	236	244	A_u
–	–	66	B_u	–	–	73	B_u	–	–	20	B_u
–	–	23	A_u	–	–	28	A_u	–	–	27	A_u

† Approximate estimate derived from the ν_sSiOGe frequency in the spectra of $Sc_2(Si, Ge)_2O_7$ solid solutions.

between the lattice vibrations and many of the deformation vibrations of the complex anion, leading to a considerable "mixing" of the vibrations, is confirmed; this means that any attempts at identifying these vibrations with the help of the quasimolecular model are devoid of physical content. Thus in the frequency range 500 to 600 cm^{-1} the lattice vibrations of thortveitite correspond simultaneously to deformations of the OSiO angles and vibrations of the Sc–O bonds. Since, apart from the complex anion, only forces acting along the shortest oxygen–cation distances (in the first coordination sphere of the latter) were taken into account in the calculation leading to approximate agreement with experimental data, the interaction of the internal vibrations of the complex anion with the lattice vibrations is predominantly a kinematic interaction. It should be noted that the relation between the force constants of the Sc–O and Yb–O "bonds" obtained by calculation appears justified.

§4. Silicates of Rare-Earth Elements in the Divalent State

A spectroscopic study of rare-earth silicates (not obtained in the form of single crystals) in the divalent state [72] was of interest in a number of respects. The ionic radii of the practically accessible cations Sm^{2+}, Eu^{2+}, and Yb^{2+}, the most stable divalent cations of the rare earths, are (according to Zachariasen) 1.11, 1.09, and 0.93Å, which almost exactly corresponds to the range of ionic radii embraced by the ions Sr^{2+}–Ca^{2+} among the alkaline-earth cations. Further, the actual fact that (according to chemical research) compounds of compositions 3:1, 2:1, 3:2, and 1:1 exist in the YbO–SiO_2 and EuO–SiO_2 systems suggests that these are very similar to the system CaO–SiO_2, which contains compounds of the same composition; in the SmO–SiO_2 system only compounds of compositions 2:1 and 1:1 are observed, as in the SrO–SiO_2 system, in which compounds of composition 3:1 and 3:2 are not formed. In the practical respect, comparison between the structures of the silicates of divalent rare-earth elements and the alkaline-earth silicates is also of considerable interest, since it enables us to estimate the suitability of crystals of the latter as "matrices" into which Ln^{2+} ions may be introduced to any extent required. Finally, the silicates of the divalent rare earths themselves have a practical value: thus europium orthosilicate Eu_2SiO_4 has already found a use owing to its high Verdet constant.

We might expect that a reduction in the charge and an increase in the dimensions of the cation would have the effect that dimensional relationships would be much more dominant than in the case of the silicates of trivalent R.E., that the part played by the partly directional char-

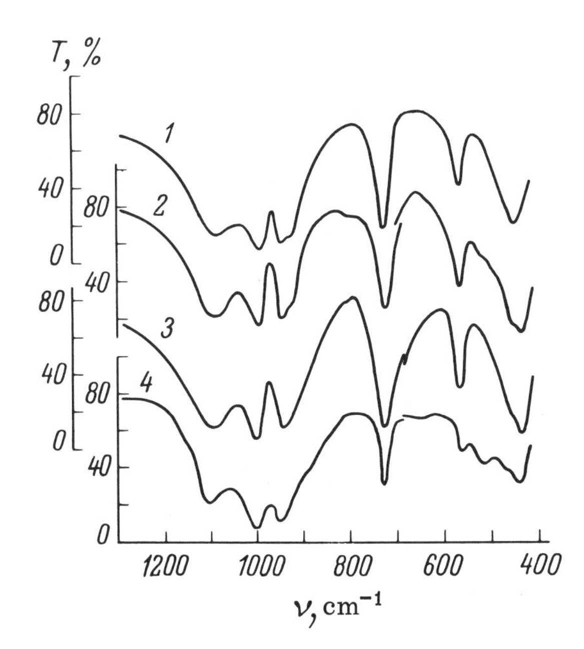

Fig. 112. Infrared spectra of the metasilicates of divalent rare-earth elements and calcium. 1) $CaSiO_3$; 2) $YbSiO_3$; 3) $EuSiO_3$, pressed in KBr, 5 mg; 4) $SmSiO_3$ in vaseline oil.

TABLE 59. Frequencies of the Absorption Maxima in the Infrared Spectra of the Metasilicates of the Rare-Earth Elements $Ln_3Si_3O_9$

$Sm_3Si_3O_9$	$Eu_3Si_3O_9$	$Yb_3Si_3O_9$	Pseudo-wollastonite	Identification of the frequencies (ion $Si_3O_9-D_{3h}$)
1160 sh.	–	–	--	–
1100 v.s.	1094 v.s.	1090 v.s.	1088	A_2'' (ν_{as} O^-SiO^-)
999 v.s.	997 v.s.	996 v.s.	994	E' (ν_s O^-SiO^-, ν_{as} SiOSi)
943 v.s.	941 v.s.	941 v.s.	943	E' (ν_{as} SiOSi, ν_s O^-SiO^-)
–	926?	924 sh.	928	
(884)	–	–	–	Impurity?
(840)	–	–	–	
–	(800)	(804)	–	
728 s.	726 v.s.	724 v.s.	723	E' (ν_s SiOSi)
(644)	–	–	–	Impurity?
562 m.	564 s.	564 s.	565 s.	A_2'' (ρO^-SiO^-)
(515)	–	(524)	–	Impurity?
–	–	(496)	–	
470 sh.	452 sh.	448 sh.	–	See text
440 v.s.	432 v.s.	434 v.s.	444	E' (δO^-SiO^-, δSiOSi)

acter of the Ln–O bonds would be smaller, and that the interaction between the lattice vibrations and the internal vibrations of the complex anions would weaken.

The results of an investigation into the infrared spectra of crystalline compounds in the $YbO-SiO_2$, $EuO-SiO_2$, and $SmO-SiO_2$ systems enabled us, firstly, to establish the chemical individuality of these compounds, and secondly, by comparison with the earlier-described spectra of the alkaline-earth elements, to elucidate the main features of their structures. Let us give more detailed attention to the spectra of some compounds of different compositions.

The metasilicates $LnSiO_3$ (Fig. 112) over the whole frequency range studied have spectra almost entirely identical with those of the high-temperature form of calcium metasilicate (pseudowollastonite) and strontium metasilicate.† (Exceptions include certain bands of impurity origin.) From this we may conclude not only that cyclical Si_3O_9 anions exist but also that the crystals are practically isostructural. Comparison of certain details of the spectra such as the width of the ν_{as} O^-SiO^- (A_2'') vibration around 1100 cm^{-1} and the splittings of the bands at 940 to 920 cm^{-1} shows that the spectra of Yb, and to some extent Eu, metasilicates are very similar to that of pseudowollastonite. The smaller width of the band around 1100 cm^{-1} and the more symmetrical contour of that at 940 cm^{-1} in the spectrum of $SmSiO_3$ makes this similar to $SrSiO_3$. The identification of the vibrational frequencies of the complex anions in Table 59 is based on the results of a calculation for $Sr_3Si_3O_9$ (Chap. IV). Except for the bands of impurity origin, there are no indications of any deviation from the maximum (D_{3h}) symmetry of the ring configuration, nor of any perturbation of the deformation vibrations of the ring by lattice vibrations. Only the appearance of a "shoulder" around 450 to 470 cm^{-1} suggests the removal of degeneracy for the ring vibration of type E' (at 430 to 440 cm^{-1}) or activation of one of the vibrations (forbidden by the selection rules) of types A_2' and A_1', the frequencies of which lie (according to calculation) in the range 480 to 500 cm^{-1}.

† All the silicates of the divalent rare earths here described were synthesized at comparatively high temperatures. The possibility that structures of the wollastonite type might occur for $LnSiO_3$ compounds at low temperatures was not studied.

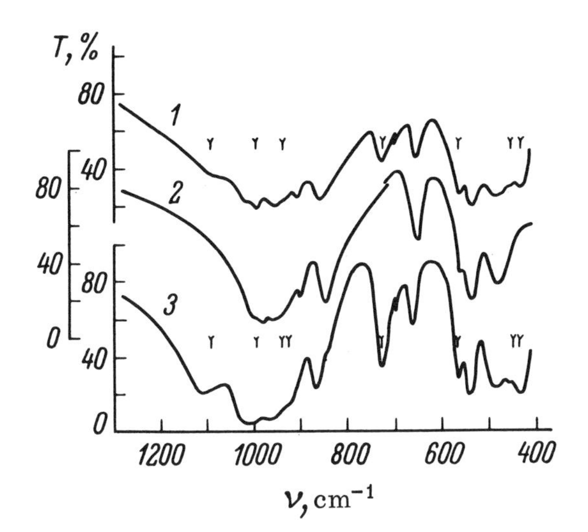

Fig. 113. Spectroscopic proof of the existence of compounds $Ln_3Si_2O_7$ similar to rankinite. Infrared spectra: 1) $Eu_3Si_2O_7$, 2) $Ca_3Si_2O_7$, 3) $Yb_3Si_2O_7$, pressed in KBr, 5 mg. In spectra 1 and 3 there are some impurity bands of the corresponding metasilicates (labeled γ).

TABLE 60. Frequencies of the Absorption Maxima in the Infrared Spectra of the Pyrosilicates of Rare-Earth Elements $Ln_3Si_2O_7$

$Eu_3Si_2O_7$	$Yb_3Si_2O_7$	$Ca_3Si_2O_7$ (rankinite)	Identification of the frequencies of the ion Si_2O_7 (C_{2v})
~ 1090*	1105*	–	
1015 v.s.	1020– v.s.	1002	
997* v.s.	1000*	980	
958 v.s.	964 v.s.	950	ν_{as} SiO_3 (2), ν'_{as} SiO_3 (2), ν_s SiO_3, ν'_s SiO_3, ν_{as} SiOSi ($3B_1$, B_2, $2A_1$, A_2)
~ 936*	940*?	–	
–	924*	–	
910 m.	–	903	
856 s.	856 s.	847	
–	(844 sh.)	–	Impurity?
727*	728*	–	–
654 m.	662 m.	654	ν_s SiOSi (A_1)
561*	566*	560	δ_{as} SiO_3 (2), δ'_{as} SiO_3 (2) (B_1, B_2, A_1, A_2)
535 v.s.	542 v.s.	538	
482 s.	487 s.	482	
448*	454*	–	–
431*	436*	–	–

*Bands due to the superposition (complete or partial) of the absorption bands of impurities of $EuSiO_3$ and $YbSiO_3$.

The existence of pyrosilicates $Ln_3Si_2O_7$ could only be established in the $YbO-SiO_2$ and $EuO-SiO_2$ systems; we see from the curves of Fig. 113 that these could not be obtained in pure form but always contained considerable amounts (probably 20 to 30%) of the corresponding metasilicates, which could easily be identified spectroscopically. At the same time, consideration of the bands not coinciding in frequency with those of any other compounds in the system considered clearly indicated the existence of compounds with a structure of the rankinite type $Ca_3Si_2O_7$. The most indicative band in this respect was the ν_s SiOSi vibration at around 650 cm^{-1} (Table 60).

The spectra of all three Ln_2SiO_4 orthosilicates exhibit considerable similarity with each other (Fig. 114), but, judging from the relative intensity of the band at 840 cm^{-1}, the number of

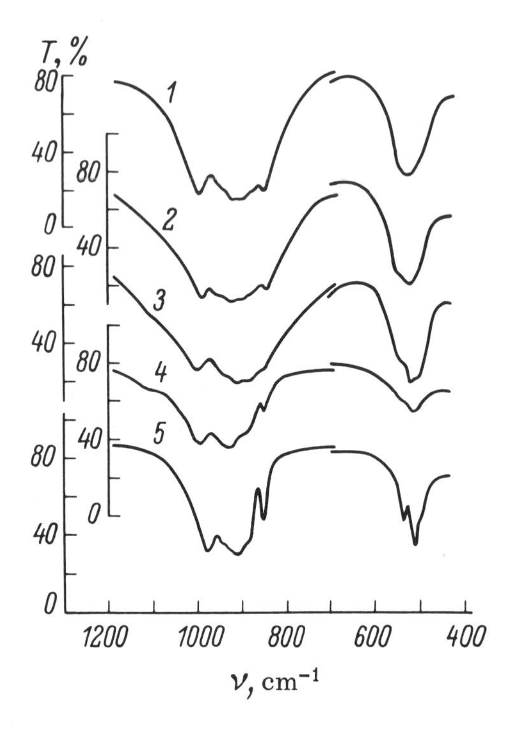

Fig. 114. Infrared spectra of the orthosilicates of divalent rare-earth elements, calcium, and strontium. 1) Ca_2SiO_4; 2) Yb_2SiO_4; 3) Eu_2SiO_4 pressed in KBr, 5 mg; 4) Sm_2SiO_4; 5) Sr_2SiO_4 in vaseline oil.

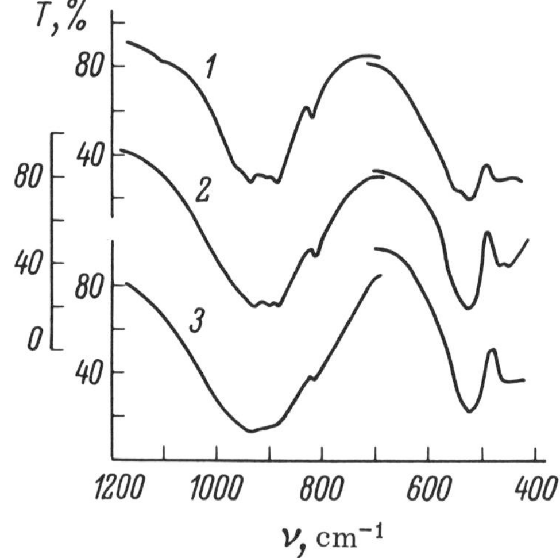

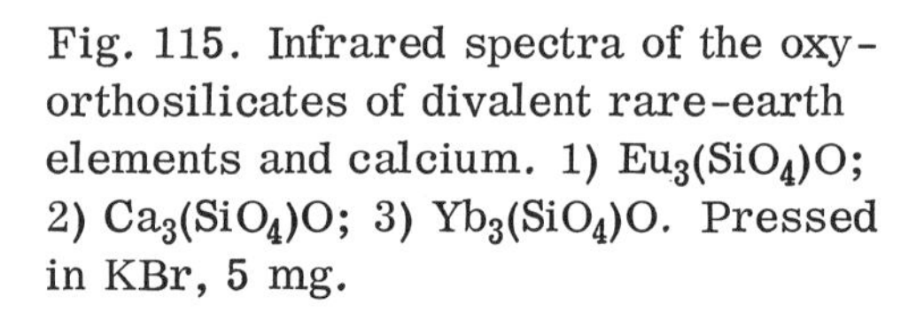

Fig. 115. Infrared spectra of the oxyorthosilicates of divalent rare-earth elements and calcium. 1) $Eu_3(SiO_4)O$; 2) $Ca_3(SiO_4)O$; 3) $Yb_3(SiO_4)O$. Pressed in KBr, 5 mg.

maxima in the complex band at 880 to 940 cm^{-1}, and the clarity of the triplet structure of the band around 500 cm^{-1}, we may consider that Sm_2SiO_4 and Eu_2SiO_4 are similar to strontium orthosilicates, while the spectrum of Yb_2SiO_4 is practically identical with that of β-dicalcium silicate. We note that good agreement was obtained in [73] between the interplane distances obtained from Debye photographs for Eu_2SiO_4 and Sr_2SiO_4 crystals [74]. (According to x-ray structural analysis of $EuSO_4$ and $EuCO_3$, the ionic radius of divalent europium is 0.03 Å smaller than that of the Sr^{2+} ion [75].)

The spectra of the oxyorthosilicates $Ln_3(SiO_4)O$, the existence of which (as in the case of the pyrosilicates) has only been confirmed for Ln = Eu and Yb, are almost indistinguishable from the spectrum of tricalcium silicate (see Fig. 115).

It follows from the results obtained that isomorphous substitutions of Ca^{2+} ions by Yb^{2+} and Eu^{2+} and of Sr^{2+} ions by Sm^{2+} and Eu^{2+} may well take place over a wide range of concentrations without any serious lattice distortions. The similarity of the ionic radii of the cations for the same formal charge, in accordance with the laws of classical crystal chemistry, constitutes a sufficient condition for the construction of similar crystal lattices for the compounds under consideration. This result is not unexpected, since on passing from compounds with Ln^{3+} cations to those with Ln^{2+} the electrostatic field of the cations weakens, which leads to a fall in the mutual polarization of the cations and anions and hence to a less important role of the covalent bonds between them, these usually constituting one of the main causes of deviations from purely "packing" laws in the structure of ionic crystals.

Comparison of the spectra of silicates with divalent rare-earth cations, on the one hand, and calcium and strontium silicates on the other also leads to a number of interesting conclusions associated with the problem of identifying the frequencies in the spectra of these crystals. In comparing the spectra of these compounds one notices the slightness of the displacement of all the bands, including those in the low-frequency range, right up to the long-wave boundary of the spectral range studied. Since on passing from the silicates of, for example, calcium to those of ytterbium the mass of the cation increases sharply, we may suppose that the observed constancy of the frequencies in the spectra indicates that all the bands relate to the internal vibrations of the complex anions. In other words, the contribution of lattice vibrations, i.e., the vibrations of the cation–oxygen bonds, to frequencies exceeding 400 cm^{-1} is insignificant. This enables us to identify not only the stretching but also some of the deformation vibrations of the silicon–oxygen anions. Thus, for example, the absorption bands observed in the spectra of all the metasilicates considered around 565 and 430 to 440 cm^{-1} should be ascribed to the deformation vibrations of cyclical Si_3O_9 anions. In accordance with the results of calculation, the first of these frequencies probably corresponds to a vibration of the A_2'' type (for symmetry D_{3h}), i.e., ρO^-SiO^-, while the lower-frequency one corresponds to the δO^-SiO^- vibrations, "mixed" with the δSiOSi of the ring, of symmetry type E'. Unfortunately, the absence of experimental data relating to the orientation of the electric moments of the transitions prevents us from verifying this assignment. In an analogous way, the bands around 560, 540, and 480 cm^{-1} in the spectra of silicates with the structure of rankinite may be ascribed to the three highest-frequency vibrations of the bent Si_2O_7 group, $\delta'_{as} SiO_3 (B_1)$ and $\delta_{as} SiO_3 (B_2, A_1)$. In the spectra of the orthosilicates the band in the low-frequency region, exhibiting a triplet structure for all compounds except β-Ca_2SiO_4, corresponds to the triply degenerate deformation vibration of the SiO_4 tetrahedron. The frequency of this vibration varies from 510 or 511 cm^{-1} in Sr_2SiO_4 and Sm_2SiO_4 to 516 cm^{-1} in Eu_2SiO_4 and 521 to 523 cm^{-1} in Yb_2SiO_4 and β-Ca_2SiO_4, not showing any dependence on the

TABLE 61. Frequencies of the Absorption Maxima in the Infrared Spectra of the Orthosilicates of Rare-Earth Elements Ln_2SiO_4

Sr_2SiO_4	Sm_2SiO_4	Eu_2SiO_4	Yb_2SiO_4	β-Ca_2SiO_4	Identification of frequencies
–	(1100)	(1100)	–	(1120)	Impurity?
968	993 v.s.	999 v.s.	991 v.s.	996	
940 sh.	?	942 sh.	950 sh.	941 sh.	
908	924 v.s.	912 v.s.	920 v.s. 900 v.s.	917 901	$\nu_{as} SiO_4$
885 sh.	890 sh.	876 sh.	878 sh.	872 sh.	
838	848 s.	846 sh.	842 v.s.	845	
532	535 sh.	538 sh.	~540 sh.	?	
510	511 s.	516 v.s.	521 v.s.	523	$\delta_{as} SiO_4$
495 sh.	–	501 sh.	?	?	

TABLE 62. Frequencies of the Absorption Maxima in the Infrared Spectra of the Oxyorthosilicates of Rare-Earth Elements $Ln_3(SiO_4)O$

$Eu_3(SiO_4)O$	$Yb_3(SiO_4)O$	$Ca_3(SiO_4)O$	Identification of frequencies
(~1100)	–	–	Impurity?
956 sh.	?	950 sh.	ν_{as} SiO_4
936 v.s.	933 v.s.	933	
908 s.	~910 sh.	901	
886 v.s.	890 v.s.	883	
818 m.	816 m.	812	ν_s SiO_4
540 sh.	538 sh.	?	δ_{as} SiO_4
522 v.s.	521 v.s.	525	
?	510 sh.	?	
~460 s.	~450 s.	464	δ_s SiO_4 and ν $(M^{2+}-O^{2-})$?
–	–	452	

mass of the cation; its value is higher than might be expected for the free $[SiO_4]^{4-}$ ion, and this may be ascribed to the influence of the internal field of the crystal, although it gives no indications as to the closeness of the characteristic frequencies of the vibrations of the cation–oxygen bonds. In the spectra of the orthosilicates, in addition to the band around 520 cm^{-1}, absorption occurs at 460 cm^{-1}, and this may presumably be ascribed to lattice vibrations involving the participation of the O^{2-} "ion."

References

1. Yu. I. Smolin and S. P. Tkachev, Kristallografiya, 14:22 (1969).
2. Yu. I. Smolin, Kristallografiya, 14:985 (1969).
3. Yu. I. Smolin and Yu. F. Shepelev, Neorg. Materialy, 5:1823 (1969).
4. Yu. I. Smolin and Yu. F. Shepelev, Neorg. Materialy, 3:1034 (1967).
5. Yu. I. Smolin and Yu. F. Shepelev, Neorg. Materialy, 4:1133 (1968).
6. Yu. I. Smolin, Yu. F. Shepelev, and T. V. Upatova, Dokl. Akad. Nauk SSSR, 187:322 (1969).

6a. Yu. I. Smolin, Kristallografiya, 15:47 (1970).

7. S. S. Batsanov, G. N. Grigor'eva, and I. P. Sokolova, Zh. Strukt. Khim., 3:338 (1962).
8. W. L. Baun and N. T. McDevitt, J. Amer. Ceram. Soc., 46:294 (1963).
9. N. T. McDevitt and W. L. Baun, Spectrochim. Acta, 20:799 (1964).
10. N. T. McDevitt and A. D. Davidson, J. Opt. Soc. Amer., 56:636 (1966).
11. H. G. Häfale, Z. Phys., 148:262 (1957).
12. F. Vratny, Appl. Spectroscopy, 13:59 (1959).
13. J. R. Ferraro, J. Mol. Spectr., 4:99 (1960).
14. A. Hadni, J. Claudel, E. Décamps, X. Gerbaux, and P. Strimer, Compt. Rend., 255:1595 (1962).
15. A. Walker and J. R. Ferraro, J. Chem. Phys., 43:2689 (1965).
16. Yu. Ya. Kharitonov and I. Z. Babievskaya, Dokl. Akad. Nauk SSSR, 168:615 (1966).
17. W. C. Steele and J. C. Decius, J. Chem. Phys., 25:1184 (1956).
18. C. E. Weir and E. R. Lippincott, J. Res. Nat. Bur. Standards, 65A:173 (1961).
19. C. E. Weir and R. A. Schroeder, J. Res. Nat. Bur. Standards, 68A:465 (1964).
20. J.-P. Laperches and P. Tarte, Spectrochim. Acta, 22:1201 (1966).
21. W. F. Bradley, D. L. Graf, and R. S. Roth, Acta Crystallogr., 20:283 (1966).
22. E. M. Levin, R. S. Roth, and J. B. Martin, Amer. Mineralog., 46:1030 (1961).

22a. S. D. Ross, J. Mol. Spectr., 29:131 (1969).

23. T. F. Tenisheva, T. M. Pavlyukevich, and A. N. Lazarev, Izv. Akad. Nauk SSSR, Ser. Khim., 1771 (1965).

24. G. Pannetier and A. Dereigne, Bull. Soc. Chim. France, 1850 (1963).
25. V. P. Surgutskii, V. I. Gaivoronskii, and V. V. Serebrennikov, Zh. Neorg. Khim., 13:1010 (1968).
26. J. R. Ferraro and J. S. Ziomek, J. Inorg. Nucl. Chem., 26:1397 (1964).
27. K. I. Petrov, V. P. Vasil'eva, and V. G. Pervykh, Zh. Neorg. Khim., 11:1837 (1966).
28. R. G. Charles, J. Inorg. Nucl. Chem., 27:1489 (1965).
29. V. T. Matveichuk, A. V. Shevchenko, and N. V. Skripchenko, Neorg. Materialy, 2:514 (1966).
30. V. N. Krishnamurthy and S. Soundarajan, Z. Anorg. Allg. Chem., 349:220 (1967).
31. V. I. Ivanov, M. B. Varfalomeev, et al., Kristallografiya, 10:509 (1965).
32. A. N. Lazarev, T. F. Tenisheva, I. A. Bondar', and N. A. Toropov, Izv. Akad. Nauk SSSR, Ser. Khim., 1220 (1963).
33. A. N. Lazarev, T. F. Tenisheva, I. A. Bondar', L. N. Koroleva, and N. A. Toropov, Neorg. Materialy, 4:1287 (1968).
34. E. A. Kuz'min and N. V. Belov, Dokl. Akad. Nauk SSSR, 165:88 (1965).
35. L. A. Harris and C. B. Finch, Amer. Mineralog., 50:1493 (1965).
36. C. Michel, G. Buisson, and E. F. Bertaut, Compt. Rend., 264B:397 (1967).
36a. B. A. Maksimov, Yu. A. Kharitonov, V. V. Ilukhin, and N. V. Belov, Dokl. Akad. Nauk SSSR, 183:1072 (1968).
37. T. F. Tenisheva, A. N. Lazarev, and T. M. Pavlyukevich, Izv. Akad. Nauk SSSR, Ser. Khim., p. 1553 (1965).
38. J. Fournier and P. Cohlmuller, Compt. Rend., 266c:1361 (1968).
39. A. G. Cockbain and G. V. Smith, Mineralog. Mag., 36:411 (1967).
40. V. B. Glushkova, I. A. Davtyan, and E. K. Keler, Neorg. Materialy, 3:119 (1967).
41. G. Buisson and C. Michel, Mater. Res. Bull., 3:193 (1968).
42. A. N. Lazarev and T. F. Tenisheva, Izv. Akad. Nauk SSSR, Otd. Khim. Nauk, p. 964 (1961).
43. A. N. Lazarev and T. F. Tenisheva, Trans. Sixth Conf. on Exp. and Tech. Mineralogy and Petrography, Izd. AN SSSR, Moscow (1962), p. 319.
44. A. N. Lazarev, T. F. Tenisheva, I. A. Bondar', and L. N. Koroleva, Izv. Akad. Nauk SSSR, Otd. Khim. Nauk, 557 (1962).
45. A. N. Lazarev, T. F. Tenisheva, and I. A. Bondar', Neorg. Materialy, 1:1207 (1965).
46. I. Warshaw and R. Roy, Progr. in Science and Technology of the Rare Earths, Vol. 1, Pergamon Press (1964), p. 203.
47. A. N. Lazarev, T. F. Tenisheva, I. A. Bondar', and L. N. Koroleva, Neorg. Materialy, 5:74 (1969).
48. D. W. J. Cruickshank, H. Linton, and G. A. Barclay, Acta Crystallogr., 15:491 (1962).
49. N. G. Batalieva and Yu. A. Pyatenko, Zh. Struktur. Khim., 9:921 (1968).
50. J. Ito and H. Johnson, Amer. Mineralog., 53:1940 (1968).
51. P. W. Ranby, D. H. Mash, and S. T. Henderson, Brit. J. Appl. Phys. Supl. 4:18 (1955).
52. A. N. Lazarev and T. F. Tenisheva, Neorg. Materialy, 5:82 (1969).
53. D. W. J. Cruickshank, J. Chem. Soc. (L.), 5486 (1961).
54. B. E. Robertson and C. Calvo, Acta Crystallogr., 22:665 (1967).
55. C. Calvo, Acta Crystallogr., 23:289 (1967).
56. N. C. Webb, Acta Crystallogr., 21:942 (1966).
57. C. Calvo, Inorg. Chem., 7:1345 (1968).
58. J.-C. Grenier and R. Masse, Bull. Soc. Frac. Mineral. Cristallogr., 90:285 (1967).
59. L.-O. Hagman, I. Jansson, and C. Magneli, Acta Chem. Scand., 22:1419 (1968).
60. A. N. Lazarev and T. F. Tenisheva, Dokl. Akad. Nauk SSSR, 177:849 (1967).
61. E. Steger and B. Kassner, Z. Anorg. Allg. Chem., 355:131 (1967).
62. A. F. Lubchenko and E. I. Rashba, Opt. i Spektr., 4:580 (1958).
63. J.-P. Labbé, Compt. Rend., 259:1822 (1964).
64. J.-P. Labbé, Thesis, Univ. of Paris (1965).

65. T. F. Tenisheva, A. N. Lazarev, I. A. Bondar', and N. V. Vinogradova, Izv. Akad. Nauk SSSR, Ser. Khim., 1764 (1965).
66. A. N. Lazarev, T. F. Tenisheva, A. N. Sokolov, A. P. Mirogorodskii, and N. A. Toropov, Dokl. Akad. Nauk SSSR, 183:352 (1968).
67. V. M. Goldschmidt, Norsk. Geol. Tidskrift, 12:247 (1931), cited in Strukturbericht, Vol. 2, (1928-1932).
68. T. Shimanouchi, M. Tsuboi, and T. Miyazawa, J. Chem. Phys., 35:1597 (1961).
69. J. J. Hopfield, Phys. Rev., 112:1555 (1958).
70. J. F. Scott and S. P. S. Porto, Phys. Rev., 161:903 (1967).
71. M. M. Elcombe, Proc. Phys. Soc., 91:947 (1967).
72. A. N. Lazarev and T. F. Tenisheva, Neorg. Materialy, 1:569 (1965).
73. M. W. Shafer, T. R. McGuire, and J. C. Suits, Phys. Rev. Lett., 11:251 (1963).
74. N. A. Toropov and Mao Chih-Ch'iung, Zh. Neorg. Khim., 7:1632 (1962).
75. I. Mayer, E. Levy, and A. Glasner, Acta Crystallogr., 17:1071 (1964).

CHAPTER VI

SPECTRA AND FLEXIBILITY OF Si—O—Si AND P—O—P BONDS

In the first stages of the x-ray structural analysis of crystalline silicates, the observation that there were only comparatively few structural types of the complex silicon–oxygen anions, and that these were reproduced without any major alterations in many different crystals, led to a general discussion regarding the rigidity of these anions. This was reflected in the basic concept of a silicon–oxygen framework, the configuration of which largely determined the structure of the crystal.

A review of this kind of concept has only recently been carried out, following the accumulation of a large number of experimental facts, all indicating a high degree of lability of the silicate anions. An important part has been played in these developments by N. V. Belov's idea (verified in many structural analyses) to the effect that the configuration of the silicon–oxygen anion is capable of adapting itself to the structural motif determined by the structure and relative dispositions of the coordination polyhedra of the cations [1].

On the other hand, research into organosilicon compounds has shown that siloxane molecules, which are very similar in structure to silicate anions, have a considerable internal mobility, in many cases exceeding the mobility of corresponding organic molecules. These characteristics of the siloxanes determine a number of their practical uses, covering a wide field in the application of high-polymer siloxanes: their low-temperature coefficient of viscosity, their high elasticity extending down to very low temperatures, their great compressibility, and so on [2]. Among the many direct experimental proofs of the low excitation energy of the internal motions in siloxane molecules we may mention the results of an investigation into the properties of siloxanes by nuclear magnetic resonance. In these compounds, in addition to the high rotational mobility of the methyl radicals (which extends down to liquid-nitrogen temperatures), the great freedom with which deformations and internal rotations take place in chains of Si–O–Si bonds down to −100°C has been known for a long time [3]. Electron- and x-ray-diffraction studies of structures containing flat trisiloxane rings have also established an anomalously high mobility of the methyl radicals; these are ascribed to the large amplitude of the internal rotations ("nonplanar" Si–O–Si deformations) in these rings [4-7].

It would be natural to suppose that these features of the siloxane bonds would also be reflected in the properties of complex silicate anions. Work in this direction has only recently been started. Here we shall consider existing data, chiefly derived from x-ray and spectroscopic investigations, regarding the internal mobility of the X–O–X oxygen bridges and its relationship with the characteristics of their electron structure. The list of compounds considered is limited to siloxanes, silicates, and the analogous phosphates. The structures of the phosphonitriles (phosphazenes with $-\overset{\diagdown\diagup}{P}{=}N{-}\overset{\diagdown\diagup}{P}{=}N{-}$ bonds) which are also distinguished by high flexibility and are isoelectronic with the siloxanes, have been considered in a number of reviews [8-10]. Considerable attention has been paid to phase transformations, both in connection with recent developments in the concept of the part played by lattice vibrations in the mechanism of solid-state transformations [11, 12], and also because any lack of rigidity in the structure under consideration may be expected to manifest itself most sharply near phase-transformation points.

§1. Spectroscopic Evidence for the Flexibility of Si–O–Si Bonds in Siloxanes

Modern theory associates the molecular mechanism of the high flexibility of organic high polymers principally with the low-energy barriers to the internal rotations. Hence a study of the mechanism underlying the flexibility of siloxane chains and others of similar nature in heteroorganic compounds, as well as in inorganic anions with similar bonds, should clearly start with an investigation into the internal rotations in these chains. It may be initially considered that the high values of the SiOSi valence angles and the great distances between the side substituents neighboring the Si atoms facilitate internal rotation in the siloxane chains. There are no grounds for supposing that the partial double-bondedness of these chains might lead to any substantial increase in the retarding potential, since each of the Si atoms has at least two d orbitals oriented favorably for interaction with the unshared pairs of electrons of a specified O atom.

On the other hand, it follows from the principles set out earlier as to the electron structure of X–O–X bridges (where X = Si, P, and sometimes Ge) that the most frequently encountered angles (XOX = 130 to 150°) constitute an intermediate case between the structures formed by "pure" σ bonds and structures characterized by the complete use of the virtual systems of π bonds. The fact that small changes in the electronegativity of the substituents, i.e., small changes in the energy of the d orbitals of the atom X, lead to substantial changes in the valence angle of the oxygen suggests that the energy of the X–O–X system will change comparatively little when the angle XOX deviates from the equilibrium value for the particular compound. In other words, the force constants of the deformation vibrations δXOX are probably small, which for large masses of the atoms should lead to a very low frequency of these vibrations. Hence the population of the excited states determined by the Boltzmann factor $e^{-h\nu/kT}$ is extremely significant even at ordinary temperatures. This mechanism may clearly constitute an additional source of flexibility for chains with X–O–X bonds in contrast to the majority of organic polymers, in which, as a rule, the deformation of the valence angles requires considerably greater expenditure of energy than the excitation of the internal rotations.

We shall now consider what conclusions as to the characteristics of the dynamics of the silicon–oxygen bridges may be drawn from a study of the simplest siloxanes.

Theoretical consideration was given to the possible spectroscopic manifestations of the low energy and great anharmonicity of the δSiOSi deformation vibration and its strong interaction with rotation (relative to an axis parallel to the Si . . . Si direction) by Thorson and Nakagawa [13]. These authors analyzed the dynamics of a triatomic "quasilinear" (their terminology) molecule characterized by a quantum of deformation vibration small compared with the stretching vibrations and a potential-energy function of the form

$$V = V_0(r) + {}^1/_2 K_a a^2 + {}^1/_2 K_s s^2 + K_a X_a a^2 \alpha^2 + K_s X_s s^2 \alpha^2 + \ldots$$

Here a and s are symmetry coordinates corresponding to the two stretching vibrations; $\alpha = {}^1/_2\,(180° - \angle \mathrm{SiOSi})$; $r = r_{\mathrm{Si-O}} \cdot \alpha$. The coefficients X_a and X_s characterize the interaction of the stretching vibrations with the deformation of the SiOSi angle, which in turn interacts strongly with the rotation; the potential $V_0(r)$ is not harmonic:

$$V_0(r) = {}^1/_2 K_r r^2 + K_B/(c^2 + r^2)$$

The choice of parameters K_B and c in the theory determines the form of the potential $V_0(r)$; for $(2K_B/K_r)^{1/2} - c^2 > 0$ the potential has a minimum of $r_0 = [(2K_B/K_r)^{1/2} - c^2]^{1/2}$ and a maximum at r = 0 (i.e., for ∠SiOSi = 180°). For a smallish potential maximum at ∠SiOSi = 180° and a minimum at a large angle such as 150° (and hence a small moment of inertia of the molecule relative to an axis parallel to Si . . . Si), we may expect a complex structure of the band of stretching vibrations, which embraces a considerable range of frequencies, and is associated with combination transitions and transitions from excited rotational and deformation sublevels. The strong interaction of deformation and rotation in the model considered also leads us to expect a considerable displacement of the rotational levels when compared with the levels of a rigid rotator, and a complex structure of the δSiOSi band.

These results explained the absence of any band corresponding to the symmetrical ν_s SiOSi stretching vibration in the infrared spectrum of disiloxane $H_3SiOSiH_3$ [14, 15], despite the bent shape of the molecule (∠SiOSi = 144° [16]) established by electron diffraction. Evidently the large number of satellites of this band, occupying a wide frequency range and comparable in intensity with the weak principal maximum, leads to the almost complete disappearance of the latter. Transitions from excited rotational and deformation sublevels explained the complex shape and the appearance of a satellite of the ν_{as} SiOSi band at approximately 100 cm^{-1} above the frequency of this vibration. Lack of success in attempts at securing the ordinary rotational P–R structure of the parallel ν_{as} SiOSi band of gaseous disiloxane was ascribed to the blurring of this structure by the same transitions from excited levels. The complex form of the ν_{as} SiOSi band and the presence of satellites is evidently also characteristic of other siloxanes; for example, a very strong satellite of the ν_{as} SiOSi band is found in the spectrum of $Cl_3SiOSiCl_3$ [17], where the SiOSi angle is particularly large. We also note that the observation of a band at 67 cm^{-1} in the long-wave infrared spectrum of $H_3SiOSiH_3$ [18], the frequency of this being unaltered by deuterization, probably constitutes a valid confirmation of the assumption as to the low δSiOSi frequency.

Hexamethyldisiloxane [19] shows two frequencies in the infrared spectrum, ν_{as} SiOSi and ν_s SiOSi (probably as a result of the value of ∠SiOSi, which is smaller than in $H_3SiOSiH_3$), and both bands have a complicated structure which may be associated with the interaction of these vibrations with internal rotations and deformations of the Si–O–Si bridge.

The theoretical calculation of the energy levels of the hexamethyldisiloxane molecule associated with internal rotation around the Si–O bond is a complex problem, since one must consider: a) the interaction of the two rotators, the $Si(CH_3)_3$ groups, with each other, b) the interaction between the internal rotations and the rotations of the molecule as a whole, and finally c) the interaction between the internal rotations and the deformation of the SiOSi angle, similar to that considered by Thorson and Nakagawa for the triatomic model, namely, the interaction of δSiOSi with the rotation of the molecule.

The inclusion of even the first of the interactions mentioned leads to a considerable complication of the energy spectrum as shown in [20] in connection with calculating the torsional vibrations of a molecule of the $H_3C{-}X{-}CH_3$ type. The interaction of the internal rotations with the external plays no significant part for small moments of inertia of the internal rotators, but undoubtedly has a considerable importance for the hexamethyldisiloxane molecule. (The moment of inertia of the $Si(CH_3)_3$ group for tetrahedral angles and r_{SiC} = 1.88 Å and r_{CH} = 1.09 Å is $277.5 \cdot 10^{-40}$ g·cm², while the total moment of inertia of the molecule relative to an axis parallel to the Si . . . Si line for ∠SiOSi = 140° and r_{SiO} = 1.63 Å equals $514.5 \cdot 10^{-40}$ g·cm².) We shall here confine ourselves to an attempt at securing a mainly qualitative interpretation of the structure of the bands of the hexamethyldisiloxane Si–O–Si stretching vibrations. More reliable information regarding the low-frequency energy levels of this and other molecules considered in this section can only be obtained by direct experimental measurements in the long-wave part of the spectrum and theoretical analysis, allowing for the factors indicated.

Before considering the possible identification of individual components of the band structure, let us examine the classification of the lowest-frequency internal motions of the framework of the $(H_3C)_3SiOSi(CH_3)_3$ molecules. These clearly include the δSiOSi deformation vibration and two torsional vibrations τ of the $Si(CH_3)_3$ groups corresponding to the rotation of these groups around the Si–O bonds in the same or opposite directions. For a symmetry of C_{2v}, the most probable symmetry for the bent configuration ($\angle$SiOSi = 135 to 140°) of the molecule, these three vibrations respectively relate to symmetry types A_1, A_2, B_2.† Hence according to the selection rules for combination transitions the δSiOSi vibration may give paired combinations, active in the infrared spectrum, with ν_{as} SiOSi(B_1) and ν_s SiOSi(A_1), while among the torsional vibrations only that of type A_2 can combine with ν_{as} SiOSi and only that of type B_2 with ν_s SiOSi. We must emphasize the substantial difference between the two torsional vibrations of the $(H_3C)_3SiOSi(CH_3)_3$ molecules: $\tau_1(A_2)$ and $\tau_2(B_2)$ (Fig. 116). From symmetry considerations we see that for the first of these the Si–O–Si bridge should be stationary, i.e., in the strict sense of the word this is a torsional vibration of the $Si(CH_3)_3$ groups. On the other hand, for the τ_2 vibration there may be a displacement of the O atom in a plane perpendicular to the SiOSi plane; the rotation of the two $Si(CH_3)_3$ groups in one direction is equivalent to the rotation of the SiOSi plane in the opposite direction. In other words, this vibration is at the same time a "nonplanar" deformation vibration (rotational oscillation) of SiOSi. This may be illustrated by comparison with the motions of an analogous molecule with $\angle$SiOSi = 180°, in which the "planar" and "nonplanar" SiOSi deformations are degenerate and only one torsional vibration remains:

Bent	Straight
$\tau_1(A_2)$	$\tau(A_1''$ or $A_{1u})$
$\tau_2(B_2)$, $\delta(A_1)$	$\delta(E'$ or $E_u)$

A comparison of the calculated (from spectroscopic data) and experimental values of the thermodynamic functions of hexamethyldisiloxane‡ carried out by a group of American authors [24] led to the conclusion that the rotation of the $Si(CH_3)_3$ groups was very nearly free (with a potential barrier not exceeding a few hundreds of cal/mole). Under these conditions the frequency τ_1 is extremely low; for a barrier of around 100 cal/mole it is no greater than a few cm^{-1} owing to the much reduced moment of inertia of the two almost coaxial $Si(CH_3)_3$ groups. For the τ_2 vibration the reduced moment is mainly determined by the small (as compared with the total moment of inertia of the two $Si(CH_3)_3$ groups) moment of inertia of the $\mathrm{Si}\overset{\mathrm{O}}{\frown}\mathrm{Si}$ bridge relative to an axis parallel to the Si . . . Si axis, which for the same height of the barrier and a moment of the order of $10 \cdot 10^{-40}$ $g \cdot cm^2$ gives a frequency of 20 to 30 cm^{-1} for τ_2. According to the Thorson and Nakagawa theory, the flexibility of the SiOSi angle may lead to an additional rise in frequency. (It follows from the foregoing that the τ_2 vibration of the $(H_3C)_3SiOSi(CH_3)_3$ molecule to a certain extent corresponds to the external rotation of the triatomic model of Thorson and Nakagawa.)

† A detailed treatment is given for the symmetry properties of molecules with internal rotations and the selection rules for optical transitions in [21-23].

‡ The potential barrier for the rotation of the CH_3 groups relative to the Si–C bond is evidently around 1600 cal/mole, as in $Si(CH_3)_4$; similar values are obtained from microwave data for H_3SiCH_3 and $H_2Si(CH_3)_2$ [25, 26].

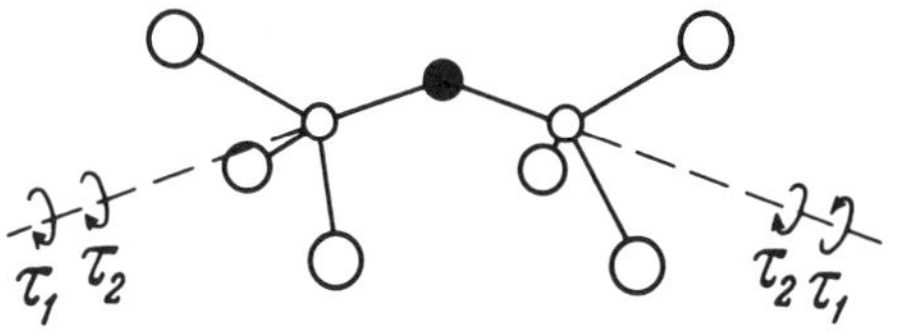

Fig. 116. Scheme of internal rotations around the Si –O bonds in the $(CH_3)_3SiOSi(CH_3)_3$ molecule with symmetry C_{2v}. Arrows indicate the direction of rotation of the $Si(CH_3)_3$ groups for the torsional vibrations τ_1 and τ_2.

In the infrared spectrum of crystalline hexamethyldisiloxane at low temperatures (120°K) there is a complex structure in the region of the ν_s SiOSi frequency (520 cm^{-1}); on the high-frequency side of the principal maximum there is a rapidly converging series of bands, on the long-wave side there are two weak maxima,† and in addition to this absorption occurs in the immediate neighborhood of ν_s SiOSi. On raising the temperature this structure becomes blurred, although it never vanishes, but forms two wings of the ν_s SiOSi band in the spectrum of the liquid and the gas; the band itself widening considerably, particularly on the low-frequency side. Hence the structure of the ν_s SiOSi band may be mainly ascribed to the interaction of this vibration with the internal (and not the external) degrees of freedom of the molecule; the part played by the latter reduces merely to a certain blurring of the whole picture. In the region of the ν_{as} SiOSi vibration (1060 cm^{-1}) there is also a complex structure, part of which may also be seen in the spectrum of the liquid and the gas.

On the grounds of the foregoing considerations the structure of the bands may be explained as being due to combinations of the $\nu_s \pm \tau_2$ type (for ν_{as} these transitions are inactive), the $\nu_{as,s} \pm \delta$ type, and transitions from excited levels ("upper-level" transitions), i.e., $\nu_{as,s} + \tau_2 - \tau_2$ or $\nu_{as,s} + \delta - \delta$. This enables us to construct (in Fig. 117) a hypothetical scheme of the energy states of the $\mathrm{Si}^{\diagup \mathrm{O} \diagdown}\mathrm{Si}$ bridge in the $(H_3C)_3SiOSi(CH_3)_3$ molecule in agreement with experimental data by characterizing this state in terms of four quantum numbers N_{as}, N_s, n, and l corresponding to vibrations ν_{as}, ν_s, δ, and τ_2. The resultant sequence of levels agrees with the assumption as to the very weakly restricted rotations; the scheme of levels shown in Fig. 117 for the fine structure of the vibrational states (with different n and l) and the variation in their separations (ΔU) are in good agreement with the conclusions of Thorson and Nakagawa if we suppose that $(-\mu/m+2X_a)>0$ and $(-\mu/m+2X_s)\leqslant 0$. (Here μ is the reduced mass of the SiOSi, m is the mass of the O atom.) In other words we find that the ν_{as} SiOSi vibration interacts more strongly with the deformation of the SiOSi angle and with the oscillations of the whole SiOSi plane than do the symmetrical vibrations; this is probably associated with the large amplitude of the displacement of the O atom. In accordance with the foregoing considerations the frequencies δ and τ_2 are low, being 130 and 37 cm^{-1} for the ground state ($N_{as} = 0$, $N_s = 0$, $n = 0$, $l = 0$).

Thus a study of the spectrum of hexamethyldisiloxane clearly confirms the assumption as to the great freedom of displacement of the O atom in the $\mathrm{Si}^{\diagup \mathrm{O} \diagdown}\mathrm{Si}$ bridges in a plane perpendicular to the Si . . . Si line. The low frequencies of the corresponding vibrations suggest the possibility of considerable changes taking place in the structure of siloxane molecules or silicate anions with changing temperature. Subsequently, by the "flexibility" of the oxygen bridges we shall understand the nonrigidity of the SiOSi angle, together with a considerable freedom of the internal rotations, i.e., rotations of the SiOSi plane. Comparison of the low δSiOSi frequencies in $H_3SiOSiH_3$ and $(H_3C)_3SiOSi(CH_3)_3$ with the far higher δCOC frequencies of the analogous hydrocarbons, 414 cm^{-1} for H_3COCH_3 [27] and 467 cm^{-1} for polyoxymethylene [28], demon-

† The lower-frequency of these maxima is much stronger and is possibly associated with combination vibrations not related to the structure of the ν_s SiOSi band here considered.

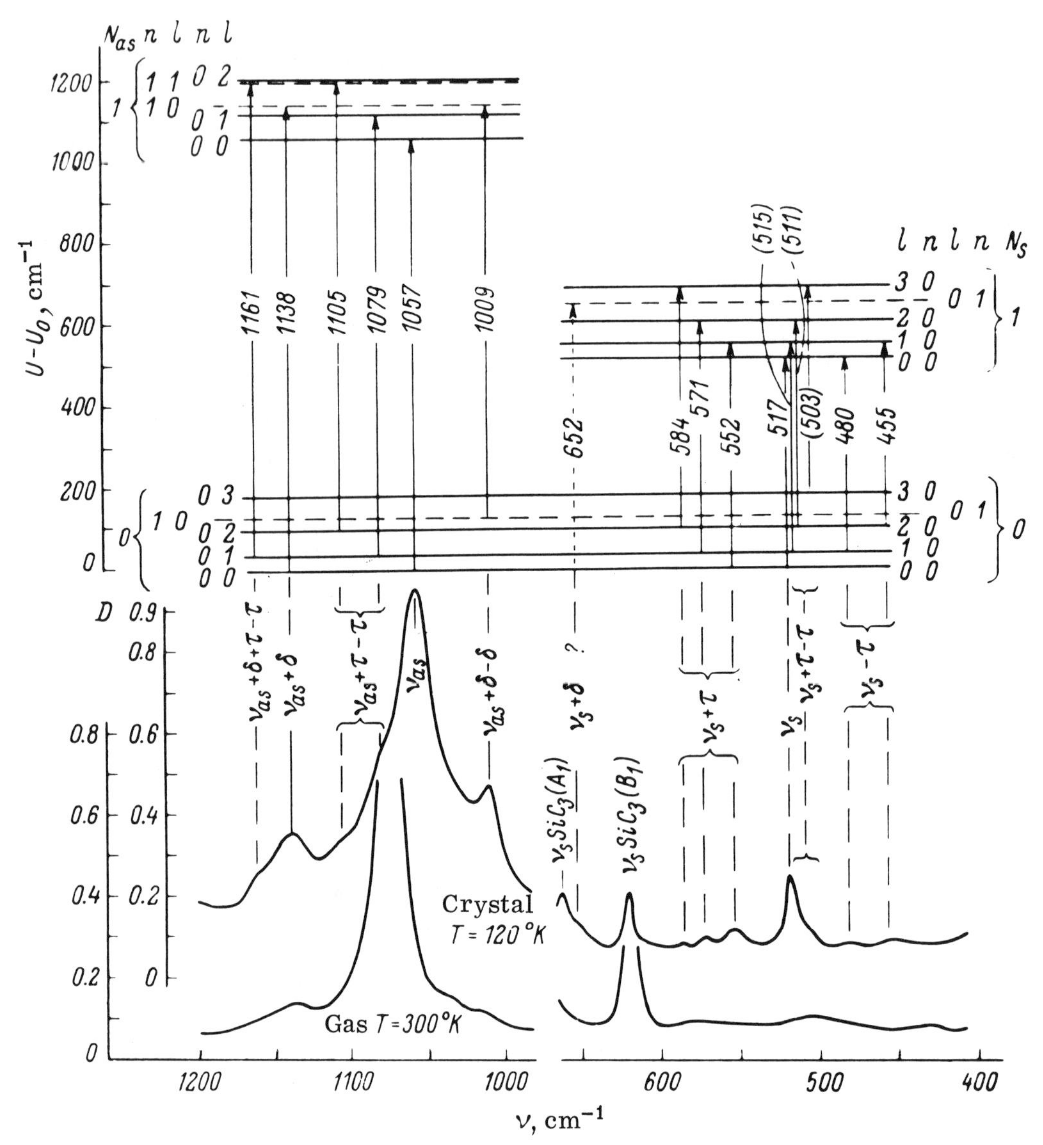

Fig. 117. Possible interpretation of the structure of the bands of Si –O –Si vibrations in the spectrum of hexamethyldisiloxane.

strates the possibility that the deformations of the SiOSi angle may contribute substantially to the flexibility of the siloxane chains, in contrast to many organic polymers.

Among the various physical manifestations of the flexibility of the siloxane molecules we must also remember the anomalously low ($\approx 0.6 \cdot 10^{-12}$ sec) dielectric relaxation time established for hexamethyldisiloxane [29], indicating the participation of internal rotations in the mechanism governing the reorientation of the molecule. We note that an analogous anomaly in dielectric relaxation also occurs for molecules such as diphenyl ether [30, 31]. This similarity may be associated not only with the similarity in the geometrical configurations of the molecules but also with the common characteristics in the electron structure of the oxygen bridges, responsible for their flexibility. It is an interesting point that in the spectrum of diphenyl ether the very low frequency of 102 cm^{-1} [32] is ascribed to the δCOC vibration.

Hexamethylcyclotrisiloxane $[OSi(CH_3)_2]_3$ [33] is the simplest representative of the class of cyclic siloxanes and has been repeatedly studied both spectroscopically [34] and

TABLE 63. Vibrations of the Framework of the $[OSi(CH_3)_2]_3$ Molecule with Intrinsic Symmetry D_{3h}

Assignments	Symmetry type	Selection rules		Frequencies, cm^{-1}
		IRS	RS	
ν_{as}SiOSi	E'	a ∥	dp	1020 IRS
	A_2'	ina	ina	—
ν_{as}CSiC	E''	ina	dp	790 RS
	A_2''	a ⊥	ina	~790 IRS
ν_sCSiC	E'	a ∥	dp	688 IRS, RS
	A_1'	ina	p	723 RS
ν_sSiOSi	E'	a ∥	dp	607 IRS
	A_1'	ina	p	586 RS
δSiOSi	E'	a ∥	dp	450 IRS
	A_1'	ina	p	454 RS
γSiC_2	E'	a ∥	dp	390 IRS
	A_2'	ina	ina	—
ρSiC_2	E''	ina	dp	287 RS
	A_2''	a ⊥	ina	309 IRS
δCSiC	E'	a ∥	dp	250 RS
	A_1'	ina		
τSiC_2	E''	ina	dp	198 RS
	A_1''	ina	ina	—
$\delta_\perp$SiOSi	E''	ina	dp	35
	A_2''	a ⊥	ina	

Note: RS taken from Kriegsmann [34], frequencies 390 and 309 cm^{-1} in the IRS from Smith [37], frequency 35 cm^{-1} from an analysis of combination transitions.

Notation: ν = stretching; δ = deformation (bending, scissoring); γ = "fanning" (wagging); ρ = "pendulum oscillations" (rocking); τ = torsional. Orientation of the moment given relative to the σ_h plane.

by x-ray and electron-diffraction methods [6, 7]. According to the results obtained, the molecule contains a plane trisiloxane ring remaining exactly the same in different states of aggregation (but see p. 276). The planar configuration of the ring is considered as a consequence of a certain stress associated with the magnitude of the SiOSi angle, which is smaller than in linear and higher cyclic methylsiloxanes; a deviation from planar configuration requires an additional reduction of this angle. In the plane ring there is an "automatic" (i.e., symmetry-induced) separation of the internal rotations around the Si–O bonds (the nonplanar deformations of the ring) from the higher-frequency deformation and stretching vibrations of the SiOSi group, which are not associated with motion of the Si and O atoms out of the plane.

Kriegsmann's identification of the vibration frequencies of the heavy atoms in the framework of the molecule [34] is presented in Table 63, with a few slight changes: the ν_{as} CSiC(A_2'') vibration in the infrared spectrum is ascribed to a "shoulder" at about 790 cm^{-1} situated on the slope of the stronger ρCH_3 band at 814 cm^{-1}; in interpreting the frequencies of the internal and external deformation vibrations of the SiC_2 groups allowance is made for Kovalev's calculations of the corresponding vibrations in molecules of the $(CH_3)_2SiX_2$ type [35, 36], and Smith's data [37] regarding the infrared spectrum of $[OSi(CH_3)_2]_3$ in the range of a CsBr prism. The frequency of the ν_{as} SiOSi vibration is lower than in the spectra of the linear siloxanes and siloxane rings of larger dimensions; this may be associated not only with the purely kinematic

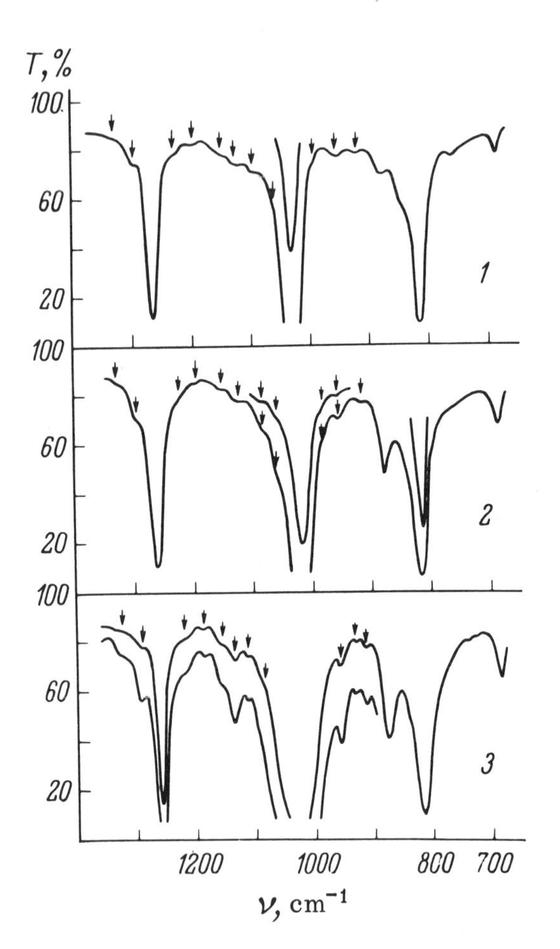

Fig. 118. Infrared spectrum of $[OSi(CH_3)_2]_3$ in various states of aggregation. 1) Vapor (path 130 mm, p ~ 3 and 0.5 mm Hg); 2) liquid (solution in CS_2, thickness 0.012 mm, concentration 5 and 0.5 mol. %); 3) crystal (polycrystalline sample obtained by cooling the melt in a layer 0.08 mm thick and in a capillary layer). Bands of combinations indicated by arrows.

effect (the reduction in the SiOSi angle) but also with the corresponding reduction in the order of the Si–O–Si bonds.

It follows from the curves presented in Fig. 118 that near the ν_{as} SiOSi (1020 cm^{-1}) and $\delta_s CH_3$ (1255 cm^{-1}) vibration bands there is a whole series of weaker maxima situated approximately symmetrically with respect to these bands.† These maxima may evidently be ascribed to combination (sum and difference) transitions involving vibrations with frequencies between 35 and 130 cm^{-1}. Since the combination bands are found without any marked changes in the spectra of hexamethylcyclotrisiloxane crystals, solutions, and vapor, the low-frequency vibrations participating in these transitions should be attributed to the internal degrees of freedom of the molecule. The only marked change observed on passing to the spectrum of the crystal lies in the splitting of the satellites into two components

$$\nu_{as}\text{SiOSi} \pm 105 \text{ cm}^{-1}$$

which may probably be associated with Fermi resonance as a result of the chance coincidence of the frequencies of the corresponding internal vibration of the $[OSi(CH_3)_2]_3$ molecule and one of the crystal lattice vibrations. In the region of the ν_s SiOSi vibration band there is an obvious satellite on the high-frequency side of the main maximum; on the long-wave side there is a "shoulder" at a much shorter distance, 25 instead of 35 cm^{-1}. This absorption may be due to

† In the $[OSi(CH_3)_2]_3$ molecule with symmetry D_{3h} the $\delta_s CH_3$ vibrations are associated with the irreducible representations A_1', A_2'', E''.

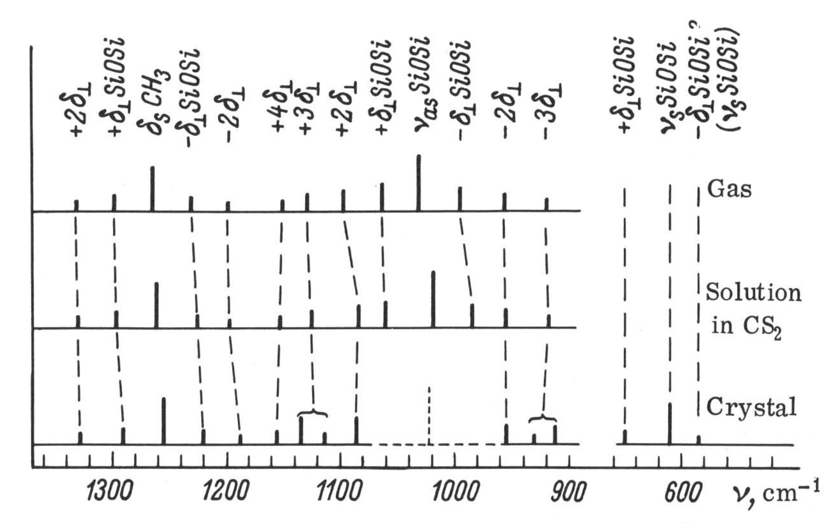

Fig. 119. Scheme for the assignment of the combination frequencies in the $[OSi(CH_3)_2]_3$ spectrum.

the appearance of a ν_s SiOSi vibration of type A_1' in the infrared spectrum as a result of the infringement of the selection rules.

Despite the difficulty of an exact determination of the frequencies of the weak (and not always well-resolved) maxima on the sides of the far stronger central bands, we may at any rate say with some conviction that the ν_{as} SiOSi and δ_s CH_3 band satellites are approximately equidistant, at least for the spectra of the gas phase. Then the observed system of satellites should be ascribed to series of the $\nu \pm n\omega$ type with n = 1, 2, 3, 4. Figure 119 presents a scheme for the assignment of the frequencies based on the foregoing principles.

The following considerations weigh against the alternative version of ascribing these frequencies to difference combinations with various internal vibrations of the $[OSi(CH_3)_2]_3$ molecule possessing frequencies at about 35, 70, 105, and 130 cm^{-1}: a) none of these frequencies has been observed in the Raman spectrum, despite special efforts (the spectrum of a melt was obtained; in this the background made it difficult to establish the presence or absence of the line at 35 cm^{-1}, but the rest should be visible); b) even for the $Cl_2Si(CH_3)_2$ molecules the frequencies of the δCSiC, ρSiC_2, and τSiC_2 vibrations exceed 150 cm^{-1}, and hence it is difficult to explain the existence of at least four fundamental vibrations of the $[OSi(CH_3)_2]_3$ molecule in the lower-frequency range; c) the ascription of all four of the frequencies in question to different deformation vibrations of the ring and SiC_2 groups leads to difficulties in the interpretation of the frequencies observed in the infrared spectrum and the Raman spectrum between 200 and 400 cm^{-1} (see Table 63).

Thus the system of band satellites observed in the infrared spectrum is associated with combinations involving the fundamental and overtones of only one low-frequency vibration of the molecule $\omega \simeq 35$ cm^{-1}. This frequency cannot be ascribed to the torsional vibrations of the CH_3 groups around the Si–C bonds Si–C(τCH_3); it follows from the frequencies of the $\nu \pm n\omega$ transitions that the anharmonicity of the ω vibration is relatively small and hence in order to estimate the frequency of the torsional vibration τCH_3 (assuming that $\tau CH_3 = \omega$) we may use the relation $\tau = m\sqrt{V_0 B_I}$, where B_I is the rotational constant. For a moment of inertia of the CH_3 group I = $5.3 \cdot 10^{-40}$ $g \cdot cm^2$ and a potential barrier to internal rotation around the Si–C bond of V_0 = 1600 cal/mole, these being fairly constant for different molecules with Si–CH_3 groupings, the expected frequency of τCH_3 has a value of the order of 170 cm^{-1}. Thus the nonplanar deformation vibration of the siloxane ring† is the only vibration of the $[OSi(CH_3)_2]_3$ molecule to which the frequency 35 cm^{-1} may be attributed, while the large thermal amplitude of the heavy atoms noted in structural investigations receives a natural explanation in the presence, among the vibrations of the framework, of such low-frequency vibrations that even at low temperatures the excited states are substantially occupied. The thermal amplitudes characterized by the temperature factors B for the Si and O atoms in the trisiloxane cycle are not only large but also very anisotropic ($B_z = 10.39 Å^2$, $B_{x,y} = 5.5 Å^2$ [7]), which indicates the considerable mobility of these atoms in a direction perpendicular to the plane of the ring. Hence the existence of a fairly strong series of combination bands involving a nonplanar deformation vibration of the ring is in turn a consequence of the strong interaction of the vibrations due to substantial deviations from equilibrium configuration. We note that in crystals formed by the corresponding organic rings of trioxymethylene $(OCH_2)_3$ the average temperature factor $B = 4.38 Å^2$ for the C and O atoms, and the motions of these atoms are isotropic [38]. The COC angles in this ring, like the OCO angles, are close to the tetrahedral values, and the molecule exists in the nonplanar conformation of a "crown," C_{3v}.

A study of the infrared spectrum of hexamethylcyclotrisiloxane at –150°C shows some reduction in the relative intensity of the bands ascribed to difference combinations, but a quantitative estimate of the changes in intensity is difficult.

One notices the fact that the frequency of the nonplanar deformation vibration of the trisiloxane ring, roughly equal to 35 cm^{-1}, coincides almost exactly with the frequency of the corresponding vibration (internal rotation around the Si–O bond) in the molecule of hexamethyldisiloxane. This only refers, of course, to the 0–1 transition, since there are fundamental differences between the higher excited states of the nonplanar deformation of SiOSi in these molecules: in the case of the ring the higher levels inevitably retain their vibrational character and no transition to free rotation occurs. These differences are associated with the characteristics of the geometrical structure of the molecules, but for small displacements the value of the vibrational quantum is almost the same, and may very well serve as a characteristic of the rigidity of the Si–O–Si bonds with respect to nonplanar deformation, provided that the similar electron structure of the bonds (mainly determined by the substituents adjoining the Si) is preserved.

§2. Phase Transformations and Conformations of Cyclopolysiloxane Molecules

Cyclopolysiloxane molecules $(OSiRR')_n$ with $n > 3$ form nonplanar rings if R and R' are aliphatic or aromatic radicals. Correspondingly different conformations of these rings may

---

† According to the selection rules, among the $\delta_\perp$SiOSi vibrations only the vibration of type E″ gives active combinations ν_{as}SiOSi $\pm n\delta_\perp$SiOSi for any n, yet for even n combinations with vibrations of the A_2'' type are also possible. We note, by the way, that among possible combinations ν_{as}SiOSi of types E′ and A_2' and $n\delta_\perp$SiOSi (E″ and A_2'') for $n = 3$ there is only one active vibration: ν_{as}SiOSi(E′) $\pm 3\delta_\perp$SiOSi(E″). Hence it is difficult to explain the splitting of this vibration in the spectrum of the crystal without having recourse to the earlier-mentioned assumption of a Fermi resonance.

According to x-ray structural analysis [7] the $[OSi(CH_3)_2]_3$ crystal belongs to the R3m $-C_{3v}^5$ space group and contains one molecule in the unit cell. The center of gravity of the molecule occupies a position with symmetry C_{3v}; hence the selection rules for the symmetry group of the free molecule D_{3h} may be broken. Thus it is possible for there to be two rotational vibrations of the molecule in the lattice, of symmetry types A_2 and E; on the basis of the symmetry properties the latter may resonate with the overtone $3\delta_\perp$SiOSi of the internal vibration of the molecule.

be expected. The structure of the siloxane ring clearly constitutes a compromise between the repulsion of groups not joined by valence bands and the requirement of maximum $d\pi - p\pi$ interaction. The relation between the structure of the rings and these factors was considered in detail by Craig and Paddock [39], who calculated the overlap integrals of the d and p (or hybrid sp) orbitals for various configurations of the rings. This established the possible existence of several ring conformations, energetically almost equally favorable, formed by atoms possessing vacant d orbitals and atoms with unshared pairs in the p orbitals. Thus for eight-membered rings, for example, in cyclotetrasiloxanes $(OSiR_2)_4$, in view of the comparatively weak $d\pi - p\pi$ interactions, configurations with symmetry S_4, C_{2h}, and C_{4v} are almost equally probable. It is therefore possible that any choice in favor of one particular configuration will be largely dependent on external factors, such as the temperature and state of aggregation, and in the case of solutions on the nature of the solvent.

On the other hand, the low rigidity of the SiOSi bond angles and the small damping of the internal rotations around the Si–O bonds suggest that the activation energy necessary for a transformation to take place from one configuration of the siloxane ring to another is relatively small and comparable with the ordinary thermal energies.

In the phosphonitrile rings, with their considerably stronger $p\pi - d\pi$ bonds, the flexibility of the molecular configuration is evidently even more considerable; this is reflected in the vibrational spectra of these compounds [40-43]. X-ray and electron-diffraction data [44-49] indicate a great variety in the configurations of the phosphonitrile rings, depending on the electronegativity of the exocyclic substituents associated with the phosphorus, the state of aggregation, etc.† As already noted, the reasons for the flexibility of the system of bonds involving $p\pi - d\pi$ interactions are of quite a general character, while the properties of the cyclic siloxanes agree closely with the principles underlying the low deformation resistance of the rings. However, direct structural methods have not yielded a great deal of information relating to the conformations of cyclosiloxane molecules, and a study of this problem by spectroscopic procedures is of undoubted interest.

In what follows we shall consider the most readily analyzed data relating to the interconversions of certain siloxane rings during phase transformations and the possible role of internal vibrational states of the molecules in preparing for these transformations.

Octamethylcyclotetrasiloxane $[OSi(CH_3)_2]_4$ [50] is the simplest of the cyclic dimethylpolysiloxanes with nonplanar rings. In crystals of this compound (t_m = 17.5°C) a phase transformation was noticed at −16.3°C; this was apparently a transformation of the first kind and was characterized by a slight jump in dielectric constant and the existence of a latent heat of transformation [51]. It was suggested that the development of restricted rotations of the molecule might possibly cause this transformation. The structure of the low-temperature form was analyzed in detail by x-ray diffraction in [52]; the space group was $P4_2/n - C_{4h}^4$ (Z = 4; a = 16.1, c = 6.47 Å); the $[OSi(CH_3)_2]_4$ molecule had symmetry C_i and formed a nonplanar ring with $\angle OSiO$ = 109° and $\angle SiOSi$ = 142.5°, this evidently being associated with the energetic unfavorability of the condition $\angle SiOSi \simeq 160°$ required for a planar configuration. The structure of the crystals above the transformation temperature was only studied partially [52]; a space group of $I4_1/a - C_{4h}^6$ was established; the dimensions of the unit cell along the a axis remained unaltered, while along the c axis the value rose to 6.83 Å. According to [52] the thermal energy at the transformation temperature was too small for any changes in the internal structure of the molecule, and the phase transformation was associated with orientational disordering of the

† It is interesting to note that the spectroscopic study of the conformations of $(PNCl_2)_4$ molecules [49a] revealed a great deal in common as regards structure (in the solid state) with the octamethylcyclotetrasiloxane about to be described.

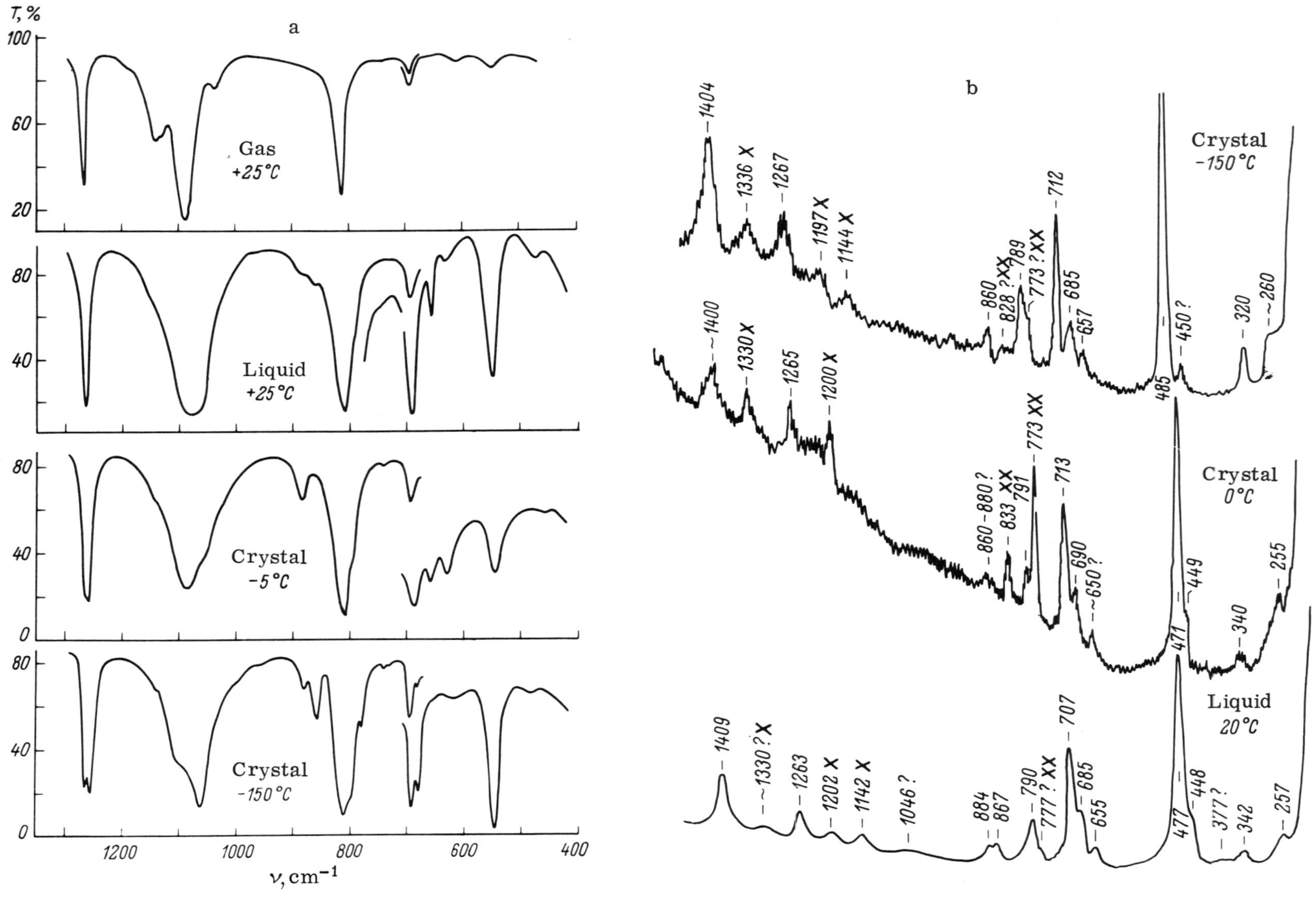

Fig. 120. Vibrational spectra of octamethylcyclotetrasiloxane in various states of aggregation. a) Infrared spectrum (thickness of the layer the same for the solid and liquid phases); b) Raman spectrum. Crosses mark the lines of νCH_3 vibrations appearing as a result of excitation by the short-wave lines of mercury; two crosses mark the Ar lines occuring in the spectrum of the lamp.

TABLE 64. Distribution of the Stretching Vibrations of the $[SiOC_2]_4$ Framework (νSiOSi and νCSiC) and the Rocking Vibrations of the CH_3 (ρCH_3) Groups with Respect to Symmetry Types for Various Configurations of the Molecule $[OSi(CH_3)_2]_4$

Characteristics and notation of the vibrations	Symmetry types of the point groups											
	D_{4h}	C_{4h}	C_{4v}	D_4	C_4	D_{2d} (V_{dII})	D_{2d} (V_{dI})	S_4	C_{2h}	C_i	C_{2v} (C_{2vI})	C_{2v} (C_{2vII})
$\nu_{11,12}$ (*as*SiOSi); $\nu_{13,14}$ (*s*SiOSi); $\nu_{15,16}$ (*s*CSiC); $\rho_{13,14}$ ($s\parallel$); $\rho_{15,16}$ ($as\perp$)	$\underline{E_u}$	$\underline{E_u}$	$\underline{\mathbf{E}}$	$\underline{\mathbf{E}}$	$\underline{\mathbf{E}}$	$\underline{\mathbf{E}}$	$\underline{\mathbf{E}}$	$\underline{\mathbf{E}}$	$\underline{2B_u}$	$\underline{2A_u}$	$\underline{\mathbf{B_1}}+\underline{\mathbf{B_2}}$	$\underline{\mathbf{B_1}}+\underline{\mathbf{B_2}}$
ρ_3 ($s\perp$)	A_{1u}	$\underline{A_u}$	A_2	$\mathbf{A_1}$	$\underline{\mathbf{A}}$	$\mathbf{B_1}$	$\mathbf{B_1}$	$\underline{\mathbf{B}}$	$\underline{A_u}$	$\underline{A_u}$	$\mathbf{A_2}$	$\mathbf{A_2}$
ν_4 (*as*CSiC); ρ_4 ($as\parallel$)	$\underline{A_{2u}}$	$\underline{A_u}$	$\underline{\mathbf{A_1}}$	$\underline{A_2}$	$\underline{\mathbf{A}}$	$\underline{\mathbf{B_2}}$	$\underline{\mathbf{B_2}}$	$\underline{\mathbf{B}}$	$\underline{A_u}$	$\underline{A_u}$	$\underline{\mathbf{A_1}}$	$\underline{\mathbf{A_1}}$
ρ_7 ($s\perp$)	B_{1u}	B_u	$\mathbf{B_2}$	$\mathbf{B_1}$	$\mathbf{B}$	$\mathbf{A_1}$	A_2	$\mathbf{A}$	$\underline{A_u}$	$\underline{A_u}$	$\mathbf{A_2}$	$\underline{\mathbf{A_1}}$
ν_8 (*as*CSiC); ρ_8 ($as\parallel$)	B_{2u}	B_u	$\mathbf{B_1}$	$\mathbf{B_2}$	$\mathbf{B}$	A_2	$\mathbf{A_1}$	$\mathbf{A}$	$\underline{A_u}$	$\underline{A_u}$	$\underline{\mathbf{A_1}}$	$\mathbf{A_2}$
$\nu_{9,10}$ (*as*CSiC); $\rho_{9,10}$ ($as\parallel$); $\rho_{11,12}$ ($s\perp$)	$\mathbf{E_g}$	$\mathbf{E_g}$	$\underline{\mathbf{E}}$	$\underline{\mathbf{E}}$	$\underline{\mathbf{E}}$	$\underline{\mathbf{E}}$	$\underline{\mathbf{E}}$	$\underline{\mathbf{E}}$	$\mathbf{2B_g}$	$\mathbf{2A_g}$	$\underline{\mathbf{B_1}}+\underline{\mathbf{B_2}}$	$\underline{\mathbf{B_1}}+\underline{\mathbf{B_2}}$
ν_1 (*s*CSiC); ν_2 (*s*SiOSi); ρ_1 ($s\parallel$)	$\mathbf{A_{1g}}$	$\mathbf{A_g}$	$\underline{\mathbf{A_1}}$	$\mathbf{A_1}$	$\underline{\mathbf{A}}$	$\mathbf{A_1}$	$\mathbf{A_1}$	$\mathbf{A}$	$\mathbf{A_g}$	$\mathbf{A_g}$	$\underline{\mathbf{A_1}}$	$\underline{\mathbf{A_1}}$
ν_3 (*as*SiOSi); ρ_2 ($as\perp$)	A_{2g}	$\mathbf{A_g}$	A_2	$\underline{A_2}$	$\underline{\mathbf{A}}$	A_2	A_2	$\mathbf{A}$	$\mathbf{A_g}$	$\mathbf{A_g}$	$\mathbf{A_2}$	$\mathbf{A_2}$
ν_5 (*as*SiOSi); ν_6 (*s*CSiC); ρ_5 ($s\parallel$)	$\mathbf{B_{1g}}$	$\mathbf{B_g}$	$\mathbf{B_1}$	$\mathbf{B_1}$	$\mathbf{B}$	$\mathbf{B_1}$	$\underline{\mathbf{B_2}}$	$\underline{\mathbf{B}}$	$\mathbf{A_g}$	$\mathbf{A_g}$	$\underline{\mathbf{A_1}}$	$\mathbf{A_2}$
ν_7 (*s*SiOSi); ρ_6 ($as\perp$)	$\mathbf{B_{2g}}$	$\mathbf{B_g}$	$\mathbf{B_2}$	$\mathbf{B_2}$	$\mathbf{B}$	$\underline{\mathbf{B_2}}$	$\mathbf{B_1}$	$\underline{\mathbf{B}}$	$\mathbf{A_g}$	$\mathbf{A_g}$	$\mathbf{A_2}$	$\underline{\mathbf{A_1}}$
$\rho\parallel$ and $\rho\perp$ (*as* and *s*) oscillations of the CH_3 groups parallel to and perpendicular to the CSiC plane	—	—	—	—	—	$C_2' \to C_2'$	$C_2'' \to C_2'$	—	$C_2 \to C_2$	—	C_2, σ_v	C_2, σ_d

Note: Types of symmetry active in the IRS underlined, those active in the RS printed bold.

molecules, namely, rotations around the c axis by angles equal to multiples of 90°, which statistically led to an effective site symmetry S_4, the shape of the actual molecules remaining the same.

A preliminary study of the spectrum of liquid $[OSi(CH_3)_2]_4$ led Kriegsmann [53] to the conclusion that the symmetry of the molecule was C_{4v} or S_4. The results of a more detailed study of the spectrum of this compound in various states of aggregation are shown in Fig. 120. Substantial changes in the spectrum arising from the solid-state transformation† hardly agree with the idea that the shape of the molecules remains constant, since in a molecular crystal with weak intermolecular interactions the spectrum in the frequency range characterizing the internal vibrations of the molecule depends only slightly on the packing of the molecules and their external degrees of freedom. Hence the changes in the spectrum may be ascribed to changes in the structures of the molecules, the absence of substantial changes in the frequencies of the Si–O and Si–C vibrations evidently indicating that the lengths of the bonds and the valence angles in the molecule remain constant. It follows from this that the change in the structure of the molecule is limited mainly to a change in conformation. Transformations of this kind frequently require very little expenditure of energy and may occur even at low temperatures.

In order to verify the foregoing considerations, let us consider a possible interpretation of the spectra of the two crystalline forms and the liquid phase. Here we confine ourselves to the frequency range of the stretching vibrations of the framework, whose characteristic nature allows us to neglect interactions with other vibrations of the molecule, at any rate to a first

† Variation of temperature over the range of stability of each form does not lead to any serious changes in the spectrum.

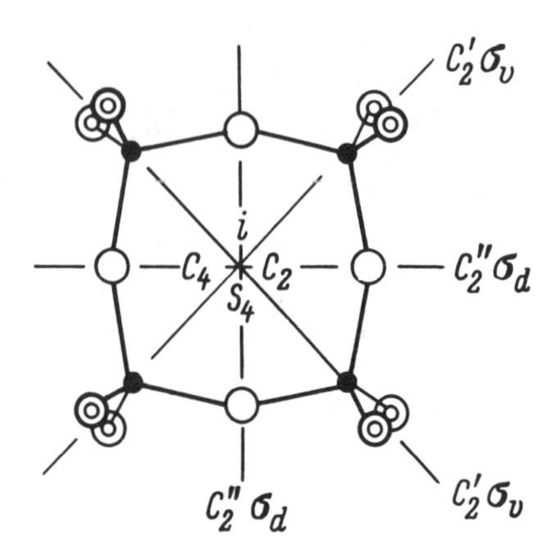

Fig. 121. Symmetry elements of the plane centrosymmetric $[OSiC_2^*]_4$ " molecule."

approximation. In the spectral range considered (1200 to 400 cm^{-1}), in addition to the stretching vibrations of the framework ν, we also find the rocking vibrations (oscillation ρ) of the CH_3 groups and these have to be considered when assigning the frequencies. Table 64 presents the distribution of these vibrations with respect to the types of symmetry of the various point groups, starting with the highest possible symmetry of the $[OSi(CH_3)_2]_4$ molecule, D_{4h}. This enables us to establish a correspondence between the normal vibrations of the molecule for any conformations. In view of the indeterminacy in the choice of notation for certain symmetry elements (C_2' and C_2'', σ_v and σ_d) of the D_{4h} group, Fig. 121 shows their selected arrangement in the molecule.

It is natural to begin identification of the frequencies from the spectrum of the low-temperature form, for which the structure of the molecule has been established by x-ray diffraction. The $[OSi(CH_3)_2]_4$ molecule forms a nonplanar ring, the four O atoms and the four Si atoms lying respectively each in a single plane, the angle between the planes being 30°. The intrinsic symmetry of the molecule is C_i, and since the center of symmetry of the molecule is the center of symmetry of the crystal, there are no grounds for expecting any disruption of the mutual exclusion rule owing to interactions between the vibrations of the molecule in the cell. On the basis of the vibration frequencies of the SiOSi and SiC_2 groups characteristic of siloxanes and similar compounds (ν_{as} SiOSi = 1000 to 1100 cm^{-1}, ν_{as} SiC_2 = 750 to 800 cm^{-1}, ν_s SiC_2 = 650 to 750 cm^{-1}, ν_s SiOSi = 450 to 600 cm^{-1}), we may derive an interpretation of the frequencies satisfying the selection rules for the C_i group (Table 65).

The transformation to the high-temperature form, in which the molecule occupies a position with symmetry S_4, suggests, on dispensing with the idea of statistical symmetry [52], that the molecule may have an intrinsic symmetry corresponding to one of the point groups D_{4h}, C_{4h}, $D_{2d}(V_d)$, S_4, a subgroup of which is the site group S_4. The first two, which give considerably fewer active frequencies than those observed experimentally, may at once be discarded, while the configuration $V_{d\,II}$ is eliminated since it only admits the appearance of one ν_{as} SiOSi band in the infrared spectrum. The choice between configurations with symmetry $V_{d\,I}$ and S_4 is not so unambiguous, but the latter group is favored by the appearance of a band at 458 cm^{-1} corresponding to the $\nu_7(\nu_s$ SiOSi) vibration in the infrared spectrum. Then the band at 860 cm^{-1}, vanishing in the infrared spectrum of the high-temperature form, may be ascribed to vibrations ρ_7 or ρ_8 (we note that ρ_1 and ρ_2 are inactive for both forms).

Thus during the solid-phase transformation in octamethylcyclotetrasiloxane the molecule changes from a conformation with symmetry C_i to one with symmetry S_4. The mechanism of this kind of rearrangement may be described, if we compare the schemes in Fig. 122, as an internal rotation in the ring (with a simultaneous displacement of the C atoms), or, if we remember the much greater rigidity of the OSiO angles as compared with the SiOSi, as a peculiar

TABLE 65. Identification of the Fundamental Frequencies of the Stretching Vibrations in the Framework of the $[OSi(CH_3)_2]_4$ Molecule

Characteristics and notation of vibrations	Crystal						Liquid		
	low-temperature form			high-temperature form			Symmetry types C_{4v}	frequencies	
	Symmetry types C_i	frequencies IRS	frequencies RS	Symmetry types S_4	frequencies IRS	frequencies RS		IRS	RS
ν_{as}SiOSi $\nu_{11,12}$	$2A_u$	1104	—	E	1088	—	E	1077	—
		1063	—						
ν_5	A_g	—	—	B	1058	—	B_1	—	1045
ν_3	A_g	—	—	A	—	—	A_2	—	—
ν_{as}CSiC ν_4	A_u	799	—	B	797	—	A_1	794	793
$\nu_{9,10}$	$2A_g$	—	789	E	— *	791	E		
ν_8	A_u	782	—	A	—	?	B_1	—	—
ν_sCSiC ν_1	A_g	—	712	A	—	713	A_1	—	711
$\nu_{15,16}$	$2A_u$	695	—	E	690	690	E	693	685
		682	685 **						
ν_6	A_g	—	657	B	658	654	B_1	656 ***	656
ν_sSiOSi $\nu_{13,14}$	$2A_u$	547	—	E	544	—	E	548	550
ν_2	A_g	484 ****	485	A	—	471	A_1	471	476
ν_7	A_g	—	450	B	458	449	B_2	—	450

*Band probably masked by stronger neighbor.

**Combination? (485 + 198; Ag × Ag = Ag).

***Combination? (471 + 185; $A_1 \times A_1 = A_1$).

****Combination? (320 + 163).

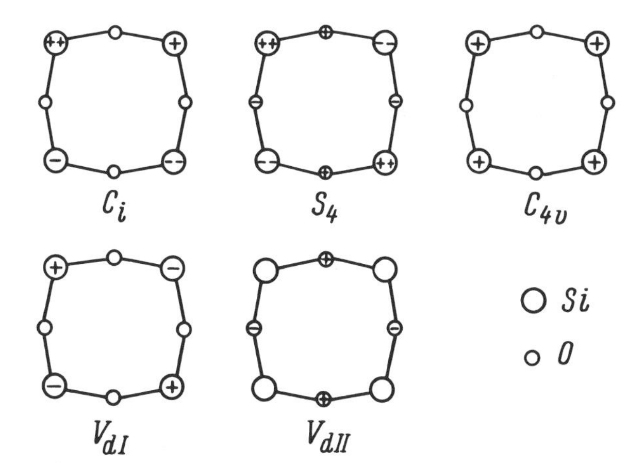

Fig. 122. Scheme of various conformations of the nonplanar tetrasiloxane ring. Signs + and - indicate the position of the atom above or below the plane of the figure.

kind of inversion (in opposite directions) of two oxygen atoms in each ring. Naturally the concept of inversion does not suggest the passage of the O atom through the Si...Si line, but only its passage through the plane in which the four Si atoms lie (with symmetry C_i). Such a mechanism of the phase transformation in no way contradicts the x-ray data, which indicate the absence of any substantial changes in the positions of the Si, O, and C atoms along the a and b axes and a 5.5% elongation along the c axis (clearly this arises from a change in the packing of the CH_3 groups in the molecules situated one on top of another). The symmetry S_4 of the molecule derived from the spectroscopic data agrees with its site symmetry arising from the space group of the high-temperature form of the crystal established by x-ray diffraction.

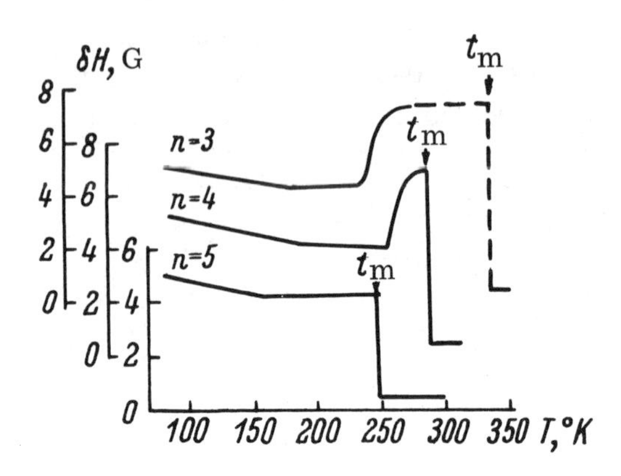

Fig. 123. Temperature dependence of the width of the NMR line of the protons in the molecules of cyclic siloxanes $[OSi(CH_3)_2]_n$.

It also should be noted that the supposition as to the relation between the mechanism of the solid-phase transformation in $[OSi(CH_3)_2]_4$ and the increase in the rotational mobility of the molecule as a whole contradicts results obtained by proton magnetic resonance [3]. The width of the line is about 5.5 G at liquid-nitrogen temperature; it falls slightly (to about 4 G) on heating to some −100°C, then remains constant until the transformation temperature, in the neighborhood of which it increases rapidly to 7 G. We may thus suppose that the increase in the width of the line during the solid-phase transformation in octamethylcyclotetrasiloxane is due to a change in the disposition and hence the vibrations of the methyl groups on changing the conformation of the molecule. However, an analogous type of temperature dependence of the line width also occurs for hexamethylcyclotrisiloxane (Fig. 123), for which the line width starts rising sharply at a temperature of around −30°C, although x-ray diffraction shows no signs of solid-state transformations right down to low temperatures. A possible explanation for this contradiction lies in the following: the planar configuration of the ring is not a consequence of the symmetry of the hexamethylcyclotrisiloxane crystals, which belong to the $R3m - C_{3v}^5$ space group [7]. Hence within the bounds of the same lattice the molecule may have also a lower symmetry than D_{3h} (but not lower than C_{3v}). If we also remember the anomalously large thermal amplitudes of the Si and O atoms in a direction perpendicular to the plane of the ring, we may suppose that the ring has a nonplanar (C_{3v}) equilibrium configuration, transforming into a statistically flat one under the influence of thermal excitations. The changes in the NMR spectrum may probably be explained on the basis of a transformation of this kind taking place without any change in the symmetry of the crystal. Hence it may be interesting to make a more careful study of the shape of the potential curve of the nonplanar deformation vibration of the trisiloxane ring.

On passing from the high-temperature form of the $[OSi(CH_3)_2]_4$ crystal to the liquid phase, considerable changes occur in the spectrum, suggesting a second change in conformation. Among the groups containing a C_4 or S_4 axis (which follows from the absence of any splitting of the degenerate vibrations), according to the selection rules the most probable are the C_4, or rather C_{4v}, for which Table 65 gives the corresponding scheme of frequencies. The absence of the band near 710 cm^{-1} from the infrared spectrum is probably due to its low intensity. The appearance of a band at 860 cm^{-1} in the infrared spectrum is evidently associated with the vibration ρ_1, which is inactive for both forms of the crystal. Passing to the spectrum of the gas phase gives no indication as to any fresh changes of conformation.

The conclusion as to the conformation with symmetry C_{4v} (or C_4) in the liquid phase agrees with the results of measurements of the dipole moments of liquid cyclosiloxanes $[OSi(CH_3)_2]_n$ [54]:

$n=$	4	5	6	7	8
μ (Debye units) =	1.09	1.35	1.56	1.78	1.96

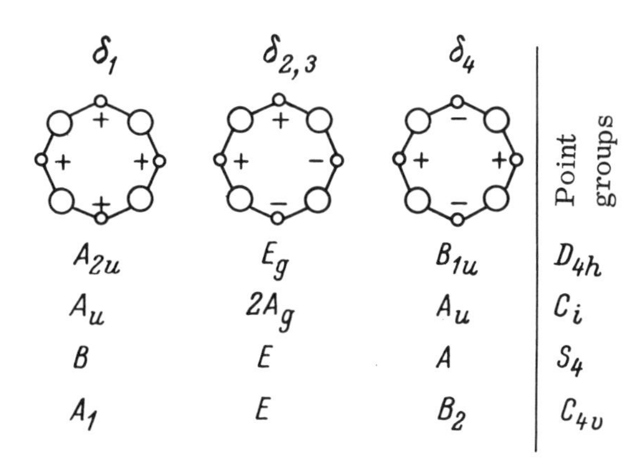

Fig. 124. "Nonplanar" deformation vibrations of the siloxane ring of the $[OSi(CH_3)_2]_4$ molecule. For simplicity the signs + and − are only used to show the directions of displacement of the O atoms.

Actually the other conformations discussed here C_i, S_4, V_d have no static dipole moment. We note also that the idea of there being very little energy differences between the conformations of the siloxane ring, particularly those with symmetry C_i (or C_{2h}) and S_4, receives indirect confirmation from the x-ray structural analysis of the octamethylcyclotetrasilazane crystal $[HNSi(CH_3)_2]_4$: in these crystals both conformations of the silazane ring coexist within the same structure [55].

Let us give some more detailed consideration to the question of the possible molecular mechanism of the transformations here described. As already noted, these transformations are closely connected with the deformations of the ring of silicon–oxygen bonds. The symmetry group of the low-temperature form is not a subgroup of the high-temperature form, and conversely: $C_i \rightarrow S_4$, $S_4 \rightarrow C_{4v}$. Hence the transformation cannot be explained simply by the excitation of the higher vibrational states, keeping the force field of the molecule constant, i.e., it is not a transformation of the second kind. We note that the transformation would then probably not exhibit the hysteresis phenomena clearly observed in [51]. However, we may suppose that the thermal excitation of the extremely low-frequency "nonplanar" deformation vibrations of the siloxane ring† ($\delta_\perp$SiOSi) have a considerable importance in overcoming the activation barrier separating possible (comparatively close in energy) configurations of the $[OSi(CH_3)_2]_4$ molecule; the stability of the molecule with respect to several forms of vibrations may fall, and it may pass to another configuration under the influence of intermolecular interactions. The role of the latter factor, i.e., of the cooperative character of the transformation process, may be very considerable and demands special study.

Figure 124 shows the $\delta_\perp$SiOSi vibrations schematically and gives their distribution with respect to the symmetry types of various point groups. It follows directly from this scheme that the preparation for the $C_i \rightarrow S_4$ transformation, requiring the simultaneous transfer of two neighboring O atoms, upward and downward respectively, through the plane of the four Si atoms, may be associated with the excitation of δ_4 vibrations. By analogy with this, the $S_4 \rightarrow C_{4v}$ transformation may be associated with the participation of the δ_1 vibration.

The $\delta_\perp$SiOSi vibrations, in contrast to the "in-plane" deformation vibrations of the ring, for which the bond lengths change considerably, are probably the lowest frequency of all the vibrations of the $[OSi(CH_3)_2]_4$ molecule, and, judging from the δSiOSi frequencies in $(H_3C)_3SiOSi(CH_3)_3$ and $[OSi(CH_3)_2]_3$, lie at about 50 to 100 cm^{-1}. In the Raman spectrum of liquid $[OSi(CH_3)_2]_4$, in the frequency range below 250 cm^{-1}, bands occur at 202 to 184 cm^{-1}

† The term "nonplanar" deformation vibration is inapplicable in this case and is only used by analogy with the classification of the motions of the planar ring so as to distinguish, from other deformations of the nonplanar ring, those which change the bond lengths very little but lead to substantial "vertical" displacements of the atoms.

(maximum at 198 cm^{-1}), a narrow line at 163.5 cm^{-1}, a double line at 136 to 147 cm^{-1}, and a band at 59 to 74 cm^{-1}. This band, in which no structure has yet been resolved, despite the high resolution of the spectral apparatus employed (3.5 Å/mm), may be ascribed to the $\delta_{\perp}$SiOSi vibrations of the tetrasiloxane ring. For symmetry C_{4v} all these vibrations are allowed by the selection rules. As in the cases considered earlier, we may seek to identify the bands due to the interaction of $\delta_{\perp}$SiOSi vibrations with the stretching vibrations of the ring in the IR spectrum. These probably include the satellite of the ν_{as} SiOSi band situated at 1140 cm^{-1}, which is weak in the spectrum of the crystal and liquid but clearly discernible in that of the gas. We may also ascribe the weaker band at 1036 cm^{-1} seen in the gas spectrum to a corresponding difference transition. The temperature dependence of the intensity ratio of the 1140- and 1036-cm^{-1} bands in no way contradicts their relation to the ν_{as} SiOSi ± $\delta_{\perp}$SiOSi vibrations, while the $\delta_{\perp}$SiOSi frequency is equal to approximately 50 cm^{-1}. In the present case we are clearly concerned with the $\delta_{2,3}$ vibration, which is capable of forming active combinations with the degenerate ν_{as} SiOSi vibration for any of the three conformations of the molecule. An analogous origin (ν_s SiOSi + $\delta_{\perp}$SiOSi) may underly the diffuse band in the infrared spectrum at about 620 cm^{-1}, which had earlier never received any interpretation. The value of the $\delta_{\perp}$SiOSi frequency derived from this band is about 70 cm^{-1}. It is not impossible, however, that the absorption around 1036 and 620 cm^{-1} may be at least partly associated with trisiloxane impurity. The fact that the frequencies of the $\delta_{\perp}$SiOSi vibrations are higher than in hexamethylcyclotrisiloxane agrees with the clearly nonplanar configuration of the tetrasiloxane ring.

$[OSi(CH_3)_2]_n$, $n > 4$ Rings [56]. The infrared and Raman spectra of these compounds have been studied repeatedly [37, 53, 57-59], but only in the liquid phase, and no attempts have been made at establishing their conformations. Only the crystalline "octamer" $[OSi(CH_3)_2]_8$ [60] has been subjected to x-ray structural analysis; the space group is $P\bar{4}2_1c-D_{2d}^4$, there are two molecules in the unit cell occupying positions with site symmetry S_4, and the configuration of the ring is clearly reminiscent of the shape of the tetrasiloxane ring in the high-temperature form of $[OSi(CH_3)_2]_4$ crystals. Even these results suggest the possibility that there may be a change in the conformation of the octasiloxane ring on melting, since in the liquid phase its molecules have a nonzero dipole moment, which is impossible for a point group of symmetry S_4 or any other containing an S_4 element.

Let us consider the results of a study of the infrared spectra of $[OSi(CH_3)_2]_n$ compounds for n = 5, 6, 7 in the liquid and solid phases. In view of the considerable differences in the behavior of the molecules with odd and even numbers of siloxane links in the rings during the phase transformation, we start by considering the spectra of dodecamethylcyclohexasiloxane given in Fig. 125. It follows from a comparison of curves 1 and 2 that the transformation from the liquid phase to the solid on cooling the sample fairly rapidly to −150°C does not lead to any great change in the number of bands observed in the spectrum. Immediately after cooling we observe a very marked narrowing and an increase in the intensity of the bands; holding the sample at −150°C (for 3 to 4 h) greatly reduces this narrowing, the changes in the spectrum usually ending after the first hour. This may be explained by the existence of a degree of freedom which is not completely frozen even at −150°C and leads to a gradual removal of the stresses arising on initial cooling (crystallization is accompanied by a sharp reduction in specific volume). The removal of the stresses also probably leads to the subsequent broadening of the bands as a result of the ordering of the interaction between these external (or certain extremely low-frequency internal) vibrations and the stretching vibrations of the molecules. In other words we may suppose that in a crystal with severe and spatially nonperiodic stresses each molecule absorbs light almost independently, and interactions of the internal degrees of freedom with the external fail to appear, precisely because of the substantial deviations from the regular lattice. This is evidently the only explanation for the (at first glance rather unexpected) broadening of the bands as the ordering of the structure increases on holding at constant temperature.

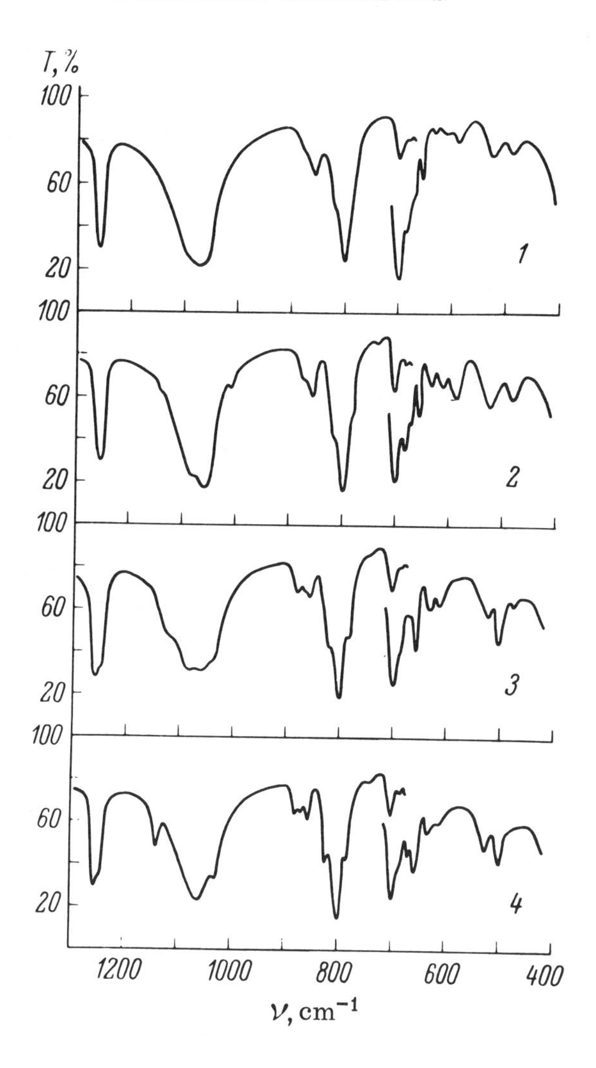

Fig. 125. Infrared spectrum of $[OSi(CH_3)_2]_6$. 1) Liquid, t = 25°C; 2) 1 h after cooling to −150°C; 3) 3 h after cooling to −150°C, then heating to −50°C for 0.5 h, then again cooling to −150°C; 4) same sample as in curve 3 after holding at −30°C for 1.5 h then cooling again to −150°C. In the KBr prism range the layer thickness is 0.012 mm, in the NaCl range the layer is capillary.

On raising the temperature to −100°C and over, considerably greater changes start appearing in the spectrum: new bands appear and others vanish. These changes take place particularly quickly on raising the temperature to about −50°C. However, they may be frozen at any stage for an indefinitely long time by cooling to a temperature below −100°C (curve 3). A moderate holding at −50°C evidently leads to the completion of the structural transformations, since no further changes in the spectrum are observed. It follows from the foregoing that the "annealing" of the structure obtained by rapidly cooling liquid $[OSi(CH_3)_2]_6$ which takes place in the temperature range above −100°C (T_g), i.e., for kT > 100 to 120 cm^{-1}, leads to the formation of a new structure differing from the previous not only in packing but also in the internal structure of the molecules.† We see in fact by comparing curves 2 and 4 (Fig. 125) that intermolecular effects cannot produce such substantial changes in the spectrum. The cooling of this new form to −150°C or heating almost to the melting point does not lead to any great changes in the spectrum; it must therefore be acknowledged as the only stable form of $[OSi(CH_3)_2]_6$ crystals differing in molecular structure from the liquid and metastable crystalline forms.

† The notation T_g is used arbitrarily here, since according to visual observation the sample has a crystalline character not only after "annealing" but also immediately after the freezing of the liquid. It is also interesting to note that the spectrum of the liquid sample taken immediately after melting, at a temperature close to the melting point, differs somewhat from that of the liquid cooled to the melting point from room temperature.

The data presented agree closely with the already-mentioned results of a study of the mobility of linear and cyclic dimethylpolysiloxanes by the NMR method [3], which indicated a substantial rotational mobility of the radicals, persisting right down to −196°C (this may also explain the changes in the infrared spectrum in the temperature range below −100°C), and a great mobility of the actual siloxane chains (internal rotations and deformations of the Si–O–Si bonds) down to −100°C.

Let us now try to find what changes in the shape of the siloxane ring characterize the two possible configurations of the $[OSi(CH_3)_2]_6$ molecule. For this purpose we mainly use the bands of the stretching vibrations of Si–O–Si in the ring, not only because these vibrations are the most sensitive to a change of conformation but also because the $\nu_{as}\,SiC_2$ vibration bands overlap the range of frequencies of the rocking vibrations (ρ) of the CH_3 groups, while the number of bands in the range of frequencies of the $\nu_s\,SiC_2$ vibrations is in all cases greater than would follow from the proposed symmetry of the molecule. Table 66 shows the symmetry types of the Si–O and Si–C stretching vibrations in the $[OSi(CH_3)_2]_n$ molecules for n = 5, 6, 7 and various symmetries; in all cases the potentially highest symmetry D_{nh} was taken as a starting point. The notation of the symmetry elements for a molecule with symmetry D_{6h} is shown in Fig. 126, since the choice of C_2' and C_2'' axes and planes σ_v and σ_d is not unambiguous, while in Table 66 on passing to point groups of lower order the symmetry elements retained are shown wherever necessary.

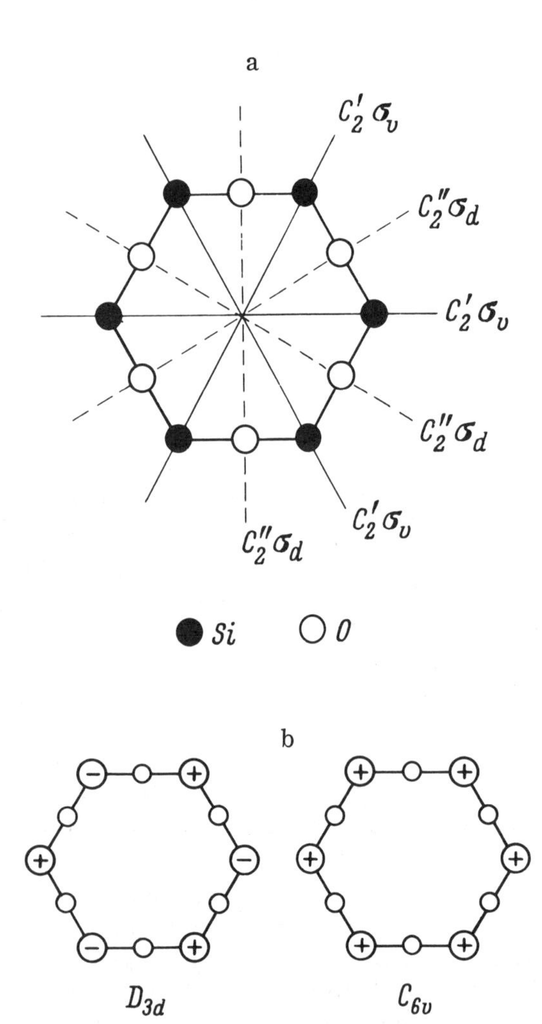

Fig. 126. Notation of the symmetry elements of the $[OSi(CH_3)_2]_6$ molecule. a) For symmetry D_{6h}; b) scheme of actually existing conformations of a ring with symmetry D_{3d} and C_{6v}.

TABLE 66. Symmetry Types of the Vibrations of the $[OSi(CH_3)_2]_n$ Molecules for Several Point Groups

Description of vibrations (and numeration for n=6)		$n=6$							$n=2r+1$			
		Point groups										
		D_{6h}	$D_{3d}(C_2')$	$D_{3d}(C_2'')$	S_6	D_6	C_{6v}	C_6	D_{nh}	D_n	C_{nv}	C_n
ν_{as}SiOSi	1, 2	$\underline{E_{1u}}$	$\underline{E_u}$	$\underline{E_u}$	$\underline{E_u}$	$\underline{E_1}$	$\underline{E_1}$	$\underline{E_1}$	$\underline{E_1'}$	$\underline{E_1}$	$\underline{E_1}$	$\underline{E_1}$
	3, 4	E_{2g}	E_g	E_g	E_g	E_2	E_2	E_2	. .	. .	. .	. .
	5	B_{1u}	A_{1u}	$\underline{A_{2u}}$	$\underline{A_u}$	B_1	B_1	B	E_r'	E_r	E_r	E_r
	6	A_{2g}	A_{2g}	A_{2g}	A_g	$\underline{A_2}$	A_2	$\underline{A}$	A_2'	$\underline{A_2}$	A_2	$\underline{A}$
ν_sSiOSi	7, 8	$\underline{E_{1u}}$	$\underline{E_u}$	$\underline{E_u}$	$\underline{E_u}$	$\underline{E_1}$	$\underline{E_1}$	$\underline{E_1}$	$\underline{E_1'}$	$\underline{E_1}$	$\underline{E_1}$	$\underline{E_1}$
	9, 10	E_{2g}	E_g	E_g	E_g	E_2	E_2	E_2	. .	. .	. .	. .
	11	A_{1g}	A_{1g}	A_{1g}	A_g	A_1	$\underline{A_1}$	$\underline{A}$	E_r'	E_r	E_r	E_r
	12	B_{2u}	$\underline{A_{2u}}$	A_{1u}	$\underline{A_u}$	B_2	B_2	B	A_1'	A_1	$\underline{A_1}$	$\underline{A}$
ν_{as}CSiC	13, 14	E_{2u}	$\underline{E_u}$	$\underline{E_u}$	$\underline{E_u}$	E_2	E_2	E_2	E_1''	$\underline{E_1}$	$\underline{E_1}$	$\underline{E_1}$
	15, 16	E_{1g}	E_g	E_g	E_g	$\underline{E_1}$	$\underline{E_1}$	$\underline{E_1}$	. .	. .	. .	. .
	17	$\underline{A_{2u}}$	$\underline{A_{2u}}$	$\underline{A_{2u}}$	$\underline{A_u}$	$\underline{A_2}$	$\underline{A_1}$	$\underline{A}$	E_r''	E_r	E_r	E_r
	18	B_{2g}	A_{2g}	A_{1g}	A_g	B_2	B_1	B	$\underline{A_2''}$	$\underline{A_2}$	$\underline{A_1}$	$\underline{A}$
ν_sCSiC	19, 20	$\underline{E_{1u}}$	$\underline{E_u}$	$\underline{E_u}$	$\underline{E_u}$	$\underline{E_1}$	$\underline{E_1}$	$\underline{E_1}$	$\underline{E_1'}$	$\underline{E_1}$	$\underline{E_1}$	$\underline{E_1}$
	21, 22	E_{2g}	E_g	E_g	E_g	E_2	E_2	E_2	. .	. .	. .	. .
	23	A_{1g}	A_{1g}	A_{1g}	A_g	A_1	$\underline{A_1}$	$\underline{A}$	E_r'	E_r	E_r	E_r
	24	B_{1u}	A_{1u}	$\underline{A_{2u}}$	$\underline{A_u}$	B_1	B_1	B	A_1'	A_1	$\underline{A_1}$	$\underline{A}$

Note. Symmetry types active in the IRS are underlined.

Both in the spectrum of the liquid and in that of the crystal (by crystal we shall in future mean the stable form) there are two bands in the ν_s SiOSi frequency range, of which only one (the higher-frequency one) remains active in both cases, while the second becomes inactive by virtue of the transformation. In the ν_{as} SiOSi frequency range there are two bands in the spectrum of the liquid (these are particularly clear on freezing the latter), but only one in the spectrum of the crystal, since the considerably weaker maxima at 1029 and 1141 cm^{-1} are probably associated with combinations of stretching and deformation vibrations of the ring (as in the spectrum of $[OSi(CH_3)_2]_4$). The fact that there are no splittings of the degenerate vibrations indicates the existence of a high-order symmetry axis, and hence on the basis of the selection rules given in Table 66 the $[OSi(CH_3)_2]_6$ molecule may be assigned to point group $D_{3d}(C_2')$ in the crystal and C_6 in the liquid. This explains the changes in the spectrum in the regions of ν_{as} and ν_sSiOSi during the transformation and the existence of a dipole moment for the molecule in the liquid.

Figure 126 gives a systematic indication of the changes in the structure of the molecule associated with the $D_{3d} - C_{6v}$ transformation, resulting in a change of ring conformation; in the crystal the $Si(CH_3)_2$ groups are turned alternately upward and downward from a plane perpendicular to the three-(six-) fold axis, while in the liquid all these groups are turned in the same direction. The fall in symmetry from C_{6v} to C_6 is probably due to the interaction of neighboring $Si(CH_3)_2$ groups. Keeping a constant value of ∠SiOSi, in the C_{nv} conformations of the $[OSi(CH_3)_2]_n$ molecules the distance between the CH_3 radicals diminishes with increasing n, and a fall in symmetry to C_n may prove favorable.

Three weak bands in the range 580 to 630 cm^{-1} may be ascribed to the combinations ν_sSiOSi + δ; the frequency differences of these bands and the three ν_sSiOSi frequencies ($\nu_{7,8}$, ν_{11}, ν_{12}) are almost identical, about 110 cm^{-1} (Table 67). Passing to the spectrum of the crystal, in which the ν_{11} vibration is inactive, also leads to the loss of the 588-cm^{-1} band. The combining vibration with frequency 110 cm^{-1} probably belongs to the set of "nonplanar" deformation vibrations $\delta_{\perp}$SiOSi, corresponding to internal rotations in the ring and interacting strongly with the stretching vibrations. The identification of the satellites of the ν_{as}SiOSi band at around 1141 and 1029 cm^{-1} in the spectrum of the crystal as corresponding to transitions ν_{as} + δ and ν_{as} + $\delta - \delta$ leads to $\delta \simeq 80$ cm^{-1}. Since difference combinations fail to appear, it is not clear whether this difference in the frequencies 80 and 110 cm^{-1} is due to their belonging to different deformation vibrations.

In conclusion we note that, despite the absence of any information regarding the spectrum of crystalline $[OSi(CH_3)_2]_8$, comparison of the x-ray structural data mentioned with the existence of two ν_{as}SiOSi bands in the infrared spectrum of the liquid [58] and the existence of a dipole moment indicates that conversion of the ring takes place during the phase transformation in accordance with an analogous scheme: $S_4 \rightarrow C_8$.

A study of the cyclical penta- and heptasiloxanes would, apart from anything else, be of particular interest in view of the fact that five- and seven-fold axes cannot exist in crystal lattices; hence, independently of the crystal symmetry, the degenerate vibrations should split and the inactive ones become active. The fact that this is not observed experimentally constitutes a direct indication of the weak influence of intermolecular interactions (symmetry of the internal field) on the vibrational spectrum of crystalline siloxanes. In other words, the observed spectrum is almost entirely determined by the internal vibrations of the molecule. It is interesting to note that, on rapid cooling to –150°C, the spectrum of $[OSi(CH_3)_2]_5$ initially exhibits splitting of the band at about 525 cm^{-1}, evidently corresponding to the doubly degenerate vibra-

TABLE 67. Frequencies of the Stretching Vibrations of the Siloxane Rings in $[OSi(CH_3)_2]_n$ Molecules and of the Combinations Involving "Nonplanar" Deformations of These Rings

$[OSi(CH_3)_2]_6$					$[OSi(CH_3)_2]_5$			$[OSi(CH_3)_2]_7$		
liquid (25° C)		crystal (–150° C)		assignments	liquid (25° C)		crystal (–150° C)	liquid (25° C)	crystal (–150° C)	assignments
RS [53]	IRS	IRS: metastable form	IRS: stable form		RS [53]	IRS	IRS	IRS	IRS	
493 (10)	489 w.	480 w.	—	ν_{11}	488 (10)	485 w.	477 w.	489 w.	496 w.	A
—	—	—	499 m.	ν_{12} } ν_sSiOSi	—	—	—	—	—	} ν_sSiOSi
—	523 m.	524 m.	525 m.	$\nu_{7,8}$	530 (0)	530 w.	525 w.	516 w.	523 w.	E_1
—	588 w.	587 w.	—		—	—	—	587 w.	—	
—	611 v.w.	611 v.w.	612 v.w.	} ν_sSiOSi + $\delta_{\perp}$	—	613 w.	609 w.	—	—	} ν_sSiOSi + $\delta_{\perp}$
—	631 v.w.	634 v.w.	632 w.		639 (1)	—	—	635 v.w.	632 w.	
—	—	1011 w.	1029 m.	—*	—	1005 sh.	1015 m.	—	—	—*
—	1070—	1059 v.s.	1064 v.s.	$\nu_{1,2}$	—	1068 v.s.	1062 v.s.	1064 v.s.	1064 v.s.	E_1 } ν_{as}SiOSi
—	–1090 v.s.	1080 v.s.	—	ν_6 } ν_{as}SiOSi	—	1090 v.s.	1083 v.s.	1090 v.s.	1100 v.s.	A
—	—	1137 sh.	1141 m.	ν_{as}SiOSi + $\delta_{\perp}$	—	—	—	—	—	

Note: *ν_{as}SiOSi + $\delta_{\perp} - \delta_{\perp}$ or ν_{as}SiOSi – $\delta_{\perp}$.

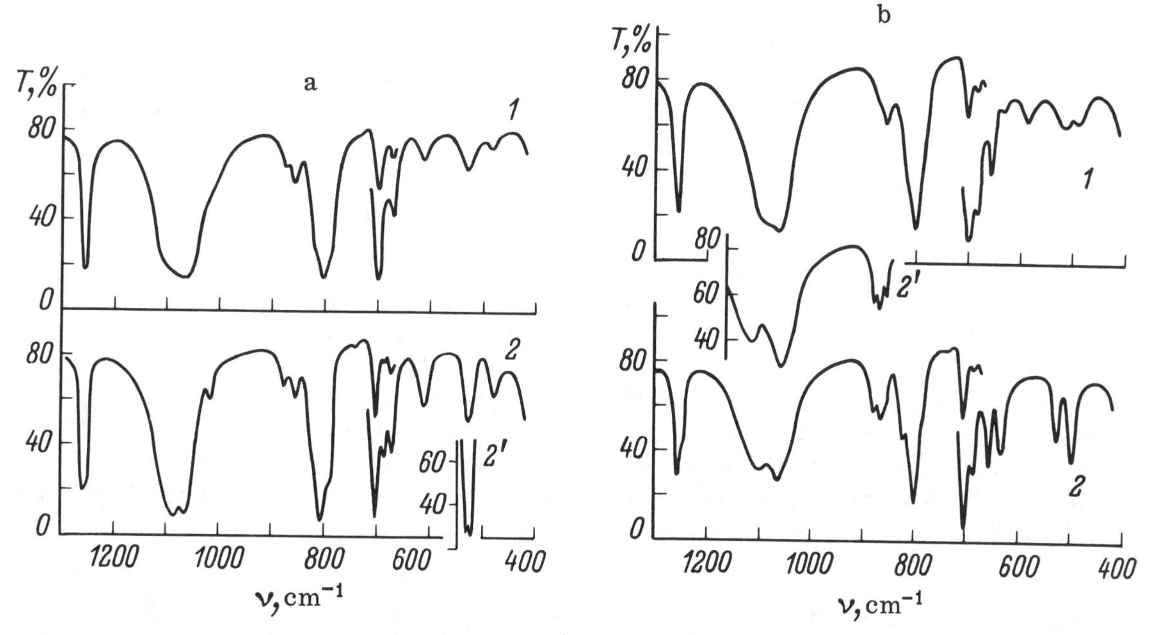

Fig. 127. Infrared spectra of cyclohexanes with an odd number of links. a) $[OSi(CH_3)_2]_5$; b) $[OSi(CH_3)_2]_7$; 1) liquid, t = 25°C, 2) crystal, t = −150°C (after holding for 2 h), 2') the same, 15 min after cooling to −150°C. In the range of the KBr prism the layer thickness is 0.012 mm; in the range of the NaCl prism there is a capillary layer.

tion ν_s SiOSi. However, this splitting vanishes after holding at −150°C for about an hour in the same way as the very pronounced narrowing of all the bands. In the spectrum of $[OSi(CH_3)_2]_7$, immediately after freezing there is an increase in the separation between the two ν_{as} SiOSi bands (Fig. 127b, curve 2'); holding at −150°C reduces this separation, which subsequently remains constant. These results show that the severe stresses arising as a result of the initial cooling may lead not only to imperfect packing but also to deformations (distortions) of the molecules themselves.

In contrast to the rings with even n, neither long holding at −150°C nor subsequent heating almost to the melting point, or alternatively slow cooling of the liquid, lead to any substantial changes in the spectrum. The spectra of the crystals differ comparatively little (apart from better differentiation of the bands) from those of the liquid phase. There is thus little doubt that the configuration of the rings remains the same in the liquid and in the crystal. A determination of the symmetry of the molecules may be based, on the one hand, on the existence of degenerate vibrations, indicating the presence of C_5 or C_7 axes, and, on the other hand, on the presence of two bands in the ν_{as} SiOSi and two in the ν_s SiOSi frequency range. It follows from Table 66 that only D_n and C_n groups satisfy these demands. The first of these may be rejected as nondipolar, and hence the molecules have symmetry C_5 and C_7, provided that the large number of bands in the νSiC_2 frequency range does not mean a lower symmetry for these groups.

The band at about 610 cm^{-1} in the spectrum of $[OSi(CH_3)_2]_5$ may probably be associated with the combinations ν_s SiOSi + $\delta_\perp$SiOSi. If we suppose that the combining frequency ν_s SiOSi is a vibration of the type E_1 (530 cm^{-1}), then the resultant value of $\delta_\perp$ = 83 cm^{-1} may be compared with that observed by Kriegsmann in the Raman spectrum at 76 cm^{-1}. In the spectrum of the "heptamer" the combinations ν_s SiOSi + $\delta_\perp$SiOSi may well represent the bands at 587 and 635 cm^{-1}, which gives $\delta_\perp$SiOSi $\simeq$ 100 to 120 cm^{-1}. The vanishing of the first of these in the spectrum of the crystal is hard to explain unless we suppose that its appearance in the spectrum of the liquid is associated with the breaking of the selection rules.

It is interesting to compare the values of the $\delta_{\perp}$SiOSi frequencies in siloxane rings of various dimensions. The nonplanar SiOSi deformation vibration in the plane cyclotrisiloxane molecule evidently corresponds to rather higher-frequency vibrations in the nonplanar $[OSi(CH_3)_2]_4$ molecule and still higher frequencies in the spectra of the larger rings of the $[OSi(CH_3)_2]_n$ type:

n	3	4	5	6	7
$\delta_{\perp}$ SiOSi	35	50–70	~80	80, 100	100, 120 cm^{-1}

There is a clear tendency for the frequencies to rise on increasing n; despite the possibility of individual errors in identifying these vibrations from the combination transitions in the spectra of the rings with n = 5 to 7, this may easily be explained by kinematic factors. For constant values of the SiOSi (and OSiO) angles (starting from n = 4), as n increases so does the deviation from the coplanar configuration of the Si–O–Si bonds in the ring, and correspondingly so does the extent to which changes in bond length take part in these vibrations. For a fairly large n, the frequencies of the vibrations considered may probably approach those of the plane deformations of the SiOSi bond in linear polysiloxanes (we remember that the δSiOSi frequency in hexamethyldisiloxane approximately equals 130 cm^{-1}).

Summarizing the results of the spectroscopic analysis of the dimethylcyclopolysiloxanes, we may now set up a general scheme for the proposed changes in the conformation of the siloxane rings in the $[OSi(CH_3)_2]_n$ molecules during phase transformations (Table 68).

The difference in the behavior of the rings with even and odd numbers of links is evidently explained by the existence of only one possible conformation for the rings with odd n, involving a comparatively high symmetry of the molecule and an almost equilibrium value of the SiOSi angles. Conformations with symmetry C_n in the crystal phase are clearly energetically unfavorable, since for even n these are either not observed at all or are unstable. This may constitute one of the reasons, apart from packing hindrances, for the anomalously low melting points of the penta- and heptasiloxanes [61]:

$n =$	3	4	5	6	7	8
t_m, °C =	64	17.5	—38	—3	—26	31.5

Thus a study of the structure of the siloxane rings and its changes during phase transformations fairly convincingly demonstrates the flexibility of the molecules with Si–O–Si bonds. Apart from the dimethylcyclopolysiloxanes considered here, siloxanes with other radicals probably have different conformations, for example, diphenyl cyclosiloxanes, the crystals of which

TABLE 68. Symmetry of the Molecules of Cyclic Dimethylpolysiloxanes

n	Crystal		Liquid
	low temperature form	high temperature form	
3	D_{3h} (C_{3v}?)		D_{3h}
4	C_i	S_4	C_{4v} (C_4?)
5	C_5		C_5
6	D_{3d} [C_6]		C_6
7	C_7		C_7
8	S_4?		C_8

Note: The metastable form given in square brackets.

form several types of structure. However, detailed information regarding the structure of these crystals is not yet available. The results of x-ray structural analysis of the silicates also indicate the existence of various equilibrium configurations of the cyclic anions: the $[Si_4O_{12}]^{-8}$ anion has symmetry D_{4h} (or more exactly S_4) in baotite [62] and C_s in kainosite [63]; the $[Si_6O_{18}]^{-12}$ ion has a flat D_{6h} ring in the beryl crystal [64] but a nonplanar configuration with symmetry S_6 in dioptase [65,66], while its symmetry in tourmaline is C_{3v} [67,68]. Up to the present, however, no solid-phase transformations associated with changes in the conformation of the cyclic anion have been described in silicates; this is probably due to the considerable rigidity of the lattice of cation coordination polyhedra. The question as to the various conformations of metaphosphate rings was discussed earlier (Chap. II, § 5).

§ 3. Mobility of the Oxygen Bridges and Phase Transformations in Phosphates and Silicates

In this section we shall consider some characteristics of crystal structures characterized by anomalously large (over 150°) valence angles of the oxygen in the X–O–X bridges. We might expect that in these cases the resistance to the internal rotation would be still more weakened. This is promoted, firstly, by the increase in the distances between the side radicals, and secondly by the reduction in the potential barrier of the internal rotation as a result of the increase in the overlapping of the d orbitals of the X atom with the p orbitals of the oxygen, the nodal plane of which is perpendicular to the XOX plane. Actually in the limiting case with $\angle XOX = 180°$ the internal rotation in the X–O–X bond is evidently completely free, i.e., it only depends on the interaction between the end groups. In the linear $R_3X{-}O{-}XR_3$ molecule the vectors of the corresponding π orbitals (p_x and p_y are the O orbitals and d_{xx} and d_{yz} the X orbitals, the z axis lying along the line of the bonds) may be oriented along any mutually-perpendicular directions (see Fig. 17). Quantum-chemical consideration of these questions was undertaken by Shustorovich, who showed that both in the chains of organic cumulenes $H_2C{=}C{=}C\ldots C{=}CH_2$ and in inorganic cumulenes the rotations of the atoms within the chains were completely free [69, 70]. In this respect the straightened oxygen bridges in silicates and phosphates differ considerably from chains of conjugate bonds, for which high barriers to internal rotation are typical.

On the other hand, the low transverse rigidity of the cumulene chains was also noted in the foregoing papers†; this applied particularly to the oxygen bridges in the chains of inorganic cumulenes. This low rigidity was largely associated with the fact that the reduction in the contribution of certain d orbitals arising from the deformation was compensated by a strengthening of the π bonds involving other d orbitals, and also with the very slight change in the energy of the system when the latter deviated from the equilibrium configuration. In other words, the mutual degeneracy of the in-plane and "nonplanar" deformations of the angle XOX taking place for $\angle XOX = 180°$ evidently arises, not only from the increase in the "rotational" quantum associated with the reduction in the moment of inertia of the O atom relative to the X...X axis, but also from a reduction in the deformation quantum (see p. 43). Thus in structures with XOX angles close to 180° both the small resistance to the internal rotation and the low rigidity of these angles lead us to expect a considerable increase in the mobility of the oxygen atoms in the plane perpendicular to the X...X axis, and especially in the direction perpendicular to the XOX plane.

The thermal amplitudes of the oxygen atoms in the Si–O–Si bonds obtained when refining x-ray analyses of the silicate structures agree closely with the foregoing considerations. Thus Cruickshank et al. [72] refined the structure of thortveitite $Sc_2Si_2O_7$

† Pitzer and Strickler [71] gave some examples of anomalously low force constants for deformation of the C=C=C angles in small molecules and considered possible interpretations of these anomalies.

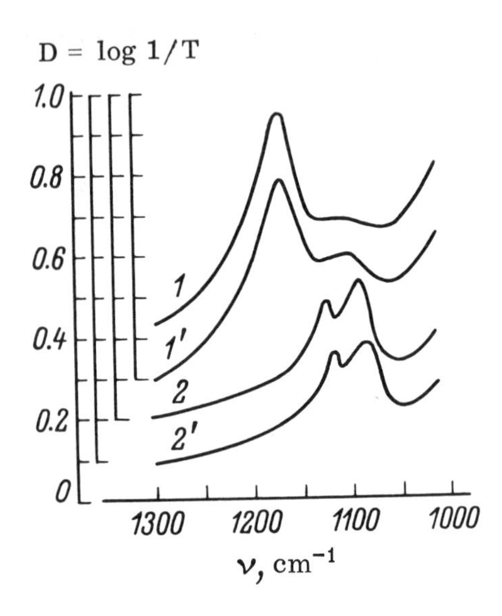

Fig. 128. Temperature dependence of the infrared spectra of scandium and yttrium pyrosilicates (low-temperature forms) in the region of the band representing the antisymmetric Si –O –Si vibration. 1) $Sc_2Si_2O_7$, T = 120°K; 1') the same at T = 300°K; 2) $Y_2Si_2O_7$, T = 120°K; 2)' the same at = 300°K.

(using the method of least squares), confirming the original conclusion of Zachariasen [73] as to the centrosymmetry (symmetry C_{2h}) of the Si_2O_7 anion, and determined the anisotropic thermal amplitudes of the atoms. It was found that the mean square amplitude of the vibrations of the bridge oxygen atom (~ 0.06 Å^2) in a direction perpendicular to the Si . . . Si axis was roughly twice that of the Si and terminal O atoms. At the same time the amplitudes of the Sc atoms were insignificant, so that the large amplitudes of the O and Si atoms could not be associated with lattice vibrations, i.e., with the vibrations of the complex Si_2O_7 anions as a whole relative to the sublattice of the Sc cations. Thus we may conclude that the linear Si–O–Si system in thortveitite is quite flexible.

Certain characteristics of the infrared spectrum of thortveitite may also be considered as an indication of the flexibility of the Si–O–Si bridge [74]. On analyzing the vibration frequencies of the complex anion in this crystal, which contains only one Si_2O_7 group in the unit cell, it was noted that there were no grounds for expecting the splitting of the nondegenerate vibrations.† Nevertheless, near the ν_{as} SiOSi vibration band (1170 cm^{-1}) a satellite was observed with a maximum at about 1100 cm^{-1}; this could not be ascribed to any other fundamental vibrations of the Si_2O_7 group. We see from Fig. 128 that the relative intensity of the band at 1100 cm^{-1} falls on reducing the temperature. This clearly suggests ascribing this band to ν_{as} SiOSi transitions from excited levels ($\nu + n\omega - n\omega$), in the same way as the satellites of the bands of the corresponding vibrations characteristic of the siloxane spectra. Since the change in the intensity of the band at about 1100 cm^{-1} is no greater than 50% on reducing the temperature from 300 to 120°K, the frequency ω associated with the factor $\exp(-h\omega/kT)$ is no greater than 100 cm^{-1}, if in order to be specific we suppose that the ν_{as} SiOSi vibration only interacts with one low-frequency vibration ω, and that the contribution of transitions with $n > 1$ to the intensity is negligible. Hence we are justified in assigning the vibrations combining with ν_{as} SiOSi to deformations of SiOSi and rotations around the Si–O bonds. For comparison the same figure shows the ν_{as} SiOSi band in the spectrum of yttrium pyrosilicate, the doublet structure of which is evidently associated with an increase in the number of Si_2O_7 groups in the unit cell (see Chap. V). The relative intensity of the two components in the spectrum of this pyrosilicate shows no temperature dependence.

† Strictly speaking this is only valid for the unexcited crystal lattice, i.e., for temperatures around 0°K, since the vibrations of the lattice change its symmetry.

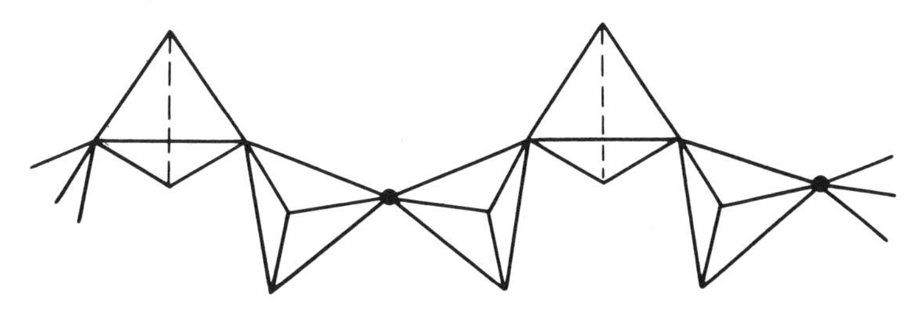

Fig. 129. Chain of silicon–oxygen $(Si_3O_9)_\infty$ tetrahedra in the structures of bustamite, wollastonite, and pectolite.

A no less characteristic example is obtained by comparing the structures of bustamite (Ca, Mn) SiO_3 and wollastonite $CaSiO_3$, which were recently refined by Buerger and colleagues [75, 77]. The structure of the two crystals is very much the same: the complex metasilicate anions are constructed in the form of infinite $(Si_3O_9)_\infty$ chains containing three silicon–oxygen tetrahedra in the identity period (Fig. 129). The three nonequivalent SiOSi angles in the chain have the following values (in degrees):

(Ca, Mn) SiO_3	$CaSiO_3$
161	149
135	139
137	140

The isotropic temperature factor B for the O atom forming the bridge straightened to ∠SiOSi = 161° in the bustamite chain (shown by a black circle in Fig. 129) was at least twice as large as that of the remaining atoms of the structure, including the O atoms forming the bent bridges. On the other hand, in the wollastonite chain, with its relatively less straightened oxygen bridge, the temperature factor of the corresponding oxygen atom is only slightly greater than the average value for all the other atoms. Thus comparison of the thermal amplitudes in these structures also agrees with the assumption that there is an increase in the flexibility of the Si–O–Si bridges on increasing the valence angle.

It should be noted that, in all the structures considered, an increased mobility of the O atom is characteristic of the Si–O–Si bridges with the largest SiOSi angles and simultaneously the shortest interatomic distances $r_{Si-O(Si)}$, which reflect the relatively high orders of these bonds. The shortening of the straightened Si–O–Si bridges cannot be associated with an increase in the ionic part of the interaction between the Si and O, since the O atoms of these bridges typically lie at greater distances from the nearest cations.

The comparison of the interatomic distances and SiOSi angles in the structures of bustamite, wollastonite, and pectolite [78] is instructive. In all three crystals the configurations of the complex anions are similar (Fig. 129); the $r_{Si-O(Si)}$ distances are on the average 0.06 Å greater than the r_{SiO^-} distances, being, respectively, 1.66 and 1.60 Å. However, the $r_{Si-O(Si)}$ distances show a clear tendency to fall as ∠SiOSi increases (Table 69).

TABLE 69

Compound	∠SiOSi (greatest of three), deg	$r_{Si-O(Si)}$, Å
Wollastonite . . .	149	1.65; 1.64
Pectolite	150	1.65; 1.61
Bustamite	161	1.62; 1.61

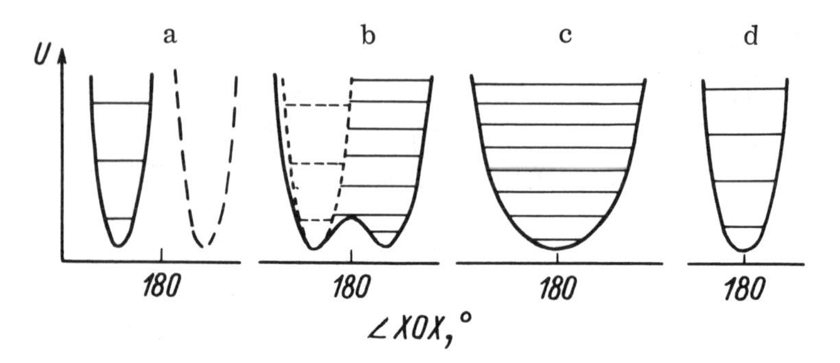

Fig. 130. Probable form of the potential curve of the δXOX vibration in X–O–X systems with various equilibrium values of ∠XOX. a) bent X–O–X bridge; b) "quasi-linear" (inversion possible); c) flexible with ∠XOX = 180°; d) ordinary linear XOX molecule. In order to simplify the scheme the interaction of the deformations with rotations is disregarded.

This tendency also appears within the bounds of a single structure, for example, buştamite, on comparing the lengths and angles in the three nonequivalent Si–O–Si bonds of the chain anion:

∠ SiOSi, deg	$r_{Si-O(Si)}$, Å
135	1.67; 1.65,
137	1.66; 1.65,
161	1.62; 1.61.

Thus a 10° increase in the SiOSi angle corresponds in these structures to a reduction of 0.02 to 0.03 Å in the length of the Si–O(Si) bond. It is interesting that the extrapolation of this relationship to the tetrahedral SiOSi angle gives an $r_{Si-O(Si)}$ distance of 1.72 to 1.75 Å, i.e., close to the length of the single Si–O bond calculated by the Schumaker–Stevenson method (1.76 Å), while on extrapolating to 180° one obtains 1.55 to 1.58 Å, coinciding with the shortest of the experimentally measured Si–O⁻ lengths.† The lengths of the Si–O(Si) bond in the pyrosilicate ion of thortveitite with ∠SiOSi = 180° is slightly greater, 1.607 ± 0.007 Å [72], although the distances r_{Si-O^-} = 1.626 ± 0.012 Å, owing to the partially covalent character of the O⁻ ... Sc bonds, are still longer. It was shown in [72] that the structure of thortveitite could be interpreted not only for the space group C_{2h}^3–C2/m but also for groups Cm and C2, achieving even a slightly better reliability coefficient; here the SiOSi angle equals 165°. These data, together with other indications as to the flexibility of the straightened Si–O–Si bridges, suggest that the potential function of the δSiOSi vibration in thortveitite may be described by schemes (Fig. 130b or c) corresponding to intermediate cases between the ordinary "rigid" bent (a) or straightened (d) triatomic molecules.

We may also present one further example of the increased thermal amplitudes of the oxygen in straightened X–O–X bridges. In the earlier-considered (p. 172) double-chain structure formed by the ribbons of the sillimanite anion, the angles of the Si–O–Al^{IV} bonds perpendicular to the direction of the ribbon exceed 170°, while the longitudinal bonds of the same type

† The correlation between the lengths of the Si–O–Si (or Si–O–Al) bonds and the values of the SiOSi angles in structures of the framework type has recently been studied in detail [77a]. The average lengths of the Si–O(Si) bonds obtained from this correlation is 1.643 Å for ∠SiOSi = 120° and 1.558 Å for ∠SiOSi = 180°.

have $SiOAl^{IV}$ angles close to the tetrahedral value. Correspondingly the interatomic distances in the straightened SiOAl bonds are shortened, while the mean square thermal amplitudes of the O atoms are sharply anisotropic, and in a direction perpendicular to the Si . . . Al line are roughly twice those of the other oxygen atoms.

Spectra and Phase Transformations in Pyrophosphates with the $Mg_2P_2O_7$ Structure [79, 80]. Let us consider possible differences between the XOX bridges with potential functions represented by schemes b and c in Fig. 130. For a small height of the potential "hump" it is difficult to distinguish these two cases by x-ray diffraction, while the securing of spectrographic evidence in favor of one model or the other demands a study of transitions (including those from nonzero levels) lying in the long-wave part of the infrared range. We may expect nevertheless that at fairly low temperatures the nonlinear configuration of the X–O–X bridge with a potential represented by scheme b will be "frozen." The spectrum of thortveitite at T = 120°C still shows no indications of a bent form of anion. Such an indication might, for example, be provided by the appearance of a ν_s SiOSi band. In certain phosphates which we shall be considering shortly, solid-phase transformations take place; these indicate a transformation from the straightened configuration of the P_2O_7 group to a bent configuration on reducing the temperature.

Magnesium pyrophosphate $Mg_2P_2O_7$ and thortveitite $Sc_2Si_2O_7$, as mentioned by Machatschki [81], constitute a typical example of heterovalent isomorphism: Sc → Mg, Si → P. Roy et al. [82] noted the existence of a reversible phase transformation in $Mg_2P_2O_7$ at 68°C, the high-temperature β-form being isomorphous with thortveitite. This was confirmed by analysis of the β-$Mg_2P_2O_7$ structure [83]. The same investigation established a doubling of the lattice parameters along the a and c axes on passing to the low-temperature α-form. When studying the specific heat in [84] it was suggested that the α-β transformation constituted a transformation of the first kind, although possible arguments were also advanced in favor of its being one of the second kind. The considerable entropy of the transformation, 2 · 13 cal/mole · deg, suggests the existence of ordering processes, independently of whether this transformation is of the first or second order.

A study of the infrared spectrum of $Mg_2P_2O_7$ enables the configuration of the P_2O_7 group in the α-form to be discussed, together with the possible mechanism of the solid-phase transformation [79]. The spectrum of α-$Mg_2P_2O_7$ (Fig. 131) clearly indicates a noncentrosymmetric ($\angle POP < 180°$) configuration of the anion: the ν_sPOP band occurs in the spectrum, and the

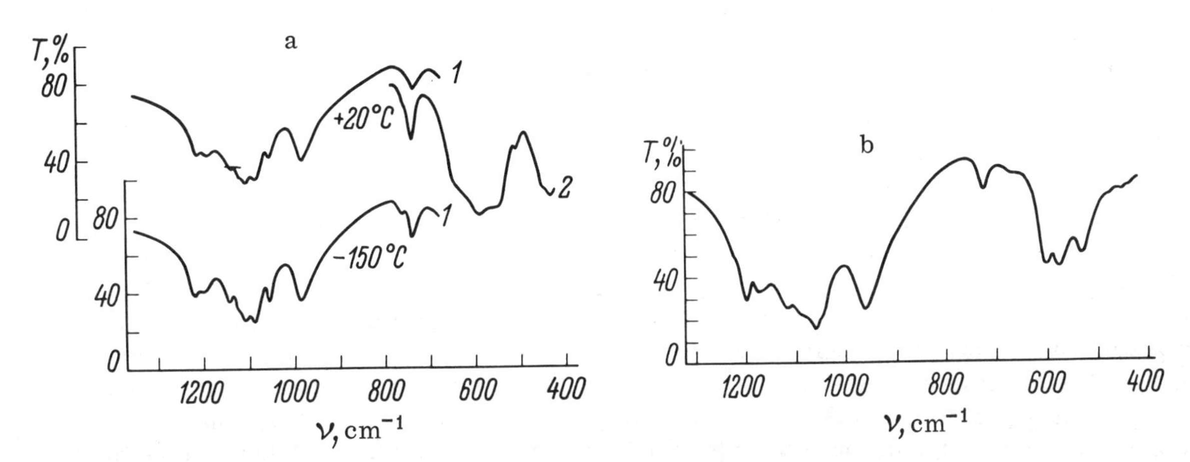

Fig. 131. Infrared spectra of low-temperature forms. a) $Mg_2P_2O_7$ [pressed in KBr: 1) 2 mg, 2) 5 mg]; b) $Zn_2P_2O_7$ (3 mg).

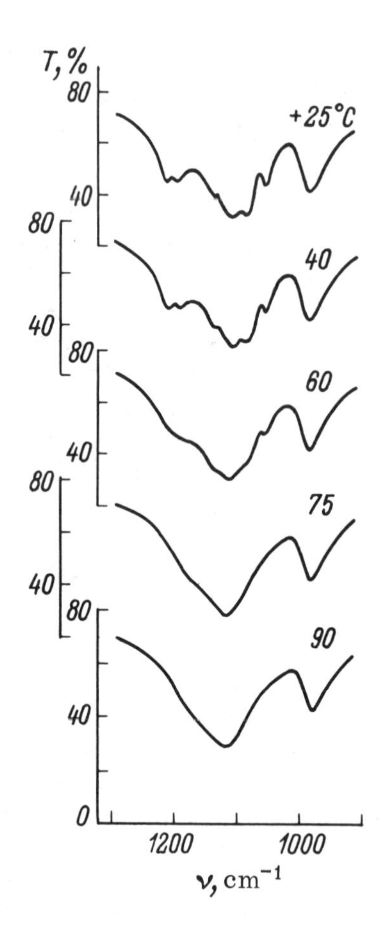

Fig. 132. Changes in the infrared spectrum of $Mg_2P_2O_7$ on raising the temperature.

large number of bands in the range 1050 to 1200 cm^{-1} indicates the splitting of the degenerate (for symmetry D_{3d}) vibrations of the PO_3. The rather considerable (~20 cm^{-1}) splitting of these vibrations and the high frequency and intensity of the ν_sPOP band suggest that the POP angle is no greater than 160°. The symmetry of the bent P_2O_7 ion cannot be greater than C_{2v}, but the appearance of a ν'_{as} PO_3 vibration band (the prime denotes the antiphase nature of the vibrations of the two PO_3 groups in the anion) in the infrared spectrum suggests a lower symmetry, C_2, C_s, or simply C_1. We note that for a large value of ∠POP, although not 180°, the trans configuration, with its maximum possible symmetry C_s, gives the greatest distance between the O atoms and the PO_3 groups for a specified ∠POP, while on increasing the angle to 180° the symmetry rises to D_{3d} without any change in the relative orientation of the PO_3 groups.

The infrared spectrum of β-$Mg_2P_2O_7$ (Fig. 132) consists of a small number of very diffuse (even at temperatures a long way from the transformation point) bands, which may be associated with the three active stretching vibrations of the P_2O_7 ion possessing intrinsic symmetry D_{3d}: ν_{as} POP (A_{2u}), ν'_{as} PO_3 (E_u), and $\nu'_s PO_3$ (A_{2u}). The identification of the frequencies of the two forms is presented in Table 70. We remember that, as in the case of the isostructural thortveitite, the reduction of the site symmetry of the anion from D_{3d} to C_{2h} makes the splitting of the degenerate vibrations potentially possible, but without leading to the removal of the mutual exclusion rule. Correspondingly, in the spectrum of the high-temperature form, the band of the ν_sPOP vibration relating to the inactive irreducible representation has a vanishingly small intensity. A departure from the rigorous application of the ban might be explained, for example, by interactions with low-frequency lattice vibrations or the influence of defects in the crystal structure.

TABLE 70. Frequencies of the Low- and High-Temperature Forms of $Mg_2P_2O_7$ and $Zn_2P_2O_7$ in the Infrared Spectrum

Δ_{av}, % freq. increase ($Mg_2P_2O_7$ vs. $Zn_2P_2O_7$)	Low temperature forms		Assignments	Symmetry types of the vibrations of the P_2O_7 ion				High temperature forms	
	$Mg_2P_2O_7$	$Zn_2P_2O_7$		C_{2v}	C_s	C_2	D_{3d}	$Mg_2P_2O_7$	$Zn_2P_2O_7$
	—	(1224 sh., w.)	—	—	—	—	—	—	—
1.45	1211 s.	1198 s.	ν_{as} PO_3	A_1	A'	A	E_g (ina)	—	—
	1193 s.	1172 s.		B_2	A''	B		—	—
	1133 s.	1118 s.	ν'_s PO_3	B_1	A'	B	A_{2u}	~1170 s.	~1150 s.
	(~1118 sh., w.)	—	—	—	—	—	—	—	—
	1105 v.s.	1090 sh.	ν'_{as} PO_3	B_1	A'	B	E_u	1118 v.s.	1103 v.s.
	1081 v.s.	1061 v.s.		A_2 (ina)	A''	A			
	1052 s.	~1046 sh.	ν_s PO_3	A_1	A'	A	A_{1g} (ina)	—	—
2.05	982 s.	961 s.	ν_{as} POP	B_1	A'	B	A_{2u}	978 s.	960 s.
	761 sh., v.w.	—	—	—	—	—	—	—	—
	740 w.	726 w.	ν_s POP	A_1	A'	A	A_{1g} (ina)	727 v.w.	—
	—	(670 v.w.)	—	—	—	—	—	—	(~675 v.w.)
4.5	638 sh.	602 s.	δ'_{as} and δ'_s PO_3 and ν $M^{2+}\ldots O$					~638 sh.	600 s.
	587 v.s.	575 s.						586 v.s.	573 s.
	558 v.s.	530 s.						~550 sh.	530 s.
	(509 v.w.)	—	—	—	—	—	—	(~508 v.w.)	—
>11.0	454 sh.	—	ρ' PO_3 and ν $M^{2+}\ldots O$					—	—
	437 v.s.	—						440 v.s.	—

However, the character of the change in the $Mg_2P_2O_7$ spectrum on increasing the temperature, as well as certain characteristics of the spectrum of the high-temperature form, can hardly be made to agree with the idea of a phase-transformation mechanism amounting to the simple straightening of the POP angle from 160 to 180°. An increase in the POP angle should lead to a fall (possibly very slight) in the ν_s POP frequency and a far greater increase in the ν_{as} POP frequency. Actually, as the temperature rises there is only a very slight fall in both frequencies, while in the neighborhood of the phase-transformation temperature there is a sharper fall in the ν_s POP frequency, by about 10 cm^{-1} (Fig. 133), with a considerable reduction in the intensity of the band. The frequency of ν_{as} POP remains practically unchanged. The frequencies of the vibrations of the terminal PO_3 groups change comparatively little on raising the temperature, but the bands corresponding to these vibrations become so diffuse that the broadening cannot be explained by the relatively slight rise in temperature without involving additional assumptions.

We may also note that, from a priori considerations, a sharp rise in the angle POP and hence in the order of the P–O–P bond on increasing the temperature, i.e., on increasing the interatomic distances, is improbable.

The foregoing characteristics of the thermal changes in the infrared spectrum of $Mg_2P_2O_7$ may be naturally explained if we suppose that the mechanism of the phase transformation does not lie in a static straightening of the P_2O_7 groups but that this transformation is associated with the thermal excitation of a process in which the O atom in the P–O–P bond (and at the same time the O atoms of the end PO_3 groups) jumps into an equivalent position relative to the P...P axis, and that for a sufficiently high temperature this leads to a statistical centrosymmetry of the P_2O_7 ions. In other words, we suppose that the potential curve of the δPOP vibration of the P_2O_7 group has two minima (Fig. 130) separated by a relatively low potential hump corresponding to the straightened form of the anion. The thermal excitation of the higher levels of the low-frequency δPOP vibration may lead to the overcoming of the barrier, which corre-

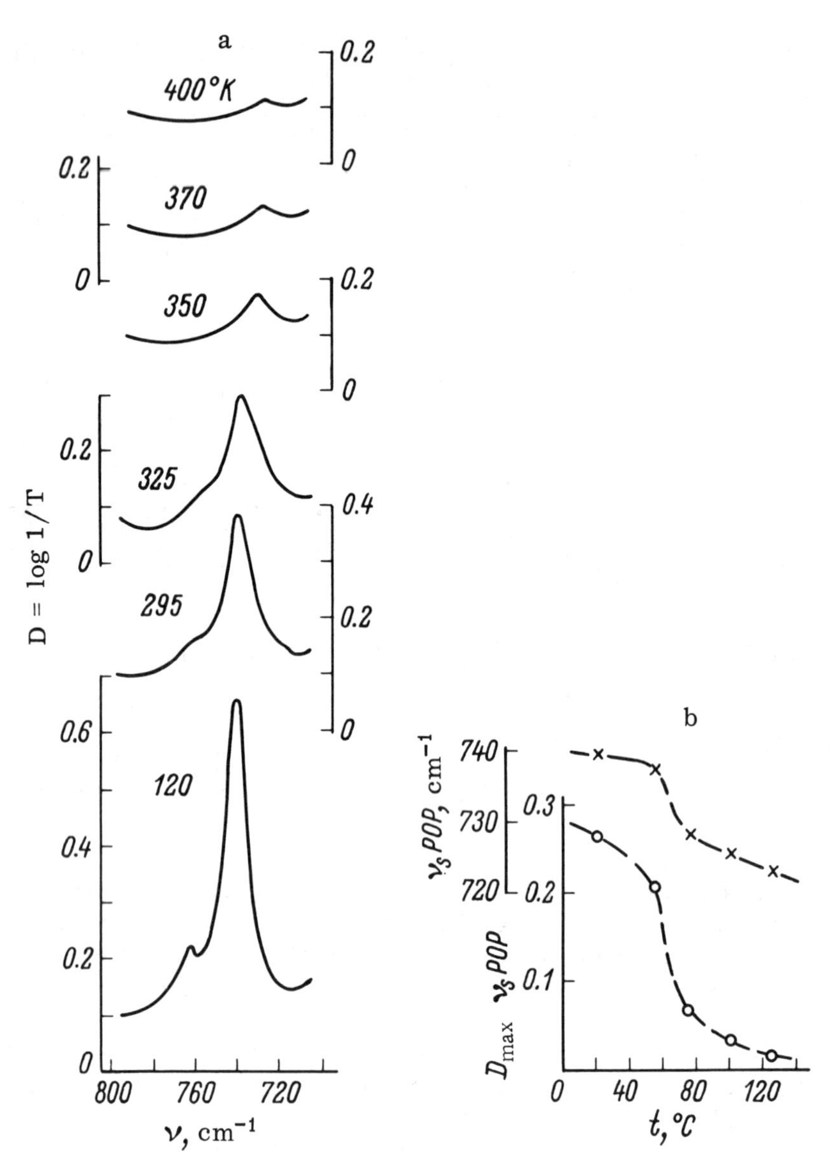

Fig. 133. Thermal changes in the ν_sPOP vibration band of $Mg_2P_2O_7$. a) The ν_s POP band at various temperatures: b) temperature dependence of the frequency and intensity of the two bands.

sponds to a transformation to the centrosymmetric configuration.† Then the equilibrium value of the angle POP (that corresponding to the potential-energy minimum) remains constant, which leads to the constancy of the stretching frequencies, while the selection rules approach those corresponding to the centrosymmetric system. The considerable amplitude of the vibra-

†Strictly speaking, the P_2O_7 ion becomes symmetrical relative not to the P ... P axis but to a line parallel to this, passing through the center of gravity of the bent P_2O_7 group. This difference is slight, since on inversion of the P_2O_7 ion the amplitude of displacement of the P atoms is smaller than in the case of the O atoms of the PO_3 groups, and particularly the O atom in the P–O–P bond. We note that the number of minima in the potential function of the δPOP vibration is given by the site symmetry C_{2h} of the P_2O_7 ion in β-$Mg_2P_2O_7$ and is equal to two if the site symmetry in α-$Mg_2P_2O_7$ is C_s or C_2 and four if the site symmetry is C_1.

tions of the O atoms in the P–O–P bridges, and partly in the PO_3 groups, is the reason for the spread of the bands. Interaction with deformations and internal rotations leads to a complex structure of the stretching vibration bands in the spectra of the "flexible" siloxane molecules. Hence the slight displacement of the ν_sPOP band to lower frequencies during the transformation may be associated with the increasing role of the excited states, for which the POP angle is on the average greater, while the existence of the satellite on the high-frequency side of the ν_sPOP band, which intensifies with falling temperature, may be ascribed to states with smaller POP angles.

The low temperature of the transformation in $Mg_2P_2O_7$ may be considered as an indication of a small (order of 0.01 eV) height of the potential hump, if we suppose that this is close to kT at the transformation point. However, this supposition ignores the cooperative character of the transformation, as a result of which the potential barrier for the transfer of an O atom in one P_2O_7 group, taken separately, may be considerably higher. In order to obtain fuller information as to the character of the potential function, direct investigations in the low-frequency part of the spectrum are desirable. We must mention once again that it is clear, even by analogy with the earlier-considered disiloxane molecules, that the actual transformation mechanism is associated not only with δPOP vibrations but also with the strongly interacting internal rotations around the P–O (P) bonds. This does not mean that there is a transposition of the terminal O^- atoms in the P_2O_7 group (improbable in the present case), but only supposes the possibility of the $P-O^-$ bonds (like the P–O(P)) "precessing" around directions corresponding to their disposition in the case of an ideally rigid centrosymmetric configuration.

An analogous transformation also occurs in zinc pyrophosphate $Zn_2P_2O_7$. The existence of two forms of these crystals transforming one into the other at 132 ± 8°C with an endothermic effect was noticed in [85]. The transformation occurs reversibly and without any marked hysteresis. X-ray data indicate a higher symmetry for the high-temperature form with a similarity to the structure of $Mg_2P_2O_7$. The infrared spectrum of the low-temperature form of $Zn_2P_2O_7$ (Fig. 131) also indicates this similarity, although probably incomplete. It is an interesting point that the infrared spectrum of $Zn_2P_2O_7$ clearly indicates octahedral coordination of the Zn^{2+} ions in these crystals, which of course is a necessary condition for their structure to be similar to that of magnesium pyrophosphate. In the majority of cases the Zn^{2+} ions (Zn being a group IIB element) are characterized by tetrahedral coordination in crystals formed by the salts of the corresponding oxy acids. Since the ionic radius of Zn^{2+} is usually given a greater value than that of Mg^{2+}, this is evidently associated with the markedly directional character of the Zn–O bonds, contrasting with Mg–O.

Let us compare the spectra of the two phosphates given in Fig. 131. Except for weak bands not belonging to the fundamental vibrations, for each band in the spectrum of $Mg_2P_2O_7$ we find a corresponding band in the spectrum of $Zn_2P_2O_7$. However, in the $Mg_2P_2O_7$ spectrum there is a strong band at 437 to 454 cm^{-1} and this is absent from the $Zn_2P_2O_7$ spectrum. The absorption in this region may be associated with Mg–O vibrations in the oxygen octahedra, or more precisely, with the mixing between these vibrations and the deformation vibrations of the P_2O_7 groups (for example $\rho' PO_3$), leading to an increase in the frequencies of the latter. The absence of this band from the $Zn_2P_2O_7$ spectrum is due to the lower frequencies of the Zn–O

Spatially these minima, approximately characterizing the position of the bridge O atom, are arranged in a general position, i.e., symmetrically with respect to the point of symmetry C_{2h} in β-$Mg_2P_2O_7$. New structural data [89, 90] confirm the assumption as to the symmetry C_1 of the P_2O_7 ion in α-$Mg_2P_2O_7$; from the point of view here set out, the existence of two forms similar to the β-form near the transformation temperature [87] may be considered as an indication that the two pairs of potential wells have different depths, i.e., an effective symmetry $C_{2h} = C_2 \times C_s$ is established in the two processes.

bonds in the octahedra; no other bands associated with Zn–O vibrations are observed in the spectrum, which is only possible for octahedral coordination of the Zn.

Table 70 shows the average displacements of the frequencies Δ_{av} of the fundamental vibrations for groups of neighboring bands on passing from $Mg_2P_2O_7$ to $Zn_2P_2O_7$. The value of Δ_{av} increases from 1.5% for the range 1200 to 1050 cm^{-1} to 4.5% for the deformation vibrations of P_2O_7 in the range 640 to 530 cm^{-1} and exceeds 11% for the lower-frequency vibrations, which mix strongly with those of the lattice. It follows from this that the changes in the vibration frequencies of the P_2O_7 groups are mainly due to kinematic interaction with the M^{2+}–O bonds. This is also indicated by the negative sign of Δ_{av} on replacing the Mg^{2+} ions by the heavier Zn^{2+}, although the proportion of covalence in the M^{2+}–O bonds is probably thereby increased.

The heating of $Zn_2P_2O_7$ in the neighborhood of the α-β transformation produces changes in structure similar to those occurring for $Mg_2P_2O_7$. The most characteristic is a marked broadening of the bands, which in no way diminishes on moving away from the transformation point to higher temperatures, and the complete loss of the ν_s POP bands. A graph similar to that shown in Fig. 133b gives the transformation temperature as about 140°C. As in the case of $Mg_2P_2O_7$, the number of bands of P–O stretching vibrations in the spectrum of the high-temperature form agrees with a centrosymmetric configuration of the anion. The considerable blurring of the spectrum and the absence of any displacement in the ν_{as} and ν_s POP bands on passing from the bent to the centrosymmetric configuration of the P_2O_7 ion suggests a statistical character of the centrosymmetry in this case also.

We should note the following characteristic features common to both $Mg_2P_2O_7$ and $Zn_2P_2O_7$ and also the earlier-discussed silicate structures, having a high mobility of the oxygen bridges (as indicated by x-ray diffraction): 1) a considerable proportion of covalence in the bonds between the cation and the O atoms of the complex anion, leading to a relatively low order of the $X-O^-(M^+)$ bonds and correspondingly a strengthening of the $p\pi-d\pi$ interaction in the X–O–X bonds; 2) a coordination number of the cation relative to oxygen with a value close to the highest possible value for the particular cation, or at least an ionic-radius ratio r_M/r_O close to the smallest value permitting the existence of the particular coordination polyhedron in question. The purport of this condition is that of ensuring the maximum deformability of the oxygen polyhedron of the cation necessary for increasing the mobility or even permitting the inversion of the X–O–X bridges in the anion.

X-ray structural investigations aimed at refining the structure of β-$Mg_2P_2O_7$ and β-$Zn_2P_2O_7$ [87, 88] confirmed the high mobility of the oxygen atoms in the P–O–P bridges. The thermal amplitudes of these atoms were very large and anisotropic; the greatest displacements occurred in a plane perpendicular to P . . . P. The interatomic distances in the P_2O_7 group of the β-$Mg_2P_2O_7$ crystal indicate a relative shortening of the bridge bonds† $r_{PO(P)}$ = 1.557 ± 0.002 Å (1.569 Å), r_{PO^-} = 1.534 ± 0.010 Å (1.556 Å) and 2 × 1.542 ± 0.009 Å (2 × 1.554 Å). The ascription of the crystals of the centrosymmetric $C2/m-C_{2h}^3$ group was confirmed by the ESR method, Mn^{2+} ions being introduced into the cation positions [89]. Moreover in $Zn_2P_2O_7$ the existence of two solid-state transformations, one of the first kind at 132° and one of the second kind at 155°, was indicated. The latter occurs without any changes in lattice parameters, but with a reduction in symmetry from C2/m to Cm (in the range 155° to 132°). In the low-temperature forms of these pyrophosphates the P–O–P bridges, as we should expect from spectroscopic data, were nonlinear [90], while the volume of the unit cell was greater in the α-form than in the β-form, by four times for $Mg_2P_2O_7$ and by six times (below 132°) for $Zn_2P_2O_7$.‡ It is an interesting fact that in α-

† Figures in parentheses refer to β-$Zn_2P_2O_7$.

‡ This by itself serves as an indication that a transformation of the first kind as well as one of the second kind occurs in $Zn_2P_2O_7$.

$Cu_2P_2O_7$, for which an α-β transformation from C2/c to C2/m occurs at 72° with a doubling of cell volume [87], the O atom of the P–O–P bridge lies on a two-fold axis [90], while in $Zn_2P_2O_7$ (between 155 and 132°) the bridge O atoms lie in a plane of symmetry. We note, however, that in a recent x-ray investigation [91] into the α-β transformation in $Cu_2P_2O_7$, which confirmed our assumption as to the statistical nature of the centrosymmetry of the β-forms, it was concluded that the statistics related to the volume (i.e., spatial) and not the vibration disorder of the β-form of $Cu_2P_2O_7$. The transformation has a heterogeneous character, although the symmetry of the α and β phase satisfies the requirements imposed on transformations of the second kind. The close connection established in [91] between the mechanism of the inversion of the P_2O_7 ion and the change in the configuration of the cation polyhedron indicates the importance of simultaneously considering both factors when developing a theory of such transformations. We note that the relaxation character of these transformations does not allow us to use a dynamic theory of solid-phase transformations similar to that developed in [11, 12].

Among pyrophosphates of the $M_2P_2O_7$ type, a structure similar to thortveitite and the high-temperature form of $Mg_2P_2O_7$ is found in the case of $Mn_2P_2O_7$ [86]. In this crystal the interatomic distances are $r_{P-O(P)} = 1.57$, $r_{P-O^-} = 1.74$ and 2×1.60 Å; although the accuracy of determining these is low, the relative shortening of the P–O–P bridge is undoubted. The octahedral coordination of the comparatively small Mn^{2+} ion allows a certain freedom in its positioning in the polyhedron. The infrared spectrum of $Mn_2P_2O_7$ is similar to the spectra of the high-temperature forms of $Zn_2P_2O_7$ and $Mg_2P_2O_7$ and may be interpreted in an analogous manner [92]. The frequencies of the deformation vibrations of the P_2O_7 groups are 612, 574, and 531 cm^{-1}, closer to the corresponding frequencies in $Zn_2P_2O_7$, and as in this crystal the absorption at 400 to 450 cm^{-1} found in $Mg_2P_2O_7$ is absent. The bands of the three P–O stretching vibrations are wide, and in the ν_s POP frequency range there are two very weak bands at 732 and 715 cm^{-1}. Reducing the temperature to −150°C does not lead to any change in the spectrum, i.e., the anion remains centrosymmetric, although only statistically.

In this connection it is interesting to note the discussion associated with the straightened X–O–X bridges. Despite the centrosymmetric form of certain structures with an O bridge atom in the center of symmetry established by x-ray diffraction, Liebau [93] expressed doubts as to the validity of these results. It was considered that these structures were only pseudocentrosymmetric, and in a number of cases refinement of the x-ray results supported this view. On the other hand, the flexibility of the SiOSi (POP) angle in the straightened bridges makes the boundary between strictly centrosymmetric and pseudocentrosymmetric bridges very tenuous. The possibility that even at low temperatures statistically symmetrical bridges may exist further complicates the problem of determining the equilibrium values of the SiOSi and POP angles close to 180°.

Cubooctahedral Pyrophosphates MP_2O_7. This group of compounds† contains phosphates with quadrivalent cations of Si, Ti, Zr, Hf, Sn, U. Levi and Peyronel [94, 95] indexed the Debye photograph of ZrP_2O_7 in the cubic system, analyzed the structure of this crystal, and came to the conclusion that it contained centrosymmetric P_2O_7 anions with approximately equal

† A cubooctahedral form is also observed for GeP_2O_7. It should be noted that the compounds $SiO_2 \cdot P_2O_5$ and $GeO_2 \cdot P_2O_5$ also form monoclinic and pseudohexagonal forms, the spectra of which were interpreted by Lecomte, Lelong, and Boullé [100-102] on the basis of the pyrophosphate formula. However, certain of these forms probably have a different nature with the same chemical composition. Thus for β-$GeO_2 \cdot P_2O_5$ (monoclinic form) Avduevskaya and Tananaev proposed a metaphosphate formula $Ge(PO_3)_2O$ [103]. The infrared spectrum of these crystals has a doublet at about 1000 cm^{-1} with closely spaced components, and five bands, of which three are reasonably strong, between 800 and 600 cm^{-1}, which tends to favor the latter view.

$r_{P-O(P)}$ and r_{P-O^-} distances (1.52 and 1.56 Å). It was also established that these crystals were only strictly cubic at high temperatures, while at room temperature they were apparently only pseudocubic (although the discrepancies were very slight). Dilatometric, thermographic, and x-ray analyses of ZrP_2O_7 revealed the existence of a reversible phase transformation at 300°C [96]. The phase transformation is accompanied by a slight but noticeable endothermic effect and a step in thermal expansion. The linear expansion coefficient at temperatures above the transformation point is much smaller than below.

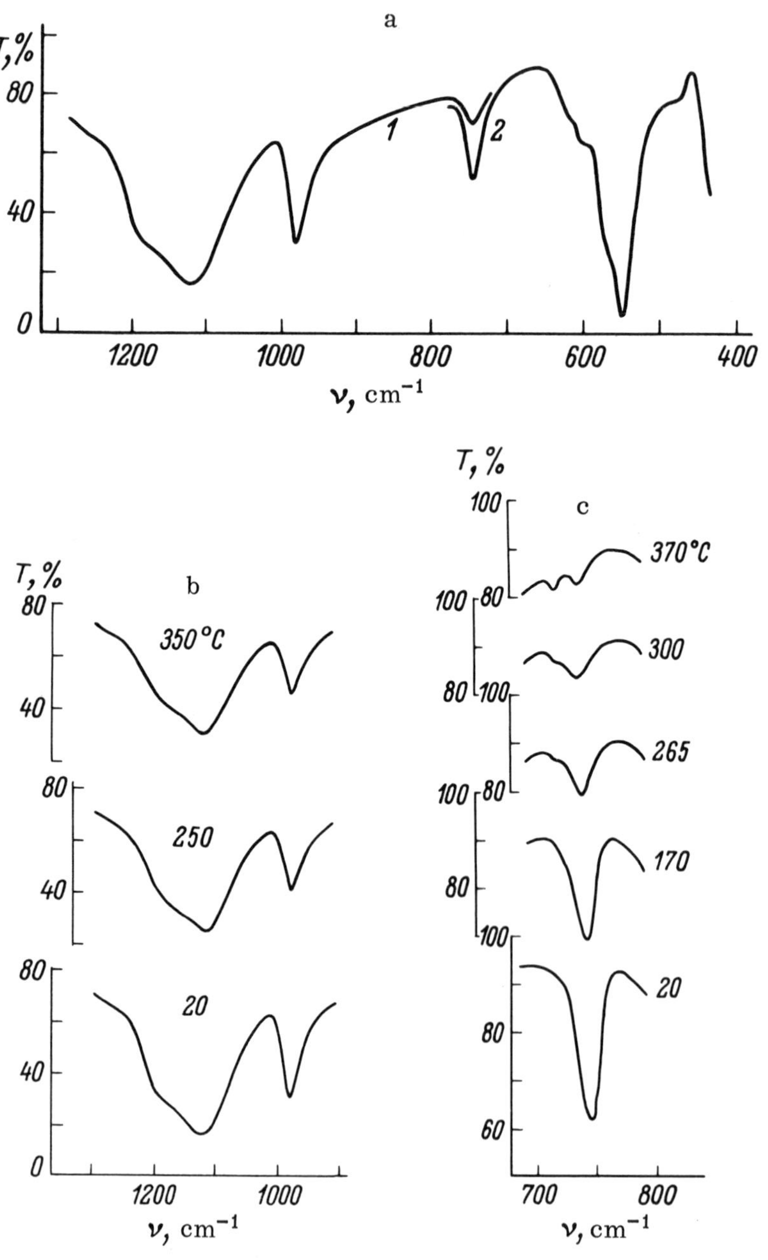

Fig. 134. Infrared spectrum of ZrP_2O_7. a) At room temperature [pressed in KBr: 1) 2.5 mg; 2) 5 mg]; b, c) spectral changes in heating.

The infrared spectra and in part the Raman spectra of cubooctahedral pyrophosphates were studied by Steger and Leukroth [97] at room temperature. The resultant curves showed strong absorption bands at about 750 cm^{-1}; these were probably due to the ν_s POP vibration of the P_2O_7 anions. The appearance of this band in the infrared spectrum cannot be made to agree with a centrosymmetric structure of the P_2O_7 anion (this was not noted in [97]), since allowing for the influence of crystal symmetry on the selection rules will not lead to the breaking of the mutual exclusion rule. The bridge O atoms coinciding with the center of symmetry of the anion lie in special positions with symmetry $\bar{3}-C_{3i}$ in the crystal (space group $Pa3-T_h^6$); the irreducible representation A_{1g} of the point group D_{3d} of the anion corresponds to the inactive representations A_g and F_g of the factor group of the crystal T_h.

All the foregoing results suggest that the low-temperature forms of these crystals are not strictly cubic owing to the slightly bent configuration of the P_2O_7 ions, while the phase transformation is associated with a transformation to a statistically centrosymmetric structure. Hence it would appear interesting to study the changes in the spectra of these crystals on raising the temperature. As a result of the foregoing we may expect the ν_s POP band to vanish near the transformation point, the frequencies of this vibration and the ν_{as} POP remaining almost constant.

The results of an investigation into the infrared spectrum of ZrP_2O_7 between room temperature and 400°C agree closely with these principles [92]. Raising the temperature (Fig. 134) has little effect on the intensity and frequency of the vibration bands of the PO_3 groups and ν_{as} POP. At the same time, as in the $Mg_2P_2O_7$ case, on raising the temperature the intensity of the ν_s POP band falls sharply, and above the temperature of the phase transformation it becomes negligibly small. As the fundamental ν_s POP band weakens an additional maximum appears on the long-wave side. Above the temperature of the phase transformation the two maxima differ little in intensity, and the spectrum of ZrP_2O_7 is reminiscent of that of $Mn_2P_2O_7$ in this frequency range. Thus the temperature changes in the ZrP_2O_7 spectrum support the view as to the possibility of an inversion of the anion in this crystal and enable us to ascribe a potential function corresponding to scheme b in Fig. 130 to the δPOP vibration. The manner in which the structure of the ν_s POP band varies in the ZrP_2O_7 spectrum enables us to associate the appearance of the long-wave satellite with the increasing role of the higher vibrational levels of δPOP and τPO_3, to which the large POP angle and lower frequency ν_s POP correspond. As in the cases considered earlier, the number of configurations of the P_2O_7 ion corresponding to the potential-energy minimum, arranged symmetrically with respect to the bridge O atom (in the centrosymmetric form), is determined by the symmetry of this point in the high-temperature form of the crystal, in the present case C_{3i}, and evidently equals six, since the bent P_2O_7 anion has no symmetry elements in common with the C_{3i} group, neither C_3 nor C_i.

It should be noted that the conditions allowing the possibility of an inversion of the pyrophosphate anion formulated in the foregoing are only partly satisfied for the Zr^{4+} cation. The bonds between Zr and O are to a considerable extent covalent, but the ionic radius is quite large, and its ratio to that of oxygen is only slightly smaller than the limiting value for octahedral coordination. It is possible, however, that the actual dimensions of the zirconium ion in this crystal are smaller than expected. This would appear to be indicated by the fact that in the denser form of ZrP_2O_7 formed at high pressure the similarity of the infrared spectrum and molecular refraction to those of the cubooctahedral form indicate that no change has occurred in the coordination number [98]. Hence the cubooctahedral form may be considered as relatively loosely packed. In the structure derived by Levi and Peyronel one notices substantial channels or cavities adjacent to the bridge oxygen atoms of the P_2O_7 groups. From this point of view the negative coefficient of thermal expansion of the isostructural UP_2O_7 crystal [99] at temperatures over 400°C is also worth noting.

The phenomena considered clearly relate to the most obvious and hence most easily interpreted manifestations of the mobility of the X–O–X bridges in structures with large XOX angles. It is extremely probable that in structures with smaller angles the internal rotations and deformations of the XOX angles are also of considerable importance in many processes, since the excitation energy of these motions remains smaller than or comparable with the ordinary thermal energy. Unfortunately, as yet no sharp changes in the structure of metasilicate or metaphosphate chains have been observed on the basis of the mechanism just described, although in x-ray investigations into complex silicate structures it has sometimes been suggested that the centrosymmetry of the oxygen bridges has a statistical character. Thus it has been suggested that in the tridimite-like lattice of nepheline $KNa_3Al_4Si_4O_{16}$ (space group $P6_3$, $Z = 2$) two of the Si–O–(Si, Al) bonds form angles statistically equal to 180° [104, 105].

The part played by internal rotations in the reconstruction of complex frameworks is excellently illustrated by the α-β transformation (at 575°C) in quartz. In 1940 Raman and Nedungadi [106] discussed the possible relation between the α-β transformation in quartz and the thermal excitation of a vibration at a frequency of 207 cm^{-1} observed in the Raman spectrum of α-quartz. Study of the temperature changes in the Raman spectrum revealed a considerable broadening of the 207 cm^{-1} line with increasing temperature, together with a considerable reduction in its frequency, and its complete loss near the transformation point [107]. This vibration, belonging to the irreducible representation A_1 [108, 109] (factor group D_3), corresponds to a displacement of the O atom in a direction perpendicular to the SiOSi plane, i.e., a "nonplanar" deformation of SiOSi, as shown by an analysis of the normal vibrations (nondegenerate types of symmetry) of the unit cell of α-quartz carried out by Kleinmann and Spitzer [110]. It is in this direction that the O atoms are displaced during the transformation from trigonal α-quartz to the hexagonal (D_6) β-quartz. The displacements of the Si atoms are much smaller, but in direction these also correspond to their displacements during the α-β transformation. A calculation of the frequencies using a valence-force model with three force constants, characterizing the rigidity of the Si–O bonds and the valence angles of the silicon and oxygen atoms, led to the conclusion that the last of these parameters was close to zero. This is also supported by an analysis of the intensities in the Raman spectrum.

Thus the nature of the α-β transformation of quartz is closely connected with the internal rotation† in the Si–O–Si bridges. Information obtained in relation to the thermal amplitudes of the atoms [111, 112] on refining an x-ray analysis of the structure of α-quartz by the method of least squares agrees closely with the foregoing considerations; the ellipsoid representing the displacements of the O atom has its longest axis in the direction of the "nonplanar" deformation of SiOSi, the next smallest lies in the SiOSi plane perpendicular to the Si . . . Si line, and the shortest in a direction parallel to Si . . . Si ($\angle$SiOSi = 144°). Unfortunately, it is difficult to study the changes in the spectrum of the β-form of quartz near the transformation temperature, since the A_1-type vibration of the α-form responsible for the transformation corresponds to a B_1-type vibration of the β-form inactive in the infrared and Raman spectra.

Later studies of the Raman spectra of quartz revealed a more complex picture of the spectrum and its thermal changes in the low-frequency range, which in addition to the 207-cm^{-1} band contained another weak band at 147 cm^{-1} (at room temperature), also belonging to symmetry type A_1. Possible explanations include the existence of a potential function with two minima for the vibration at 207 cm^{-1} [112a], or resonance between single-phonon (optical phonon with $k = 0$) and two-quantum (acoustic phonons at the edge of the zone) transitions [112b]. However, within the framework of the harmonic model the relation between the type-A_1 vibra-

† However, in contrast to the cases considered earlier, the greater freezing of these motions in the three-dimensional polymerized lattice probably leads to a relatively large value of the vibrational quantum and higher transformation temperatures.

tion at 207 cm^{-1} and the mechanism of the transformation is in no doubt, and a further calculation of the quartz spectrum [112c] agrees with these principles. We also note that the frequency 207 cm^{-1} exhibits the greatest displacement under the influence of pressure; this was established by studying the Raman spectrum of quartz at a pressure of 40 kbar [112d].

According to the foregoing, the α-β transformation in quartz is due to the thermal excitation of a totally symmetric vibration of the α-form. For strict harmonicity of this vibration the symmetry of the excited states cannot change, and hence a phase transformation involving a change of symmetry D_3 to D_6 can only arise from the anharmonic character of the vibration.† Such transformations as the α-β transformation in quartz constitute phase transformations of the second kind or are at least similar to these and have a resonance character. They differ considerably in this respect from the phase transformations of the first kind considered earlier in octamethylcyclotetrasiloxane ($C_i \rightarrow S_4 \rightarrow C_{4v}$, one group is not a subgroup of another), in which there are clear hysteresis phenomena near the transformation temperature. The possible role of vibrations (in the present case internal rotations) in transformations of this kind, as already mentioned, simply amounts to preparation for the transformation, but there is no direct link between the vibrational states of the system above and below the transformation point.

Among the spectroscopic manifestations of the internal rotations of the Si–O–Si bridges in organic crystals we must mention the fine structure of the absorption band of an oxygen impurity in pure silicon. According to the results of Hrostowski and Alder [113], this structure is related to internal rotations by virtue of the existence of several energetically equivalent positions of the oxygen atom in the interstices of the lattice.

In conclusion we should mention that many authors have tried to explain a number of anomalous properties of vitreous silica (specific heat at low temperatures, negative thermal expansion, low-temperature maxima of the dielectric and mechanical losses) on the basis of the high mobility of the Si–O–Si bridges [114-118]. This mobility is probably aided by the large (compared with quartz) SiOSi angle in vitreous silica‡ (this agrees with the lower density of the glass, while still retaining the same mode of condensation of the silicon–oxygen tetrahedra as in the crystal), and the existence of a distribution of the configurations of the Si–O–Si bridges deviating much further from the mean than that of the crystal. A study of the frequency dependence of the position of the low-temperature maximum of the acoustic and dielectric losses§ indicates the existence of a relaxation process with an activation energy of the order of $E = 1300$ cal/mole and a frequency $\nu_{max} \simeq 10^{12}$, i.e., about 30 cm^{-1} ($\omega = \nu_{max} \exp[-E/kT]$). The observation of a whole range of scattering with low frequencies, from 10 to 50 cm^{-1} [121], in the Raman spectrum of vitreous SiO_2 may indicate the existence of corresponding vibrations of the optical type.

---

† For nontotally symmetric vibrations the symmetry of the excited states may differ from that of the ground state. Hence the existence of anharmonicity is in this case not obligatory in order to give a change of symmetry, but then the symmetry of the high-temperature configuration will be a subgroup of the symmetry group of the low-temperature form. However, in the majority of cases a rise in temperature leads to an increase in symmetry.

‡ There have been contradictory estimates of the SiOSi angle in vitreous silica in the literature [119, 120]. According to recent data [120a], the most probable value of this angle is 144°, as in the crystal, but individual deviations may be considerable and embrace a range of 120 to 180°, i.e., the distribution is considered to be rather drawn out in the direction of large angles.

§ We note that in the $Mg_2P_2O_7$ crystal there is also apparently a low-temperature maximum of the dielectric losses, since as temperature falls from 20 to –150°C the value of tan δ increases.

References

1. N. V. Belov, Crystal Chemistry of Silicates with Large Cations, Izd. AN SSSR, Moscow (1961).
2. C. Eaborn, Organosilicon Compounds, Butterworths, London (1960).
3. E. G. Rochow and H. G. LeClair, J. Inorg. Nucl. Chem., 1:92 (1955).
4. W. L. Roth, J. Amer. Chem. Soc., 69:474 (1947).
5. W. L. Roth and D. Harker, Acta Crystallogr., 1:34 (1948).
6. E. H. W. Aggarwal and S. H. Bauer, J. Chem. Phys., 18:42 (1950).
7. G. Peyronel, Atti Accad. Naz. dei Lincei, Rendiconti Classe Sci. Fis. Matem. e Nat., 15:402 (1953); 16:78, 231 (1954).
8. V. P. Davydov and M. G. Voronkov, Polyphosphazenes [in Russian], Izd. AN SSSR, Moscow–Leningrad (1962).
9. R. A. Shaw, S. W. Fitzsimmons, and B. C. Smith, Chem. Revs., 62:247 (1962).
10. N. L. Paddock, in the book edited by M. F. Lappert and G. J. Leigh, Developments in Inorganic Polymer Chemistry, Elsevier Publ., Amsterdam–London–New York (1962), p. 87.
11. P. Anderson, in: Physics of Dielectrics [Russian translation], Izd. AN SSSR, Moscow–Leningrad (1960), p. 290.
12. W. Cochran, Adv. Physics., 9:387 (1960).
13. W. R. Thorson and I. Nakagawa, J. Chem. Phys., 33:994 (1960).
14. R. C. Lord, D. W. Robinson, and W. C. Schumb, J. Amer. Chem. Soc., 78:1327 (1956).
15. J. R. Aronson, R. C. Lord, and D. W. Robinson, J. Chem. Phys. 33:1004 (1960).
16. A. Almenningen, O. Bastiansen, V. Ewing, K. Hedberg, and M. Traetteberg, Acta Chem. Scand., 17:2455 (1963).
17. A. N. Lazarev, M. G. Voronkov, and T. F. Tenisheva, Opt. i Spektr., 5:365 (1958).
18. D. W. Robinson, W. J. Lafferty, J. R. Aronson, J. R. Durig, and R. C. Lord, J. Chem. Phys., 35:2245 (1961).
19. A. N. Lazarev and T. F. Tenisheva, Opt. i Spektr., 18:217 (1965).
20. K. D. Moller and H. G. Andresen, J. Chem. Phys., 37:1800 (1962).
21. R. J. Myers and E. B. Wilson, J. Chem. Phys., 33:186 (1960).
22. H. C. Longuet-Higgins, Mol. Phys., 6:445 (1963).
23. A. J. Stone, J. Chem. Phys., 41:1568 (1964).
24. D. W. Scott, J. F. Messerly, S. S. Todd, G. B. Guthrie, J. A. Hossenlopp, R. T. Moore, Ann Osborn, W. T. Berg, and J. P. McCullough, J. Phys. Chem., 65:1320 (1961).
25. R. W. Kilb and L. Pierce, J. Chem. Phys., 27:108 (1957).
26. L. Pierce, J. Chem. Phys., 31:547 (1959).
27. H. Herzberg, Vibrational and Rotational Spectra of Polyatomic Molecules, Van Nostrand, New York, 1945.
28. H. Tadokoro, A. Kobayashi, Y. Kawaguchi, S. Sobajima, S. Murahashi, and Y. Matsui, J. Chem. Phys., 35:369 (1961).
29. R. S. Golland and C. P. Smith, J. Amer. Chem. Soc., 77:268 (1955).
30. F. K. Fong, J. Chem. Phys., 40:132 (1964).
31. R. D. Nelson and C. P. Smith, J. Phys. Chem., 69:1006 (1965).
32. J. A. Katon, W. R. Feairheller, and E. R. Lippincott, J. Mol. Spectr., 13:72 (1964).
33. A. N. Lazarev, Opt. i Spektr., 18:792 (1965).
34. H. Kriegsmann, Z. Anorgan. und Allgem. Chem., 298:223 (1959).
35. I. F. Kovalev, Opt. i Spektr., 8:315 (1960).
36. I. F. Kovalev, in: Physical Problems of Spectroscopy, Izd. AN SSSR, Moscow (1962), p. 360.
37. A. L. Smith, Spectrochim. Acta, 19:849 (1963).
38. V. Busetti, M. Mammi, and G. Garazzolo, Z. Kristallogr., 119:310 (1963).

39. D. P. Craig and N. L. Paddock, J. Chem. Soc. (L.), 4418 (1962); see also D. P. Craig and K. A. R. Mitchell, J. Chem. Soc. (L.), 4682 (1965).
40. L. W. Daasch, J. Amer. Chem. Soc., 76:3403 (1954).
41. A. C. Chapman, N. L. Paddock, D. H. Paine, H. T. Searle, and D. R. Smith, J. Chem. Soc. (L.), 3608 (1960).
42. E. Steger and R. Stahlberg, Z. Anorgan. und Allgem. Chem., 326:243 (1964).
43. I. C. Hisatsune, Spectrochim. Acta, 21:1899 (1965).
44. A. Wilson and D. F. Carrol, J. Chem. Soc. (L.), 2548 (1960).
45. H. McD. McGeachin and F. R. Tromans, J. Chem. Soc. (L.), 4777 (1961).
46. H. J. Becher and F. Seel, Z. Anorgan. und Allgem. Chem., 305:148 (1960).
47. R. Hazekamp, T. Migchelsen, and A. Vos, Acta Crystallogr., 15:539 (1962).
48. G. J. Bullen, Proc. Chem. Soc. (L.), 425 (1960).
49. M. W. Dougill, J. Chem. Soc. (L.), 5471 (1961).
49a. I. C. Hisatsune, Spectrochim. Acta., 25A:301 (1969).
50. A. N. Lazarev and T. F. Tenisheva, Izv. Akad. Nauk SSSR, Ser. Khim., 1168 (1964).
51. J. D. Hoffman, J. Amer. Chem. Soc., 75:6313 (1963).
52. H. Steinfink, B. Post, and I. Fankuchen, Acta Crystallogr., 8:420 (1955).
53. H. Kriegsmann, Z. Anorgan. und Allgem. Chem., 298:232 (1959).
54. R. O. Sauer and D. J. Mead, J. Amer. Chem. Soc., 68:1794 (1946).
55. G. S. Smith and L. E. Alexander, Acta Crystallogr., 16:1015 (1963).
56. A. N. Lazarev and T. F. Tenisheva, Izv. Akad. Nauk SSSR, Ser. Khim., 983 (1966).
57. Ya. M. Slobodin, Ya. E. Shmulyakovskii, and K. A. Rzhendzinskaya, Dokl. Akad. Nauk SSSR, 105:958 (1955).
58. N. Wright and M. J. Hunter, J. Amer. Chem. Soc., 68:803 (1947).
59. E. Richards and H. W. Thomson, J. Chem. Soc. (L.), 124 (1949).
60. L. K. Frevel and M. J. Hunter, J. Amer. Chem. Soc., 67:2275 (1945).
61. W. Noll, Chemie und Technologie der Silicone, Verl. Chemie, Weinheim (1960).
62. V. I. Simonov, Kristallografiya, 6:544 (1960).
63. G. F. Volodina, I. M. Rumanova, and N. V. Belov, Dokl. Akad. Nauk SSSR, 149:173 (1963).
64. W. L. Bragg and J. West, Proc. Roy. Soc., 111A:691 (1926).
65. N. V. Belov, V. P. Butuzov, and N. I. Golovastikov, Dokl. Akad. Nauk SSSR, 83:953 (1952).
66. H. G. Heide, K. Boll-Dornberger, E. Thilo, and E. M. Thilo, Acta Crystallogr., 8:425 (1955).
67. N. V. Belov and E. N. Belova, Dokl. Akad. Nauk SSSR, 69:185 (1949).
68. G. E. Donnay and M. J. Buerger, Acta Crystallogr., 3:379 (1950).
69. E. M. Shustorovich, Zh. Strukt. Khim., 4:642 (1963).
70. E. M. Shustorovich, Zh. Strukt. Khim., 4:743 (1963).
71. K. S. Pitzer and S. J. Strickler, J. Chem. Phys., 41:730 (1964).
72. D. W. J. Cruickshank, H. Linton, and G. A. Barclay, Acta Crystallogr., 15:491 (1962).
73. W. H. Zachariasen, Z. Kristallogr., 73:1 (1930).
74. A. N. Lazarev, T. F. Tenisheva, and I. A. Bondar', Izv. Akad. Nauk SSSR, Neorg. Mat., 1:1207 (1965).
75. M. J. Buerger and C. T. Prewitt, Proc. Nat. Acad. Sci., 47:1884 (1961).
76. D. R. Peacor and M. J. Buerger, Z. Kristallogr., 117:331 (1962).
77. D. R. Peacor and C. T. Prewitt, Amer. Mineralogist, 48:588 (1963).
77a. G. E. Brown, G. V. Gibbs, and P. H. Ribbe, Amer. Mineralogist, 54:1044 (1969).
78. C. T. Prewitt and D. R. Peacor, Amer. Mineralogist, 49:1527 (1964).
79. A. N. Lazarev and T. F. Tenisheva, Izv. Akad. Nauk SSSR, Ser. Khim., 242 (1964).
80. A. N. Lazarev, T. F. Tenisheva, and M. A. Petrova, Izv. Akad. Nauk SSSR, Neorg. Mat., 1:1379 (1965).
81. F. Machatschki, Fortschrift Mineral., 20:48 (1936).
82. R. Roy, E. T. Middleswarth, and F. A. Hummel, Amer. Mineralogist, 33:458 (1948).

83. K. Lukaszewicz, Roczniki Chemii, 35:31 (1961).
84. F. L. Oetting and R. A. McDonald, J. Phys. Chem., 67:2737 (1963).
85. E. L. Katnack and F. A. Hummel, J. Electrochem. Soc., 105:125 (1958).
86. K. Lukaszewicz and R. Smajkiewicz, Roczniki Chemii, 35:741 (1961).
87. C. Calvo, Canad. J. Chem., 43: 1139 (1965).
88. C. Calvo, Canad. J. Chem., 43:1147 (1965).
89. J. G. Chambers, W. D. Datars, and C. Calvo, J. Chem. Phys., 41:806 (1964).
90. K. Lukaszewicz, Seventh Internat. Cong. on Crystallography, Abstracts, Izd. Nauka, Moscow (1966), p. 59.
91. B. E. Robertson and C. Calvo, Acta Crystallogr., 22:665 (1967).
92. A. N. Lazarev and T. F. Tenisheva, Izv. Akad. Nauk SSSR, Ser. Khim., 403 (1964).
93. F. Liebau, Acta Crystallogr., 14:1103 (1961).
94. G. R. Levi and G. Peyronel, Z. Kristallogr., 92:190 (1935).
95. G. Peyronel, Z. Kristallogr., 94:311 (1936).
96. D. E. Harrison, H. A. McKinstry, and F. A. Hummel, J. Amer. Ceram. Soc., 37:277 (1954).
97. E. Steger and G. Leukroth, Z. Anorgan. und Allgem. Chem., 303:169 (1960).
98. G. B. Sclar, L. C. Carrison, and C. M. Schwarz, Nature, 204:573 (1964).
99. H. P. Kirchner, K. M. Merz, and W. R. Brown, J. Amer. Ceram. Soc., 46:137 (1963).
100. B. Lelong, Ann. Chim., 9:229 (1964).
101. J. Lecomte, A. Boullé, C. Doremieux-Morin, and B. Lelong, Compt. Rend., 258:131 (1964).
102. J. Lecomte, A. Boullé, C. Doremieux-Morin, and B. Lelong, Compt. Rend., 258:1447 (1964).
103. K. A. Avduevskaya and I. V. Tananaev, Zh. Neorg. Khim., 10:366 (1965).
104. M. J. Buerger, G. E. Klein, and G. Donnay, Amer. Mineralogist, 39:805 (1954).
105. Th. Hahn and M. J. Buerger, Z. Kristallogr., 106:308 (1955).
106. C. V. Raman and T. M. K. Nedungadi, Nature, 144:147 (1940).
107. P. K. Narayanaswamy, Proc. Indian Acad. Sci., A28:417 (1948).
108. B. D. Saksena, Proc. Indian Acad. Sci., A22:379 (1945).
109. B. D. Saksena and H. Narian, Proc. Indian Acad. Sci., A30:128 (1949).
110. D. A. Kleinmann and W. G. Spitzer, Phys. Rev., 125(2):16 (1962).
111. R. A. Young and B. Post, Acta Crystallogr., 15:337 (1962).
112. G. S. Smith and L. A. Alexander, Acta Crystallogr., 16:462 (1963).
112a. S. M. Shapiro, D. C. O'Shea, and H. Z. Cummins, Phys. Rev. Letters, 19:361 (1967).
112b. J. F. Scott, Phys. Rev. Letters, 21:907 (1968).
112c. M. M. Elcombe, Proc. Phys. Soc., 91:947 (1967).
112d. J. F. Asell and M. Nicol, J. Chem. Phys., 49:5395 (1968).
113. H. J. Hrostowski and B. J. Alder, J. Chem. Phys., 33:980 (1960).
114. O. L. Anderson and H. E. Bommel, J. Amer. Ceram. Soc., 88:125 (1955).
115. H. T. Smith, J. W. Londeree, and G. E. Lorey, J. Amer. Ceram. Soc., 36:238, 327 (1953).
116. H. T. Smith, H. S. Skogen, and W. B. Harsell, J. Amer. Ceram. Soc., 38:140 (1955).
117. J. Volger and J. M. Stevels, Philips Res. Rep., 11:452 (1956).
118. O. L. Anderson and G. J. Dienes, Coll.: Noncrystalline Solids, John Wiley and Sons, New York–London (1960), p. 449.
119. J. Zarzycki, Verres et Refractaires, 11:3 (1957).
120. R. J. Breen, R. M. Delaney, P. J. Persiani, and A. H. Weber, Phys. Rev., 105:520 (1957).
120a. R. L. Mozzi and B. E. Warren, J. Appl. Crystallogr., 2:164 (1969).
121. P. Flubacher, A. Leadbetter, J. Morrison, and B. Stoicheff, J. Phys. Chem. Solids, 12:53 (1959).